APPLICATIONS OF INFORMATION AND CONTROL SYSTEMS

Applications of Information and Control Systems

Volume III of a selection of papers from INFO II, the Second International Conference on Information Sciences and Systems, University of Patras, Greece, July 9-14, 1979

Edited by

D. G. LAINIOTIS
Dept. of Electrical Engineering,
State University of New York at Buffalo, Amherst, N.Y., U.S.A.

and

N. S. TZANNES
University of Patras, Chair of Information Theory,
School of Engineering, Patras, Greece

D. Reidel Publishing Company

Dordrecht : Holland / Boston : U.S.A. / London : England

Library of Congress Cataloging in Publication Data

CIP

International Conference on Information Sciences and Systems, 2d, University of Patras, 1979.

A selection of papers from INFO II, the Second International Conference on Information Sciences and Systems, University of Patras, Greece, July 9–14, 1979.

Includes indexes.

Contents: v. 1. Advances in communications. v. 2. Advances in control. 3. Applications of information and control systems.

1. Electronic data processing—Congresses. 2. Telecommunication—Congresses. 3. Automatic control—Congresses. 4. Pattern recognition systems—Congresses. I. Lainiotis, Demetrios G. II. Tzannes, Nicolaos, 1937– III. Title.

QA75.5.I58 1979 001.64 80–18203
ISBN 90–277–1140–2 (v. 1)
ISBN 90–277–1129–1 (v. 2)
ISBN 90–277–1143–7 (v. 3)

Published by D. Reidel Publishing Company
P.O. Box 17, 3300 AA Dordrecht, Holland

Sold and distributed in the U.S.A. and Canada
by Kluwer Boston Inc., Lincoln Building,
160 Old Derby Street, Hingham, MA 02043, U.S.A.

In all other countries, sold and distributed
by Kluwer Academic Publishers Group,
P.O. Box 322, 3300 AH Dordrecht, Holland

D. Reidel Publishing Company is a member of the Kluwer Group

Printed in The Netherlands

CONFERENCE COCHAIRMEN

D.G. Lainiotis N.S. Tzannes

SPONSORED BY

The Ministry of Culture & Sciences of Greece
The University of Patras
The IEEE Automatic Control Society
The IEEE Information Theory Group
The IEEE Systems, Man and Cybernetics Society
The World Organization of General Systems and Cybernetics
The IEEE Greece
The Greek Society for Cybernetics and Systems

STEERING COMMITTEE

D.G. Lainiotis,	U.S.A.
N.S. Tzannes,	Greece
J. Cruz,	U.S.A.
G.F. Ferrate-Pascual,	Spain
K. Kerner,	Austria
J.L. Lions,	France
Y. Sunahara,	Japan
C. Tapiero,	Israel
C.P. Tsokos,	U.S.A.
M. Younis,	Egypt

TECHNICAL PROGRAM COMMITTEE

D.G. Lainiotis,	U.S.A.
J. Aguilar-Martin,	France
E. Baptista,	Brazil
A. Bensoussan,	France
E.J. Davison,	Canada
A. Ephremides,	U.S.A.
M. Gevers,	Belgium
H. Kwakernaak,	Holland
E. Mosca,	Italy
K. Najim,	Morocco
G. Saridis,	U.S.A.
P.D. Scott,	U.S.A.
C. Tapiero,	Israel
C.P. Tsokos,	U.S.A.
N.S. Tzannes,	Greece
M.I. Younis,	Egypt

LOCAL ARRANGEMENTS COMMITTEES

GREECE

A. Sissouras, Chairman
G. Megalokonomos
P. Panagiotopoulos

U.S.A.

C. Nikias, Chairman
E. Cosmidis
P. Groumpos
C. Holtz
S. Thomopoulos

CONFERENCE SECRETARIES

Mrs. Niki Sarantoglou, Greece
Mrs. Helen Alexandridis, Greece
Miss Ismini Kariofili, U.S.A.

THE CONFERENCE HAS BEEN AIDED FINANCIALLY BY

The Ministry of Culture & Sciences of Greece
The University of Patras
A.B.E.E.A. Co.
A.B.E.X. Co.
ACHAIA-CLAUSS Co.
Astir Hotel
BIOSOL Co.
DINAMIKI Co.
ELLENIT Co.
KAZA Co.
Kalogerakis Co.
Koutsadonis N.
Piraiki-Patraiki Co.
Pirelli Hellas S.A.
Titan Cement Co.

TABLE OF CONTENTS

PREFACE

This volume contains a selection of papers presented at INFO II, the Second International Conference on Information Sciences and Systems, held at the University of Patras, Greece, from July 9th to 14th, 1979. Additional selections of papers from this Conference are published in separate volumes, also by D. Reidel and are entitled "Advances in Communication" and "Advances in Control", respectively.

INFO II, was hosted by the University of Patras and sponsored by, The Ministry of Culture & Sciences, IEEE Automatic Control Society, IEEE Information Theory Group, IEEE Systems -Man and Cybernetics Society, World Organization of General Systems and Cybernetics, IEEE GReece, and the Greek Society for Cybernetics and Systems. It was the second such Conference held at the University of Patras, the first one having taken place from August 19th to 24th, 1976, with its papers published by the Hemisphere Publishing Corporation in a three volume set, entitled "Applications and Research in Information Sciences and Systems".

INFO II, then, was the second high level Conference on these important topics in the Eastern Mediterranean Region. The reason that these conferences have been established is because we feel strongly that the topics covered have reached a high degree of maturity in this region, both from the point of view of theory and application. The applications areas are especially important to the Mediterranean region in view of the Common Market connections of such countries as France, Italy and Greece and the anticipated admission of Spain. In view of this, a major objective of the conferences is to focus attention on these important technical subjects and to facilitate the exchange of information and cross-fertifization of ideas between researchers of the Mediterranean region and those of the rest of the world, thereby promoting research and development by the region's engineers, researchers and academicians.

The Conference keynoted by Prof. J.L.Lions, France, Prof.L.A. Zadeh, U.S.A., and Prof. M.Dertoyzos, U.S.A., was highly successful as evidenced by the attendance of over 250 scientists and engineers from approximately 30 countries, and by the presentation of over 200 technical papers by well-known scientists and engineers from around the world.

The technical program of the Conference consisted of technical sessions that covered all-important aspects of control, information processing/communications, and computers. In particular, the-

D. G. Lainiotis and N. S. Tzannes (eds.), Applications of Information and Control Systems, xiii-xiv.

re were numerous technical sessions covering all-important aspects of control, such as, control theory, control applications, estimation, identification, adaptive systems, linear systems, stability, cybernetics, computational methods, and simulation. The program also included many sessions on information processing/communications such as, information theory, coding, signal analysis, signal processing, communication theory, satellite communications, pattern recognition and image processing. Moreover, there were several sessions on computers, in particular on computer systems, computer communication networks, and automata. Finally, the technical program included numerous sessions on important applications of systems technology, such as, power, energy modeling and planning, earth resources, transportation, economics and management, and physiological systems.

In view of the broad scope of the Conference technical program and the extensive coverage of many important aspects of systems theory and applications by internationally known researchers, we hope that this collection of papers will be a useful supplement to the published literature and textbooks used for research and teaching.

For the success of the Conference we are indebted to a great many people and institutions, primarily to the authors of the papers, without whom the conference would not have existed, and this book would not have materialized. We are particularly indebted to the great majority of them who paid some or all of their own expenses. We are also indebted to the institutions throughout the world that gave permission to the authors to attend and covered their expenses.

We are greatly indebted to the University of Patras for hosting the Conference, and for its financial support, as well as to the various other institutions and companies, whose financial support was essential for the success of the Conference. We would also like to thank the above mentioned scientific societies for adding their name as conference co-sponsors.

Finally, in a peronal vein, we would like to thank the session chairmen, and the members of various conference committees, for their enthusiasm and dedication with which they worked in making INFO II a reality.

Nicolaos S. Tzannes Demetrios G. Lainiotis

SECTION I

IMAGE PROCESSING

FAST LINEAR METHODS FOR IMAGE FILTERING

Franz Pichler

Institut für Systemwissenschaften
Arbeitsbereich Systemtheorie
Johannes Kepler Universität Linz/Austria

Filtering Operators for two-dimensional signals are realizable by spatial domain methods and also by spectral domain methods. It is known that in both cases there exist for linear filters fast algorithms for computer implementation. The paper discusses such methods for linear filters which are realizable by convolution operators defined for images on a spatial domain which is a direct product of finite abelian groups. It is shown that any such operator is quasi-separable, that is that the operator is representable by a finite sum of separable operators.
On the other hand - by the existence of the character functions of the spatial-domain group - such convolution operators can be effectively realized via the spectral-domain representation computed by the related Fourier transform. In the special case of the two-dimensional dyadic group as spatial domain the spectral-domain realization is given by the fast Walsh transform.

D. G. Lainiotis and N. S. Tzannes (eds.), Applications of Information and Control Systems, 3-11.

1. INTRODUCTION

The application of the theory of Fouriertransform has a long history in signal processing. It serves there as a tool for the analysis as well as synthesis of signals and systems. The development of computers opened new roads for signal processing, but the discovery of the Fast Fouriertransform algorithm (FFT-algorithm) maintaince the importance of the classical methods. This raises the question of a general framework of Fouriertransform theory from which - depending on the type of the given problem - fast algorithms for computer implementation can be derived.

Our paper presents such a general framework and demonstrates the applicability of it to filtering of images (= two-dimensional signals). Some parts of our paper cannot claim to be original but as a pedagogical discussion. In this respect it compares with other work e.g. ANDREWS [1], AHMED-RAO [2] and GONZALEZ-WINTZ [3]. But we do claim that we were able to derive some new results: In chapter 3 it is proved that the kernel of a two-dimensional convolution-operator can be represented as a sum of separable kernels. In case of an (ideal) convolution-filter we even have a separable kernel. This result seems to open new possibilities for "fast" spatial-domain realizations.

Chapter 5 introduce the concepts of "manifold-filtering" and of the "Gibbs-differential operators". It seems that these concepts hold some promise for image-processing.

2. TWO-DIMENSIONAL FILTERING OPERATORS

It is known that any linear operator $\mathbb{L} : \mathcal{L} \to \mathcal{L}$, where $\mathcal{L}$ denotes the linear space of images $B \in \mathcal{L}$ of the kind $B : N{\times}N \to \mathbb{R}$ ($N := \{0,1,2,\ldots,n-1\}$, $\mathbb{R}$ is the real numbers), can be uniquely representend by their weighting-function (=kernel)

$$L : N^2 \times N^2 \to \mathbb{R} : (x,y,\xi,\eta) \mapsto L(x,y,\xi,\eta) \qquad (1)$$

When the spatial-domain $\mathcal{R} := N{\times}N$ has the algebraic structure of an abelian group $(N{\times}N, \boxplus)$ the convolution of images $H,B \in \mathcal{L}$ is defined by

$$H*B(x,y) := \sum_{(\xi,\eta)} H((x,y) \boxminus (\xi,\eta))\ B(\xi,\eta) \qquad (2)$$

($\boxminus (\xi,\eta)$ denotes the inverse element of (ξ,η) in $(N{\times}N, \boxplus)$. Any linear operator $\mathbb{L}$ which has a weighting function of convolution type, i.e.

$$L(x,y,\xi,\eta) = H((x,y) \boxminus (\xi,\eta)) \qquad (3)$$

is called a convolution operator. The image H is derived by application of $\mathbb{L}$ to the "impuls-image" E $(E(x,y) = 1$ for $(x,y) = (0,0)$ and $E(x,y) = 0$ for all $(x,y) \neq (0,0))$. H is called the impulse-response of $\mathbb{L}$.

We mention now two important properties of two-dimensional convolution operators.

(P1) All eigen-functions of a convolution operator $\mathbb{L}$ are characters $\chi_{ij}(i,j \in N)$ of the spatial-domain group $(N \times N, \boxplus)$.

(P2) The weighting-function L of $\mathbb{L}$ can be represented by

$$L(x,y,\xi,\eta) = \sum_{(i,j)} \lambda_{ij}\, \chi_{ij}((x,y) \boxminus (\xi,\eta)) \tag{4}$$

Here χ_{ij} denotes the eigenvalues of the eigenfunction χ_{ij}.

Taking $\chi_{ij}((x,y) \boxminus (\xi,\eta)) = \chi_{ij}(x,y)\ \chi_{ij}(\boxminus (\xi,\eta))$

and $\chi_{ij}(\boxminus (\xi,\eta)) = \overline{\chi_{ij}(\xi,\eta)}$ into account we can write for equation (4)

$$L(x,y,\xi,\eta) = \sum_{(i,j)} \lambda_{ij}\, \chi_{ij}(x,y)\ \overline{\chi_{ij}(\xi,\eta)} \tag{5}$$

The spatial case that the spatial-domain group $\mathcal{R} = (N \times N, \boxplus)$ is given by a direct product $(N \times N, \boxplus) = (N, \boxplus_1) \otimes (N, \boxplus_2)$ of abelian groups $(N, \boxplus_1)$ and $(N, \boxplus_2)$ is of special interest to us. Then every character χ_{ij} of $\mathcal{R}$ is given by the (usual) product of the individual characters χ_i and χ_j of $(N, \boxplus_1)$ and $(N, \boxplus_2)$, respectively:

$$\chi_{ij} = \chi_i \chi_j \qquad (i,j \in N) \tag{6}$$

Therefore $\chi_{ij}(x,y)\overline{\chi_{ij}(\xi,\eta)} = \chi_i(x)\overline{\chi_j(y)}\chi_i(\xi)\overline{\chi_j(\eta)}$ and equation (5) can now be represented in the form

$$L(x,y,\xi,\eta) = \sum_{(i,j)} \lambda_{ij} p_i(x,\xi) q_j(y,\eta) \tag{7}$$

where $p(x,\xi) := \chi_i(x)\ \overline{\chi_i(\xi)}$ and $q(y,\eta) := \chi_j(y)\ \ \overline{\chi_j(\eta)}$.

Equation (7) shows that the kernel L of a convolution operator $\mathbb{L}$ which is defined on a spatial-doamin group of product form $\mathcal{R} = \mathcal{R}_1 \otimes \mathcal{R}_2$ is represented by a sum of separable kernels $L_{ij}(x,y,\xi,\eta) := p_i(x,\xi) q_j(y,\eta)$.

Let L, P_i and Q_j denote the matrix representations of $L(x,y,\xi,\eta)$, $p_i(x,\xi)$ and $q_j(y,\eta)$, respectively.

Then equation (7) reads

$$\mathbb{L} = \sum_{(i,J)} \lambda_{ij} P_i \otimes Q_j^T \tag{8}$$

Here Q_j^T denotes the transpose of Q_j and $P_i \otimes Q_j^T$ is the Kronecker product of P_i and Q_j^T.

3. REALIZATION OF CONVOLUTION OPERATORS

By realization we mean a "blue print" for the actual construction of a system in hardware or software. For convolution operators of the kind discussed in chapter 2 we distinguish between the following two kinds of realizations:

a) the realization is based on the representation of the weighting function as given in equation (7) or (8).
   In this case we call it a <u>spatial domain realization</u>.
b) the realization makes use of a change of the basis-vectors in the image-space and uses the representation of equation (5) of the weighting-function.
   This is the case of a <u>spectral domain realization</u>.

To get a spatial-domain realization for $\mathbb{L}$ we use (8). If B denotes now the image represented by the N×N matrix $B = [B(x,y)]$ it can be shown, that

$$\mathbb{L}B = \sum_{(i,j)} \lambda_{ij} P_i B Q_j \tag{9}$$

For a spectral-realization of $\mathbb{L}$ we have to introduce some notations. Let b denote the $N^2 \times 1$ matrix, which is derived from the image B by scanning it row by row; then with the $N^2 \times N^2$ weighting matrix $L = [L(x,y,\xi,\eta)]$ we can compute $\mathbb{L}B$ by

$$\mathbb{L}B = Lb \tag{10}$$

Let X and $\bar{X}$ denote the $N^2 \times N^2$ matrices defined by the characters in the following way: $X := [X_{kl}]$ and $\bar{X} := [\bar{X}_{kl}]$ where $X_{kl} := \chi_{ij}(x,y)$ and $\bar{X}_{kl} := \chi_{ij}(\xi,\eta)$ for $l \doteq ij$ and $k \leftrightarrow (x,y)$ and $k \leftrightarrow (\xi,\eta)$, respectively.

Let Λ denote the diagonal-matrix $\Lambda = \mathrm{diag}(\lambda_{ij})$ built up by the eigenvalues λ_{ij} of $\mathbb{L}$. Then using equation (5) the weighting matrix L can be respresented in the form

$$L = X \Lambda \bar{X}^T \tag{11}$$

and $\mathbb{L}B = Lb$ can be computed by

$$\mathbb{L}B = X\Lambda X^{-T}b \qquad (12)$$

Equation (12) gives the basis for an effective realization of $\mathbb{L}$ in the spectral-domain. Since $X^{-T}X = I$ and therefore $X^{-T} = X^{-1}$ we have also

$$\Lambda = X^{-T}LX \qquad (13)$$

4. FAST ALGORITHMS

For both methods of realization as described basically by equations (9) and (12), respectively, there exist fast algorithms for hard- or software implementation. These fast methods are founded on the possibility to factorize the matrices P_i and Q_j (in case of spatial-domain realization) and the matrices X and X^{-T} (in case of spectral-domain realization). Generally, these methods are known as "fast fourier transform methods". One of the first contributions in which the general framework of finite abelian groups was used seems to be the article by NICHOLSON [5]. Another good source, covering this topic is the recent contribution by KUNZ [4] .

5. SPECIAL TYPES OF CONVOLUTION OPERATORS

There are many ways to specialize convolution operators. For our purpose it seems to be appropiate to discuss convolution operators which have special spectral properties and-as another class - those defined on special spatial domain groups.

(a) Classification with respect to the spectral domain

(a1) Rectangular convolution filters

A convolution operator $\mathbb{L}$ with the property that all eigenvalues λ_{ij} have value 1 shall be called a convolution filter. The set $\mathcal{D} \subseteq N\times N$ which is defined by $\mathcal{D} = \{(i,j): \lambda_{ij} \neq 0\}$ is called the domain of a convolution filter. We call a convolution filter $\mathbb{L}$ rectangular if its domain is rectangular, that is if $\mathcal{D}$ is given by $\mathcal{D} = \mathcal{D}_1 \times \mathcal{D}_2$ with $\mathcal{D}_1 \subset N$ and $\mathcal{D}_2 \subset N$.

Let $\mathbb{L}$ be assumed to be a rectangular convolution filter. Then by equation (7) we have

$$L(x,y,\xi,\eta) = \sum_{(i,j)\in\mathcal{D}} p_i(x,\xi)q_j(y,\eta)$$

$$= \Big(\sum_{i\in\mathcal{D}_1} p_i(x,\xi)\Big(\sum_{j\in\mathcal{D}_2} q_j(y,\eta)\Big)\Big)$$

or with $P(x,\xi) := \sum_{i\in\mathcal{D}_1} p_i(x,\xi)$ and $Q(y,\eta) := \sum_{j\in\mathcal{D}_2} q_j(y,\eta)$

we also have

$$L(x,y,\xi,\eta) = P(x,\xi)Q(y,\eta) \tag{14}$$

Equation (14) shows that rectangular convolution filters are separable. Therefore such a filter has a rather "nice" spatial domain realization based on the form

$$\mathbb{L}B = P\ B\ Q^T \tag{15}$$

where P and Q denoted the N×N matrices corresponding to the functions $P(x,\xi)$ and $Q(y,\eta)$, respectively.

(a2) Multiplicity Filters

For each character χ_i and χ_j of $(N, \boxplus_1)$ and $(N, \boxplus_2)$ respectively, the multiplicity $M(\chi_i)$ and $M(\chi_j)$ can be defined by the number of generalized Rademacher-functions, required to define (by their multiplication) χ_i and χ_j respectively (for more detail see the article by WEISS [8]).

Let M denote a subset of the natural numbers (including 0): Then a convolution filter with domain $\mathcal{D}$ given by

$$\mathcal{D} = \{(i,j):M(i) + M(j) \in M\} \tag{16}$$

is called a <u>M-multiplicity filter</u>.

Up to this definition the concept of a M-multiplicity filter looks quite artificial. But the applicability of this concept derives from certain mathematical results which relate the concept of multiplicity of characters to approximation properties. We state one result with respect to the approximation of polynomials:

Theorem (WEISS [8] , PICHLER [6]):

The character-expansion of a polynomial-function of degree n contains only summands $\alpha_i\ \chi_i$ for which $M(\chi_i) \le n$.

As a consequence of this theorem a {n} -multiplicity filter (to state a rather simple example) would produce as output only polynomial images B of the form

$$B(x,y) = a_o x^{i_o} y^{j_o} + a_1 x^{i_1} y^{j_1} + \ldots + a_k x^{i_k} y^{j_k}$$

with $i_o + j_o = i_1 + j_1 = \ldots = i_k + j_k = n$.

Therefore M-multiplicity filters can be considered as comb-filters with special approximation properties which are related to polynomial approximation. Experimental work done with 1-dimensional M-manifold filters refer to ROTH [7].

(a3) Logical Differential operators
By extension of the concepts of GIBBS' [9] from the 1-dimensional case to two dimensions, a convolution-operator $\mathbb{L}$ with the property that

$$\mathbb{L} x_{ij} = |i|\,|j|\, x_{ij} \tag{17}$$

(where $|i|$ and $|j|$ denotes an evaluation of the elements i and j in the groups $(N, ⊞_1)$ and $(N, ⊞_2)$, respectively, is called a logical differential operator.
There exists a basic theory for logical differential operators (e.g. a theory of logical differential equations and methods for their solution). Therefore this specialisation fo a convolution operator may hold out some hope of applicability to image processing.

(b) Classification with respect to the spatial domain

If we select the component groups $(N, ⊞_1)$ and $(N, ⊞_2)$ respectively, as special abelian groups we derive a special case of convolution operators. The most familiar case is obtained when we take $(N, ⊞_1) = (N, ⊞_2) = (N, +\ (\mathrm{mod}\ n))$ (= the cylic group Z_n of order n).
Then $\mathcal{R} = Z_n \otimes Z_n$ and the relevant convolution operators are those most often realized in the spectral-domain by the (finite discrete) Fast Fouriertransform (GOOD [10], COOLEY-TUKEY [11]). Another familiar type of a convolution operator is obtained if we select $(N, ⊞_1) = (N, ⊞_2) = (N, \oplus)$, where $\oplus$ denotes "addition modulo 2" of the numbers 0,1,...,n-1. This specialization results in dyadic convolution filters which have been fully investigated (e.g. PICHLER [6] , [12]). The related characters are the Walsh-functions. Here the Fast Walshtransform gives an efficient spectral-domain realization. "Between" these two (to some extent) extreme classes of convolution operators there exists a spectrum of other classes which are "nice" to realize and might stay "nice" also for certain applications (Compare e.g. with CASPARI [13] and KUNZ [4]).

6. CONCLUSION

Our paper presents an introduction to the "world" of 2-dimensional convolution operators. We investigated the case of a spatial-domain which is the direct product of abelian groups. Then the realization of such operators in the spatial-domain as well as in the spectral domain seems to be "nice". The principles, by which special classes of convolution operators can be constructed have been presented. The usual convolution operators - usually realized by FFT-methods - and the dyadic convolution operators - represented via the FWT - turned out to be within these classes.

All methods which can be derived from the framework discussed are - to borrow a phrase frome KUNZ [4] - "optimal in respect to method" (methoden-optimal). This not only because of the fast realization algorithms available but also because we have a well established theory (e.g. a theory of optimal Wiener filtering).

The concept discusses make a "top down design" of realizations possible. It is therefore to be hoped that they are also helpful in deriving methods, which are "optimal in respect to the problem" (problemoptimal). This view ancouraged us to present this general framework.

LITERATURE

Andrews, H.C.: 1975, Two Dimensional Transforms, in: "Picture Processing and Digital Filtering" (ed. T.S.Huang), Springer Verlag Berlin, pp. 21-68

Ahmed, N. and Rao, K.R.: 1975, Orthogonal Transforms for Digital Signal Processing, Springer Verlag Berlin

Gonzalez, R.C. and Wintz, Paul: 1977, Digital Image Processing, Addison Wesley Publishing Comp., Reading, Mass.

Kunz, H.: 1977, Approximation optimaler linearer Transformationen durch eine Klasse schneller, verallgemeinerter Fourier-Transformationen, Dissertation ETH 5832 Zürich, Juris Druck + Verlag Zürich

Nicholson, P.J.: 1971, Algebraic Theory of Finite Fourier Transforms Journal of Comp. and Syst., Sc.5, pp. 524-547

Pichler, F.: 1972, Dyadische Faltungsoperatoren zur Beschreibung linearer Systeme, Sitzungsberichte der Österr. Akademie der Wissenschaften, Smn 18o-12, pp. 69-87

Roth, D.: 1973, Digitale Vielfaltfilter, Archiv für Elektronik und Übertragungstechnik, 27 (1973), pp.72-78

Weiss, P.: 1967, Zusammenhang von Walsh-Fourierreihen mit Polynomen, Mh. für Mathematik 71, pp. 165-179

Gibbs, J.E. and Millard, M.J.: 1969, Some methods of solution of linear ordinary logical differential equations, National Physical Laboratory, Teddington, Middlesex, England, DES Report No.3

Good, I.J.: 1958, The Interaction Algorithm and Practical FOURIER Analysis, Journal of the Royal Statistical Society, Series B, Vol.B2o, No.2, pp. 361-373

Cooley, J.W. and Tukey, J.W.: 1969, An Algorithm for the Machine Calculation of Complex FOURIER Series, Mathematics of Computation, Vol.19, pp. 297-3o1

Pichler, F.: 1973, Walsh Functions: Introduction to the Theory, in: Signal Processing (ed.J.W.R.Griffiths, P.L. Stocklin and C. van Schooneveld), Academic Press London and New Yor, pp. 23-41

Andrews, H.C. and Caspari, K.L.: 1970, Generalized Technique for Spectral Analysis, IEEE Trans. on Computers, Vol.C-19, No.1, Januar 197o

RECURSIVE PROCESSING FOR POSITION CORRELATION IN NOISY IMAGES

Dan Andrée and Åke Wernersson

Optimization and Systems Theory
Royal Institute of Technology
S-100 44 Stockholm 70, Sweden

ABSTRACT

A method is given for estimating the position of a noisy sub-image relative to a larger reference image or a map containing some kind of contour. The first step is to estimate where the contours are crossing each scan in the smaller sub-image. This scan processing is done using a recursive state estimation method. A SNR of unity gives approximately an uncertainty of two pixels along the scan. The estimated edges are finally to be correlated with the stored contour of the larger reference image. The correlation is done using a 2-dim maximum likelihood estimate. The number of operations and the memory requirements are proportional to size of the registered sub-image. The size of the larger reference image is of less importance. The method is tested on Landsat images.

1. INTRODUCTION – THE PROBLEM

In this paper we summarize a method for extracting position information from images, i.e. to find the position of a noisy sub-image relative a larger reference image.

The physical situation to be considered is the following: A given scene is described by a reference image. The scene is assumed to contain some kind of well defined contour(s), e.g. between land and water. Since the position information in an image is essentially determined by the contours/edges, the reference image is stored as contour positions. A portion of the scene is registrated giving a noisy sub-image. The problem is to estimate

D. G. Lainiotis and N. S. Tzannes (eds.), Applications of Information and Control Systems, 12-19.

the (x,y)-position of the sub-image relative to the larger reference image - compare the illustration in Fig. 1.

Problems of this type are relevant in several cases of 'image feed-back' control systems such as: updating a fully automated navigational system or controlling an industrial robot etc.

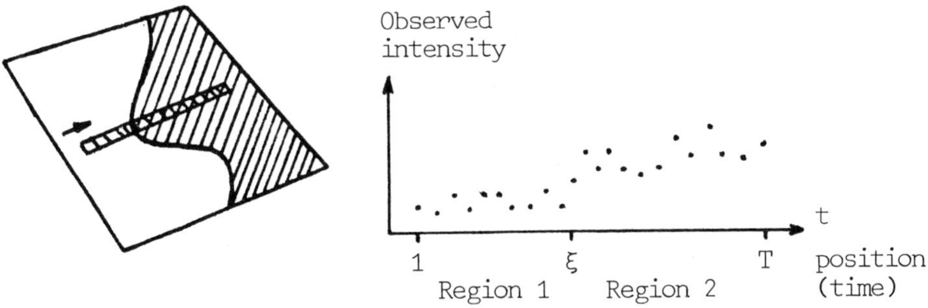

FIG. 1. From one scan we get observations as illustrated in the figure. The first step in the processing is to estimate, for each scan, the position ξ of the edge. The second step is to use these edge estimates to make a position correlation with the reference image.

2. THE STRUCTURE OF THE RECURSIVE SCAN PROCESSING

The purpose of the scan processing is to obtain an estimate where each scan in the smaller sub-image is crossing the contour(s), compare figure 1.

The structure of the scan processing is shown in figure 2. The position index t along the scan is below referred to as time. The observations are first processed forward in time to obtain the <u>filter estimate</u> $\hat{e}(t)$, where

$$\hat{e}(t) = \Pr(\text{index } t \text{ is in region } 2 | \text{observations } Z(\tau) \text{ with } \tau \le t)$$

Hence, $\hat{e}(t)$ is a one-sided estimate. Of course, we are interested in the two-sided <u>smoothing estimate</u>

$$\hat{e}(S|T) = \Pr(\text{index } S \text{ is in region } 2 | \text{ all observations } Z(\tau) \text{ with } \tau \le T).$$

The smoothing estimate, $\hat{e}(S|T)$, is also obtainded recursively, but backwards in time using the filter estimates - the last row in figure 2.

Measurements:		Z(1)		Z(t+1)	Z(T)
		↓		↓	↓
Likelihood function:		L(·)		L(·)	L(·)
		↓		↓	↓
Filter:	$\hat{e}(0)$ →	$\hat{e}(1)$ →...→	$\hat{e}(t)$ →	$\hat{e}(t+1)$ →....→	$\hat{e}(T)$
		↓	↓		↓
Smoothing:		$\hat{e}(1\|T)$ ←..←	$\hat{e}(t\|T)$ ←	$\hat{e}(t+1\ T)$ ←..←	$\hat{e}(T\|T)$

FIG. 2: The structure of the recursive scan processing. For each observation Z(t+1) a likelihood ratio is calculated which is weighted non-linearly with $\hat{e}(t)$ to give $\hat{e}(t+1)$. $\hat{e}(0)$ equals the initial 'guess' that Z(1) is taken from region 2. The smoothing estimates $\hat{e}(t|T)$ are computed recursively using $\hat{e}(t)$, but backwards.

The non-linear recursive method is based on recent results from state estimation of Markov chains - c.f. ref. [1,3]. If we exclude more than one edge in the model for the scan, i.e. the Markov chain is absorbing, $\hat{e}(t|T)$ is the conditional distribution function of the edge. Given the smoothing estimate, the natural edge estimate $\hat{\xi}$ is the point where $\hat{e}(t|T)$ crosses the 50%-level - i.e. the conditional median.

The behaviour of the filter and smoother is shown in figure 3. The curves are the simulated observations and the corresponding filter and smoothing estimate, respectively. In the observations some 'spikes' have been added to demonstrate the robustness of the method.

It should be mentioned that several of the parameters, i.e. the levels h_1, h_2 in fig. 1, and the standard deviations have been estimated using a two dimentional clustering algorithm, see [6] pp. 94 and [1].

3. POSITION CORRELATION

From the scan processing summarized above we get position estimates of the edge/contour in the sub-image. To obtain the final position estimate these estimated edge positions are to be correlated with the pre-stored contour positions.

The 2-dimensional Maximum Likelihood Estimate is used as the position estimate. More precisely, the estimated edge positions of the sub-image are taken as the feature vector. According to the experience gained in [1,2] it is reasonable to model the fluctua-

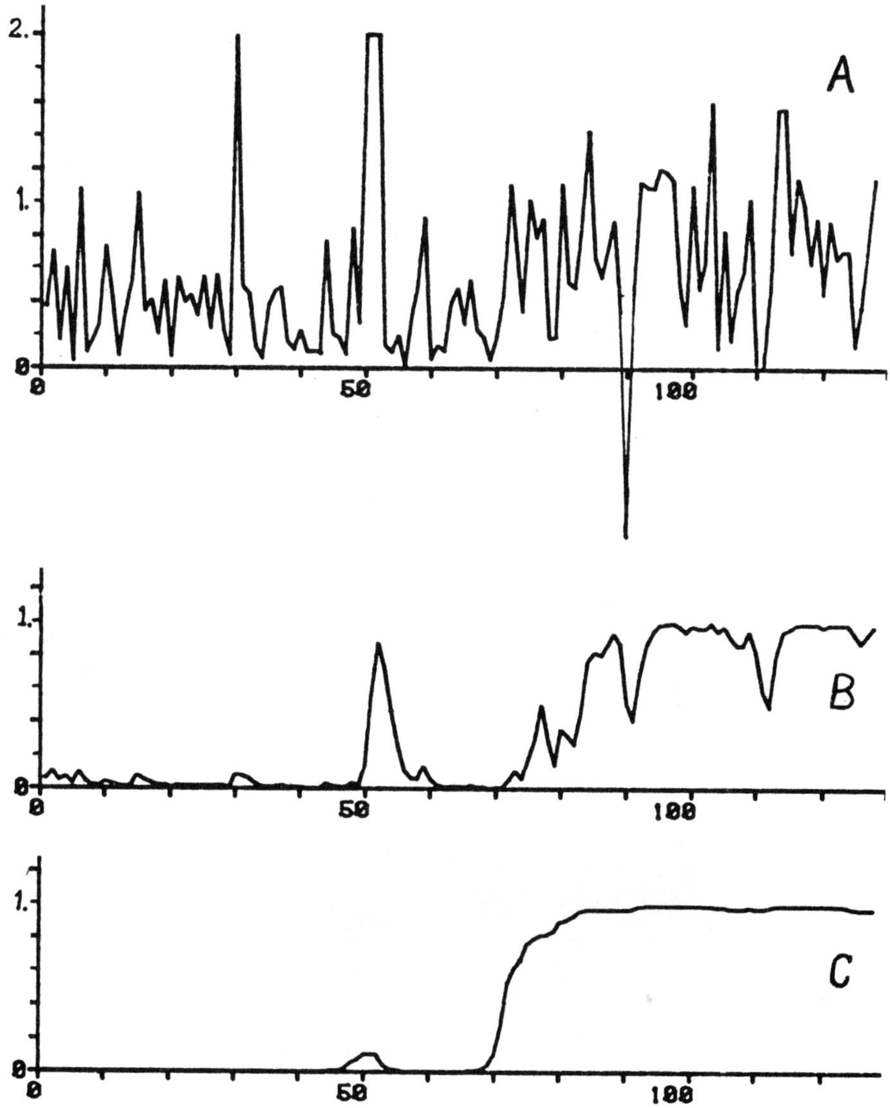

FIG. 3. Simulated noisy observations with added 'spikes' is shown in curve A. The 'spikes' give a response in the 'one-sided' filter estimate - curve B. Only the second 'spike' at index 50 gives a small response on the 'two-sided' smoothing estimate - curve C.

tions in these feature vectors by a symmetric exponential density. As pattern classes, c.f. [4,5] and [2], we choose all possible (x,y)-positions.

Using these assumptions, it turns out that the MLE is obtained by a one-dimensional contour 'sliding' along the prestored contour. The best fit is where the absolute difference between the estimated edge positions and the pre-stored positions is minimal. It should be observed that the MLE is 2-dimensional, but it is possible to handle one direction analytically. Hence, it is only necessary to make a one-dimensional numerical search.

In the minimization above the position of the registrated sub-image is assumed to have uniform a priori distribution. It is straight forward to include a non-uniform a priori information. Other types of errors in the feature vector give other functions to minimize. Multiple contours of modest complexity can also be handled.

4. A TEST ON LANDSAT IMAGES

In this section the method is illustrated, using Landsat images. Images from three spectral regions have been used, infrared (band 7 and band 6) and red (band 5). In figure 4 an image from band 7 is shown. The image consists of 128×128 pixels and the size of each pixel is 59×78 m. The signal to noise ratio (SNR), between land and water, for the three bands are approximately 3.2, 3.1, 0.7 respectively.

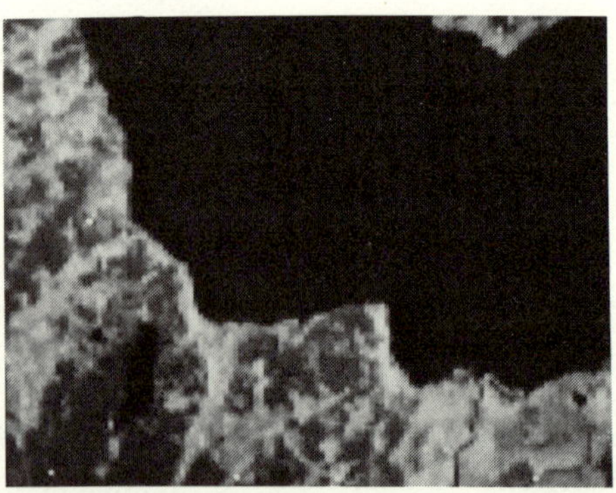

FIG. 4. A Landsat image from band 7.

Due to the practical difficulties to get the 'pre-stored' contour from a 'real' map we here use the smoothing estimate of the image in figure 4 instead of a map. The resulting 'smoothing estimate is shown in figure 5. In the processing, no limitations have been imposed on the number of edges. From the smoothing estimate in figure 5 we get the 'pre-stored' contour shown in figure 6.

FIG. 5. The smoothing estimate of the image in fig. 4.

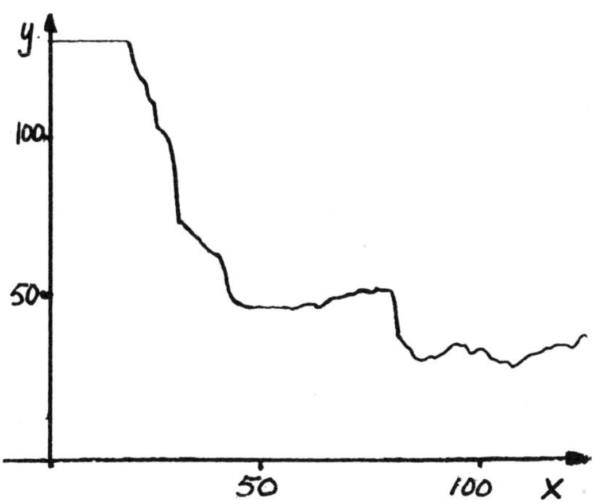

FIG. 6. The pre-stored contour obtained from fig. 5 - 128 edge positions.

To illustrate the position correlation method we next choose two sub-images, one from band 6 and one from band 5. The sub-images, of the size 16×80 pixels, are shown in figure 7. The problem is to find where these two sub-images fit along the contour in fig. 6. These sub-images are to be processed using the scan processing described in section 2. The discrimination function from the position correlation, described in section 3, is shown in figure 8A for sub-image 7A. The sharp peak value, at x = 100, is at the true position. It should be mentioned that the margin to the smallest but one peak value is more than 10 standard deviations.

For the sub-image in fig. 7B, band 5, four adjacent scans have been averaged before the scans are processed. The SNR in the resulting scans is approximately 1.4 compared with 0.7 before the averaging. The discrimination function is shown in fig. 8B. The sub-image from band 5 does not give the same sharp peak value, compared with band 6, bacause the position correlation is based on only four averaged edge position.

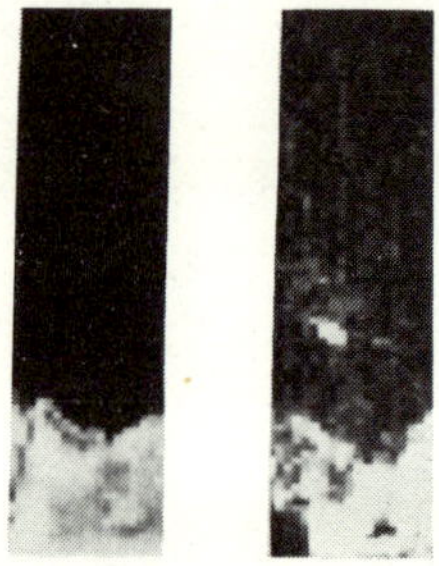

FIG. 7. Two sub-images, of size 16×80 pixels, from band 6 and band 5, respectively. The problem is to find where the two images fit the contour in fig. 6.

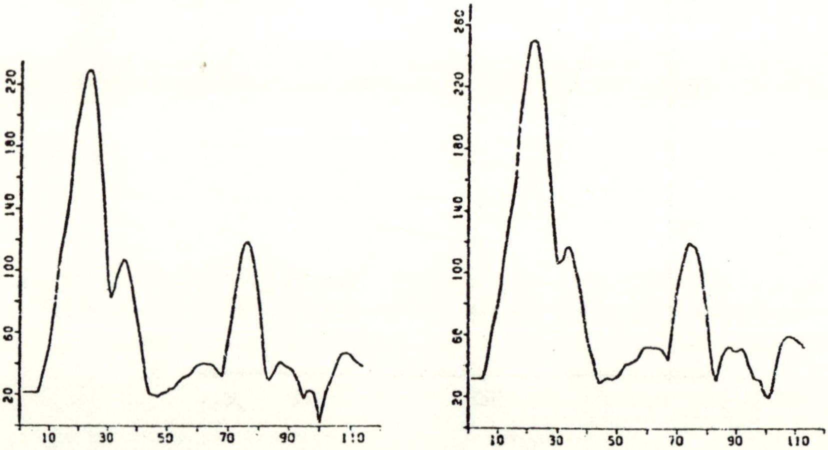

FIG. 8. Discrimination curves for the images in fig. 7. The minimizing value gives the x-estimate of the position. The y-estimate is obtained as described in section 3.

5. SUMMARY

The main ideas have been presented for a two-stage image processing method for position correlation. The method summarized here, should in several aspects be contrasted with the match correlation methods studied in, say, [7,8]. The described method is computationally simpler and gives a much sharper discrimination than matching provided the selected reference scene does not contain an 'arbitrary' number of contours.

ACKNOWLEDGEMENTS

The authors wish to thank; T. Orhaug for suggesting the problem and for many fruitful discussions; S. Nyberg for providing us with the image outputs and the Research Inst. of National Defence for financial support.

REFERENCES

1. D. Andrée, Å. Wernersson; Recursive Scan Processing for Estimating Edges in Noisy Images, FOA rep. C 30138-E1, National Defence Research Institute, 1978.

2. D. Andrée, Å. Wernersson; An Image Processing Method for Position Correlation, (to appear as a FOA report).

3. Å. Wernersson; Some State Estimation Problems Connected with Discrete Time Markov Chains, Gregson (ed.); Recent Theoretical Developments in Control, Academic Press 1978.

4. K.S. Fu; Sequential Methods in Pattern Recognition and Machine Learning, Academic Press 1968.

5. E.A. Patrick; Fundamentals of Pattern Recognition, Prentice Hall Electrical Eng. Series, 1972.

6. J. Tou, R.C. Gonzales; Pattern Recognition Principles, Addison-Wesley, 1974.

7. R.Y. Wong; Sequential Scene Matching Using Edge Features, IEEE Trans. Areosp. Electron. Syst., Vol. AES-14,pp. 128-140, Jan. 1978.

8. L. Novak; Correlation Algorithms for Radar Map Matching, IEEE Trans. Aerosp. Electron. Syst., Vol. AES-14, pp. 641-648, July 1978.

COMPRESSION OF BINARY PICTURES USING PARALLEL OPERATIONS

C. Arcelli, L. P. Cordella, S. Levialdi

Laboratorio di Cibernetica C.N.R. - Arco Felice, Naples, Italy.

ABSTRACT. - Problems of image description and classification are often faced by considering, instead of the digital image A, a compressed version S of it, which has unitary thickness everywhere. Generally S is obtained by deleting from A those elements adjacent to the background which are not necessary either to preserve the connectedness of the image or to identify the protruding regions of A. The process is iterated until deletable elements no longer exist in A. In this communication we present a different approach involving a step by step propagation of the background over the image as far as the whole A has been covered. Within every step, we extract from A and assign to S, those elements which are embedded in significant protrusions or ensure that a connected S is obtained. Since no deletion operations are used, it becomes possible to perform a transformation of parallel nature which is isotropic and involves only operations on the nearest neighbors of every element.

1. INTRODUCTION

The reduction of the amount of data to be processed, by eliminating that part of information which is not needed for the given task, is an important problem in many fields. Specifically in image recognition, it seems often convenient to transform elongated digital images such as typed or handwritten characters, chromosomes, fingerprints etc. into images of unitary thickness (1,2): in fact the features characterizing an elongated image like an alphanumeric symbol for instance, are generally not dependent on the thickness of the sign representing it. On the other hand, both for recognition and storage purposes, it may be useful to have a thinning transformation which could be applied also to non-elongated images. In this case,

D. G. Lainiotis and N. S. Tzannes (eds.), Applications of Information and Control Systems, 20-28.

however, the thinned image should preserve the information about the width and contour shape of the original input.

In principle it would be interesting to have so versatile algorithms that, applied to the digitized version of an input image, allow to derive from it a thinned image which is: i) representative of the original image (in the sense that it preserves the wanted discriminant information), ii) connected when deriving from a connected image, iii) independent of rotation, enlargement and contraction of the input, iv) reversible, so that the original image can be recovered starting from its thinned version. In practice, due to the properties of the digital plane, it is very difficult not only to get all these features together, but even, in some cases, to assign to them an unquestionable meaning.

Generally thinning algorithms (3-7) of binary digital images are based on the deletion of some contour elements according to rules involving nearest neighbors and preserving connectedness. Furthermore, to avoid the disappearance of digital arcs, the deletion of the elements having only one neighbor is inhibited. The process is iterated until there are no more elements satisfying the deletion conditions. Such transformations appear as essentially parallel in nature, so that they can be achieved using only operations which cause the simultaneous change of the state of every image element as a function of the state of itself and of its nearest neighbors. This fact and the consideration that parallel hardware is expected to play a very important role in the near future of picture processing, because of the reduction of computation time, (8) has made parallel thinning algorithms very popular in the literature. However, to avoid that certain configurations either disconnect or vanish, these algorithms must be devised as a sequence of four phases, each of which considering as candidates to deletion only those pixels 4-adjacent to the background in a specified direction, i.e., the four phases respectively involve north, south, east and west border pixels. The transformation achieved in this way is not completely parallel: in fact only subsets of the contour are processed simultaneously and the sequence according to which the phases are executed influences the shape of the thinned image, thus leading to an anisotropic behavior.

Another approach to thinning, based on the ideas suggested by Blum (9) for binary images in the Euclidean plane, is that of preserving all those image points lying in positions where at least two wavefronts of a fireline propagating with constant speed from the contour towards the inside of the image, meet during propagation (quench points). This approach, which has not been much studied in the digital plane, is more suitable when detailed information on the image contour shape has to be preserved. In this communication we present the outline of a parallel algorithm based on this last approach and discuss the consequences of shape distorsions occurring to a digital fireline during propagation.

2. SKELETONIZING

According to our algorithm, in analogy with the fireline propagation, the background $\bar{A}$ (set of 0-elements) of a picture propagates step by step over the image A (set of 1-elements) including at every step all and only the elements of A which are 4-adjacent to it (see Fig. 1).

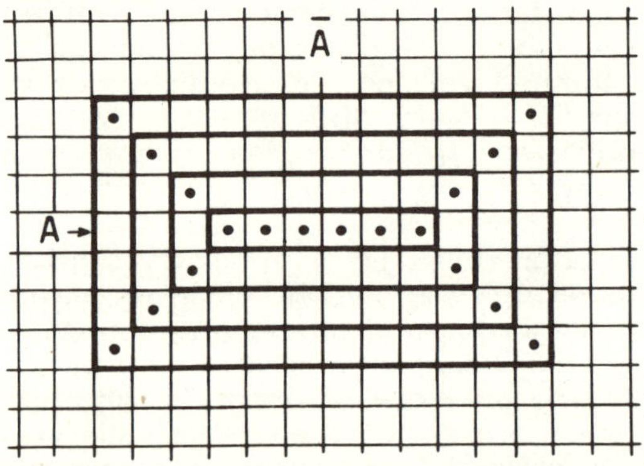

Fig. 1 - Step by step propagation of $\bar{A}$ over A. The set of dotted elements is the skeleton of A.

The process causes the shrinking of A, originating shorter and shorter contours. On every current contour, some suitable elements are selected and their locations stored in a memory area. The process terminates when all the elements of A have been included into $\bar{A}$. The union of the selected elements constitutes the thinned image of A and will be called skeleton of A. For what has been already said, the skeleton elements should satisfy the condition of lying in quenching positions of interacting wavefronts. From an operative point of view, the parts of the contour where they can be found are either convex arcs or arcs which are repeating or "brush past" (10) (i.e., they are element to element adjacent to another contour arc). In the latter case all the elements belonging to the specified arcs should be attributed to the skeleton, while in the former case only one element, lying in the position of higher curvature, has to be selected (refer to Fig. 2).

It can be shown that the local maxima (11) obtained by applying a 4-distance transform to A, are a subset of the elements, satisfying the above operative definition (Fig. 3).

Moreover local maxima constitute the smallest subset of A from which the whole A can be reconstructed by a reverse 4-distance transform (11). The skeleton could then be made by the union of

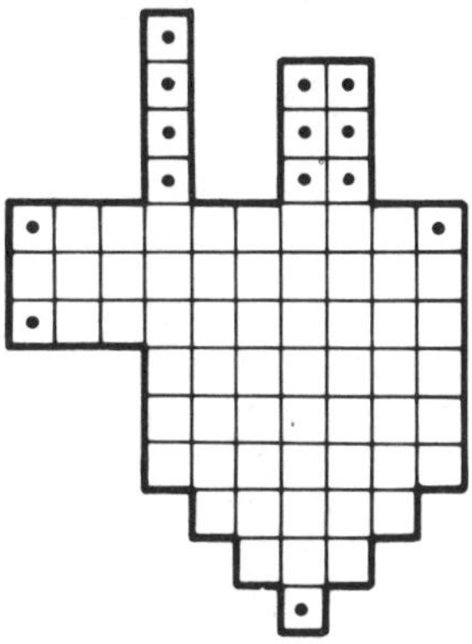

Fig. 2

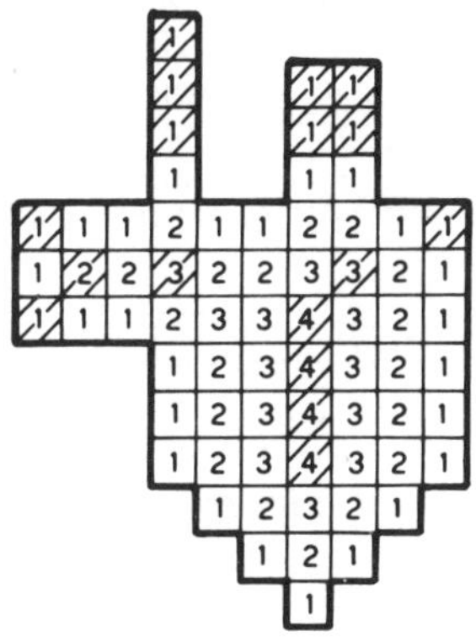

Fig. 3

Fig. 2 - Examples of elements (dots) which should be selected as skeleton elements at the first propagation step.

Fig. 3 - 4-distance transform of A. The local maxima are dashed.

local maxima and of some other elements which, although non local maxima, belong to convex or protruded regions of the image. However, two main problems arise: first of all it is difficult in practice to establish an operative rule to decide whether convexity exists on a digital image, especially when gentle contour variations occur (12,13) ; secondly, due to the distorsions of the contour of A during the propagation process, it may happen that to a contour arc detectable as convex according to a given criterion for detecting convexity, a non convex (or non detectable as convex) contour arc succeeds (see Fig. 4). This may happen whatever the criterion adopted for detecting convexities and the metric used for the propagation of the background. The consequence is that the skele-

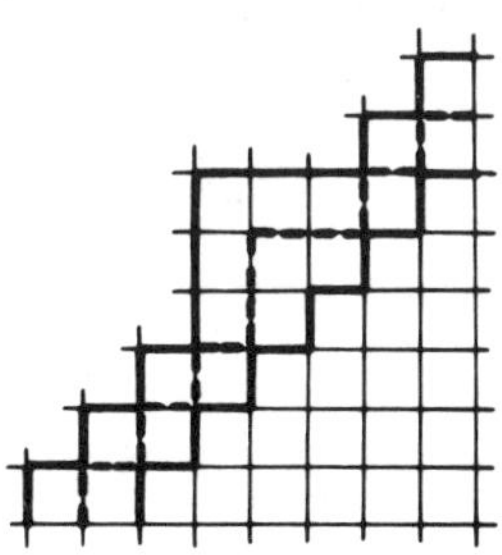

Fig. 4

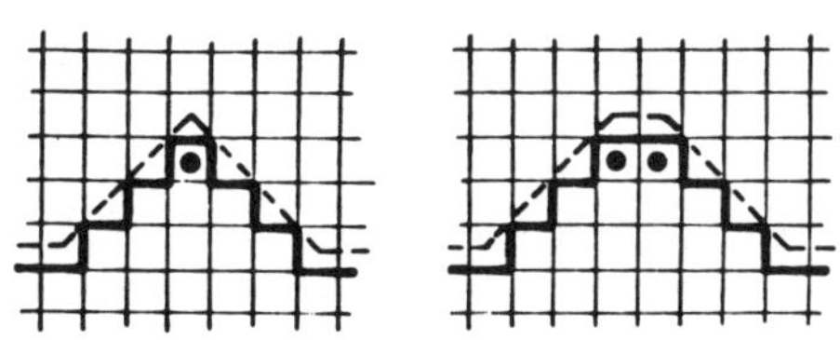

Fig. 5

Fig. 4 - A convex configuration which, after some steps of propagation, is transformed in a non convex one. Successive contours are alternatively outlined by continuous and dashed lines.

ton of a connected image could be disconnected. It is then necessary to study the situations where disconnections appear and find a strategy for preserving skeleton connectedness.

3. CONNECTING STRATEGIES

For the sake of simplicity and so as to have an algorithm especially tailored for running on a parallel computer, we will limit here ourselves to consider the case that only some elements identifying sharp convexities are added to local maxima. Specifically, if the amplitude of the angle formed by two interacting wavefronts of the fireline is taken as the parameter characterizing the sharpness, then the values smaller than or equal to $\pi/2$ are assumed as identifying detectable convexities. This implies that, besides local maxima, only the dotted elements in Fig. 5 are added to the skeleton. The usefulness of the algorithm is however not strongly limited since, in a wide number of practical cases, the shape of an image is mainly characterized by convexities as strong as the ones so detected.

Once indicated by A_i, A_{i+1}, etc. the successive sets of 1-elements obtained after i, i+1, etc. steps of propagation of $\bar{A}$ and by C_i, C_{i+1}, etc. their respective 4-connected contours, it is easy to verify that the so obtained skeleton S is 8-connected only if the following conditions are both verified:

α) every element of S detected on C_i is 8-connected to A_{i+1}, i.e., it is 8-adjacent to A_{i+1} or it is connected to A_{i+1} through a path made by elements of S.

β) at least one element belonging to S and 8-adjacent to elements of S detected on C_i, exists on A_{i+1}.

Since such conditions are not always satisfied (Fig. 6), our connecting strategy imposes that, as soon as a local maximum or an element of convexity has been detected on a current contour C_i, at

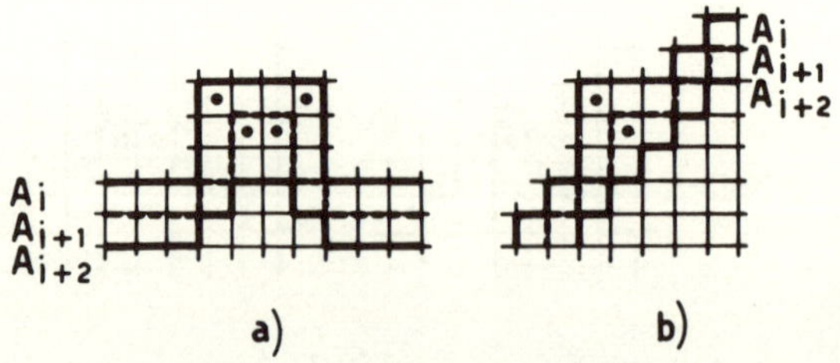

Fig. 6

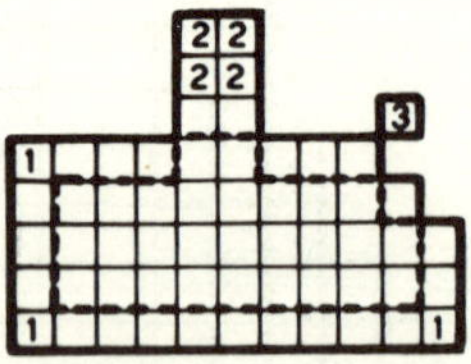

Fig. 7

Fig. 6 - Cases in which condition α (a) and condition β (b) are not verified.

least one element is generated which is 8-adjacent to it and nearer to the kernel, i.e. to the set of elements belonging to A and bearing the greater 4-distance from $\bar{A}$. The location of the successor(s) of a skeleton element detected at step i, depends on the structure of the configuration in which the element is embedded.

In the hypotheses we made, skeleton discontinuity may take place only in the 3x3 neighborhood of a local maximum. In fact the elements not local maxima that we limited ourselves to consider as individuating convexities, belong to geometrical configurations which either are not distorted by the type of propagation process we use, or are distorted in such a way that an 8-adjacent local maximum is originated on A_{i+1}.

The local maxima extracted at step i can be distinguished into three types, depending on their metric properties with respect to A_{i+1}; we will indicate by m_k, (k = 1,2,3) a local maximum, according to its type. Examples of the three types are labeled 1,2,3 in Fig. 7. Without entering in too many details, it can be seen that the m_1's satisfy condition α but not always condition β . The possible contour distorsions are shown in Fig. 8 a,b: in both cases an element b 8-adjacent to an m_1 existing on C_i can be individuated on A_{i+1} by using only parallel operations (namely: diagonal expansion of m and intersection of the so obtained set with A_{i+1}). Such local operations are applied to all the local maxima, but they

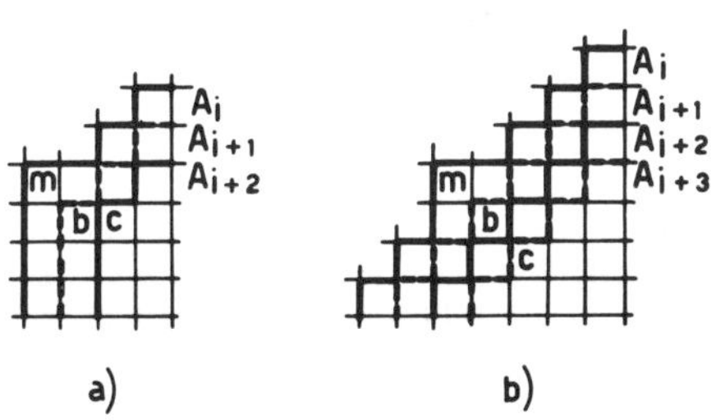

Fig. 8

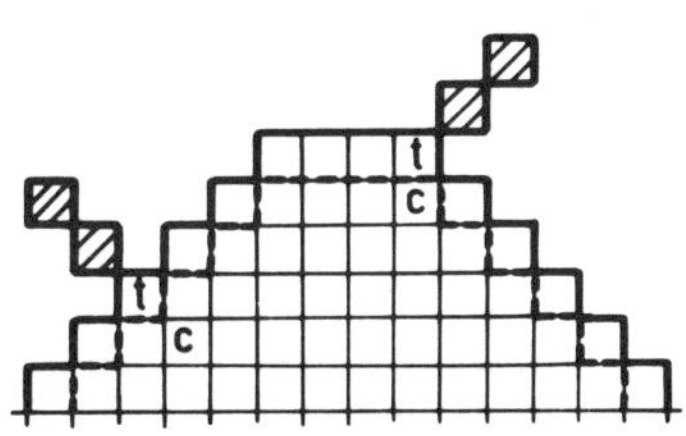

Fig. 9

are active only on the m_1's. The elements b are attributed to S and also used for generating successors on A_{i+2}. According to the geometry of the contour arc in which b is embedded it appears convenient that the skeleton branch passing through b has the diagonal, horizontal or vertical direction. Namely, in terms of local properties, if b is 4-adjacent to only one element of A_{i+2} its successor c is located on A_{i+2} and is 4-adjacent to b; if b is 4-adjacent to two elements of A_{i+2}, c must be 8-adjacent to b and belonging to A_{i+3}. At successive propagation steps, the elements c generate successors with the same technique now illustrated for b.

The m_2's never satisfy both conditions α and β . Differently from the case of the m_1's, it is then necessary to separate the m_2's from the remaining local maxima before applying to them the connecting process: in fact the procedure we devised for generating the successors of the m_2's is not sufficiently selective, so that if applied to some of the m_1's or m_3's, would generate spurious elements. Note that at every step the m_2's are a subset of the difference set W_i between A_i and the 8-expansion of the set of elements of A_i having all 1-elements in their 8-neighborhood (8-core of A_i). The set W_i also contains the convexity elements of the type of Fig. 5.

Moreover the parallel operations by which W_i is extracted, are the simplest ones for extracting such elements. In conclusion, W_i is added to S and to satisfy the requirements of condition α , the set L_i of elements 8-adjacent both to W_i and to A_{i+1} is also added to S. To fulfill in any case also the requirements of condition β such last elements are 4-expanded and the intersection of the obtained set with A_{i+1} added to S. The so obtained elements must also be added to L_{i+1}, in order that memory is retained of the already existing skeleton path. It can be proved that the application of the whole process to the elements of W_i which are not m_2's, never generates unexpected branches.

To guarantee the connectedness to the kernel of the third type of local maxima, the same procedure employed for the m_1's can be mainly used: the diagonal expansion of the whole set of local maxima (and then of the m_3's) is intersected with C_i and the obtained elements t_i treated exactly as the elements of the type b found in the case of the m_1's (refer to Fig. 9).

4. CONCLUSIONS

The skeleton we obtain is connected, reversible (this is guaranteed by the presence of local maxima), isotropic in the sense that the detectability of a configuration as a convex one does not depend on its orientation. Finally it is sufficiently representative of the original image, as it can be seen from the examples given in Fig. 10 . A closer representativity could be obtained by allowing branches to start from all the meaningful convexities of the image, but this is material for further work.

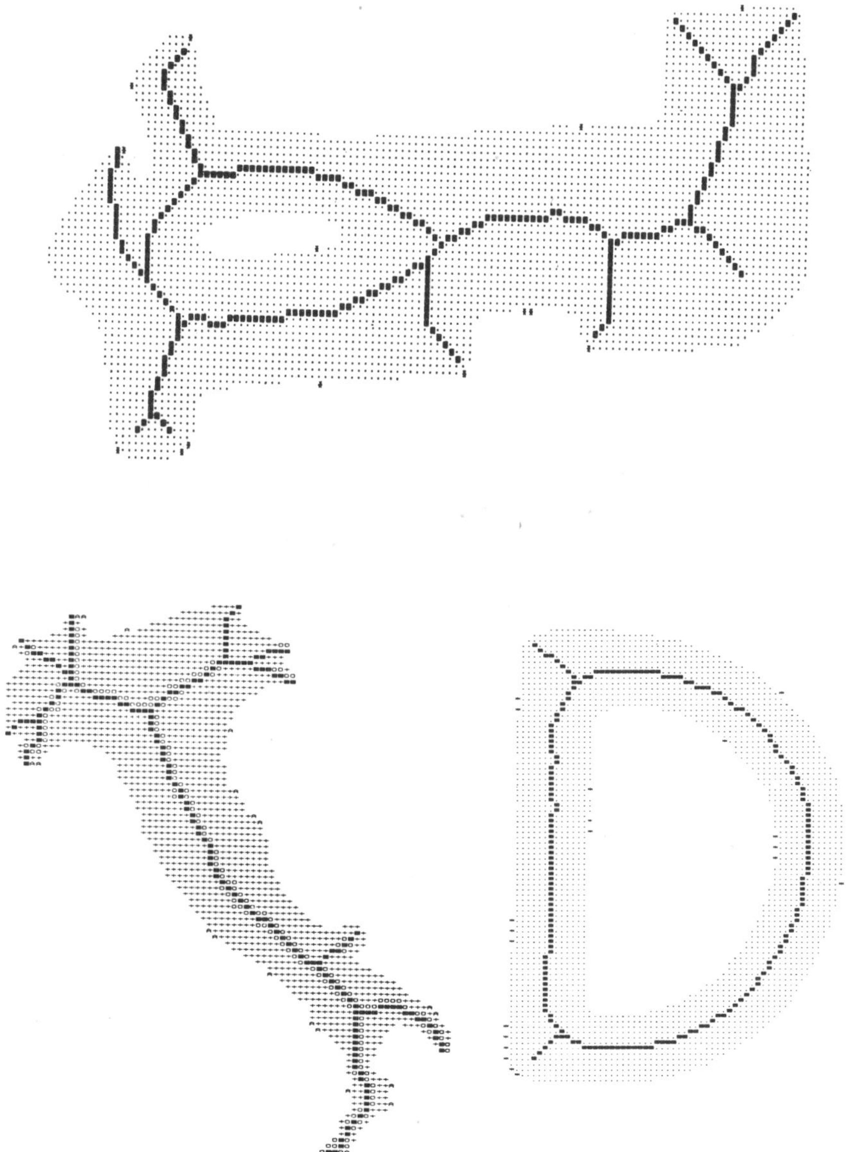

Fig. 10 – Examples of skeletons.

REFERENCES

(1) Rosenfeld, A., and Kak, A.C.: Digital Picture Processing, 1976, Academic Press, New York.

(2) Pavlidis, T.: "A review of algorithms for shape analysis" 1978, Computer Graphics and Image Processing, 7, pp. 243-258.

(3) Hilditch, C.J.: "Linear skeletons from square cupboards", 1969 in Machine Intelligence IV, B. Meltzer and D.Michie, Eds. Edinburgh University Press, pp. 403-420.

(4) Deutsch, E.S.: "Thinning algorithms in rectangular, hexagonal, and triangular arrays", 1972, Comm. ACM, 15, pp. 827-837.

(5) Stefanelli, R., and Rosenfeld A.: "Some parallel thinning algorithms for digital pictures", 1971, J. ACM, 18, n.2, pp. 255-264.

(6) Rosenfeld A.: "A characterization of parallel thinning algorithms", 1975, Information and Control, 29, pp.286-291.

(7) Arcelli, C., Cordella, L., and Levialdi, S.: "Parallel thinning of binary pictures", 1975, Electronics Letters, 11, n.7, pp. 148-149.

(8) Cordella, L., Duff, M.J.B., and Levialdi, S.: "An analysis of computational cost in image processing" - a case study", 1978, IEEE Trans. Computers, C-27, pp. 904-910.

(9) Blum, H.: "A transformation for extracting new descriptors of shape", in Models for the perception of speech and visual form, 1967, W. Wathen-Dunn, Ed. Cambridge, MA., MIT Press, pp. 362-380.

(10) Rosenfeld, A.: "Arcs and curves in digital pictures", 1973, J. ACM, 20, no.1, pp. 81-87.

(11) Rosenfeld, A. and Pfaltz, J.L.: "Sequential operations in digital pictures processing", 1966, J. ACM, 13, no.4, pp. 471-494.

(12) Arcelli, C. and Cordella L.: "Concavity point detection by iterative arrays", 1974, Computer Graphics and Image Processing, 3, pp. 34-47.

(13) Arcelli, C., and Massarotti, A.: "Regular arcs in digital contours", 1975, Computer Graphics and Image Processing, 4, pp. 339-360. Erratum 1976, vol. 5, p. 289.

CONSTRAINT OPTIMIZATION TECHNIQUES FOR FAN BEAM PROJECTION IMAGING

C.E. Goutis* and T.S. Durrani**

*Department of Electrical and Electronic Engineering, University of Newcastle upon Tyne NE1 7RU, U.K.
**Dept. of Electronic Science and Telecommunications, University of Strathclyde, Glasgow G1 1XW, U.K.

ABSTRACT. The reconstruction from diverging ray projections is formulated and solved as a constraint optimization problem, using the projections as constraints, and adopting a general cost function which depends on derivatives. The application of this solution to several specific cost criteria such as energy, entropy and derivative norm, created a number of relationships (models). The fan beam 'additive' and 'multiplicative' ART derived from the energy and entropy models respectively are presented. ART associated to new criteria are also given. The reconstructions of the algorithm produced using the minimum energy criterion are compared with the solutions of the convolution algorithm whose approximations are discussed.

1. INTRODUCTION

There is a growing number of disciplines where reconstruction of an object from its projections is required. These are as diverse as computerised axial Tomography (1) where the competition for fast and high resolution scanners is very strong, Radioastronomy (2) and Ultrasonics (3).

Projection data may be obtained using parallel rays or diverging rays (fan beam) depicted in Fig.1. The absorption of the X-ray beam passing through the (O_kAB) strip depends on the structure of the slice (cross section). A projection point is given by $In(I_o/I)$, where I_o and I denote the strip flux before and after passing through the slice respectively. For each position of the X-ray source, the detector measurements give a complete projection. Rotating the source produces different

D. G. Lainiotis and N. S. Tzannes (eds.), Applications of Information and Control Systems, 29-37.

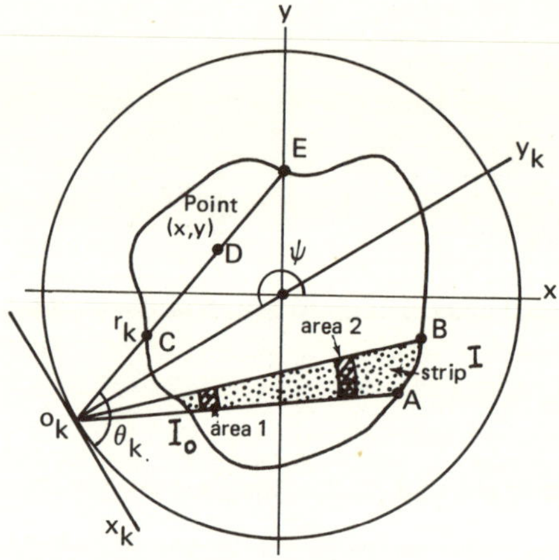

Fig.1 Fan beam geometry,
$\ell_1(\Theta_k) = C_k C$, $\ell_2(\Theta_k) = O_k E$, and $r_k = O_k D$

projections. The aim of Tomography is to reconstruct the full image of the cross-section from the projections at various orientations. Normally, the structure of the whole 3-dimensional (3-D) object may be found by portioning it into a number of parallel cross-sections. An extension to three dimensions of the scanning scheme given in Fig. 1 results in cone beam projections.

The main solution methods for reconstruction from parallel projections are: (a) ART and other related methods (4), (b) convolution algorithm (5), (c) Fourier methods (6), and (d) constrained optimization methods (7) (8). In the following, the fan beam constrained optimization method, which includes derivative dependent cost criteria, will be introduced. An algorithm giving the solution of the minimum energy criterion will also be given in Section 2. The natural extension of ART techniques to divergent ray projections will be presented in Section 3. Note that the ART for twin-cone projection data given in (9) are not considered as the true extension. This will be justified in Section 3. Computer simulation will be included in Section 4.

2. CONSTRAINT OPTIMIZATION

Reducing the strip (O_kAB) to a line, the continuous projection at angle Ψ_k is given by

$$g_k(\Theta_k) = \int_{\ell_2}^{\ell_1} f(x,y)\, dr_k \qquad (1)$$

where f(x,y) represents the slice for reconstruction r_k, Θ_k are the polar coordinates with original at O_k, and ℓ_1, ℓ_2 are as in Fig. 1. This reduction facilitates the optimization analysis and can readily produce the results for the discrete projections.

Using the projections as constraints and employing the Euler-Lagrange method, the solution of the fan beam problem is the function f(x,y) which optimizes the functional integral

$$J = \iint_D C\left(f\,;\,\frac{\partial f}{\partial x}\,...\,\frac{\partial^n f}{\partial x^n}\,;\,\frac{\partial f}{\partial y}\,..\,\frac{\partial^n f}{\partial y^n}\,;\,x,y\right)dxdy$$

$$+\sum_{k=1}^{N}\int_{o}^{2\pi} g_k(\Theta_k)\left[g_k(\Theta_k) - \int_{\ell_1}^{\ell_2} f(x,y)dr_k\right]d\Theta_k \qquad (2)$$

where C is the cost criterion dependent on f, its derivatives and (x,y); D is the domain of definition (area) of the slice, $\{g_k(\Theta_k)\}$ are the continuous 1-D Lagrange multipliers; and r_k, Θ_k, ℓ_1 ℓ_2, are as Fig. 1.

Optimizing J by setting its first variation with respect to f equal to zero and employing the Green's theorem gives

$$\frac{\partial C}{\partial f} + \sum_{i=o}^{n}(-1)^i\left(\frac{\partial^i C_{io}}{\partial x^i} + \frac{\partial^i C_{io}}{\partial y^i}\right) - \sum_{k=1}^{n}\frac{g_k(\Theta_k)}{r_k} = 0 \qquad (3)$$

where $C_{io} = \dfrac{\partial C}{\partial\left(\dfrac{\partial^i f}{\partial x^i}\right)}$, $C_{oi} = \dfrac{\partial C}{\partial\left(\dfrac{\partial^i f}{\partial y^i}\right)}$, $i = 1, 2 .. n$

and n is a positive integer. Table 1 lists the relationship (models), between the image and the Lagrange multipliers, produced by substitution for C in Eqn.(3) the associated cost criteria. All models in Table 1 can be extended to true 3-D models by substituting for $g_k(\Theta_k)$, the corresponding 2-D Lagrange multiplier.

TABLE 1

No.	Cost Criterion	Model, $(x,y) \in D$
1	$\frac{1}{2} f^2$ Energy	$f = f_1 = \sum_{k=1}^{N} \frac{g_k(\Theta_k)}{r_k}$
2	$-f. \ln f$, Entropy 1	$f = \exp(-1+f_1)$
3	$\ln f$, Entropy 2	$f = 1/f_1$
4	f^a	$f = (\frac{2}{a} f_1)^{1/(a-1)}$
5	$\frac{1}{2} [(\frac{\partial f}{\partial x})^2 + (\frac{\partial f}{\partial y})^2]$	$-\nabla^2 f = -(\frac{\partial^2 f}{\partial x^2} \quad \frac{\partial^2 f}{\partial y^2}) = f_1$
6	$\frac{1}{2}(1-a)f^2 + \frac{1}{2}a[(\frac{\partial f}{\partial x})^2 + (\frac{\partial f}{\partial y})^2]$	$(1-a) \cdot f - a \nabla^2 f = f_1$
7	$\frac{1}{2}f^2 + \frac{1}{2}\sum_{m=1}^{n}[a_m(\frac{\partial^m f}{\partial x^m})^2 + b_m(\frac{\partial^m f}{\partial y^m})^2]$	$f + \sum_{m=1}^{n}(-1)^m[a_m \frac{\partial^{2m} f}{\partial x^{2m}} + b_m \frac{\partial^{2m} f}{\partial y^{2m}}] = f_1$

Utilising the minimum energy model, the Lagrange multipliers can be found from the projection data, viz,

$$\underline{\lambda} = [U \otimes I] \, \Lambda^{+} \, [U^{*} \otimes I] \, \underline{g} \tag{4}$$

where $\underline{g}$ and $\underline{\lambda}$ are the vectors produced from the projection data and the discretization of the Lagrange multipliers, U is the discrete Fourier transform matrix, $\otimes$ denotes Kronecker product, * denotes transposition, † denotes pseudoinversion and Λ is the block eigenvalues of the matrix relating $\underline{\lambda}$ and $\underline{g}$. Detailed steps of the algorithm based on (4) are given in (10). The performance of the algorithm is examined in Section 4.

3. EXTENSION OF ART

The fan beam summation method is produced by substituting for $\{\lambda_k(\Theta)\}$ in the minimum energy model of Table 1 the associated projections, viz,

$$f(x,y) = \sum_{k=1}^{N} \frac{g_k(\Theta_k)}{r_k}$$

where r_k, Θ_k as in Eqn. (1).

The generation of a 2-D function from the 1-D projection is achieved by assigning to all image points with the same Θ_k the value $g_k(\Theta_k)$, and then weighting by 1/rk. This is called weighted angular back projection. A physical interpretation of the factor $1/r_k$ may be obtained from Fig. 1. Taking equal areas 1 and 2 (Fig.1) and also f(x,y)- constant and equal in these areas, the contribution of area 1 to the measurement associated to the divergent strip (O_kAB) is twice as much as the contribution of area 2. In very loose terms, we may say that the "sampling rate" of area 1 is double the "rate" of 2. Therefore, the corresponding measurement must affect the value of the reconstruction in area 1 twice as much as the value in area 2. This is exactly what the factor $1/r_k$ does.
Note that in the true 3-D situation, this factor becomes $1/r_k^2$.

Employing the weighted angular backprojection, the fan beam "additive" ART may be implemented by the following steps:

1. Adopt an initial estimate of the reconstruction. The solution of the summation method using the normalised projection, i.e. $g_{ki} = g_{ki}/(\ell_2 - \ell_1)$, for all k,i, where g_{ki} is the ith measurement of the kth projection, may be taken as such an estimate since it is very near to the image for reconstruction.

2. Compute the projection $\hat{g}_{ki}$, i=1, ...P which the present estimate would give (see Section 2). To speed up the computation, the discrete image $\{f_{mj};\ m,j=1,2..P\}$ may be used, where each square pixel has its intensity concentrated at its centre. Calculate the error

$$\varepsilon_{ki} = \frac{g_{ki} - \hat{g}_{ki}}{\ell_2 - \ell_1}, \text{ for all i}$$

3. Update the present image estimate by backprojecting the error, i.e. update all image pixels whose centres lie within (ki) strip

$$f_{mj}^{n+1} = f_{mj}^{n} + \lambda \frac{\varepsilon_{ki}}{r_k} \tag{5}$$

where r_k refers to the pixel centre, and γ is the relaxation factor. Repeat it for all i.

4. Repeat steps 2 and 3 for the other projections using them once each, preferably in random order. When all

projections have been used a cycle is complete. Repeat the cycle until the error is acceptably small.

Non linear methods may be obtained by employing the models Nos. 2, 3 and 4 in Table 1. This is achieved by only substituting for f_{mj} in Eq.(5) the following

(a) $\log f_{mj}$ maximum Entropy 1

(b) $1/f_{mj}$ maximum Entropy 2

(c) $f^{\mu-1}$ criterion No. 4.

It is worth noting here that a previous attempt to extend ART to twin-cone beam scanning (9) assumes (i) that the area 1 and 2 in Fig. 1 give the same contribution, for constant image, to the measurement of the associated strip; this is not correct for diverging rays and (ii) that the extension of the back-projection concept does not include the weighting factor.

4. COMPUTER SIMULATIONS AND DISCUSSION

In general, the convolution algorithm is very popular, has very low computational requirements, and give reasonable sharp but noisy reconstruction. Its backprojects 1-D functions, corresponding to the Lagrange multipliers, to obtain the solution. The vector produced by discretization of these functions is related to $\underline{g}$, by a linear system whose matrix is approximated by various block diagonal matrices. This means that each 1-D function for backprojection depends only on the corresponding projection.

It is natural to extend the convolution algorithms to diverging ray geometries using the weighted angular backprojection (see Section 3). However, the algorithm introduced in (11) uses the weighting factor $1/rk^2$. This might mean that the approximation of the associated matrix by block diagonal matrices in the fan beam geometry is not as useful as in the parallel ray case.

The non recursive algorithm of Section 2 and the convolution algorithm in (11) were implemented and their results compared. The point spread function (psf) of these linear algorithms is space variant. An image with only one non zero point (point image) is adopted and its N=15 equispaced projections, $\{\Psi_k = 2\pi(k-1)/N,\ k=1,..N\}$ (see Fig. 1) are computed. From these projections the point image is reconstructed giving the associated psf of the projecting-reconstructing system.

Picts. 1a and b show the psf's corresponding to the convolution and non recursive algorithms respectively. Ideally,

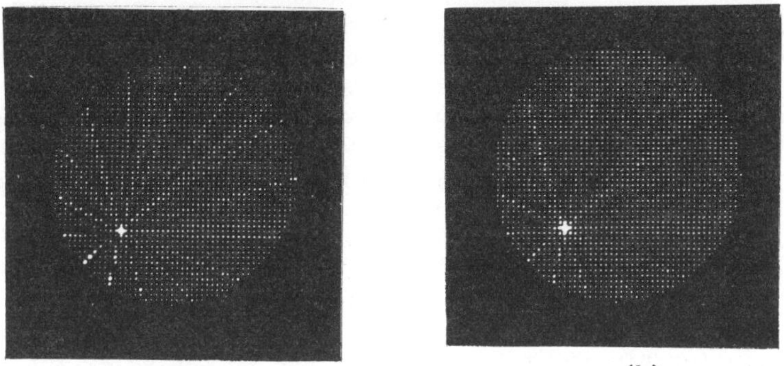

(a) (b)

Fig. 1. Reconstruction of a point image by (a) convolution algorithm and (b) non recursive algorithm. No. of projections N = 15, No. of Points per projection P = 49.

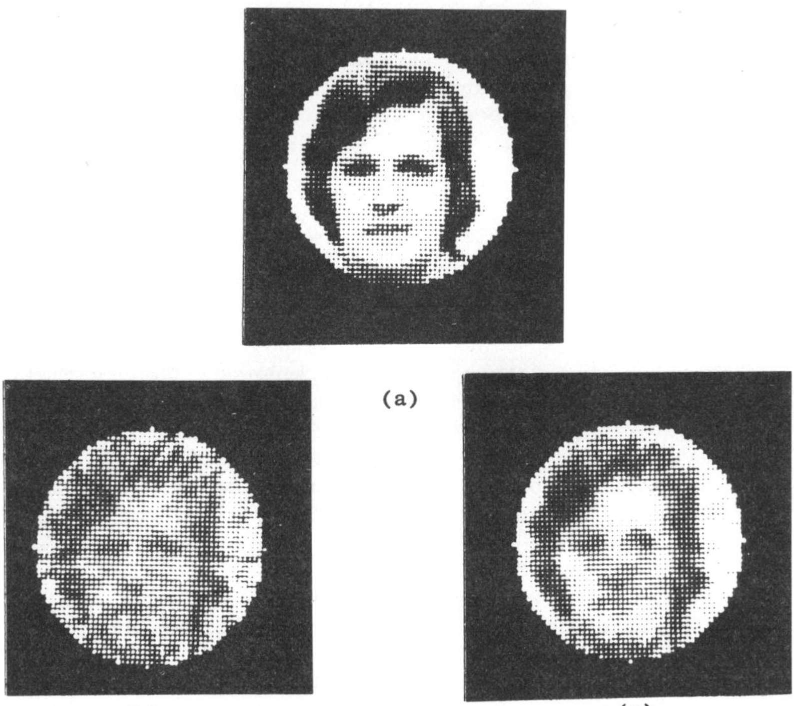

(a)

(b) (c)

Fig.2 (a) Original; and reconstructions by (b) convolution algorithm and (c) non recursive algorithm. N = 15, P = 49.

the psf should be the same as the point image, as the reconstructing system must be the inverse of the projecting system. In practice, a finite number of projections is available and the original image cannot be fully recovered. As in the parallel ray geometry these psf's are 'star' type. This is due to the back-projection model adopted as the last reconstruction step by both methods and it cannot be avoided. Observing Pict.1., is can be seen that the 'star' branches and the variations, in general, outside the 'star' centre are weaker in the psf associated with the non recursive algorith than that associated with the convolution algorithm. It seems that this drawback of the convolution algorithm is partially due to the weighting factor $1/r_k^2$.

Picts. 2a, b and c are the original and its reconstructions from N=15 equispaced projections and P=49, computed by using an approximation of (1). The reconstruction of the convolution algorithm is very noisy while the non recursive algorithm gives a smooth solution. This is in accordance with the psf's in Pict. 1. 2-D interpolations to increase the sharpness of Fig. 2c shall be given in (12). The natural extension of the Fourier methods to the diverging ray geometries may be achieved by using the new projection slice theorem introduced in (10).

REFERENCES

1. Scudder, H.J. "Introduction to computer aided tomography", 1978, Proc. IEEE, v.66, No. 6, pp. 628-637.
2. Bracewell, R.N., and Riddle, A.C. "Inversion of fan-beam scans in radio astronomy", 1967, The Astrop.J. v. 150, pp. 427-434.
3. Mueller, R.K., Kavel, M., and Wade, G. "Reconstructive tomography and applications to ultrasonics", 1979, Proc. IEEE, v.67, No. 4, pp. 567-587.
4. Herman, G.T., Lent, A, and Rowland, S.T. "ART: mathematics and applications", 1973, J. Theor. Biol., v. 42, pp. 1-32.
5. Shepp, L.A. and Logan, B.F. "The Fourier reconstruction of a head section", 1974, IEEE Trans. on Nuclear Science, v. NS-21, pp. 21-43.
6. Mersereau, R.M., and Oppenheim, A.V. "Digital reconstruction of multidimensional signals from their projections", 1974, Proc. IEEE, v.62, No. 10, pp. 1319-1338.
7. Goutis, C.E. "Constrained optimization approach to the reconstruction from projection problems",1976, Southampton University Report, U.K.
8. Goutis, C.E. and Durrani, T.S. "Constraint optimization for reconstruction from projections; a recursive algorithm", submitted to IEEE Trans. on Pattern analysis and Machine intelligence, available from IEEE Computer Soc. Repository.

9. Schlindwein, M. "Iterative three-dimensional reconstruction from twin-cone beam projections", 1978, IEEE Trans. on Nucl. Sci., v. NS-25, No. 5, pp. 1135-1145.
10. Goutis, C.E. and Durrani, T.S. "New algorithms for image reconstruction for fan beam projections", 1978, in Pattern Recognition and Signal Processing" edited by C.H. Chen, Sijthoff and Noordhoff, Publ., pp. 505-519.
11. Herman, G.T., et al., "Reconstruction using divergent-ray shadowgraphs", 1977, in "Reconstruction Tomography in Diagnostic Radiology and Nuclear Medicine", ed. Terpogossian, M.M., et al., University Park Press, Baltimore.
12. Goutis, C.E., and Durrani, T.S. "Constraint optimization and 2-D interpolation for tomography", to be published.

ALGORITHMS FOR DETECTING SYMMETRIES IN IMAGES

N.A. Alexandridis
Dept. of Digital Systems
& Computers,Nat.Techn.Univ.,
Athens, Greece

J.G. Kollias
Dept. of Computer
Science, Univ. of
Patras, Patras, Greece

M. Hadzopoulos
Unit of Applied Math.,Univ. of Athens
Athens, Greece

ABSTRACT: The problem of extracting the dominant structural information from an object in a scene is of key importance in the fields of digital image processing, pattern recognition, on-line or interactive graphics, image approximation and transmission, etc. The gross informational parameter treated in this paper is directional symmetries in the object. Symmetries are examined along four basic directions, i.e. horizontal, vertical, left-diagonal and right-diagonal, and ALGOL-type global picture processing algorithms are presented for detecting such symmetries in two-dimensional data sets. The use of information concerning symmetries can improve computer analysis of pictures or pattern perception by automatic means.

1. INTRODUCTION

The problem of extracting the dominant structural information from an object in a scene is of key importance in the fields of digital image processing, pattern recognition, on-line or interactive graphics, image approximation and transmission, etc. This study considers that the object in a scene has been digitized (through on electro-optical scanner) and stored in the computer main memory. This digitized object is viewed as a two-dimensional array of picture elements or "pixels", where each pixel is an average of local light intensity in the picture around a point.

D. G. Lainiotis and N. S. Tzannes (eds.), Applications of Information and Control Systems, 38-45.

To extract information from the stored picture the following (top-down) regular decomposition scheme is assumed (1,2): Initially, the entire picture, P, is considered as a <u>quadrant</u> which is partitioned into four <u>(sub)quadrants</u>, P_a, P_b, P_c and P_d. The pixel intensities belonging to the (sub)quadrants are used to find the total intensities of P and P_* (where $*$= a,b,c and d). The ratio

$$d(P,P_*)=\frac{\text{intensity of suquadrant } P_*}{\text{intensity of quadrant } \quad P}$$

measures the importance of a quadrant P_* in picture P. The quantidy $d(P,P_*)$ characterizes the subquadrant as "informative", "noninformative" and "not sure". "Not sure" subquadrants become quadrants and the regular decomposition process described above proceeds to the next <u>level.</u> " Non informative" subquadrants are dropped from consideration, while "informative" subquadrants are not further refined. Alexandridis and Klinger (3) propose ways of measuring the "informativeness" of subquadrants and desribe alternative regular decomposition schemes (ie. bottom-up).

In (3) it is assumed that a picture has been regularly decomposed and stored as a tree data structure, where each tree-node corresponds to a (sub)quadrant. A proposition for a recursive algoritm is given which specifies the tree-nodes that must be paired to identify symmetries in a picture along four basic directions, i.e., horizontal, vertical, right-diagonal and left-diagonal. This study considers a picture as a two-dimensional data-set. The data-set may correspond either to the pixels of a picture or to a regularly decomposed picture at some level ℓ. The notation in (3) is modified in this paper(in Section 2) in a manner which allows the presentation of ALGOL-like descriptions of the algorithms which detect either horizontal/vertical (Section 3) or right/left-diagonal symmetries (Section 4).

2. NOTATION

The entire picture is viewed as a two-dimensional array $P[i,j]$ of N x N pixels, where $N=2^\lambda$. Each quadrant is referred to by the triplet (x,y,z), where x,y denote the x,y co-ordinotes of the upper left point of the quadrant and z meassures the <u>length</u> of the quadrant (4). The entire picture is expressed by the triplet (1,1,N). The four (sub)quadrants resulted at the first level of decomposition (Section 1) are defined as follows:

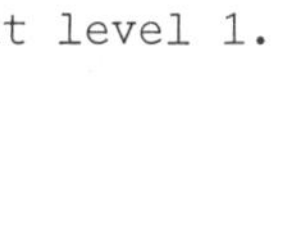

P_a =(1,1,N/2)
P_b =(1+N/2,1,N/2)
P_c =(1,1+N/2,N/2)
P_d =(1+N/2,1+N/2,N/2)

Fig. 1. Picture at level 1.

At level ℓ of decomposition ($\ell \leq \lambda$) the specific $P[i,j]$ quadrant is expressed by the triplet $(1+(i-1)N/2^{\ell},\ 1+(j-1)N/2^{\ell},\ N/2^{\ell})$. At level 2 of decomposition the picture P is partitioned into the following subquadrants P_{aa}, P_{ab}, ..., P_{cd}, P_{dd}.

	0	2	1	2	0
0					
	P_{aa}	P_{ab}	P_{ba}	P_{bb}	
2					
1	P_{ac}	P_{ad}	P_{bc}	P_{bd}	
2	P_{ca}	P_{cb}	P_{da}	P_{db}	
0	P_{cc}	P_{cd}	P_{dc}	P_{dd}	

Fig. 2. Picture at level 2.

Another convention used in this study is that axes (lines) in the picture are numbered as follows:

- The lines in the boundaries take the number 0.
- The other axes take the number ℓ which shows the level of decomposition at which the axes have been introduced.

This implies that at level ℓ, the picture has 2 horizontal and 2 vertical lines with the number 0, and $2^{\ell-1}$ horizontal and $2^{\ell-1}$ vertical axes with the number ℓ.

Two quadrants, lying symmetrically opposite with respect to a given line constitute a set $s_i^{\ell}(d)$ where i is the set number, and d denotes the direction along which the quadrants lie. The sets contain as elements the classes $S_{i,j}^{\ell}(d)$, j=1,2,..., which denote the pair of quadrants which are symmetric. In this study, d may take the following values:

d=h(v) : denotes quadrants lying symmetrically opposite a vertical (horizontal) line.

d=r(ℓ) : denotes quadrants lying symmetrically opposite a left-diagonal (right-diagonal) line.

3. HORIZONTAL AND VERTICAL SYMMETRIES

In the case of horizontal and vertical symmetries there are 2^{ℓ} sets $s_i^{\ell}(h)$. At level 2 (Fig. 2) the sets are:

$$s_1^2(h)=\{S_{1,1}^2(h),\ S_{1,2}^2(h)\}$$

$$s_2^2(h)=\{S_{2,1}^2(h),\ S_{2,2}^2(h)\}$$

$$s_3^2(h)=\{S_{3,1}^2(h),\ S_{3,2}^2(h)\}$$

$$s_4^2(h)=\{S_{4,1}^2(h),\ S_{4,2}^2(h)\}$$

where the classes are given as $S^2_{1,1}(h)=\{P_{aa},P_{bb}\}$, $S^2_{1,2}(h)=\{P_{ab},P_{ba}\}$, $S^2_{2,1}(h)=\{P_{ac},P_{bd}\}$, etc. Similarly are defined the $s^2_i(v)$ sets.

The problem is to determine the quadrants which belong to the sets $s^\ell_i(h)$ and $s^\ell_i(v)$ at any level of decomposition. The following algorithm determines these sets.

```
PROCEDURE HORVER(ℓ,d);
m ← 2 ↑ (ℓ-1) ;
k ← 2 ↑ ℓ ;
z = N/k ;
DO i=1  TO k ;
y ← 1+(i-1)*z;
      DO j=1  TO m;
      x ← 1+(j-1)*z;
      IF d=h
      S^ℓ_ij(h) = { (x,y,z),(1+(m-j)*z,y,z)}  ELSE
      S^ℓ_ij(v) = { (y,x,z),(y,1+(m-j)*z,z) } ;
      END;
END;
```

4. RIGHT-DIAGONAL AND LEFT DIAGONAL SYMMETRIES

The characteristic of the right-diagonal (left-diagonal) symmetry is that every quadrant which belongs to the upper right (left) quarter of the image finds its symmetric quadrant in the lower left (right) quarter of the image. (This implies that there are $2^{\ell-1}$ classes $S^\ell_{ij}(r)$. The problem is again the determination of the quadrants which belong to the $s^\ell_i(r)$ and $s^\ell_i(l)$ sets.

To determine the $s^3_i(r)$ sets the following two quarters of the picture are used:

0	0	3	2	3	1
3	P_{caa}	P_{cab}	P_{cba}	P_{cbb}	
2	P_{cac}	P_{cad}	P_{cbc}	P_{cbd}	
3	P_{cca}	P_{ccb}	P_{cda}	P_{cdb}	
0	P_{ccc}	P_{ccd}	P_{cdc}	P_{cdd}	

0	1	3	2	3	0
3	P_{baa}	P_{bab}	P_{bba}	P_{bbb}	
2	P_{bac}	P_{bad}	P_{bbc}	P_{bbd}	
3	P_{bca}	P_{bcb}	P_{bda}	P_{bdb}	
1	P_{bcc}	P_{bcd}	P_{bdc}	P_{bdd}	

Fig.3. Lower left and upper right quarters at level 3.

There are the following 16 classes (3):

$$S^3_{1,1}(r) = \{P_{bbb}, P_{ccc}\} \quad S^3_{1,2}(r) = \{P_{bbc}, P_{ccb}\} \quad S^3_{1,3}(r) = \{P_{bcc}, P_{cbb}\}$$
$$S^3_{1,4}(r) = \{P_{bcb}, P_{cbc}\}$$
$$S^3_{2,1}(r) = \{P_{bba}, P_{cca}\} \quad S^3_{2,2}(r) = \{P_{bca}, P_{cba}\}$$
$$S^3_{3,1}(r) = \{P_{bbd}, P_{ccd}\} \quad S^3_{3,2}(r) = \{P_{bcd}, P_{cbd}\}$$
$$S^3_{4,1}(r) = \{P_{bab}, P_{cac}\} \quad S^3_{4,2}(r) = \{P_{bac}, P_{cab}\}$$
$$S^3_{5,1}(r) = \{P_{baa}, P_{caa}\}$$
$$S^3_{6,1}(r) = \{P_{bad}, P_{cad}\}$$
$$S^3_{7,1}(r) = \{P_{bdb}, P_{cdc}\} \quad S^3_{7,2}(r) = \{P_{bdc}, P_{cdb}\}$$
$$S^3_{8,1}(r) = \{P_{bda}, P_{cda}\}$$
$$S^3_{9,1}(r) = \{P_{bdd}, P_{cdd}\}$$

The above 16 classes make up the following nine sets:

$$s^3_1(r) = \{S^3_{1,1}(r), S^3_{1,2}(r), S^3_{1,3}(r), S^3_{1,4}(r)\},$$
$$s^3_2(r) = \{S^3_{2,1}(r), S^3_{2,2}(r)\},$$
$$s^3_3(r) = \{S^3_{3,1}(r), S^3_{3,2}(r)\},$$
$$s^3_4(r) = \{S^3_{4,1}(r), S^3_{4,2}(r)\},$$
$$s^3_5(r) = \{S^3_{5,1}(r)\},$$
$$s^3_6(r) = \{S^3_{6,1}(r)\},$$
$$s^3_7(r) = \{S^3_{7,1}(r), S^3_{7,2}(r)\},$$
$$s^3_8(r) = \{S^3_{8,1}(r),$$
$$s^3_9(r) = \{S^3_{9,1}(r)\}.$$

The determination of the $s^3_i(r)$ and $S^3_i(r)$ suggests the following generalization of the right-diagonal symmetries at any level of decomposition ℓ. At level ℓ there are $2^\ell-1$ right-diagonal lines at the upper quarter of the picture. By crossing (right-diagonally) all these lines, one moves to the quadrant on the next line and column of the picture. Each move crosses an horizontal axis and enters a vertical axis. If these axes have the same number, then the quadrants belong to the same set; otherwise, they belong to different sets.

The following function uses the convention that the rows of a quarter of a picture are numbered from 1 to $2^{\ell-1}$ strarting from the botton. The function accepts as inputs the number of the row and ℓ and gives as output the axis number which is crossed when a quadrant moves to its right-diagonal adjacent quadrant.

```
FUNCTION  F(i,ℓ);
r ← 1;
IF i=0  THEN  DO;  F←0  END; ELSE
DO WHILE  F=0;
s ← i-(i/2↑r);
IF s=r  THEN DO; F ← ℓ-r+1; END;
ELSE r=r+1; END; END;
END;
```

In the following algorith the variable s counts the sets, r1 counts the quadrants of a diagonal line, while the variables r2 and r3 show respectively the horizontal and vertical axes crossed by a quadrant. The variables r and S have as arguments the axis number. The values r(axis number)=k and s (axis number) =t, signify that the examined quadrant belongs to the k-th set and it is the t-th class of this set. The following algorithm determines both the $s_i^{\ell}(r)$ and $s_i^{\ell}(\ell)$ sets:

```
PROCEDURE RIGLEF(ℓ,d);

m ← 2↑ℓ;
k ← m/2;
z ← N/m;
s ← 0;
DO i=1  TO k;
r1 ← k-i+1;
  DO j=1  TO r1;
  x ← k*z+1+(j-1)*z;
  y ← 1+(r1-j+1)*z;
  r2 ← F(j-1,ℓ);
  IF r(r2)=0  THEN DO;
  IF d=r  THEN  DO;
  S_{s+1,1}   ={(x,y,z), (y,x,z)}
  S_{s+2,1}   ={(N-y-z+2,N-x-z+2,z),(N-x-z+2,N-y-z+2,z)} END;
  ELSE
  S_{s+1,1}   ={(y,k*z+1-(j-1)*z,z),(x,(k+i+j-2)*z+1,z),z)}
  S_{s+2,1}   ={(y,k*z+1-(j-1)*z,z),(N-(k+i+j-1)*z,N-x-2+z,z)}END;
  r3 ← F(i+j-1,ℓ);
```

```
s ← s+2;  s(r3)←2;   END;
ELSE
IF d=r  THEN  DO;
S_r(r2),s(r2)  ={(x,y,z),(y,x,z)}
S_r(r2)+2,s(r2)  ={(N-y-z+2,N-x-z+2,z),(N-x-z+2,N-y-z+2,z)}
END; ELSE
S_r(r2),s(r2)={(k*z+1-(j-1)*z,y,z),(x,(k+i+j-2)*z,1,z)}
S_r(r2)+1,s(r2)={(y,k*z+1-(j-1)*z,z),(N-(k+i+j-2)*z+1,N-x-z+
2,z}
END;
r3 ← F(i+j-1,ℓ);
r(r3) ← r(r2);
S(r3)←S(r2)+1;
r(r2)←0;   S(r2)←0;
END;
END;
END;
```

5. EXTENSIONS

In some applications, images are segmented and coded as tree data structures. Such representations are usually unbalanced (leading to reduced storage or transmission bandwidth requirements) by not specifying refinements for the blank or otherwise uninteresting regions on pictures (Section 1). The symmetry detecting algorithms presented can be applied at any level of the tree, and they can be easily modified to incorporate tests which accelerate their performance by not examining non-informative quadrants.

Quite recentrly Klinger and Rhodes (5) examine the organization and access of image data stored in a disk. Such an organization is justifiable for real pictures which contain over 10^6 pixels (military reconnaisance data, radar multitarget data, etc); The authors are currently investigating the modifications needed to the algorithms to detect symmetries of disk-stored images. Their effort is directed towards a less machine-dependent description of the algorithms (i.e. a description based on relation algebra (6).

REFERENCES

1. Klinger, A.:1973, Proc. of the First Intern. Joint Conf. on Pattern Recognition, Washington, D.C.
2. Klinger, A.:1974, Proc. of the Intern. Conf. on Systems, Man, and Cybernetics, pp. 307-310.
3. Alexandridis, N., and Klinger, A.:1978, Comp. Graphics and Image Processing 8, pp. 43-77.
4. Horowitz, S., and Pavlidis, T.:1976, J. ACM 23, pp. 368-388.
5. Klinger, A., and Rhodes, M.:1979, IEEE Trans. on Pattern Analysis and Machine Intelligence 1, pp. 50-60.
6. Codd, E,;1970, Commun. ACM 13, pp. 337-347.

SECTION II

ESTIMATION AND DETECTION

MODELS OF MAN-MACHINE TASKS BASED ON NONUNIFORMLY SAMPLED SYSTEMS

George D. Kontopidis

Dept. of Electrical and Computer Engineering
University of New Hampshire, Durham NH 03824

Prediction of the closed loop performance of manually controlled systems requires adequate models for human operators. An appropriate description of the man-machine interaction based on nonuniformly sampled systems is presented in this paper.

Modern estimation and identification theory is used to establish a framework for the analysis of nonuniformly sampled systems. The irregularly sampled version of a time invariant, linear, stochastic dynamic system is used for quantitative study. Then, the human operator is modelled as a state estimator. When the parameters of the supervised plant are partially known, the irregularly spaced observations are used for parameter identification.

I. INTRODUCTION

The basic goal in developing models of the human operator is to systematically and easily predict pilot behavior and system performance in complex control tasks. After the model has been derived, properties of the composite man-machine system can be investigated. In this introductory part we summarize the major studies in manual control area. This selected summary is then used to explain the reasons for using nonuniformly sampled system models for the operator. Typical manually controlled closed loop systems are illustrated in Fig. 1.

The main approaches in modeling the human operator have been the frequency response describing function as applied by McRuer and Weir [17] and the optimal control approach used by Kleinman, Baron and Levison [9]. The helmsman of a large ship was modelled and simulated successfully by Veldhuyzen [28]. A very interesting

D. G. Lainiotis and N. S. Tzannes (eds.), Applications of Information and Control Systems, 49-59.

Tracking System Regulator System

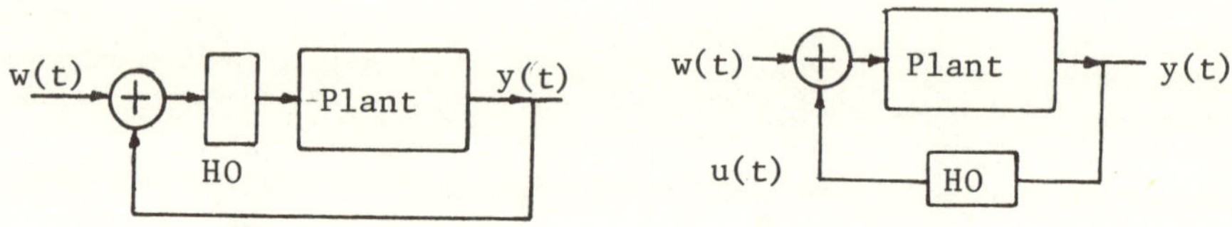

w(t): Noise & Disturbances
u(t): Manual Controls
y(t): Displayed output

HO: Human Operator

Fig. 1: Manually Controlled Systems

study in modeling the human operator as a supervisor of a distributed computer network was done by Tuga and Sheridan [27]; this is a typical example of modeling the hierarchial control of large scale systems including a human as a decision maker. Rouse's [22] dissertation and his later studies [23] in modeling the human by using modern estimation theory approaches and hypothesis testing, is a good theoretical background for improving existing models. Recent studies in the area of improving models of humans by using the position of the eye were presented by Repperger and Hartzell [21]; the effect of work load on the variances of the observation and the motor noise [9] is still an open area for research.

In the following, the assertion that the human operator can be modelled as a state estimator of a nonuniformly sampled system is introduced. The basic idea behind this assertion is that the operator does not observe the output of the system "continuously" but samples the output at discrete time instants; at the same time, he manually controls the plant in a discrete manner.

The first indication of the desirability of considering nonuniformly sampled models for the plant comes from the Kleinman's et al. [9] model. The human operator controls a linear, time invariant, stochastic system optimally, subject to his inherent limitations. In their model the estimator is designed by using a continuous time Kalman filter or a discrete time invariant one. If we assume that the visual sampling rate (of the displayed outputs) is not constant, the estimator needs modifications to take into account such changes.

The second indication of the irregularity of the sampling comes from Rouse [23], Kok [10] and Kvalselh [8]. They underline the hypothesis that the human supervisor acts partially as an observer/reconstructor and partially as a controller/decision maker. The following hypotheses are also used:

- The model of the observer depends on the supervised system, the display and the noise parameters.
- The controller part depends only on the supervised system and can be considered independent of the observer part.

Now, if we impose the hypothesis that the human supervisor acts as a discrete process with a unique processing unit, the observation rate depends on the time it takes to process a given piece of information and the time it takes to realize the controling action.

Another indication of sampling irregularity comes from systems containing computers as well as humans. In large scale systems (eg. computer networks) the time between the lower priority tasks is usually irregular because of the asynchronous interrupt configuration. For modeling such systems, a nonuniformly sampled model is appropriate. Another example of such an interrupted control is the pilot task discussed in [4].

The above reasons explain why the controlled plant is actually observed and controlled nonuniformly. Thereafter the following question arise: What modifications are necessary to existing models if we assume nonuniform sampling? In this paper we deal with a part of the above subject. The class of systems and the assumptions for the following presentation are:

- The controlled plant is a multiple input-output, linear, time invariant observable and controllable. The parameters of the plant are not necessarily known "exactly"; this means that parameter identification also takes place.
- Additive noise with known statistics corrupts the measurements the controlled plant and the control input.
- Both the control inputs and the observed outputs are sampled simultaneously at t_n and the intervals $T_n = t_{n+1} - t_n$ are assumed to be known explicitly.

In reality, the observation sequence can be found by sensing the position of the eye with respect to the display and the controling sequence can be found by sensing the changes of the actuators. We realize that the above schemes are not necessarily the best ones (eg. the human might be observing the display and performing his control task as in [4]). Generalization to a broader class of systems and relaxation of the last restrictions ("simultaneously" and "intervals known explicitly") are subjects of future work.

The results of the analysis of nonuniformly sampled systems are presented in the following manner. In Sec. II the linear system is sampled nonuniformly and modelled. In Sec. III we refer to state estimators for the nonuniformly sampled systems. In Sec. IV we assume that the original linear system is partially known; then we try to estimate the states of the system and to identify the parameters using irregularly taken samples. Finally in Sec. V we indicate areas of future work and possible extentions. Most of the proofs have been omitted but the interested reader can find them in [11].

II. MODELS AND SIMULATION ALGORITHMS

In this section the nonuniform sampling process is studied. A linear, time invariant, stochastic model is sampled at time instants t_n. The objective here is to find the difference equations which relate the input and output samples.

The state space representation of the physical system (plant) is assumed to be the following stochastic differential equation:

$$dx=Fxdt+G_1udt+G_2dw \qquad (S)$$

$$dy=C_sxdt+D_sudt+dv \qquad C_s=CF; \quad D_s=CG$$

where F is an nxn real matrix, G_1 and G_2 are $n \times l_1$ and $n \times l_2$ matrices respectively, C_s is an observation matrix, and D_s is a feed-forward gain matrix. The n-, m-dimensional stochastic processes $\{x(t)\}$ and $\{y(t)\}$ are the states and the observations (outputs) respectively. The plant noise and the measurement noise are modelled by l_2- and m-dimensional Wiener-Levy uncorrelated processes with incremental covariances $E\{dwdw^T\}=Qdt$ and $E\{dvdv^T\}=Rdt$ respectively. The control input u(t) is assumed to be an l_1-deterministic signal.

The nonuniform sampling process is studied separately for the deterministic and the stochastic component of the states x. The sampled version of the first can be found by solving the state equations from t_n to t_{n+1}. The result is a time varying discrete model given by:

$$x_1(t_{n+1})=A(n)x_1(t_n)+B(n)u(h_n) \qquad (1)$$

$$y_1(t_n)=Cx_1(t_n)$$

where h_n is a particular point between t_n and t_{n+1}, and the matrices A(n) and B(n) are given by

$$A(n)=\exp FT_n \qquad T_n=t_{n+1}-t_n \qquad (2)$$

$$B(n)=\int_0^{T_n} \exp F(T_n-s)ds\ G_1 \qquad (3)$$

The sequence $\{u(h_n)\}$ cannot be found if no assumption on the class of admissible functions u(t) is imposed. In many practical situations, where u(t) is piecewise continuous, a set of k+1 previous samples $\{u(t_{n-k}),\ldots,u(t_{n-1}),u(t_n)\}$ can be used to evaluate $u(h_n)$ by extapolation methods. In [11] a generalization of Newton-Gregory extrapolation [6] in the nonuniformly spaced data case is used for such evaluation. Also in [18] several interpolation methods are presented. For simplicity, the assumption that u(t) is piecewise constant ($u(t)=u(t_n)$ for $t\varepsilon[t_n,t_{n+1}]$) is imposed. This assumption is not very crude; there are situations where this is true eg. bang-bang control, pulse width modulation control, helmsman's task [28] etc. Then, $u(h_n)=u(t_n)$ for all n.

The sampled version of the stochastic component of the states can be found by using an averaging sampler [26, 12] as a sampling device. The result is a stochastic defference equation of the form:

$$x_2(t_{n+1})=A(n)x_2(t_n)+w_2(n) \tag{4}$$
$$z(n)=C_2(n)x_2(t_n)+v_2(n)$$

where $z(n)=y_2(t_{n+1})-y_2(t_n)$, $A(n)$ is given in (2), $C_2(n)$ is

$$C_2(n)=C_s\int_0^{T_n}\exp F(T_n-s)ds \tag{5}$$

and the processes $\{w_2(n)\}$ and $\{v_2(n)\}$ are described statistically by:

$$E\{w_2(n)\}=0;\quad \mathrm{cov}\{w_2(n),w_2(m)\}=Q_2(n)\delta_\kappa(n-m) \tag{6}$$
$$E\{v_2(n)\}=0;\quad \mathrm{cov}\{v_2(n),v_2(m)\}=R_2(n)\delta_\kappa(n-m) \tag{7}$$
$$\mathrm{cov}\{w_2(n),v_2(m)\}=R_w(n)\delta_\kappa(n-m) \tag{8}$$

with,

$$Q_2(n)=\int_0^{T_n}A(T_n-s)\ G_2Q\ G_2^T\ A^T(T_n-s)\ ds \tag{9}$$

$$R_2(n)=C_s\int_0^{T_n}B_2(T_n-s)QB_2^T(T_n-s)\ dsC_s^T+T_nR \tag{10}$$

$$R_w(n)=\int_0^{T_n}A(T_n-s)\ G_2QB_2^T(T_n-s)\ dsC_s^T \tag{11}$$

$$A(T_n-s)=\exp F(T_n-s) \tag{12}$$

$$B_2(T_n-s)=\int_0^{T_n-s}A(T_n-\tau)\ d\tau G_2 \tag{13}$$

The proof of the above relationships can be found in [11] and [1]. Eqs (7) and (10) show that the processes $\{w(n)\}$ and $\{v_2(n)\}$ are nonstationary and correlated. Simulation of the system (4) by a digital computer implies fast calculation of the integrals (9), (10), (11) and (13). This is discussed in Sec. III.

The conclusion of this section is that the plant (S) can be simulated by superimposing systems (1) and (4). The simulated states and outputs are:

$$x(t_n)=x_1(t_n)+x_2(t_n)\ \text{and}\ y(t_n)=y_1(t_n)+y_2(t_n) \tag{14}$$

(The subscripts *1* and *2* denote the deterministic and the stochastic component of the variables respectively.)

III. STATE ESTIMATION OF NONUNIFORMLY SAMPLED SYSTEMS

Estimation theory is often employed to develop a memory representation of how the human gradually forms an internal model. In extensive studies on human decision making by Rouse [4,22,23] the subject is required to estimate the state of a dynamic process. The behavior of the human supervisor as a state estimator is widely used [9,23]. The theory on short term and long term adaptation [23] is based on modeling the controlled plant as a time invariant discrete process and the human as a discrete estimator-identifier. This assumption can be crude in the following sense: visual scanning of the plant output display in general is not regular due to attention allocation caused by "control interrupts" and because of the nonconstant duration of the decision making process. Therefore the following idea is proposed: to use a discrete state estimator of the nonuniformly sampled plant.

In Fig. 2 we relate the nonuniform estimator assertion to the model proposed by Kleinman et al. [9]. The vehicle dynamics, the motor noise and the observation noise are assumed to be part of the plant. For this section, we assume that the delays and the noise covariances (τ_N, V_y, V_m) are known. The objective is to design the equalization network or more specifically the estimator based on the sampling sequence $\{t_n\}$. When V_y and V_m are given by Eqs. (21) and (22) of their paper [9], the adjustment of the model can be done by following their procedure but with a different estimation algorithm.

In the following we state the estimation problem in the nonuniformly sampled data case and a given solution.

We are given Eq. (S) which is assumed to describe the plant dynamics, the neuromotor dynamics, the motor and the observation noise. Based on the sampling sequence $\{t_n\}$ and the system (S) we can find the discrete equations (1) and (4). The objective is to estimate the states $x(t_n)$ based on the past observations $y(t_n)$.

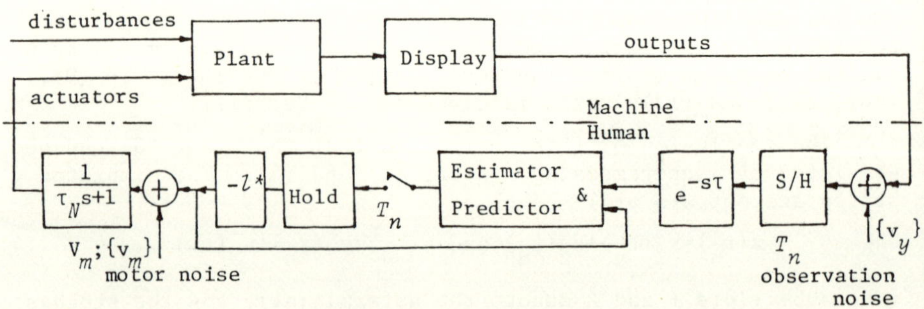

Fig. 2: Modifications of the Model of Human Operator

The extended Kalman filter [24] (for time varying systems with nonstationary noises and correlated disturbances) can be applied for the above models. The particular difficulty for such an application is that the estimator has to update from step to step the system parameters A(n), B(n) and $C_2(n)$ and evaluate the integrals (9)-(13). These computational difficulties have been reduced to a minimum by transforming the system equations to the modified Jordan canonical form [11]. A real time procedure realizing the time varying estimator has been implemented and tested succesfully. The algorithm uses as "input" the parameters F, G_1, G_2, C, and R, Q and produces as "output" the estimate of the state $x(t_n)$ based on the past observations up to $y(t_n)$.

IV. PARAMETER IDENTIFICATION USING IRREGULARLY SPACED SAMPLES

Having fixed the overall description of the controlled plant, the parameters of the human model can be found by an identification scheme. In many other cases the model of the plant is not exactly known and a composite man-machine model is desired. Generalizing the identification procedures in the nonuniformly sampled data case we look for a model of the man-machine task. The statement of the problem follows.

We are given the structure of a system S_p (which represents the controlled plant and the human operator) in terms of a vector of unknown parameters p. Mathematically the system is assumed to be described by:

$$dx=F(p)xdt+G_1(p)udt+G_2dw \qquad (S_p)$$
$$dy=C_s(p)xdt+D_s(p)udt+dv$$

where p might contain unknown elements of F, G_1 and C_s. Because of the nature of the system or because of our inability to observe the system regularly, we assume that the irregularly spaced samples of u(t) and y(t) at $\{t_n\}$ are given. The objective is to find an estimate of the vector p. Additional assumptions are imposed on the above system:

- The noises v(t) and w(t) are described statistically as Wiener Levy processes with known covariances Q and R.
- The initial state x(0) of the system S_p and the vector p are assumed to be Gaussian random vectors with mean x_o an p_o and variances S_x and S_p respectively. This restriction is not severe; for example, if we let $S_p=\omega I$ with ω a large number we do include a large class of practical problems with "unknown" p.
- The sampling sequence in not unrestricted. In [11] it is shown that there exist sequences $\{t_n\}$ under which the sampled version of S_p is unobservable even if S_p is observable; thus, ill conditioned sequences must be excepted from the identification procedure.

The parameter identification problem can be solved by a generalization of Young's [28,20] instrumental variables method [11] or by formulating the problem as a nonlinear state estimation problem. The first method has a disadvantage: it does not take into account the statistics of the system and measurement noise; the method is primarily useful for systems for low level noise. In this paper, we outline a more general approach, referencing proofs for brevity.

Step 1: Formulation of the parameter identification problem as a nonlinear, time varying augmented system state estimation problem.

The sampled version of equations Sp can be found by using the results of Sec. II; equations similar to (1) and (4) can be derived containing the parameter vector p. By defining

$$x_a(n)=[x_1^T(t_n)\ x_2^T(t_n)\ p^T\ ]^T$$
$$y_a(n)=y(t_{n+1})-y(t_n)$$

the above discrete state equations take the form

$$x_a(n+1)=A_a(x_a,n)+B_a(n)w_a(n) \tag{15}$$
$$y_a(n)=C_a(x_a,n)+v_a(n)$$

and the functions $A_a(.)$, $B_a(.)$ and $C_a(.)$ are defined properly in terms of $A(p;T_n)$, $B(p;T_n)$ and $C_2(p;T_n)$ [26,25]

Step 2: Cost function selection.

A cost function to minimize the initial state error, the output error and the trajectory error between real world samples and the output of the model (15) can be developed by using the maximum aposteriori method or the maximum likelihood method or the least squares method. The results based on the weighted least squares are given in this paper. The cost function is assumed to be

$$JLS=\tfrac{1}{2}[x_a(0)-x_o]^T S_a^{-1}[x_a(0)-x_o] + \tfrac{1}{2}\sum_{j=0}^{N-1}[y_a(j+1)-C_a(x_a,j+1)]^T R_a^{-1}(j+1)\cdot$$
$$\cdot[y_a(j+1)-C_a(x_a,j+1)] + w_a^T(j)Q_a^{-1}(j)w_a(j) \tag{16}$$

where S_a, R_a and Q_a are given apriori positive semidefinite matrices which do not depend on x_a.

If the assumptions under which minimization of (16) does not correspond to exact model matching are not met, cost functions based on maximum likelihood can be used. The following steps do apply to more general cost functions than (16).

Step 3: Minimization of the cost function leading to a two point boundary value problem.

The trajectory of the states x_a that minimizes the cost function JLS can be found by using dynamic programming [3] or by using the maximum principle. Let us simplify the notation (and make the procedure more general) by using the notation

$$\phi[\ x_a(i),\bar{w}_a(i)]=A_a(x_a,i)+B_a(i)\bar{w}_a(i) \tag{17}$$

$$\Theta_o[x_\alpha(0)] = \tfrac{1}{2}[x_\alpha(0)-x_o]^T S_\alpha^{-1}[x_\alpha(0)-x_o] \tag{18}$$

$$\Theta[x_\alpha(i),\bar{w}_\alpha(i)] = \tfrac{1}{2}\bar{w}_\alpha^T(i)Q_\alpha^{-1}\bar{w}_\alpha(i) + \tfrac{1}{2}[y_\alpha(i+1)-C_\alpha(\phi(x,\bar{w}),i+1)]^T R_\alpha^{-1}(i+1)\cdot \cdot[y_\alpha(i+1)-C_\alpha(\phi(x,\bar{w}),i+1)] \tag{19}$$

The necessary conditions for minimization of JLS subject to the system equalities are:

$$\partial H/\partial x_\alpha(n)=\lambda(n);\quad \partial H/\partial\lambda(n+1)=x_\alpha(n+1);\quad \partial H/\partial\bar{w}_\alpha(n)=0 \tag{20}$$

where H is the hamiltonian

$$H[x_\alpha(i),\lambda(i),\bar{w}_\alpha(i)]=\Theta[x_\alpha(i),\bar{w}_\alpha(i)]+\lambda^T(i+1)\phi(x_\alpha(i),\bar{w}_\alpha(i)) \tag{21}$$

and $\lambda(i+1)$ is a vector (costate) of the Lagrange multipliers assosiated with $x_\alpha(i)$. The result of the minimization is the following TPBVP:

$$x_\alpha(j+1)=A_\alpha(x_\alpha,j)-B_\alpha(j)Q_\alpha(j)B_\alpha^T(j)A_\alpha'^{-1}\lambda(j) \tag{22}$$

$$\lambda(j+1)=A_\alpha'^{-1}\lambda(j)+C_\alpha'R_\alpha^{-1}(j+1)[y_\alpha(t_{j+1})-C_\alpha(x_\alpha,j+1)] \tag{23}$$

$$\lambda(0)=S_\alpha^{-1}[x_o-x_\alpha(0)] \qquad \lambda(N)=0 \tag{24}$$

where

$$A_\alpha' = \partial A_\alpha(x_\alpha,j)/\partial x_\alpha(j);\quad C_\alpha' = \partial C_\alpha(x_\alpha,j+1)/\partial x_\alpha(j+1). \tag{25}$$

Observe that if the dimension of the vector x_α is n_α Eqs. (22)-(24) consist of a system of $(2N+1)n_\alpha$ nonlinear equations with $2N+1$ times n_α unknowns.

Step 4: Off line solution of the TPBVP.

In the literature [25] there exist several numerical methods to solve the nonlinear TPBVP. For example first order gradient, second order conjugate gradient and quasilinearization can be used. Here a modification od the steepest descent method is presented. The basic idea behind the method is to try to satisfy Eq. (20) and to try iteratively to satisfy the boundary conditions (24). In [11] it is proved that if (20) are satisfied the first variation of JLS (say ΔJLS) is given by:

$$\Delta JLS = \frac{d\Theta_o^T[x_\alpha(0)]}{dx_\alpha(0)}\,x_\alpha(0) - \lambda(N)x_\alpha(N) \tag{26}$$

Application of the steepest descent method gives that the proper steps $\Delta x_\alpha(0)$ and $\Delta x_\alpha(N)$ should be given by

$$\Delta x_\alpha(0)=-k_\chi d\Theta_o^T[x_\alpha(0)]/dx_\alpha(0);\qquad \Delta x_\alpha(N)=k_\lambda\lambda(N) \tag{27}$$

where k_χ and k_λ are properly selected positive numbers. Then, we solve (22),(23) backwards starting from an arbitrary vector $x_\alpha(N)$ and update $x_\alpha(0)$ using (27a); in the same manner, we solve (22), (23) starting from the new update $x_\alpha(0)$ up to step N where we update $x_\alpha(N)$ using (27b). The major advantage of this algorithm is the low memory requirements.

Step 5: On line solution of the identification problem.

The sequential solution of the TPBVP as is was given before is a current area of research. A brief review of the available techniques follows:

- Linearization of the Kalman filter equations about a nominal point. This method is based on the linear Kalman filter estimator of the states and parameters around a apriori trajectory $x_a(t_n)$. The selection of the nominal trajectory can be difficult [26].
- Selection of a linear model out of a set of precalculated models (partitioned algorithms). The basic idea of the method is to linearize the model about several predetermined operating points and then use a Kalman filter for each point; a decision rule is then used to select the model that gives the best estimate [14,16].
- Linearization of the invariant imbedding equations. This technique is widely used to solve the TPBVP in a sequential manner [2,25,13]. An application of the method in the nonuniformly sampled system identification area can be found in [11].

V. CONCLUSION

Several reasons for using nonuniformly sampled systems as models for humans motivated this research. A comprehensive theory of the irregular sampling process is the starting point of the whole development; then, the nonuniformly sampled models are used for solving two fundamental problems: the state estimation and the parameter identification. Applicability of the results relies on the computational studies of the presented algorithms.

There are several trends and extentions of this work. Practical manual control problems are only a subclass of the application of the theory. Each individual case requires special treatment and futher study. One direction of our work is towards solution of the nonlinear estimation problem real time by using partitioned algorithms; another, is towards solution of the estimation problem in the case of statistically known intervals. Sampling sequences with given stochastic characteristics need futher study.

REFERENCES

[1] Astrom K.J., Introduction to Stochastic Control Theory. New York: Academic Press Inc., 1970.

[2] Astrom K.J., P. Evkhoff, "System Identification. A Survey," Automatica, vol. 7, pp. 123-162, 1971.

[3] Bellman R., Introduction to the Mathematical Theory of Control Processes. New York: Academic Press Inc., vol. I&II, 1967.

[4] Enstrom K.D., W. B. Rouse, "Real Time Determination of How a Human Has Allocated His Attention Between Control and Monitoring Tasks." IEEE Trans. on Systems, Man and Cybernetics, vol. 7, pp. 153-161, March 1977.

[5] Goodwin G.C., R.L. Payne, Dynamic System Identification: Experiment Design and Data Analysis. New York: Academic Press Inc., 1977.
[6] Hamza M.H., M.A. Sheirah, "Polynomial Extrapolation in Sampled Data Control Systems," IEEE Trans. Aut. Cont. vol. 16, pp. 102-104, Feb. 1971.
[7] Hsia T.C., System Identification, Mass: Lexington Books, 1977.
[8] Kvalseth T.O., "A Decision-Theoretic Model of the Sampling Behavior of the Human Process Monitor," IEEE Tran. of Systems Man and Cybernetics, Vol. 7, pp. 810-812, Nov. 1977.
[9] Kleinman D.L., S. Baron, W.H. Levison, "A Control Theoretic Approach to Manned-Vechicle System Analysis," IEEE Trans. on Autom. Control, vol. 16, Dec. 1971.
[10] Kok J.J., R.A. Wijk, "A Model of the Human Supervisor," Proc. 13th Annual Conf. on Manual Control, pp. 210-217, June 1977.
[11] Kontopidis G.D., Nonuniformly Sampled Systems. MSEE Thesis, University of New Hampshire, Dec. 1978.
[12] Kontopidis G.D., "Computer Controlled Systems Using Multiplexed I/O" IECI Conference March 1980.
[13] Laundau I.D., "Unbiased Recursive Identification Using Model Reference Adaptive Techniques," IEEE Trans. Autom. Control, vol. 21, pp. 194-200, April 1976.
[14] Lainiotis D.G., "Optimal Adaptive Estimation: Stucture and Parameter Adaptation," IEEE Tran. Autom. Control, vol. 16, pp. 160-170, April 1972.
[15] Luenberger D.C., Optimization by Vector Space Methods. New York: John Wiley & Sons, 1969.
[16] Magill D.T., "Optimal Adaptive Estimation of Sampled Stochastic Processes," IEEE Tran. Autom. Cont. vol 10, Oct. 1965.
[17] McRuer D., D.H. Weir,"Theory of Manual Vechicular Controls," IEEE Tran. Man-Machine Systems, vol. 10, pp. 257-291, 1969.
[18] Morrison N., Introduction to Sequential Smoothing and Prediction. New York: McGraw Hill, 1969.
[19] Palmer E.,"Interrupted Monitoring of a Stochastic Process," Proc. 13th Annual Conf. on Manual Control, pp. 237-245, 1977.
[20] Pearson E.,"Finite Time Interval Linear System Identification," Automatica, vol. 12, pp. 577-587, 1976.
[21] Repperger D.W.,E.J. Hartzell, "A Relationship Between Eye Movement and Performance," 13th Annual Manual, pp. 152-165, 1977.
[22] Rouse W.B., "Cognitive Sources of Suboptimal Human Prediction," Phd Dissertation, MIT, 1972.
[23] Rouse W.B., "A Theory of Human Decision Making" IEEE Trans. on Systems Man and Cybernetics, vol. 7, pp. 274-283, 1977.
[24] Sage A.P., J. Melsa, Estimation Theory. New York McGraw Hill.
[25] Sage A.P., J. Melsa, System Identification. McGraw Hill, 1971.
[26] Schweppe F.C., Uncertain Dynamic Systems. Englewood Cliffs, N.J.: Prendice Hall, 1973.
[27] Tulga M.K., T.B. Sheridan, "Supervisory Dynamic Decision Making," 13th Annual Manual, pp. 199-209, June 1977.
[28] Young P.C., "An Instrumental Variable Method for Real Time Identification," Automatica, vol. 6. pp. 271-287, 1970.

ON THE ESTIMATION OF THE VARIOGRAM IN SPATIAL INTERPOLATION METHODS USED IN GROUNDWATERFLOW MODELLING

Michel GEVERS and Georges BASTIN

Université Catholique de Louvain
Laboratoire d'Automatique et d'Analyse des Systèmes
Bâtiment Maxwell
1348 Louvain-la-Neuve, BELGIUM

ABSTRACT

The solution of the groundwaterflow modelling problem requires the estimation of the piezometric head, considered as a two-dimensional (2-D) random process, at all nodes of a grid from measurements at some points of the spatial domain. This can be achieved using a linear minimum variance unbiased estimation method called "kriging".

Kriging requires that a model be estimated for the variogram of the piezometric head, Var $\{h(x,y)-h(x',y')\}$, as a function of the distance d between the points (x,y) and (x',y') using increments of available measurements. The choice of these increments raises several problems, which are the main object of this paper.

I. MOTIVATION : THE GROUNDWATER FLOW MODELLING PROBLEM

The steady state behavior of the groundwaterflow in an isotropic unconfined aquifer is described by the following two-dimensional flow equation :

$$\frac{\partial}{\partial x}\left(T(x,y)\,\frac{\partial h}{\partial x}\right) + \frac{\partial}{\partial y}\left(T(x,y)\frac{\partial h}{\partial y}\right) + q(x,y)=0 \qquad (1)$$

where :
$(x,y) \in \Omega$ are the spatial coordinates in the studied domain Ω
$h(x,y)$ is the piezometric head (= "the state")
$T(x,y)$ is the transmissivity (= "the parameter")
$q(x,y)$ is the aquifer recharge (or discharge) rate (= the input).

D. G. Lainiotis and N. S. Tzannes (eds.), Applications of Information and Control Systems, 60-68.

The identification problem consists in estimating $T(x,y)$ in the whole spatial domain Ω from available measurements of $h(x,y)$ at certain points in Ω. Various assumptions can be made concerning the recharge $q(x,y)$.

Often $q(x,y)$ can be assumed constant over the whole spatial domain and this constant value can be estimated [1].

The identification problem as described here has been studied by a number of authors and a variety of solutions have been proposed under various assumptions [2]- [6]. In most cases these solutions require a large amount of available piezometric measure points. In [1] various methods have been discussed and compared, and a procedure has been proposed for the case where few measure points are available.

In the proposed procedure $T(x,y)$, $h(x,y)$ and $q(x,y)$ are considered as spatial (2-dimensional) stochastic processes, and the flow equation is a stochastic equation. The method requires a discretization of the spatial domain. The solution is then obtained in two steps. First estimates of the piezometric head are obtained at all nodes of the discretization grid from the available measurements. These estimates are minimum variance unbiased estimates obtained through a procedure known as "kriging" which will be detailed later in this paper. Then estimates of the transmissivities are obtained by minimizing a uniformity criterion on $T(x,y)$ subject to satisfying the state constraints, in which the true piezometric heads are replaced by their estimated values.

This procedure has the advantage that the estimated transmissivities can be given a stochastic interpretation which is consistent with experimental geophysical findings, namely that the logarithm of the transmissivities behave like a 2-D Wiener process. In addition the kriging procedure produces covariance estimates for the estimated piezometric heads. From these error covariance estimates for $\hat{h}(x,y)$ it is possible to derive confidence intervals for the estimated transmissivities $T(x,y)$ (see [7]-[8]).

Kriging requires a model for the variogram $\mathrm{Var}\,\{h(x,y)-h(x',y')\}$ of the piezometric head as a function of the distance d between the points (x,y) and (x',y'). A crucial step of the kriging procedure consists in estimating this variogram from increments of available measurements of h. This is normally done by least squares regression using a "large number" of increments ; however no systematic procedure or theory exists as to how to choose these increments. This choice raises several questions, in particular : what is the effect of statistically dependant increments on the estimated variogram ? how many increments should one chooze ? These questions are the main object of this paper. Partial answers are given and some guidelines are proposed.

Due to space limitation we shall not elaborate on the identification of the transmissivities in the sequel, even though the conference version has largely dealt with this problem. We shall limit ourselves to the presentation of the kriging procedure for the estimation of h, and to the estimation of the variogram that is required for this procedure. We refer the reader to [1] or [8] for the application to groundwaterflow modelling.

II. STOCHASTIC INTERPOLATION IN TWO-DIMENSIONAL SPACE

Stochastic hypotheses for piezometry.

The piezometry h(x,y) is assumed to be a twodimensionnal non stationary stochastic process, which is decomposed as follows :

$$h(x,y) = m(x,y) + z(x,y) \tag{2}$$

m(x,y) is the mean of the process, also called the drift. It is assumed to vary slowly in space and to be of the form

$$m(x,y) = E\,\{h(x,y)\} = \sum_{l=1}^{L} a_l\, f_l(x,y) \tag{3}$$

The functions $f_l(x,y)$ are chozen a priori (e.g. they are monomials in x and y of order $\leqslant k$), and the a_l are unknown coefficients. For simplicity of notations we shall from now on designate the location (x_k, y_k) by the single variable u_k.

We now introduce the "variogram" γ_{kj} of the process h ; it is defined as the variance of the increments of h :

$$\gamma_{kj} = \frac{1}{2}\,\mathrm{Var}\,\{h(u_k)-h(u_j)\}\ ,\quad u_k, u_j \in \Omega \tag{4}$$

The reason for introducing the variogram will become clear later, in the context of spatial interpolation.

It will be assumed that the variogram γ_{kj} of the process h depends only on the distance d_{kj} between the points u_k and u_j :

$$\gamma_{kj} = \gamma(d_{kj}) \tag{5}$$

This is an intuitively plausible assumption for many physical processes, of which Wiener processes are a special case.

The actual piezometry is viewed as a realization of a stochastic process that is measured at n data points $h(u_1)$, ..., $h(u_n)$, which do most often not coincide with the grid nodes.

Kriging.

Within the framework of the above hypotheses one can compute unbiased linear minimum variance estimates of the piezometric head at the grid nodes without knowing or even computing the coefficients a_l of the mean m(x,y) (see (3)). Such an optimal interpolation method, called "kriging", is a twodimensional generalization of the Wiener filter and has been studied and developed by geostatisticians |9| -|11|.

Let u_o be the coordinates of an arbitrary grid node. Then it is desired to obtain a linear estimate

$$\hat{h}(u_o) = \sum_{k=1}^{n} \lambda_k h(u_k), \tag{6}$$

where $u_1, \ldots, u_n$ are the measure points, such that

$$\text{i) } E\{\hat{h}(u_o)-h(u_o)\} = 0 \tag{7a}$$

$$\text{ii) } \sigma_o^2 \triangleq \text{Var}\{\hat{h}(u_o)-h(u_o)\} \text{ is minimized} \tag{7b}$$

With $\lambda_o = -1$, one can write

$$\hat{h}(u_o) - h(u_o) = \sum_{k=0}^{n} \lambda_k h(u_k) \tag{8}$$

The solution of the kriging problem is as follows.

Since the a_l's are unknown, (7a) leads to the following L universality conditions on λ_k :

$$\sum_{k=0}^{n} \lambda_k f_l(u_k) = 0, \; l=1,\ldots, L \tag{9}$$

Furthermore σ_o^2 can be rewritten as

$$\sigma_o^2 = - \sum_{k=0}^{n} \sum_{j=0}^{n} \lambda_k \lambda_j \gamma_{kj} \tag{10}$$

Minimizing (10) subject to (9) gives :

$$-2 \sum_{j=0}^{n} \lambda_j \gamma_{kj} + \sum_{l=1}^{L} \mu_l f_l(u_k) = 0, \; k=1,\ldots,n \tag{11a}$$

$$\sum_{j=0}^{n} \lambda_j f_l (u_j) = 0 \quad , \qquad l=1,\ldots,L \tag{11b}$$

This so-called "kriging system" is a set of n+L linear equations in the n+L unknowns $\lambda_j (j=1,\ldots,n)$ and $\mu_l (l=1,\ldots L)$. μ_l are Lagrange coefficients.

Let $\hat{\lambda}_j$ and $\hat{\mu}_l$ be the solutions. Then the optimal estimate is

$$\hat{h}(u_o) = \sum_{k=1}^{n} \hat{\lambda}_k h(u_k) \tag{12}$$

The variance of the estimation error is

$$\hat{\sigma}_o^2 = \sum_{k=0}^{n} \hat{\lambda}_k \gamma_{ok} - \sum_{l=1}^{L} \hat{\mu}_l f_l(u_o) \tag{13}$$

Comments.

1) We have given here only a very brief account of the Kriging procedure. More details can be found in [1] and [9]-[11].

2) In all the kriging formulas derived above, it has been assumed that the variogram γ_{kj} is known. In practice this is not so, and the functional relation (5) between γ_{kj} and d_{kj} must be estimated from the data.

III. ESTIMATION OF THE VARIOGRAM

The estimation of the variogram from the data is a difficult problem. It consists in three different steps.

1) The estimation of the structure of the variogram, as a function of d. Some admissible structures for $\gamma(d)$ are :

$$\alpha d \; ; \; \alpha \log d (d>1) \; ; \; \alpha d^{\beta} (0<\beta<2) \; ; \; \alpha\{1-\exp(\beta d)\} \tag{14}$$

Notice that the first three expressions apply to processes of unbounded variation.

2) The estimation of the unknown parameters (α,β) from the observed data.

3) The validation of the estimated variogram.

We shall not deal here with the estimation of the structure and with the validation of the variogram (see e.g. [11] or [1]). Rather we shall concentrate on the estimation of the unknown parameters of the variogram, and only under the following very simple assumptions :

a) the average piezometry is uniform on the studied domain :

$$E\{h(u)\} = m \quad u \in \Omega \tag{15a}$$

m is unknown.

b) the increments of h are Gaussian :

$$h(u_i) - h(u_j) = \text{Gaussian} \quad u_i, u_j \in \Omega \tag{15b}$$

c) the variogram is linear :

$$\gamma(d) = \frac{1}{2}\alpha\, d \tag{15c}$$

α is an unknown parameter, to be estimated. This very simple case is chozen to illustrate the difficulties. Given n data points $h_1, \ldots, h_n$, a standard way of estimating α is to construct a large number N of increments

$$\varepsilon_k \stackrel{\Delta}{=} h_i - h_j, \quad i,j \in \{1,\ldots,N\},\ k=1,\ldots,N \tag{16}$$

Notice that

$$E\{\varepsilon_k\} = 0$$

$$E\{\varepsilon_k^2\} = 2\gamma(d_k) = \alpha\, d_k \quad (\text{with } d_k = d_{ij}) \tag{17}$$

Then α is computed by least squares regression :

$$\underset{\alpha}{\text{Min}} \sum_{k=1}^{N} (\varepsilon_k^2 - \alpha\, d_k)^2 \tag{18}$$

Even in this very simple case, the estimation of α poses several problems.

- Should one chooze all possible increments ? Notice that $N_{max} = \binom{n}{2}$. If not, how many ?
- Assuming N increments have been used to estimate α. What is the effect on the variance of $\hat{\alpha}$ of adding a new increment that is a linear combination of others. (e.g., see Fig.1, is it useful to use ε_3 if ε_1 and ε_2 have already been used ?)
- What is the effect of correlated increments ? Notice that most increments will normally be correlated.

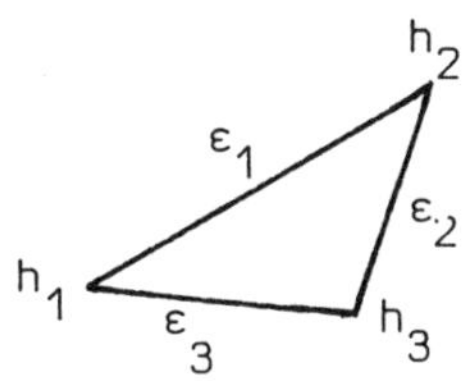

Fig.1

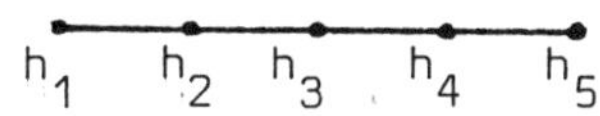

Fig.2

To illustrate these problems, consider a one-dimensional process where 5 equispaced measurements have been made : $h_1, \ldots, h_5$ (see fig.2). The following 10 increments can be defined :
$\varepsilon_1 = h_2 - h_1$, $\varepsilon_2 = h_3 - h_2$, $\varepsilon_3 = h_4 - h_3$, $\varepsilon_4 = h_5 - h_4$, $\varepsilon_5 = h_3 - h_1$, $\varepsilon_6 = h_5 - h_3$, $\varepsilon_7 = h_4 - h_2$, $\varepsilon_8 = h_4 - h_1$, $\varepsilon_9 = h_5 - h_2$, $\varepsilon_{10} = h_5 - h_1$.

We have : $\varepsilon_i^2 = \alpha d_i + v_i$ with $E\{v_i\} = 0$, $i = 1, \ldots, N$ (19)

Here $N = N_{max} = 10$. Table 1 gives the variance of the normalized least-square estimate $\hat{\alpha}_{LS}$ obtained for different choices of increments :

Table 1.

Increments used for computing $\hat{\alpha}_{LS}$	$\text{Var}\{\frac{\hat{\alpha}_{LS} - \alpha}{\alpha}\}$
$\varepsilon_5, \varepsilon_6$	1
$\varepsilon_5, \varepsilon_6, \varepsilon_9$	1.2
$\varepsilon_5, \varepsilon_6, \varepsilon_{10}$	1.5
$\varepsilon_5, \varepsilon_6, \varepsilon_7$	0.9
$\varepsilon_1, \varepsilon_2, \varepsilon_3, \varepsilon_4$	0.5

As table 1 demonstrates, adding a third increment to ε_5 and ε_6 can either increase or decrease the variance of the estimation error. Note that in this case all overlapping increments are correlated. In addition ε_{10} is the sum of ε_5 and ε_6. The lowest variance is achieved when the 4 uncorrelated increments $\varepsilon_1, \varepsilon_2, \varepsilon_3, \varepsilon_4$ are chozen for the computation of α. Adding other increments will never improve this bound.

The fact that adding new increments to the data base can deteriorate the estimate is disturbing. The reason is that the increments are often correlated. This fact remains true for 2-D processes where almost any two increments will be correlated. This leads us to consider the use of Markov estimates. Consider n data points $h_1, \ldots, h_n$ and N increments $\varepsilon_i = h_k - h_l$ in a 2-dimensional domain Ω. Then (19) can be rewritten in vector notations :

$$E^2 = \alpha D + V \qquad \text{with } E\{V\} = 0 \tag{20}$$

E^2 is defined as $E^2 = \left[\varepsilon_1^2, \varepsilon_2^2, \ldots, \varepsilon_N^2\right]^T$. D and V are similarly defined. The linear minimum variance unbiased (or Markov) estimate $\hat{\alpha}_M$ is obtained by minimizing

$$J(\alpha) = (E^2 - \alpha D)^T R^{-1} (E^2 - \alpha D) \text{ with } R = E\{VV^T\} \tag{21}$$

By the Gaussian nature of the increments, it is possible to compute R. Indeed let $\varepsilon_i \triangleq h_k - h_l$ and $\varepsilon_j \triangleq h_m - h_n$. Then for Gaussian processes one can show that

$$E\{v_i v_j\} = 2(E\{\varepsilon_i \varepsilon_j\})^2 \tag{23}$$

$$= 2(\gamma_{kn} + \gamma_{lm} - \gamma_{km} - \gamma_{ln})^2 \tag{24}$$

$$= \frac{\alpha^2}{2} (d_{kn} + d_{lm} - d_{km} - d_{ln})^2 \tag{25a}$$

$$E\{v_i^2\} = 2\alpha^2 \, d_{kl}^2 \tag{25b}$$

Notice that the increments ε_i and ε_j are uncorrelated (and hence $E\{v_i v_j\} = 0$) if and only if the 4 measure points are such that

$d_{kn} + d_{lm} = d_{km} + d_{ln}$ (see Fig.3)

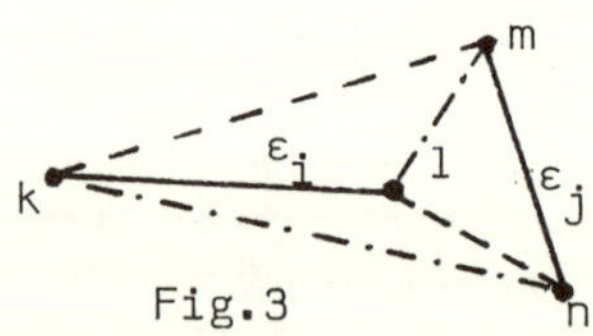

Fig.3

From (25) it is clear that one can write $R = \alpha^2 M$, where M is entirely known and depends only upon the geometrical configuration of the measure points. The Markov estimate and its normalized variance are then given by :

$$\hat{\alpha}_M = (D^T M^{-1} D)^{-1} D^T M^{-1} E^2 \tag{26}$$

$$\mathrm{Var}\{\frac{\hat{\alpha}_M-\alpha}{\alpha}\} = (D^T M^{-1} D)^{-1} \qquad (27)$$

The following results can now be proved.

<u>Result 1</u>: For a given set of N increments, $\mathrm{Var}\{\hat{\alpha}_M\} \leqslant \mathrm{Var}\{\hat{\alpha}_{LS}\}$. This result is well known.

<u>Result 2</u>: Let $\hat{\alpha}_M(N)$ be the Markov estimate obtained from N increments $\varepsilon_1, \ldots, \varepsilon_N$, and let $\hat{\alpha}_M(N+1)$ be the Markov estimate obtained by adding one more increment to the previous set. Then

$\mathrm{Var}\ \{\hat{\alpha}_M(N+1)\} \leqslant \mathrm{Var}\ \{\hat{\alpha}_M(N)\}$

<u>Result 3</u>: For a one-dimensional process, with n measure points $h_1, \ldots, h_n$, $\mathrm{Var}\ \{\frac{\alpha_M-\alpha}{\alpha}\}$ is minimum, and equal to $\frac{2}{n-1}$, when one choozes the n-1 uncorrelated increments $\varepsilon_1, \ldots, \varepsilon_{n-1}$ (as indicated on Fig.4) to estimate α. Adding other increments does not diminish the variance.

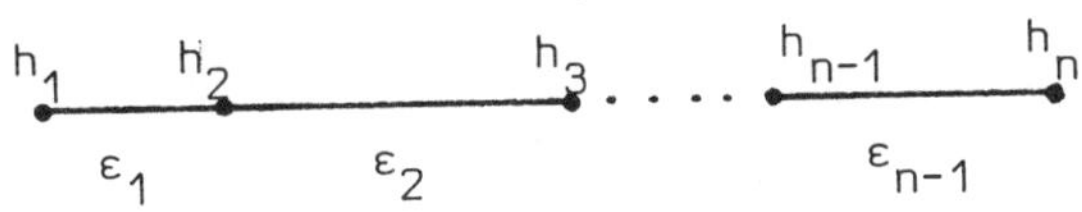

Fig.4

<u>Comments</u> :

1) Result 1 is of practical interest only when a Markov estimate can be computed, i.e. when R can be computed. This is possible for the various types of variograms proposed in (14) provided the process is Gaussian and β is known.

2) Result 2 is very important and extends to the same class of variograms as the first one. It shows that the Markov estimate, unlike the ordinary least squares estimate, is consistent. Adding a new increment will never deteriorate the quality of the estimate.

3) Result 3 gives a rule for the selection of increments for a one-dimensional process. However this result is very difficult to extend to other types of variograms or to 2-D processes (because, except for pathological cases, two increments are always correlated in 2-D space).

IV. CONCLUSIONS

Spatial interpolation of 2-D random processes is important for the identification of distributed parameter systems, in particular groundwaterflow processes. In this paper we have studied one aspect of a spatial interpolation technique, known as kriging, namely the estimation of the variogram. The main difficulties have been pointed out, and some preliminary results have been obtained.

REFERENCES

[1] BASTIN G. Approches déterministe et stochastique pour l'identification de systèmes hydrogéologiques. Thèse de Doctorat, U.C.L. 1979.

[2] NEUMAN S.P. Calibration of Distributed Parameter Groundwater Flow Models. - Water Resources Research, 9(4), Aug.1973.

[3] COOLEY R.L. A Method for Estimating Parameters and Assessing Reliability for Models of Steady-State Groundwater Flow.- Water Resources Research, 13(2), April 1977.

[4] BASTIN G., GEVERS M. Joint Use of Space Interpolation and Optimization Methods for Steady-State Aquifer Modelling with Scarce Data - in "Modelling, Identification and Control of Environmental Systems" - Van Steenkiste Ed., North Holland Publ., 1978.

[5] CAROTENUTO L. et al. A Regularized Solution of the Identification Problem for the Distributed Parameter Model of an Underground Aquifer. - IFAC Triennal World Congress, Niami Ed., June 1978.

[6] DE MARSILY G. De l'identification des systèmes hydrogéologiques. Thèse de Doctorat d'Etat en Sciences Naturelles Appliquées. Université Paris VI, 1978.

[7] BASTIN G. Stochastic Parameter Estimation of a Steady-State Groundwater Flow Model. - IMACS 79 Congress on Simulation of Systems, Italy, September 1979.

[8] BASTIN G. and GEVERS M. Stochastic Identification of Steady-State Groundwater Flow Systems. - Paper in preparation.

[9] MATHERON G. La théorie des variables régionalisées et ses applications. - Cahiers du Centre de Morphologie Mathématique Fasc.5. Fontainebleau, 1970.

[10] DELFINER P. and DELHOMME J.P. Optimum Interpolation by Kriging in "Display and Analysis of Spatial Data" - Davis and Mc Cullough Ed., J. Wiley, 1975.

[11] DELFINER P. Linear Estimation of Non-Stationary Spatial Phenomena.- in "Advanced Statistics in the Mining Industry". D. Reidel, 1976.

OPTIMAL FILTER FOR A DISCRETE-TIME DISTRIBUTED PARAMETER SYSTEM AND ITS APPLICATION TO ENVIRONMENTAL DATA PROCESSING

Sigeru OMATU, Hirosige NAGAMINE, and Takasi SOEDA

Department of Information Science and Systems Engineering
Faculty of Engineering, University of Tokushima,
Tokushima, 770, Japan

Abstract

Based on the Wiener-Hopf theory, the optimal filtering problem for a linear discrete-time distributed parameter system is considered. Using the least squares estimation error criterion, the Wiener-Hopf equation for the discrete-time distributed parameter system is derived. Then the basic equations for the optimal filtering estimate and the minimum filtering error covariance function from which we derive the recursive forms of the optimal filter and the error covariance function. Finally, we apply the filtering results derived here to predict the concentration levels of the pollutants of air pollution.

1. INTRODUCTION

In recent years much research has been devoted to study the optimal filtering problems for distributed parameter systems. The techniques used for generating the filtering estimate for continuous-time distributed parameter system include orthogonal projection theories[1]-[8], maximum likelihood approaches combined with variational inequality or dynamic programming methods[9]-[12], and the Bayesian approach[13].

On the other hand, the filtering results for the discrete-time distributed parameter system has not been discussed except for those of Tzafestas[14]. Tzafestas[14] used the Gaussian measure for the distributed parameter system whose validity has not been proved up to now. Furthermore, the derivation used the differential rule of the functional derivative whose chain rule has not been proved.

D. G. Lainiotis and N. S. Tzannes (eds.), Applications of Information and Control Systems, 69-79.

In this paper, we derive the optimal filter for the discrete-time linear distributed parameter system based on the Wiener-Hopf theory. An advantage of the present method over that of Tzafestas [14] is that the present approach does not use the concept for the Gaussian measure for the distributed parameter system and the differential rule for the functional derivative. Furthermore, we can interpret the derivation of the optimal filter from the physica viewpoint since the method is based on the Wiener-Hopf theory.

Then we consider the real-time estimation problem of air pollutant concentrations in an urban atmosphere based on observed data taken intermittently at a finite set of monitoring stations. An urban airshed can be regarded as the complex distributed parameter system on which measurements are made at discrete location points. The filtering theory for the distributed parameter system is quite well-developed as mentioned above, although the application of the filtering theory for the distributed parameter system are largely confined to simple examples. Bankoff and Hanzevack[15] developed an algorithm to estimate parameters characterizing the atomospheric mixing based on pollutant monitoring data and applied the method to modeling sulfur dioxide concentrations in Chicago. Koda and Seinfeld[16] and [17] developed a square-root distributed parameter filter for a general class of dynamic urban area. Desalu et al.[18] discretized a mathematical air pollution model by the difference approximation method and the distributed parameter system is approximated by a lumped parameter system. Then they applied Kalman filtering technique for a lumped parameter system to estimate the concentration levels of the pollutants of air pollution. Although Desalu et al.[18] used a discrete form of Chandrasekar-type equations to avoid the high dimensionality, the calculation of high dimensional matrix can not be avoided in estimation and smoothing of the system states. Furthermore, in order to identify the diffusion coefficients, they used the maximum likelihood method. But they did not considered how to determine the statistical properties of the noises. Since the algorithm of identification in [18] used repeatedly both the smoothed form of the maximum likelihood identification procedure and Kalman filter with the high dimensional matrix, much computational time is required.

In this paper, we derive the optimal filtering equation for such a mathematical air pollution model as the system preserves the distributed character. Since we can derive Kalman filter in lumped parameter system from the results of the distributed parameter system by using the Fourier expansion method, the approach adopted here is mathematically rigorous. Compared with the results of Desalu et al.[18], we use the method to determine the statistical properties of the system and measurement noises by using the algorithm developed by Sage and Husa[19]. For the identification of the diffusion coefficient, the gradient method is used and the computational time required for the prediction estimate ie less than Desalu et al.[18] since the two-stage iterative procedure may not be necessitated in the present paper.

In this paper, we apply the discrete-time filtering results derived here to predict the concentration levels of the pollutants of air pollution at Tokushima in Japan. SO_2 data are adopted at four points, that is, Aizumi, Kitajima, Kawauchi, and Matushige, in Tokushima prefecture. The covariance functions of the system and measurement noises are determined by using Sage and Husa's method[19]. The identification of the diffusion coefficient is obtained by the gradient method. Then one hour in advance prediction values for the SO_2 data are determined by using Kalman filter derived here.

2. A DISCRETE-TIME DISTRIBUTED PARAMETER SYSTEM

2.1. Statement of the problem

Let D be a bounded open domain of an r-dimensional Euclidean space with smooth boundary S. The spatial coordinate vector will be denoted by $x=(x_1,\cdots,x_r) \in D$. Consider a linear distributed parameter system which is described by the equation

$$U(k+1,x) = \mathcal{L}_x U(k,x) + B(k,x) W(k,x), \quad \forall x \in D \qquad (2.1)$$

where $U(k+1,x) = \mathrm{Col}[\, u_1(k+1,x),\cdots, u_n(k+1,x) \,]$ is an n-dimensional state vector function of the system, $W(k,x)$ is a vector-valued white Gaussian process, $\mathcal{L}_x$ is a linear spatial matrix differential operator and $B(k,x)$ is a known matrix function.
The initial and boundary condition- are given by

$$U(k_0,x) = U_0(x) \qquad (2.2)$$

$$\Gamma_\xi U(k,\xi) = 0, \qquad \forall \xi \in S$$

$$\Gamma_\xi[\cdot] = \alpha(\xi)[\cdot] + (1 - \alpha(\xi))\, \partial[\cdot]/\partial \mathbf{n} \qquad (2.3)$$

where $\mathbf{n}$ is the exterior normal vector to the surface S at a point $\xi \in S$ and $\alpha(\xi)$ is a function of class C^2 on S satisfying

$$0 \leqq \alpha(\xi) \leqq 1.$$

Assume that the initial condition $U_0(x)$ is a Gaussian random function whose mean and covariance function are given by

$$E[\, U_0(x) \,] = 0, \qquad E[\, U_0(x)\, U_0'(y) \,] = P_0(x,y) \qquad (2.4)$$

where $E[\cdot]$ denotes the expectation operator and " ' " denotes the transpose of the vector or matrix. Let the measurement data be taken at fixed m points $x^1, x^2,\cdots, x^m \in \bar{D} = D \cup S$ and let an m-dimensional column vector $U_m(k)$ be defined by

$$U_m(k) = \mathrm{Col}[\ U(k,x^1),\cdots,\ U(k,x^m)\ ]. \tag{2.5}$$

Assume that the measurement equation is given by

$$Z(k) = H(k)\ U_m(k) + V(k) \tag{2.6}$$

where $Z(k)$ is a p-dimensional measurement vector at the points $x^1, \cdots, x^m \in \bar{D}$, $H(k)$ is a known $p \times mn$ matrix, and $V(k)$ is a p-dimensional vector-valued white Gaussian process.
The white Gaussian processes $W(k,x)$ in (2.1) and $V(k)$ in (2.5) are assumed to be stochastically independent of each other and also independent of the stochastic initial condition $U_0(x)$.
The mean and covariance functions of them are given by

$$E[\ W(k,x)\ ] = 0, \quad E[\ V(k)\ ] = 0 \tag{2.7a}$$

$$E[\ W(k,x)\ W'(s,y)\ ] = Q(k,x,y)\ \delta_{ks}, \ \forall x,y \in D \tag{2.7b}$$

$$E[\ V(k)\ V'(s)\ ] = R(k)\ \delta_{ks} \tag{2.7c}$$

where δ_{ks} is the Kronecker delta, $Q(k,x,y)$ is a symmetric positive semi-definite matrix and $R(k)$ is a symmetric positive definite matrix.

2.2. Wiener-Hopf Equation
Let $\hat{U}(k,x)$ be the optimal filtering estimate based on the measurement data $Z(s)$, $s=0,\cdots,k$ and given by

$$\hat{U}(k,x) = \Sigma_{s=0}^{k}\ F(k,x,s)\ Z(s) \tag{2.8}$$

where $F(k,x,s)$ is an $n \times p$ matrix kernel function.
We adopt the performance measure J given by (2.8) which is to be minimized,

$$J = E[\ \|\tilde{U}(k,x)\|^2\ ] \tag{2.9}$$

where $\tilde{U}(k,x)$ is an estimation error given by

$$\tilde{U}(k,x) = U(k,x) - \hat{U}(k,x)$$

and $\|\cdot\|$ denotes the Euclidean norm. Then it is necessary to find a matrix function $F(k,x,s)$ which minimizes the performance J of (2.9). Now the following theory which is called the Wiener-Hopf theorem is proved.
[*Theorem 2.1*] (Wiener-Hopf theorem)
A necessary and sufficient condition for the estimate (2.8) to be optimal is that the Wiener-Hopf equation

$$\Sigma_{s=0}^{k}\ F(k,x,s)\ E[\ Z(s)\ Z'(\tau)\ ] = E[\ U(k,x)\ Z'(\tau)\ ] \tag{2.11}$$

holds for $\tau = 0, \cdots, k$, and $x \in \bar{D}$.
Equivalently, (2.11) means

$$E[\ \tilde{U}(k,x)\ Z'(\tau)\ ] = 0, \quad \tau = 0,\cdots,k, \text{ and } x \in \bar{D}. \tag{2.12}$$

The following corollary can be obtained immediately from (2.12).
[*Corollary 3.1*] The following relation holds.

$$E[\ \tilde{U}(k,x)\ \hat{U}'(k,y)\ ] = 0 \quad \text{for all } x,y \in \bar{D}. \tag{2.13}$$

Now substituting (2.6) into (2.11) and utilizing the independence of $U(k,x)$ and $V(k)$ yields

$$E[\ Z(\sigma)\ Z'(\tau)\ ] = H(\sigma)\ M_m(\sigma,\tau)\ H'(\tau) + R(\sigma)\delta_{\sigma\tau} \tag{2.14}$$

$$E[\ U(k,x)\ Z'(\tau)\ ] = M(k,x,\tau)\ H'(\tau), \tag{2.15}$$

where

$$M_m(\sigma,\tau) = E[\ U_m(\sigma)\ U_m'(\tau)\ ] \text{ and } M(k,x,\tau) = E[\ U(k,x)\ U_m'(\tau)\ ] \tag{2.16}$$

Let us now introduce the following notations;

$$\mathbf{F}(k,x) = [\ F(k,x,0),\cdots,\ F(k,x,k)]$$

$$\mathbf{M}_m(k) = \begin{bmatrix} M_m(0,0) & \cdots & M_m(0,k) \\ \vdots & & \\ M_m(k,o) & \cdots & M_m(k,k) \end{bmatrix} \qquad M^*(k,x) = [\ M(k,x,0),\ ..,M(k,x,k)]$$

$$\mathbf{R}(k) = \text{diag}[R(0),\cdots,R(k)] \text{ and } \mathbf{H}(k) = \text{diag}[H(0),\cdot\cdot,H(k)].$$

According to the above notations, (2.11) is described by

$$\mathbf{F}(k,x)[\ \mathbf{H}(k)\mathbf{M}_m(k)\mathbf{H}'(k) + \mathbf{R}(k)] = M^*(k,x)\ \mathbf{H}'(k). \tag{2.17}$$

In what follows, we shall denote $\mathbf{L}_m(k)$ as follows;

$$\mathbf{L}_m(k) = [\ \mathbf{H}(k)\ \mathbf{M}(k)\ \mathbf{H}'(k) + \mathbf{R}(k)\ ]. \tag{2.18}$$

Then the kernel function vector $\mathbf{F}(k,x)$ is described as follows;

$$\mathbf{F}(k,x) = M^*(k,x)\ \mathbf{H}'(k)\ \mathbf{L}_m^{-1}(k). \tag{1.19}$$

Then the following theorem holds.
[*Theorem 3.2*] The optimal filtering estimation error covariance function $P(x,y,k\ k)$ satisfies

$$P(x,y,k|k) = M_k(x,y) - M^*(k,x)\ \mathbf{H}'(k)\ \mathbf{L}_m^{-1}(k)\ \mathbf{H}(k)\ M^{*\prime}(k,y) \tag{2.20}$$

where

$$M_k(x,y) = E[\ U(k,x)U'(k,y)\ ], \qquad P(\xi,y,k|k) = 0,\ \xi\varepsilon\ S,\ y \in \bar{D}$$

Then the following theorem holds.
[*Theorem 3.3*] The recursive forms of the optimal filtering estimator and the optimal filtering error covariance function are given by

$$\hat{U}(k,x) = {}_x\hat{U}(k-1,x) + P^*(x,k|k-1)H'(k)[\,R(k) + H(k)$$
$$P_m(k|k-1)H'(k)]^{-1}[\,Z(k) - H(k)\ {}_x^*\hat{U}_m(k-1)\,], \qquad (2.21)$$
$$P(x,y,k|k) = P(x,y,k|k-1) - P^*(x,k|k-1)H'(k)[\,R(k)$$
$$+ H(k)P_m(k|k-1)H'(k)]^{-1}\ H(k)P^{*'}(y,k|k-1). \qquad (2.22)$$

The optimal filter for a discrete-time distributed parameter system was derived by Tzafestas based on the Bayesian approach[14]. However, the derivation is vague and does not make clear from the mathematical viewpoint because the method necessitates the Gaussian measure for the distributed parameter system and the differential rules of the functional derivative whose validities have not been proved up to now. Compared with the derivation of Tzafestas[14], the present approach neither necessitates the Gaussian measure nor the functional derivative. Furthermore, we can easily see the validity of the derivation from the physical viewpoint since the present method is based on the Wiener-Hopf theory. Therefore, the approach proposed here is not only a mathematically rigorous method but also a unified approach for the derivation of the optimal estimators based on the Wiener-Hopf theory.

3. APPLICATION TO ENVIRONMENTAL SYSTEMS

3.1. Prediction for Air Pollution

The models for predicting the concentration levels of the pollutants of air pollution are classified in two categories. One is the nonphysical model and the other is the physical model. The former has been already applied in many environmental systems. Although the latter has been mathematically well-developed, the applications to environmental problems are virtually unexplored. In this section, we consider the predictions of air pollution levels by using the results which we have derived in the preceding section. In the studies of the prediction and control of air pollution levels by physical models, we must consider some situations. We consider the following three problems:

(i) How to specify the suitable initial and boundary conditions,

(ii) How to find the statistical properties of the system and measurement noises,

and

(iii) How to determine the measurement points.

By solving three problems stated above, we discuss the prediction problem for the concentration levels of the pollutant of air pollution. In this paper, we study the prediction problem of air pollution at Tokushima prefecture in Japan. Here, wé use the SO_2 data measured each one hour in December 1-31, 1975 from four monitoring stations in Tokushima prefecture as shown in Fig.1. Since the data have a 24-hr daily cycle, we use the residual between the raw data and the monthly average at each time as the processes data to eliminate a 24-hr daily cycle. Furthermore, we assume that

(i) one-dimensional horizontal flow exists,
(ii) the boundary on the spatial domain is 0 and 1, that is, $\xi = 0$ or 1, $\xi \in S$,

and

(iii) the truncation number $N = 2$.

Since during this period(i.e., December 1-31, 1975) there is almost every day a wind blowing from west and we can observe that the diffusion of air pollution to the southern or northern direction is uniformly, we may assume the system as one-dimensional folw model.

3.2. The Performance of the Boundary Condition

It is well-known that the system describing the dynamics of air pollutant concentrations consists of distributed parameter systems. We will apply the filtering theory for the discrete-time distributed parameter system to the air pollution problem. Then, the filtering estimator for the distributed parameter system must be solved under the suitable initial and boundary conditions. On the west boundary point (i.e. $\xi = 0$), there is almost every day a wind blowing from the west and there is not such a source as influences at the boundary. Furthermore, on the east boundary point (i.e. $\xi = 1$), its boundary is on the coastline and its district is located on an open space. Under consideration mentioned above, we use four boundary conditions for the real data processing:

$$\text{(i)}\quad \partial U(k,\xi)/\partial\xi\big|_{\xi=0} = 0, \qquad \partial U(k,\xi)/\partial\xi\big|_{\xi=1} = 0. \tag{3.1a}$$

$$\text{(ii)}\quad \partial U(k,\xi)/\partial\xi\big|_{\xi=0} = 0, \qquad U(k,\xi)\big|_{\xi=1} = 0. \tag{3.1b}$$

$$\text{(iii)}\quad U(k,\xi)\big|_{\xi=0} = 0, \qquad \partial U(k,\xi)/\partial\xi\big|_{\xi=1} = 0. \tag{3.1c}$$

$$\text{(iv)}\quad U(k,\xi)\big|_{\xi=0} = 0, \qquad U(k,\xi)\big|_{\xi=1} = 0.$$

Assuming that the SO_2 data are the diffusion processes, we adopt here the differential operator $\mathcal{L}_x$ as follows:

$$\mathcal{L}_x = 1 - a\partial^2/\partial x^2$$

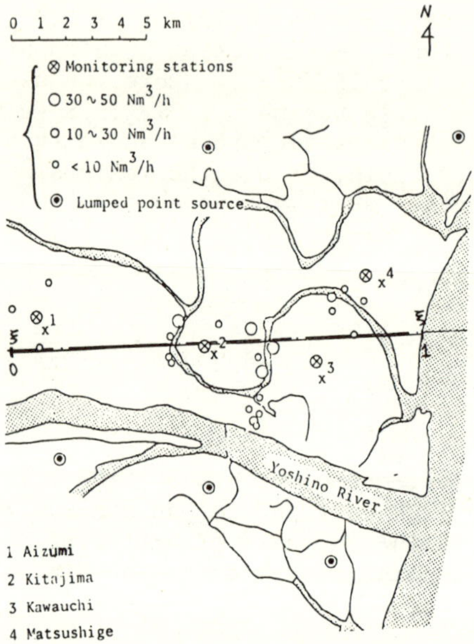

Fig.1. Map of Tokushima City

We apply the Fourier expansion method to solve the optimal filter for the distributed parameter system. The eigenvalues and eigenfunctions for four conditions are given as follows:

Condition 1.

$$\lambda_i = 1 - a\,(i\pi)^2$$
$$\phi_i(x) = \sqrt{2}\cos i\pi x. \quad (3.2a)$$

Condition 2.

$$\lambda_i = 1 - a((2i-1)\pi/2)^2$$
$$\phi_i(x) = \sqrt{2}\cos((2i-1)/2)\pi x. \quad (3.2b)$$

Condition 3.

$$\lambda_i = 1 - a((2i-1)\pi/2)^2$$
$$\phi_i(x) = \sqrt{2}\sin((2i-1)/2)\pi x. \quad (3.2c)$$

Condition 4.

$$\lambda_i = 1 - a\,(i\pi)^2, \quad \phi_i(x) = \sqrt{2}\sin i\pi x. \quad (3.2d)$$

where a is determined by using the gradient method.

The monitoring stations are selected at most 4 points (x^1, x^2, x^3, x^4) where x^1= Aizumi, x^2=Kitajima, x^3= Kawauchi, and x^4=Matsushige as sown in Fig.1. As a measure of prediction accuracy, we consider the mean square error of the predicted values and the measurement values at each time:

$$PA(i) = 1/24\left(\Sigma_{k=1}^{24}\left(Z_i(k) - Z_i(k|k-1)\right)^2\right), \; i=1,2,3,4 \quad (3.3)$$

where i denotes the i-th monitoring station.
Furthermore, as an overall performance function we take

$$J = \Sigma_{m=1}^{4} PA(m). \quad (3.4)$$

In order to determine the parameter a, we use the gradient method for the overall performance shown above. Fig.2 shows the curve of the overall performance function versus parameter a for Condition 2. Table I shows some of parameters and performances obtained by the gradient method for each of conditions. From Table I, it follows that parameter a = 0.01 for conditions 1 and 2 and parameter a = 0.02 for conditions 3 and 4 give the minimum values of the overall performance function. We also represent the mean square error of measurement values per month.

In order to obtain the covariance matrices for the noises, we

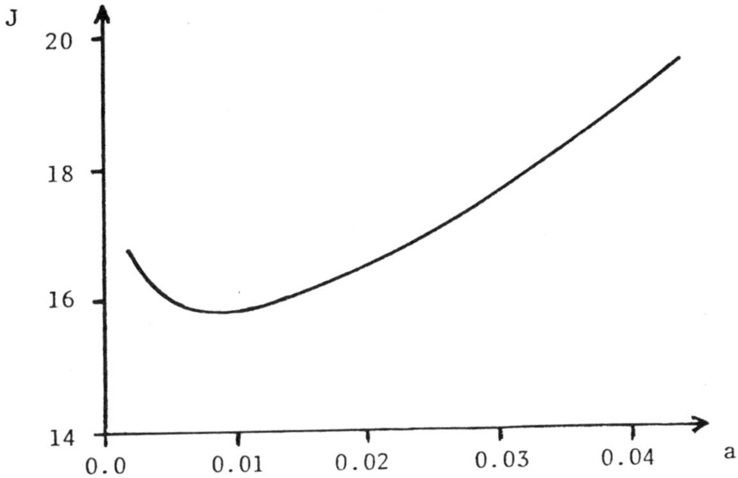

Fig.2. The curve of the overall performance function versus parameter a for Condition 2.

TABLE I. The prediction accuracy for four-points measurement case and monthly average.

condition	parameter a	Aizumi x^1	Kitajima x^2	Kawauchi x^3	Matsushige x^4	overall J
(i)	0.02	1.13	6.88	10.93	5.88	24.83
	0.01	4.26	8.64	8.07	2.73	23.70
	0.005	9.03	2.84	7.30	7.87	27.04
(ii)	0.04	1.09	4.98	7.53	5.38	18.98
	0.02	1.02	3.95	6.60	5.02	16.59
	0.01	1.00	4.54	4.59	5.51	15.84
	0.005	0.96	5.33	4.94	4.76	15.99
(iii)	0.04	1.87	3.47	4.36	7.13	16.83
	0.02	2.09	1.81	9.06	2.96	15.92
	0.01	2.17	1.77	9.04	2.98	15.96
	0.005	5.11	3.38	91.99	5.00	105.48
(iv)	0.02	2.12	1.97	7.16	4.44	15.69
	0.01	1.57	4.80	5.08	4.25	15.70
	0.005	1.29	6.63	11.24	4.08	23.24
monthly average		2.11	6.59	9.58	5.80	24.08

obtain the following results after 672 iterations; $q_{11} = 2.06$, $q_{22} = 2.41$, $r_{11}=6.70$, $r_{22}=11.10$, $r_{33}=1.30$, and $r_{44}=6.73$ where q_{ii} and r_{ii} denote the (i,i)-th component of the covariance functions of the system and measurement noises, respectively.

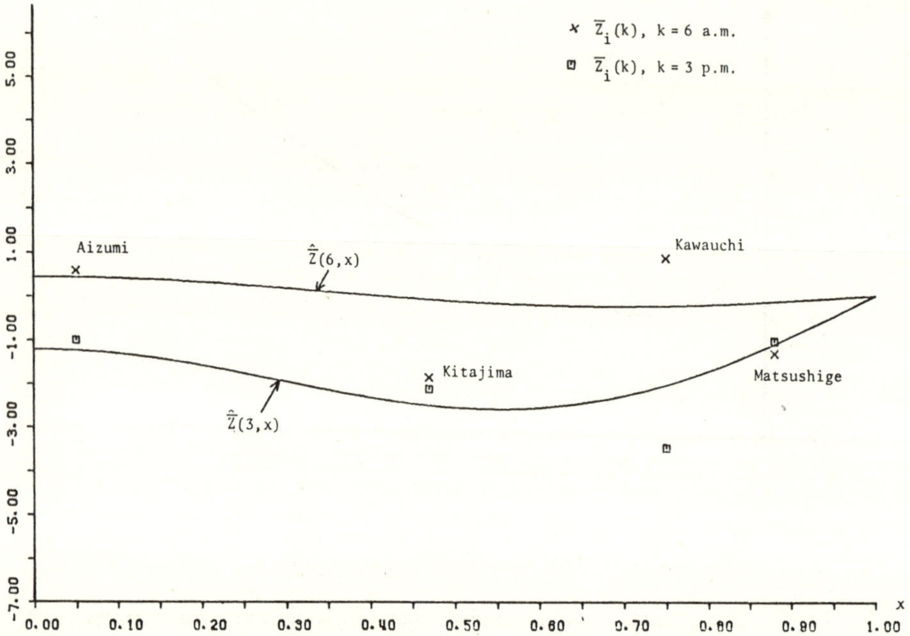

Fig.3. The distribution of the predicted concentration levels for four points measurement case.

The optimal sensor location for the one point measurement system is Kitajima which consists with the earlier results of the sensor location[7]. Finally, Fi.3 shows the one hour ahed prediction values and true values with regard to the spatial coordinate x.

4. CONCLUSIONS

In this paper, we derived the optimal filter for the discrete-time distributed parameter system based on the Wiener-Hopf theory. The results by Tzafestas[14] were verified from the mathematically rigorous way. The the results obtained here were applied to the air pollution levels. In this case, the covariance matrices of the system and observation noises were determined by using Sage and Husa's method[19]. The remaining problems are how to find the truncation number of the eigenfunctions and the initial and boundary conditions.

REFERENCES

[1] R.F.Curtain, SIAM J.Control, Vol.13, 89-103, 1975.
[2] R.F.Curtain, Rep. No.40, Control Theory Centre, Univ. Warvick, 1-35, 1975.
[3] P.L.Falb, Information and Control, Vol.11, 115-127, 1972.
[4] Y.Sakawa, Int. J. Control, Vol.16, 115-127, 1972.
[5] S.Omatu and T.Soeda, Proc. Int. Symp. on SDE, Kyoto, 1978.
[6] S.Omatu and T.Soeda, Int. J. Sys. Sci., Vol.9, 753-771, 1978.
[7] Y.Sa-aragi, T.Soeda and S.Omatu, Lec. Notes in Control and Inf. Sci. , Vol.11, Springer Verlag, 1978.
[8] S.G.Tzafestas and J.M.Nightingale, Proc. IEE, Vol.115, 1027-1093, 1968.
[9] A.Bensoussan, Filtrage Optimal des Systemes Lineaires, Dunod, Paris, 1971.
[10] A.Bensoussan, SIAM J. Control, Vol.13, 904-926, 1975.
[11] J.S.Meditch, Automatica, Vol.7, 315-322, 1971.
[12] S.G.Tzafestas and J.M.Nightingale, Proc. IEE, Vol.116, 1085-1093, 1969.
[13] H J.Kushner, SIAM J. Control, Vol.8, 346-359, 1970.
[14] S.G.Tzafestas, Int. J. Control, Vol.15, 273-295, 1972.
[15] S.G.Bankoff et al., Atmo . Env.,Vol.9, 793-808,1975.
[16] M.Koda and J.H.Seinfeld, Automatica, Vol.14, 583-595, 1978.
[17] J.H.Seinfeld and M.Koda, Lec. Notes in Control and Inf. Sci., Vol.1, 42-69, 1976.
[18] A.A.Desalu et al., IEEE Trans. AUt. Control, Vol.AC-19, 904-910, 1974.
[19] A.P.Sage and G.W.Husa, JACC, 760-769, 1969.

RADAR EMITTER RECOGNITION USING PULSE REPETITION INTERVAL

R.M. Hawkes

Department of Defence, Electronics Research Laboratory
Defence Research Centre Salisbury, South Australia.

A method for estimating radar emitter pulse repetition interval parameters from intercepted emissions is presented. Maximum likelihood parameter estimates based on certain measurements are derived. The measurements are obtained by a readily implementable observation procedure. The approach can deal with emitters with variable pulse repetition interval in a dense environment of emitters.

1. INTRODUCTION

In this paper a method for identifying radar emitters by their pulse repetition interval (PRI) is presented. The method is based on statistical techniques.

To deal with the complex signal environments likely to be encountered one emitter parameter alone can not be relied on to achieve positive identification. Thus, in this paper it is assumed that some analysis of the environment using such parameters as radio frequency and direction of arrival will be carried out in conjunction with PRI analysis.

The electromagnetic environment can be divided into two classes of emitters, namely, those emitters whose intercepted pulse amplitude is heavily modulated (such as search radars) and those emitters whose intercepted pulse amplitude has a modest amount or no modulation (such as tracking radars). The range of PRI to be encountered will be from about 10^2 μs to 10^4 μs. Search radars tend to have longer PRI while tracking radars usually have relatively short PRI.

D. G. Lainiotis and N. S. Tzannes (eds.), Applications of Information and Control Systems, 80-87.

Standard methods for emitter recognition using pulse repetition interval include the pulse train sorter and the time of arrival method. Experience has revealed certain limitations for these methods. The main limitations are in handling a variable PRI and in extending their use to a dense environment.

It is important to be able to recognize emitters with variable pulse repetition interval. Variable PRI is used in radar emitters to overcome the blind speed problem with moving target indication (MTI) and as an antijamming technique. In the MTI application the PRI may vary typically $\pm$ 10% whilst in the antijamming application $\pm$ 50% has been considered.

The recognition methods in this paper can cope with variable PRI and offer a simple, systematic approach to analysing the dense environment problem.

2. OBSERVATION PROCEDURE

In this section a readily implementable observation procedure is introduced for the determination of radar emitter PRI from intercepted emissions. Suppose that the IF of a typical electronic surveillance receiver is examined during the time interval from s_0 to s_T. This interval, which will be called the observation interval, is partitioned as $s_0, s_1, \ldots, s_{T-1}, s_T$ where $s_t - s_{t-1} = \tau$ for all t. The quantity τ will be called the gate time and the subintervals of length τ will be called the gate intervals. Note that $T\tau = s_T - s_0$. The observation procedure is to determine the number of pulses which occur during each gate interval and set n_t equal to the number of pulses which occur during gate interval t. Thus, a sequence of numbers $n_1, n_2, \ldots, n_T$ is obtained. Suppose that during the observation interval the maximum and minimum number of pulses detected in any gate interval is N_{max} and N_{min} respectively. The observation x_t at discrete-time point t is defined by

$$x_t = n_t - N_{min} \tag{1}$$

3. ASSUMPTIONS

The assumptions made throughout this paper about the radar emitters and the observation procedure are stated in this section.

A mathematical model is developed for the observed data. The observation procedure introduced generates pseudo random sequences of numbers. The equations derived for the mathematical model are based on certain assumptions concerning these sequences. The validity of these assumptions should be verified either by theoretical analysis or by empirical tests. (See references 1 and 2 for a discussion of standard random number generators).

Theoretical tests would give much more understanding of the method than the results of empirical tests. In this paper conditions relating the PRI of the emitters and the observation procedure will be given which are believed to be sufficient for the mathematical model to adequately represent the observed data. The results of simulations of the proposed PRI estimation technique which have been done justify this belief. Theoretical proofs of the results will be given in reference 6.

The basic assumptions made are as follows:

(a) The pulse trains of the emitters are unsynchronized.
(b) The pulse trains of the emitters are assumed to be unsynchronized with the gate interval partition of the observation procedure.
(c) If overlapping pulses occur it is possible to detect how many pulses occurred.
(d) The environment does not change during the observation interval.

These assumptions are not overly restrictive. For emitters with fixed PRI assumptions 3(a) and 3(b) are tantamount to assuming that the PRIs of the radar emitters are incommensurable (i.e. not divisible without remainder by the same quantity) and that the PRIs of the radar emitters and τ are incommensurable. In practice, radar emitters with commensurable PRIs will be quite rare and can possibly be treated as a single emitter with variable PRI if present together. Assumption 3(b) can always be satisfied by suitable choice of τ. It is believed that assumptions 3(c) and 3(d) are realistic.

4. EMITTERS WITH FIXED PRI

The results of this section apply to emitters with fixed PRI in a dense environment. Let $x_1^k, x_2^k, \ldots, x_T^k$ denote the observed sequence when only the kth emitter is present. If y_k is the PRI of the kth emitter and

$$(R_k - 1)y_k < \tau < R_k y_k \tag{2}$$

where R_k is a positive integer, then during any gate interval either $R_k - 1$ or R_k pulses will be observed. Thus, the observations, x_t^k, are modelled as a realization of the stochastic process, X_t^k, $t = 1,2\ldots, T$ which takes the values 0 and 1 with probability p_k and $1 - p_k$ respectively.

The probability, p_k, can be related to the parameters y_k, R_k and τ as follows. Let θ_t^k be the time measured from the start of gate interval t to where the next pulse occurs. Note that $0 < \theta_t^K < y_k$. Furthermore, if assumption 3(b) is invoked then θ_t^k may be considered to be a sample value of a random variable uniformly distributed on the interval 0 to y_k. The proof will be given in reference 6 and the results of computer simulation support this assertion. For $0 < \theta_t^k < \{\tau - (R_k-1)y_k\}/y_k$, $x_t^k = 1$, and for $\{\tau-(R_k-1)y_k\}/y_k < \theta_t^k < y_k$, $x_t^k = 0$. Thus, the expression for p_k is

$$p_k = (R_k\ y_k\ -\tau)/y_k \tag{3}$$

Simple algebraic manipulation yields

$$y_k = \tau/(R_k - p_k) \tag{4}$$

In a dense environment the observed sequence, x_1, x_2, ..., x_T is modelled as a realization of the stochastic process, $\{X_t,\ t = 1,2,\ldots,\ T\}$. The observation x_t is related to the observations for the individual emitters comprising the environment x_t^k via

$$x_t = \sum_{k=1}^{K} x_t^k \tag{5}$$

The process $\{X_t\}$ takes values from the set$\{0,1,\ldots,K\}$ where $K = N_{max} - N_{min}$ is the number of emitters present. In order to derive an expression for the probability that the dense environment process is in state j (denoted $P_{j,K}$) in terms of the quantities p_k the processes $\{X_t^k\}$ associated with the individual emitters present are assumed to be mutually statistically independent. It is believed that assumption 3(a) guarantees mutual statistical independence and simulation results have been obtained to support this statement. In reference 6 it is proved that, given the quantities $P_{j,K}$, the probabilities p_k are zeroes of the polynomial, $f_K(\lambda)$, of degree K in λ given by

$$f_K(\lambda) = \lambda^K + a_{K-1,K}\ \lambda^{K-1} + \ldots + a_{1,K}\lambda + a_{0,K} \tag{6}$$

where the coefficients can be expressed as

$$a_{r,K} = (-1)^{K-r} \sum_{j=0}^{r} P_{j,K} \binom{K-j}{r-j} \tag{7}$$

and $(\cdot)$ denotes the combinatorial symbol.

The following example illustrates this result.

<u>Example 1</u> $(K = 3)$ Starting with the expressions for $P_{j,K}$

$$P_{0,3} = p_1 p_2 p_3$$

$$P_{1,3} = (1-p_1)p_2p_3 + (1-p_2)p_1p_3 + (1-p_3)p_1p_2$$

$$P_{2,3} = (1-p_1)(1-p_2)p_3 + (1-p_1)(1-p_3)p_2 + (1-p_2)(1-p_3)p_1$$

Simple algebraic manipulation yields,

$$p_1p_2p_3 = P_{0,3}$$

$$p_1p_2 + p_1p_3 + p_2p_3 = P_{1,3} + 3P_{0,3}$$

$$p_1 + p_2 + p_3 = P_{2,3} + 2P_{1,3} + 3P_{0,3}$$

Recall that the coefficients of a polynomial with zeroes p_1, p_2 and p_3 can be computed by multiplying out the expression

$$(\lambda - p_1)\ (\lambda - p_2)\ (\lambda - p_3)$$

$$= \lambda^3 - (p_1+p_2+p_3)\ \lambda^2 + (p_1p_2+p_1p_3+p_2p_3)\lambda - p_1p_2p_3$$

This verifies the above result for $K = 3$, viz.,

$$a_{0,3} = P_{0,3}$$

$$a_{1,3} = P_{1,3} + 3P_{0,3}$$

$$a_{2,3} = -(P_{2,3} + 2P_{1,3} + 3P_{0,3})$$

Each emitter has a unique y_k and, once τ is specified, a unique p_k from (3). The significance of the above result is that in a dense environment each possible combination of emitters which can be present will have a unique set of probabilities $P_{j,K}$, $j = 0,1,\ldots,K$. The dense environment of emitters is said to be identifiable by observing the process $\{X_t\}$.

The relative frequencies of occurrence of the states of the process, P_q, $q = 0,1 \ldots, k$, during the observation interval can be calculated via

$$P_q = N_q/T \tag{8}$$

$$N_q = \sum_{t=1}^{T} N_q^t \tag{9}$$

$$N_q^t = \begin{cases} 1, & \text{if } x_t = q \\ 0, & \text{otherwise} \end{cases} \tag{10}$$

From reference 3, p392, Ex 12.23 the maximum likelihood estimate of $P_{j,K}$, denoted $\tilde{P}_{j,K}$, is given by

$$\tilde{P}_{j,K} = P_j \tag{11}$$

Using the invariance property of maximum likelihood estimates (reference 4, Ch VII, Sec 2.2) the maximum likelihood estimates of the p_k, denoted $\tilde{p}_k$, are zeroes of the polynomial $f_K(\lambda)$ of (6) with

$$a_{r,K} = (-1)^{K-r} \sum_{j=0}^{r} \tilde{P}_{j,K} \binom{K-j}{r-j} \tag{12}$$

Again using the invariance property, the maximum likelihood estimate of y_k, denoted $\tilde{y}_k$, is given from (3) by

$$\tilde{y}_k = \tau/(R_k - \tilde{p}_k) \tag{13}$$

For discussion of the properties of maximum likelihood estimates see reference 5, Ch 4. In (13) R_k is assumed to be known. This can usually be arranged if a rough idea of the PRI range to be encountered is known. For example, if $\tau<y_k$ then $R_k = 1$.

5. EMITTERS WITH VARIABLE PRI

Recognition methods for radar emitters with variable PRI in a dense environment are discussed now. In this section emitter k is assumed to have PRIs which are independent samples of a random variable taking values between α_k and β_k ($\alpha_k < \beta_k$) with probability density function $p_k(u)$. The independence assumption does not restrict the class of emitter considered as it turns out that the formulae derived with this assumption in force also apply to the dependent case. The observation procedure of section 2 is adopted and the assumptions of section 3 are invoked again. The statistical properties of the resulting observed data are analysed and their use for emitter identification investigated. It will be assumed that $\beta_k < 2\alpha_k$ and $\tau < 2\alpha_k$. These restrictions on τ, α_k and β_k are introduced to simplify the analysis, but the results presented can readily be generalized.

Let $x_1^k, x_2^k, \ldots, x_T^k$ denote the observed sequence when only the kth emitter is present. The observed sequence is modelled as a realization of the stochastic process, $\{X_t^k, t = 1,2,\ldots, T\}$ which takes the value i with probability $q_k(i)$. These probabilities can be determined using equation (3) for a single PRI value and averaging over the range of PRI α_k to β_k. To form this average the probability density function $u\, p_k(u)\, \Delta_k^{-1}$ must be used where Δ_k is the average PRI given by

$$\Delta_k = \int_{\alpha_k}^{\beta_k} p_k(u)\, u\, du. \tag{14}$$

This takes into account the fact that a shorter PRI takes up less gate intervals than a longer PRI on the average. The resulting formulae for $q_k(i)$ for various possible ranges of τ are given in Table 1.

TABLE 1

Properties of models for emitters with variable PRI

	$\tau < \alpha_k$	$\alpha_k < \tau < \beta_k$	$\tau > \beta_k$
$q_k(0)$	$(\Delta_k - \tau)/\Delta_k$	$\Delta_k^{-1} \int_{\tau}^{\beta_k} p_k(u)(u-\tau)du$	$(2\Delta_k - \tau)/\Delta_k$
$q_k(1)$	τ/Δ_k	$\Delta_k^{-1}\left\{\int_{\alpha_k}^{\tau} p_k(u)(2u-\tau)du + \int_{\tau}^{\beta_k} p_k(u)\tau\, du\right\}$	$(\tau - \Delta_k)/\Delta_k$
$q_k(2)$	0	$\Delta_k^{-1} \int_{\alpha_k}^{\tau} p_k(u)(\tau - u)du$	0

The dense environment analysis for variable PRI emitters also involves factoring polynomials. If τ is less than the minimum α_k then the analysis of section 4 is applicable with $q_k(o)$ replacing p_k and Δ_k replacing y_k. It is possible to generalize the analysis for other values of τ. For example if $\alpha_k < \tau < \beta_k$ the linear factor in the polynomial corresponding to emitter k is replaced by a quadratic factor. It is possible to set up algorithms for estimating the parameters α_k, β_k and $p_k(u)$ for each emitter present based on the observed relative frequencies of occurrence of the states for a set of values of τ.

The identification method of this paper requires a given polynomial to be factored. Normally this entails a considerable amount of computational effort. However, if changes to the environment only need be determined the problem simplifies because in splitting a polynomial into two factors the second factor is easily found once the first factor is specified. The identification method based on this concept is called adaptive identification and achieves improved accuracy compared with non adaptive methods. Experimental results obtained to date suggest that for 1 second observation intervals and for environments of up to 5 emitters PRI estimates within less than 1% of the true values can be achieved for fixed PRI emitters and reliable recognition can be achieved for variable PRI emitters.

REFERENCES

(1) Lewis, T.F.: 1975, "Distribution Sampling for Computer Simulation", Heath.

(2) Knuth, D.E.: 1969, "The Art of Computer Programming", Vol. 2, Addison Wesley.

(3) Wilks, S.S.: 1962, "Mathematical Statistics", Wiley.

(4) Mood, A.M., Graybill, F.A., and Boes, D.C.: 1974 "Introduction to the Theory of Statistics", McGraw Hill.

(5) Silvey, S.D.: 1975, "Statistical Inference", Chapman and Hall.

(6) Hawkes, R.M.: "Radar Emitter Recognition using Pulse Repetition Interval" (Technical Report in preparation).

NONSTATIONARY EEG ANALYSIS BY REALTIME REGRESSION ALGORITHMS

A.Cumani, R.Del Bello,G.U.Righini

Gruppo Sistemi e Circuiti- Istituto Elettrotecnico Nazionale "G.Ferraris" - Torino (Italy)

Abstract:The paper is concerned with real-time parameter estimation of autoregressive models used in the analysis of non-stationary electroencephalographic signals. Two algorithmically similar approaches are considered,the Kalman filter and the weighted least squares method in recursive form. The two algorithms yield pratically the same results for what concerns the tracking of spectral changes and the detection of transients, while the later is computationally more economical when implemented in factored form to improve numerical stability. Results on some test cases are presented and discussed.

1.INTRODUCTION

Autoregressive models have been found useful in electroencephalographic signal analysis both to describe the spectral properties of the stationary "background" activity and to detect short-time transients (e.g. spikes). In order to allow tracking of non-transient changes in the background activity, some researchers have proposed an extension of the autoregressive model to which Kalman filtering can be applied (1),(2),(3). In this work we propose a different approach to the tracking problem. The parameters of the autoregression are estimated by minimizing an exponentially weighted least-squares criterion. Fast recursive algorithms for solving this problem are available, so making feasible a real-time signal analysis which enables both to characterize the significant changes of background activity and to detect short-time transients by comparing the prediction error against a suitable threshold. This algorithm forms part of a system for automatic analysis of EEGs

D. G. Lainiotis and N. S. Tzannes (eds.), Applications of Information and Control Systems, 88-95.

aimed to resolve some problems in EEG clinic, such as feature extraction for diagnostic purpose, patient monitoring, drug tracking.

2.MATHEMATICAL DESCRIPTION OF THE EEG

A model which has been found useful by many authors (1),(4),(5) for describing stationary EEG segments sampled at a constant rate 1/T is the following:

$$y_n+\sum_{i=1}^{N} a_i y_{n-i} = e_n + k \qquad \text{(eq.2.1)}$$

where $y_n = y(nT)$, $E[e_n]=0$, $E[e_n e_m] = \delta_{nm} R$

and k is a term introduced to take into account the DC component of the signal. By defining

$$\theta^T = [a_1 a_2 \ldots a_N k] \qquad \text{(eq.2.2)}$$

$$H_n = [-y_{n-1} - y_{n-2} \ldots - y_{n-N} 1] \qquad \text{(eq.2.3)}$$

the signal model can be put into the form

$$y_n = H_n \theta + e_n \qquad \text{(eq.2.4)}$$

Equation (2.1) defines a stationary signal with spectrum

$$S(z) = R \,/\, A(z)\, A(1/z) \qquad \text{(eq.2.5)}$$

where $A(z) = 1+ \sum_{i=1}^{N} a_i z^{-i}$

Some authors (6),(7) have proposed to consider, instead of a pure AR model as equation (2.1), the more complete ARMA model

$$y_n + \sum_{i=1}^{N} a_i y_{n-i} = \sum_{j=0}^{M} c_j e_{n-j} + k \qquad \text{(eq.2.6)}$$

whose spectrum is

$$S(z) = R \, \frac{C(z)\, C(1/z)}{A(z)\, A(1/z)} \qquad \text{(eq.2.7)}$$

There is some inconvenient, however, in introducing MA parameters, since their estimates are not easy to compute in a recursive (on-line) way. As a matter of fact, in all cases where on-line estimation is required, the c-parameters are either not considered (pure AR model) or preassigned, so that the EEG signal can be prefiltered through $c^{-1}(z)$ and equation (2.1) applies to the prefiltered samples. For these reasons, in this work we shall consider only the pure AR model of (eq.2.1). When the stationarity assumption is dropped, the autoregression parameters must be considered as functions of the time index n, by substituting θ_n for θ in (eq.2.4). In this case, a signal spectrum in the ordinary sense cannot be defined; however, a useful characterization of the signal can still be obtained through a "time variable spectrum" defined as

$$S(z \mid n) = R \,/\, A(z|n)\, A(\tfrac{1}{z} \mid n) \qquad \text{(eq.2.8)}$$

where $A(z|n)$ is the polynomial of equation (2.5) computed with the values of the a-parameters at time nT.

3. ESTIMATION METHODS

3.1. Kalman filter . If one assumes that the time variations in the EEG characteristics may be described by a "random walk" of the autoregression coefficients, then a suitable model for the EEG signal is (3):

$$\begin{aligned} \theta_{n+1} &= \theta_n + w_n \\ y_n &= H_n\, \theta_n + e_n \end{aligned} \qquad \text{(eq.3.1.1)}$$

where $E[w_n] = 0$, $E[w_n w_m] = \delta_{nm} Q_n$.

It can be noticed that (eq.3.1.1)are the dynamic equations of a system with "state" θ_n and output y_n . This suggest the application of a standard Kalman filter for the estimation of the "state" (that is, of the autoregressive parameters). The filter equations are:

$$\begin{aligned} \hat{\theta}_{n+1/n} &= \hat{\theta}_{n/n-1} + G_n\, e_{n/n-1} \\ e_{n/n-1} &= y_n - H_n\, \hat{\theta}_{n/n-1} \\ G_n &= P_{n/n-1}\, H_n^T\, \alpha_n^{-1} \\ \alpha_n &= H_n\, P_{n/n-1}\, H_n^T + R_n \\ P_{n+1/n} &= P_{n/n-1} - G_n\, \alpha_n\, G_n^T + Q_n \end{aligned} \qquad \text{(eq.3.1.2)}$$

It must be remarked that, although the "measurement matrix" H_n depends upon the previous y values and therefore upon the system state, nevertheless it is known at time n, which is sufficient for the applicability of Kalman filtering (3). The filter equations (3.1.2) are computed recursively for each sampling instant, yielding a running estimate $\hat{\theta}_{n/n-1}$ of the autoregression parameters. Spike detection is accomplished by augmenting the filter equations (3.1.2) with the following "consistency test": at each time instant, compute the ratio of the squared innovation to its theoretical variance

$$r_n = e^2_{n/n-1} / \alpha_n \qquad \text{(eq.3.1.3)}$$

then, if r_n is less than a suitable threshold γ^2, the data point is accepted and processing continues with the next data point ;otherwise, the point is considered exceptional and a spike recognition routine is initiated. It must be remarked that test $r_n < \gamma^2$ may occasionally fail even if there are no true abnormalities in the signal; for example, letting $\gamma^2=9$ and assuming thet the model is correct, the test should fail about once every 370 data points (3σ rule).

3.2.Weighted least squares . Let us assume, for a moment, that the parameters be constant with time. Define

$$J_n = \sum_{k=N+1}^{n} \varphi^{2(n-k)} \bar{e}^2_k \qquad \text{(eq.3.2.1)}$$

where $\bar{e}_k = y_k - H_k\theta$

then, if $\varphi^2 = 1$, the parameter vector $\hat{\theta}_n$ minimizing J_n is the usual least-squares (LS) estimate of θ which can be computed, fixed n, by well-known algorithms (Levinson,etc.)(8). For $\varphi^2<1$, the fitting errors $\bar{e}_k$ are weighted by an exponential window which assign more weight to the recent signal history, so allowing the parameter estimates $\hat{\theta}_n$, when computed for each time instant n, to track the true parameters if they are slowly varying. This technique get different names ("ageweighting","discounting",etc.)(9) and has been succesfully applied in various fields; it will be referred to as "exponentially weighted least squares"(EWLS) method. The interesting fact is that the solution $\hat{\theta}_n$ of the minimization of J_n can be put in a recursive form very similar to that of the Kalman filter (10):

$$\begin{aligned}
\hat{\theta}_n &= \hat{\theta}_{n-1} + G_n \tilde{e}_n \\
\tilde{e}_n &= y_n - H_n \hat{\theta}_{n-1} \\
G_n &= P_{n-1} H^T_n \alpha^{-1}_n \\
\alpha_n &= H_n P_{n-1} H^T_n + \varphi^2 \\
P_n &= \varphi^{-2} [P_{n-1} - G_n \alpha_n G^T_n]
\end{aligned} \qquad \text{(eq.3.2.2)}$$

The quantity $\varphi^2 < 1$ is a design parameter which controls the window length and therefore the tracking speed of the filter, and is assumed constant. The consistency test for spike detection can be accomplished by comparing with a fixed threshold the quantity r'_n defined as

$$r_n' = \frac{\varphi_n^2 \vartheta_{n-1} - N}{\varphi_n^2 + H_n P_{n-1} H_n^T} \Lambda_{n-1}^{-1} \tilde{e}_n^2 \quad \text{(eq.3.2.3)}$$

where Λ_n, ϑ_n are defined recursively through:

$$\Lambda_n = \varphi_n^2 (\Lambda_{n-1} + \tilde{e}_n^2 / \alpha_n)$$
$$\vartheta_n = \varphi_n^2 \vartheta_{n-1} + 1 \quad \text{(eq.3.2.4)}$$

4. COMPUTATIONAL ASPECTS

Experience has shown that the direct application of the "covariance update" (last equation in 3.1.2 and in 3.2.2) is numerically unstable and the computed gains are often unreliable(11). This is particularly important in realt-time applications where computational accuracy must be sacrificed in favour of speed (e.g., by using 16-bit fixed point arithmetic). To circumvent these difficulties, we have implemented the EWLS algorithm by using the UDU factorization (7),(11):

$$P_n = U_n D_n U_n^T$$

where U_n is upper triangular and D_n is diagonal. The algorithms in (7) are directly applicable to the filter equations (3.2.2) with the only addition of the "time update"

$$D_n \leftarrow \varphi^{-2} D_n$$

It must be remarked that the application of the UDU^T factorization to the Kalman filter (eq.3.1.2) is less efficient than in the EWLS case, since the time update $P \leftarrow P + Q$, in factored form, requires N one-dimensional updatings of the UD factors even in the case where Q is diagonal. All the algorithms have been implemented in Fortran on a HP2100 minicomputer with floating point. Actual tests have been done off-line, after storing the EEG samples on a disk, since the actual program speed was not sufficient for real-time computations; however, real-time functioning was simulated by the program getting the samples from the disk one-at-a-time.

5. TEST RESULTS

A number of test cases was evaluated, in order to assess the performance of the proposed methods. In addition to samples of phisiological EEG signals, some digitally simulated test samples were used as well-defined references for investigating the accuracy of the tracking algorithm. Figure 1 shows one of these latter, characterized by the sudden appearance and disappearance of an α-rythm over a low-frequency background (track 1). Such a pattern could represent,e.g.,the recording from leads O1-A1 of an adult healthy subject alternately closing and opening his eyes (occipital α-rythm is blocked with eyes open). The upper part of the figure shows a logarithmic density plot of $\hat{S}(e^{j\omega T}|n)$ when evaluated through the EWLS algorithm with N=7, $\varphi^2=0.99$; tracking of spectral changes is quite evident. Track 2 of the fig.1 plots the sequence r'_n of equation(3.2.3). This plot shows the inadequacy of the prediction error alone for EEG segmentation; indeed, while the onset of the α-run could be easily detected through the appearance of large residuals, this criterion would fail to detect the end of the α-run.Figure 2 shows the same processing of fig.1 when applied to a segment of an actual EEG recording.

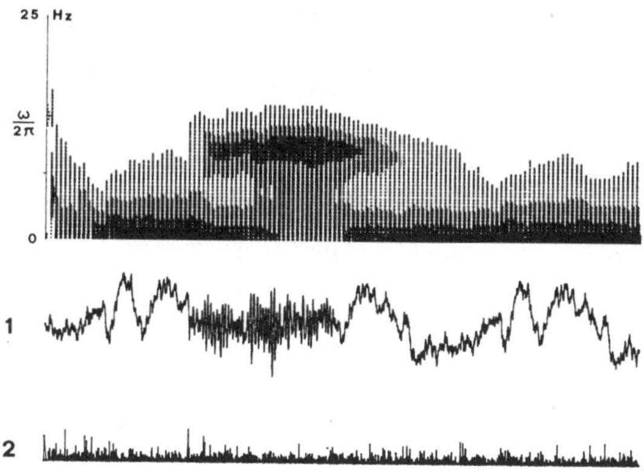

Figure 1.Example of signal processing:simulated sample

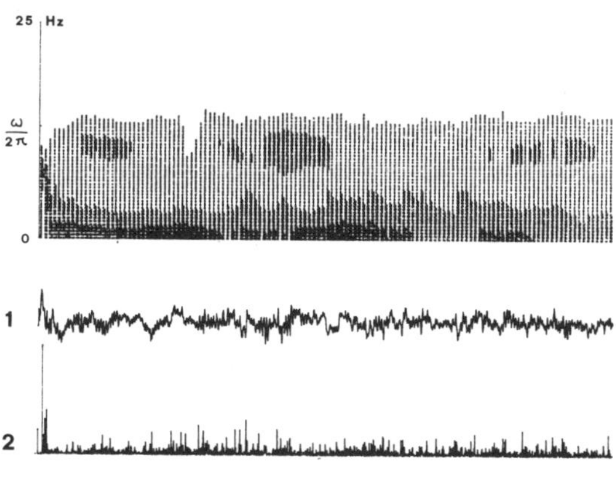

Figure 2.Example of signal processing:actual EEG record

The prediction error criterion seems nevertheless sufficient for the detection of some fast transients as ca be seen from figure 3. Here the signal (track 1) shows "runs" of fast activity (mio-eletric artefacts) over a low-frequency background. Track 2 is the sequence r'_n generated by EWLS with N=7 and φ^2=0.99; track 3 is the prediction error $\tilde{e}_n$ and the arrows mark the instants at which the algoritm detected a change in the parameters (N=7, threshold γ^2=10).

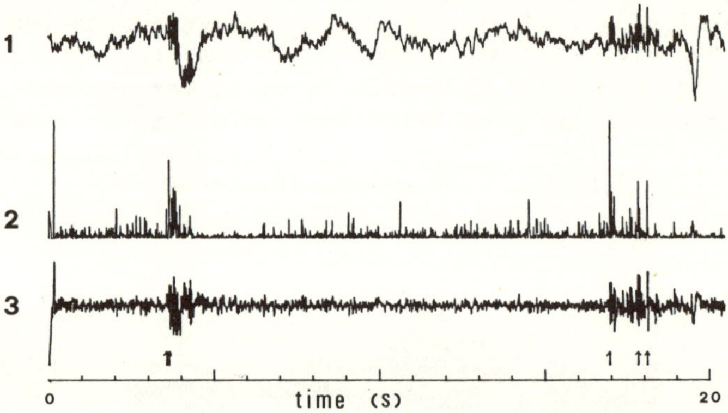

Figure 3. Fast transient detection

Spectral tracking through the KF approach has also been evaluated on these test signals, but the results are not reported here since they were very similar to those of EWLS method. This result should not be surprising, since the two algorithms are very similar except for the "covariance updating" formulae; a look at these formulae, however, reveals that both contain a mechanism (the additive factor Q in eq.3.1.2 and the multiplicative one, φ^2, in eq.3.2.2) which prevents the filter gain from converging to zero with increasing n (this would happen with Q=0 in 3.1.2 and with φ^2=1 in 3.2.2). This fact clearly involves a greater statistical variability of the estimates, but at the same time allows them to track the changes in the signal dynamics, at least if they are not too fast.

REFERENCES

(1) Gersch,W., Mathematical Biosciences,vol.7,pp.205-222,1970
(2) Lopez da Silva,F.H., Dijk,A., Smits,H., CEAN:Computerized EEG analysis,G.Dolce and H.Kunkel (ed.),Fischer,Stuttgart, 1975,pp.180-199
(3) Bohlin,T., System Identification: Advances and Case Studies, R.K.Mehra and D.G.Lainiotis (ed.),Academic Press, New York, 1976,pp.441-518

(4) Rappelsberg,P., Petsche,H., CEAN: Computerized EEG analysis, G.Dolce and H.Kunkel (ed.),Fischer,Stuttgart,1975, pp.27-40

(5) Bodenstein,G., Praetorius,H.M., Proc. IEEE, vol. 65,n. 5, May 1977,pp. 642-652

(6) Wennberg,A., Zetterberg,L.H., EEG Clin. Neurophysiol., n. 31, 1971, pp. 457-468

(7) Bohlin,T.,"Analysis of stationary EEG-signals by the maximum likelihood and generalized least-squares method", IBM SDD Nordic Laboratory, Sweden, Tech. Paper 18.200,1971

(8) Makhoul,J., Proc. IEEE, vol. 63, n. 4,Apr. 1975,pp.561-580

(9) Jazwinski, A.H.,Stochastic processes and filtering theory, Academic Press, New York, 1970

(10) Peterka,V.,"Subjective probability approach to real-time identification", Preprints IV IFAC Symp. Identification and System Parameter Estimation, Tbilisi (USSR),Sept.1976, Paper 14.1,pp.83-99

(11) Bierman,G.J., IEEE Trans. Automat. Contr.,vol. AC-23,n. 5, Oct. 1978, pp. 901-907

GIMBALESS INERTIAL SYSTEMS: ERROR ANALYSIS AND DESIGN

A.J. GRAMMATICOS - G. BITSORIS

Dynamical Systems and Measurements Laboratory
University of Patras
Patras - Greece

ABSTRACT: In a Gimbaless inertial system the translational and rotational motions are measured along vehicular coordinates. It is necessary however, to refer these motions to inertial coordinates so that the attitude and position and velocity of the vehicle are established with respect to the latter coordinates.

Mechanizations of the appropriate differential equations generate direction cosines to produce attitude information and the required transformation of coordinates, and integrate linear acceleration to produce position and velocity information. Measurement errors enter the mechanizations and then their propagation through these mechanizations is analyzed; then, appropriate design measures are proposed for improved system error performance.

1. INTRODUCTION

In a gimbaless inertial system, the sensors that measure linear acceleration and angular velocities are rigidly attached to the vehicle; hence the measured quantities are refered to vehicular coordinates. It is necessary, however, to refer these quantities to an inertial frame. The transformation of the measured quantities from vehicular coordinates to inertial coordinates requires the knowledge of the vehicle attitude which in this paper, is described by a direction-cosine matrix establishing the relative orientation of the two frames. Then, integration of appropriate defferential equations, generate the altitude, the position and the velocity of the vehicle with respect to inertial coordinates. The integration takes place in simple mechanizations of the above equations. The integration however is not free of errors, because initial condition errors and measurement errors enter the com-

D. G. Lainiotis and N. S. Tzannes (eds.), Applications of Information and Control Systems, 96-104.

puters and appear at the outputs. The error analysis shows that the output errors have oscilating and/or diverging behaviour. So, the design of appropriate systems for damping the output errors is necessary.

2. THE DYNAMIC EQUATIONS

We assume that the space vehicle in interplanetary mission is subject to gravitational forces and to the thrust of its engines.

Two reference frames are considered. The first one, called vehicular frame is rigedly attached to the vehicle with its origin at the center of mass. The second frame, called inertial frame, has its origin at the center of the sun and its x,y,z axes fixed with respect to fixed stars.

The attitude of the vehicle relative to the inertial frame is described by the direction-cosine matrix

$$C_i^v = \begin{bmatrix} c_{xx} & c_{yx} & c_{zx} \\ c_{xy} & c_{yy} & c_{zy} \\ c_{xz} & c_{yz} & c_{zz} \end{bmatrix}$$

where

$$c_{kl} = \cos(vk,il) \qquad k,l = x,y,z$$

It can be shown [1] that the direction-cosine matrix C_i^v satisfies the matrix differential equation

$$\dot{C}_i^v = -C_i^v \underset{\sim}{\Omega}_v \tag{1}$$

where

$$\underset{\sim}{\Omega}_v = \begin{bmatrix} 0 & -\omega_{vz} & \omega_{vy} \\ \omega_{vz} & 0 & -\omega_{vx} \\ -\omega_{vy} & \omega_{vx} & 0 \end{bmatrix}$$

and ω_{vx}, ω_{vy}, ω_{vz} are components of the vehicular angular velocity in vehicular coordinates which can be measured by linear accelerometers [2]. It is well known that only three of the nine equations are independent. If $C_x = \left[c_{xx}, c_{yx}, c_{zx} \right]^T$, then

$$\dot{C}_x = -\underset{\sim}{\Omega}_v C_x \tag{2}$$

is a set of independent equations.

The direction-cosine matrix is also required for the determination of the velocity and position of the vehicle with respect to the inertial frame. If $R = [x,y,z,]^T$ is the position vector of the vehicular center of mass reffered to the inertial frame, then,

$$\ddot{R} = g+a \qquad (3)$$

where

$$g = [g_x \; g_y \; g_z]^T$$

is the gravity acceleration, and

$$a = [a_x \; a_y \; a_z]^T$$

is the acceleration due to the thrust of the engines of the vehicle. The gravity acceleration g is obtained from the universal law of gravitation

$$g = -\frac{\gamma m}{R^3} R$$

which is simulated by the gravity computer; γ denotes the gravity constant and m is the attracting mass. The inertial acceleration a is obtained by the equation

$$a = C_i^v a_v \qquad (4)$$

where

$$a_v = [a_{vx} \; a_{vy} \; a_{vz}]^T$$

is the inertial acceleration of the vehicular center of mass referred to the vehicular frame.

3. IMPROVING COMPUTER PERFORMANCE

From the error analysis of direction-cosine and navigation computers, it can been seen that even with accurate measurements of the angular velocity and the linear acceleration, the output errors due to inacurate settings of initial conditions may be considerable because of their oscillatory or diverging behavior. Therefore it is necessary that these errors should be damped. The proposed method requires auxiliary information for the design of appropriate feedback compensation loops. This external information can be obtained for example, from a stellar tracking unit [3]. In the subsequent section we assume the problem of obtaining auxiliary information to be solved.

3.1 Improving Derection-Cosine Computer Performance

Let c^*_{xxs}, c^*_{yxs}, c^*_{zxs} be the direction cosines obtained from auxiliary information. We consider the following controlled system for the direction-cosine computation:

$$\dot{C}_x^* = -\underline{\Omega}_v^* C_x^* + HU \tag{5a}$$

$$U = C_{xs} - C_x^* \tag{5b}$$

where $C_x = \left[C_{xx}, C_{yx}, C_{zx}\right]^T$: $C_{xs} = \left[C_{xxs}, C_{yxs}, C_{zxs}\right]^T$ and H is a 3x3 real matrix.

A mechanization of equations (5) is shown in the following figure:

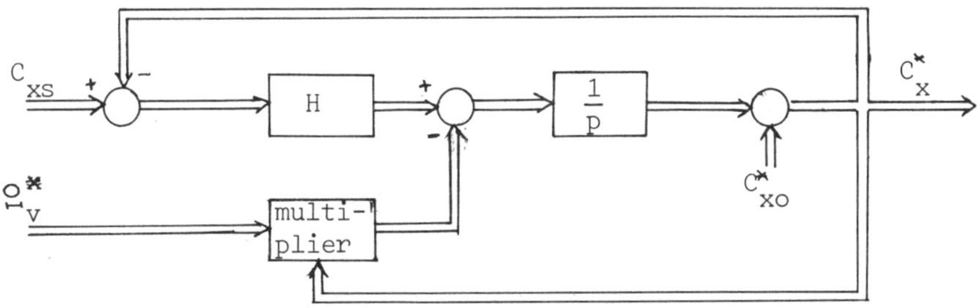

Figure 1. A mechanization of the controlled direction-cosine equations.

Let C_x be the true value of the direction-cosine vector and

$$C_x^* = C_x + \delta C_x^*$$

$$C_{xs} = C_x + \delta C_{xs}$$

Then, by substituting in (5), we obtain

$$\delta \dot{C}_x^* = -(\underline{\Omega}_v^* + H)\delta C_x^* + H\delta C_{xs}^* - \delta \underline{\Omega}_v^* C_x^* \tag{6}$$

The problem of damping the oscillations of the errors δC_x^*, due to the initial conditions errors δC_{xo}^*, consists in determining the element of H so that the real parts of the eigenvalues of $-\underline{\Omega}_v^* - H$ to be upper bounded by negative real numbers. It is not difficult to see that such a matrix is given by

$$H = \begin{bmatrix} h & 0 & 0 \\ 0 & h & 0 \\ 0 & 0 & h \end{bmatrix}$$

where h is a positive real number, because the eigenvalues of $-\underline{\Omega}_v - H$ are

$$\lambda_1 = -h$$

$$\lambda_2 = -h + j\omega_o$$

$$\lambda_3 = -h - j\omega_o$$

It must be noted that the gains h of the feedback loop are time invariant. We can now establish an estimation of the improved direction-cosine computer. From (6) we get [4]:

$$\|\delta C_x^*(t)\| \leq \exp\left[\int_{to}^{t}\mu(-\Omega_v^*-H)d\tau\right]\|\delta C_{xo}^*\| +$$
$$+\int_{to}^{t}\exp\left[\int_\tau^t\mu(-\Omega^*-H)ds\right]\|H\delta C_{xs}-\Omega_v^* C_x^*\|d\tau$$

where $\|\cdot\|$ denotes the Euclidean norm and $\mu(\cdot)$ the induced matrix measure. For the case considered, we have $\mu(-\Omega_v^*-H)=-h\,.S_o$, by a simple computation we get

$$\delta C_x^*(t) \leq \exp\left[-h(t-t_o)\right]\|\delta C_{xo}^*\| +$$
$$+(1-\exp\left[-h(t-t_o)\right]).\sup_{\tau\in[t_o,t]}(\|\delta C_{xs}-\frac{1}{h}\delta\Omega_v^* C_x^*\|), \qquad (7)$$

By inspection of the above inequality, it is seen that errors due to initial conditions errors δC_{xo}^* are damped, but it remains an error due to the errors δC_{xs} of the auxiliary information and $\delta\Omega_v^*$ of the angular velocity measurements.

3.2 Improving Navigation Computer Performance

Let R_t be the position vector of the vehicular center of mass, obtained from auxiliary information. For improving the navigation computer performance we consider the following augmented navigation system:

$$\dot{R}^* = \dot{R}^* + H_1 U_1 \qquad (8a)$$

$$\ddot{R}^* = a+g+H_2U_2 \qquad (8b)$$

$$U_1 = R_t - R^* \qquad (9a)$$

$$U_2 = \dot{R}_t - \dot{R}^* \qquad (9b)$$

where $\dot{R}_t$ is obtained by integration of R_t, and H_1, H_2 are undermined real matrices.

Let R and $\dot{R}$ be the true values of the position and velocity vectors of the vehicular center of mass and

$$R^* = R+\delta R^*$$
$$\dot{R}^* = \dot{R}+\delta\dot{R}^*$$
$$R_t = R+\delta R_t$$
$$\dot{R}_t = \dot{R}+\delta\dot{R}_t$$

By combining the above relations with eq. (8) - (9), we get

$$\begin{bmatrix} \delta\dot{R}^* \\ \delta\ddot{R}^* \end{bmatrix} = \begin{bmatrix} \delta\dot{R}^* \\ \delta g+\delta a \end{bmatrix} + \begin{bmatrix} H_1 & \mathbb{O} \\ \mathbb{O} & H_2 \end{bmatrix} \begin{bmatrix} \delta R_t-\delta R^* \\ \delta\dot{R}_t-\delta\dot{R}^* \end{bmatrix}$$

Now, taking into account that $\delta g \approx G\delta R$ and rearranging the terms of the right hand member of this equation we obtain

$$\begin{bmatrix} \delta\dot{R}^* \\ \delta\ddot{R}^* \end{bmatrix} = \begin{bmatrix} -H_1 & \mathbb{O} \\ G & -H_2 \end{bmatrix} \begin{bmatrix} \delta R^* \\ \delta\dot{R}^* \end{bmatrix} + \begin{bmatrix} H_1\delta R_t \\ H_2\delta\dot{R}_t+\delta a \end{bmatrix} \tag{10}$$

It is clear, from eq. 10, that for damping the diverging output errors in the navigation computer due to initial condition errors δR^*_o and $\delta\dot{R}^*_o$, it is necessary to determine the elements of matrices H_1, H_2, so that system (10) to be bounded input - bounded output (B.I.B.O.) stable. We shall stud y this problem using the comparison principle of Matrozov [5] and some new results on the stability analysis of non-linear comparison systems, [6-9]. This will enable us to reduce the study of the six order system (10), to a two order comparison system.

Let us consider the vector positive definite function [6],[7]

$$V(\delta R,\delta\dot{R}) = \begin{bmatrix} v_1(\ \delta R\ ) \\ v_2(\ \delta\dot{R}\ ) \end{bmatrix} = \begin{bmatrix} \|\delta R\| \\ \|\delta\dot{R}\| \end{bmatrix} \tag{11}$$

where $\|\cdot\|$ is the Euclidean norm. The total time derivative $V(t)_{(10)}$ of $V(\delta R,\delta\dot{R})$ with respect to the system (10) is computed as follows:

$$\begin{bmatrix} \dot{v}_{1(21)}(t) \\ \dot{v}_{2(21)}(t) \end{bmatrix} = \begin{bmatrix} \frac{1}{v_1}\delta R^T H_1\delta R+\frac{1}{v_1}\delta R^T\delta\dot{R}+\frac{1}{v_1}\delta R^T H_1\delta R_t \\ \frac{1}{v_2}\delta\dot{R}^T G\delta R-\frac{1}{v_2}\delta\dot{R}^T H_2\delta\dot{R}^T+\frac{1}{v_2}\delta\dot{R}^T(H_2\delta\dot{R}_t+\delta a) \end{bmatrix} \tag{12}$$

Let

$$H_1 = h_1\mathbb{1}$$

$$H_2 = h_2\mathbb{1}$$

where h_1, h_2 are undermined constants. Then from (12) follows

$$\begin{bmatrix} \dot{v}_1(t) \\ \dot{v}_2(t) \end{bmatrix} \leq \begin{bmatrix} -h_1v_1(t)+v_2(t)+h_1\|\delta R_t\| \\ 2\omega_o^2v_1(t)-h_2v_2(t)+\|h_2\delta\dot{R}_t+\delta a\| \end{bmatrix} \tag{13}$$

because $G = 2\omega^2$.

We can now establish the comparison system as follows:

$$\begin{bmatrix} \dot{w}_1 \\ \dot{w}_2 \end{bmatrix} = \begin{bmatrix} -h_1 & 1 \\ 2\omega_o^2 & -h_2 \end{bmatrix} \begin{bmatrix} w_1 \\ w_2 \end{bmatrix} + \begin{bmatrix} \| h_1 \delta R_t \| \\ \| h_2 \delta R_t + \delta a \| \end{bmatrix} \qquad (14)$$

The relation of the solution of eq. (10) with those of the associate comparison system (14) is established by the following result ([6], [7]).
If

$$\begin{bmatrix} \| \delta R_o \| \\ \| \delta \dot{R}_o \| \end{bmatrix} \leq \begin{bmatrix} w_{1o} \\ w_{2o} \end{bmatrix}$$

then

$$\begin{bmatrix} \| \delta R(t, t_o, \delta R_o, \delta \dot{R}_o) \| \\ \| \delta \dot{R}(t, t_o, \delta R_o, \delta \dot{R}_o) \| \end{bmatrix} \leq \begin{bmatrix} w_1(t, t_o, w_{1o}, w_{2o}) \\ w_2(t, t_o, w_{1o}, w_{2o}) \end{bmatrix} \qquad \begin{matrix} (15) \\ (16) \end{matrix}$$

for all $t \leq t_o$

Therefore the problem of determination of h_1, h_2 for damping the errors of the navigation computer, is reduced to the determination of h_1, h_2 in order the two dimensional comparison system (14), to be B.I.B.O. stable.

The b.i.b.o. stability of (14) is guaranteed if the real parts of the eigenvalues of the matrix

$$A = \begin{bmatrix} -h_1 & 1 \\ 2\omega_o^2 & -h_2 \end{bmatrix}$$

are upper bounded by negative real nembers. This condition is satisfied if

$$h_1 > \varepsilon_1$$

$$h_2 > \varepsilon_2$$

$$h_1 h_2 - 2\omega_o^2 > \varepsilon_3$$

where ε_1, ε_2, ε_2 are positive real numbers.

As a result of the above analysis we can design the mechanization of the controlled navigation equations, represented in figure 2.

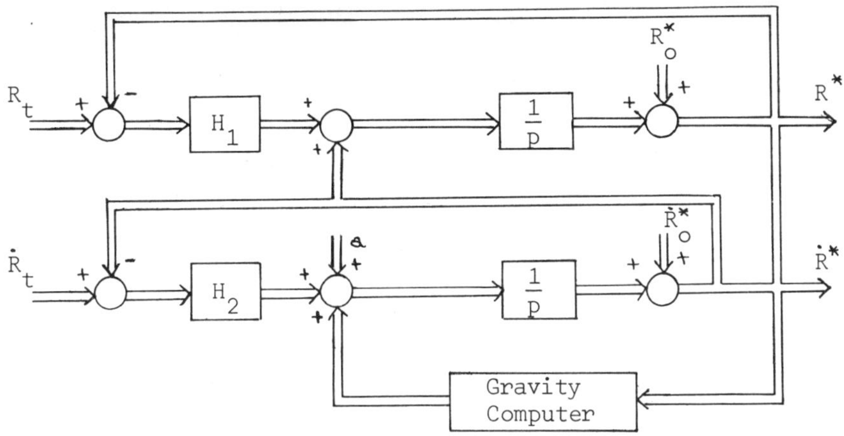

Figure 2. A mechanization of the controlled navigation equations

4. CONCLUSIONS

The problem of designing a computational system for establishing the attitude, the position and the velocity of a space vehicle, has been considered. This problem had been studied in [3] by the utilisation of classical mathematical tools of analysis.

In this paper the same problem has been studied by using state space analysis, vector positive definite functions and the comparison principle. This approach enabled us to overcome some simplifications as the hypothesis of slow varying angular velocities, made in previous papers.

REFERENCES

1. C. Broxmeyer, "Inertial Navigation Systems" Mc Graw-Hill Co. New York, 1964
2. Schuler, A.R., Grammaticos, A. and Fegley, K.A., "Measuring Rotational Motion with Linear Accelerometers". IEEE Trans. Aerosp. Electron. Syst., vol. AES-3, No 3, p.p. 465-472, 1967.
3. Grammaticos A., Schuler, A.R. and Fegley, K.A., "Damping Gimbaless Inertial Navigation Systems" IEEE Trans. Aerosp. Electron. Syst., vol. AES-3, No 3, p.p. 481-494, 1967.
4. Vidyasagar M., "Nonlinear Systems Analysis", Prentice-Hall, New Jersey, 1978.
5. Matrozov, V.M., "On the theory of Stability of motion" Prinkl. Mat. Mekhan., 26,p.p. 992-1002, 1962.
6. Bitsoris, G., "Principe de comparaison et stabilité des systè-

mes complexes" Thèse d Etat, Univ. Paul Sabatier de Toulouse 1978.

7. Bitsoris, G., "Vector positive definite functions and stability analysis of complex dynamicalsystems" Application and Research in Information Systems and Applications: Proc. 1st Int. Conf. on Inf. Sc. Syst. Patras, Greece, vol. 3, p.p. 802-804, 1976.
8. Bitsoris, G., "Utilisation des fonctions vectorielles definies positives dans l' étude de la stabilité des systémes dynamiques complexes" Technical Report L.A.A.S. 76 I 29, p. 1-96 Mai 1976
9. Bitsoris, G., "A new comparison principle for the stability analysis of complex systems". Proc. of IFAC 2nd Symp. on Large Scale Systems and Applications, Toulouse - France, 1979.

JTIDS RELATIVE NAVIGATION OVERVIEW

Leonard Chin

U.S. Naval Air Development Center
Warminster, Pennsylvania 18974

ABSTRACT

The goal and principle of relative navigation as well as new concepts which are not found in "absolute" navigation are discussed. The current Joint Tactical Information Distribution System (JTIDS) relative navigation development status is reported, and several technical areas needed near future research and development efforts are outlined.

I. INTRODUCTION

Position location and navigation within a relative frame of reference is a natural phenomenon in certain categories of scientific, commercial and military operations. For example, space ship rendezvous, aircraft collision avoidance and military grid-lock formation are considered to be relative navigation problems. The need for solutions to these problems is increasing as the space exploration program advances, commercial air traffic continues to congest and tactical warfare becomes more and more sophisticated.

The goal of relative navigation is to determine one vehicle's position and velocity with respect to a non-fixed frame of reference, which is a coordinate system established by another non-stationary vehicle. Therefore the basic requirements for the construction of a relative navigation system is that it must be capable to determine, maintain and transmit accurate position, velocity and azimuth relative to the grid coordinates. These requirements imply that a structured community must be formed such that communications among users are effective, and a navigation net architecture must be created capable of relaying messages to all community members (users) including those who are out of line-of-sight (LOS). Also, a mechanism must be devised to maintain the accuracy and stability of the grid coordinate orientation.

D. G. Lainiotis and N. S. Tzannes (eds.), Applications of Information and Control Systems, 105-114.

In principle, each user is required to access a communication link that provides navigation data (P-message) and time-of-arrival (TOA) range measurements between any two users, a dead reckoner that is capable of extrapolating navigation information in between intervals when TOA data is not available and a digital computer that is capable of providing solutions to the hybrid navigation equations. If it is desirable that the system be capable of processing absolute navigation information computed in geographic coordinates, a dual-grid (relative and absolute) hybrid navigation system must be designed. It should be pointed out that for pure relative navigation operation, stationary (geographic) users and geodetic position references are not required. In practice however, either or both of these data are sometimes available. Therefore the system design always take advantage of the situation to improve the relative navigation performance (by using geodetic data to bound the relative grid error). For this reason geographic navigation information is always processed along with and integrated into the relative navigation performance, and each user determines its navigation data in both relative and geographic grids. Therefore, it is apparent that each user is required to perform a coordinate transformation at specified intervals to convert the dead reckoner triad into the geographic and relative grid triads. In relative grid operation, it is required to estimate, in addition to position and velocity, the azimuth angle of the user's platform with respect to the established north direction. The design goal is to enable all members of the community to navigate in such a manner that they all have the same error. This goal can be achieved by a technique called "grid maintenance," which consists of three steps of operation: grid synchronization, grid acquisition and grid update. These tasks can be accomplished without any earth-fixed member participating in the exercise. Under this condition, all members of the community accumulate an identical (ideally) time increasing error. However, if one of the members has a geographic reference update capability, the increasing error can be bounded.

In the relative navigation performance, effectiveness is measured by the ability of the system to determine the range vector between any two users and the bearing angle of each platform with respect to the relative grid. Errors associated with the range vector are called "translational" and error associated with the bearing angle is referred to as "rotational." Hence, minimization of the translational and rotational errors becomes the fundamental design of the relative navigation algorithm. Usually the Extended Kalman filter theory or a variation of it is applied.

This paper is intended to discuss the current status of and future research and development (R&D) areas in relative navigation. Because of its short history (since 1965) and its limited variation of system configuration, the discussion will be focused on one mature system: the JTIDS, which is a joint project participated by all U.S. military services. JTIDS is a merger of technologies which were under development by separate services. The Air Force project, PLRACTA[1], which emphasized a secure digital information distribution system, and the Navy project, ITNS[2], which emphasized relative navigation functions. Since these projects were driving toward the same goal of fulfilling the command and control requirements, the Department of Defense directed the merger in 1975.

JTIDS is an integrated communication-navigation-identification system, which provides, among other information, precise distance measurements between all JTIDS equipped members of an operational community. A computer on the user member processes the distance information along with the dead reckoner data to form a hybrid navigation system, capable of determining both relative and geodetic position and velocity of the user member.

II. CURRENT STATUS

The current phase of JTIDS relative navigation is being developed to demonstrate a set of fundamental concepts through the use of a JTIDS transceiver (advanced development model). As such, a simplified algorithm has been implemented under the assumption that each user is equipped with a generic inertial navigation system (INS) and a dedicated computer, as well as a general computer that performs relative navigation filtering and communication functions. A JTIDS transceiver for each user is available to provide TOA range measurements and P-messages containing navigational data and associated qualities. In addition, TACAN and other position references, as well as stationary users are assumed to be available to participate in the demonstration. All transceivers can be accurately synchronized so that precise TOA range measurements are possible. For example, each user emits a pulse train at a precise time within its own pre-assigned time slot. With all clocks synchronized, one member (user) can obtain range vectors relative to other members in a proper time-order. The time-slot method of obtaining range measurement as well as achieving communication is called Time Division Multiple Access (TDMA). A different architecture using Distributed Time Division Multiple Access (DTDMA) is also under development. The TDMA architecture uses code orthogonality and frequency orthogonality to achieve separation between communication nets; the DTDMA architecture adds statistical orthogonality in time to provide additional separation of communication nets. In other words, DTDMA uses time channel for communication and TDMA uses time slots. It can be viewed that the time-structured TDMA as a special case of the time-distributed DTDMA process. Despite the fact that two different types of JTIDS terminal are being developed concurrently, the implementation of relative navigation functions remains identical for the Navy procurement. However, a certain class of the Air Force JTIDS terminals implements a slightly different relative navigation algorithm. The community structure, which is common to all terminals, is based on a hierarchy of roles assigned to each user, such that the accuracy and stability of the grid orientation can be maintained. Knowledge of the grid accuracy and the ability to stabilize it are critical issues of the current relative navigation design since the relative grid serves as a common basis for synchronization, and it is the only reference frame by which users can process TOA data and inter-lock their relative positions.

On top of the grid hierarchy is a member who has the highest relative position quality. This member is designated as the Navigation Controller who establishes the location and orientation of the grid triad. Also on top of the grid hierarchy is a member who has the highest quality in clock accuracy. This member is designated as the Time-Base Controller, who is referenced for synchronization by all members. Below the Navigation Controller and Time-Base Controller are primary and secondary users, whose roles are determined according to their location, accuracy and connectivity of the community. Primary users are members who are permitted to utilize active clock synchronization (that is, Round-Trip-Time (RTT) technique). Secondary users are permitted to utilize only passive synchronization (that is, one way receiving without the benefit of RTT). Hierarchy rules in the geographic grid are such that a Position Reference is assigned to the member who has the highest quality in geographic position accuracy. Each member, regardless of its role in the community, computes its position and velocity accuracies with respect to the geographic and relative grids, as well as the accuracy of the relative azimuth angle. The quality levels (accuracy) of the information are transmitted to the entire community and they are used dynamically as the basis for re-assigning roles of each member. These "quality levels" are also used as a basis of establishing a source (measurement) selection criterion for the relative navigation filter input. The function of the filter is to process dead reckoning and TOA range measurement data such that the "best" estimates of relative and geographic navigational information can be obtained. However, decisions have to be made whether geographic or relative coordinates are to be used for a given situation. For example, if the origin of the relative grid is not accurately known with respect to the earth (due to the lack of geographic reference data), then the use of relative grid to locate a point on earth in the geographic coordinate system is not recommended. To locate a point given by a geographic coordinate system, the geographic grid estimates should be used. Conversely, if a position is identified by a member using its on-board sensors, then navigation data would be best estimated using the relative grid. The dual-grid operating procedures together with the hybrid navigation data processing concept (to be discussed next) form the fundamental structure of the current JTIDS relative navigation design.

The relative navigation data processor consists of an Extended Kalman filter, which is the key component. Other segments include a synchronization filter, a source (measurement) selection routine, interfaces and input/output subroutines. The state vector of the Extended Kalman filter consists of elements including position, velocity, inertial platform tilts and misalignments. The filter was designed with a uniform update interval, but the INS will not reset at each update interval because it is desirable to have the INS operating in the pure inertial mode so that nonessential flow of data will be minimized. Between update intervals, the state vector is extrapolated by digital integration using INS inputs (acceleration and torquing rate data).

Mathematical descriptions of JTIDS relative navigation filter are well documented by Danik [1] and Fried [2]. In general, the filter is constructed based on the first order linear approximation of the nonlinear relative navigation system and observation, which are commonly represented by

$$\dot{x} = f(x) + g(u) \tag{1}$$

$$y_n = h(x_n) + v_n \tag{2}$$

where u and v_n are white noise process and white noise sequence respectively.

The transition matrix is computed from $\dot{\phi} = F(t)\,\phi$ using a Taylor series expansion, from which F(t) is the Jacobian of f(x). The Extended Kalman filter approach yields a "best" state vector estimate in the form of

$$\hat{x}_n = \bar{x}_n + K_n\,(y_n - h\,(\bar{x}_n))$$

where

$$\bar{x}_n = \hat{x}_{n-1} + \int_{t_{n-1}}^{t_n} f\left(\hat{x}_{n-1}, \Delta V_x, \Delta V_y, \Delta V_z, t\right) dt$$

ΔV_x, ΔV_y and ΔV_z are changes in velocity supplied by the INS. Subscripts x, y and z represent inertial platform accelerometer input axes, which are related to the geographic coordinates by

$$\begin{pmatrix} V_x \\ V_y \\ V_z \end{pmatrix} = \begin{pmatrix} \cos\alpha & -\sin\alpha & 0 \\ \sin\alpha & \cos\alpha & 0 \\ 0 & 0 & 1 \end{pmatrix} \begin{pmatrix} V_N \\ V_E \\ V_Z \end{pmatrix} \tag{3}$$

where subscripts N, E and Z represent geographic north, east and up respectively; α is the azimuth axis platform misalignment.

K_n is a time-varying gain, which is a function of the geometrical configuration of the community, trajectories of community members, sampling data rate and of course, the noise covariances of u and v_n.

The velocity vector in the relative grid coordinates (u, v and w) is related to that in the geographic coordinates through the following transformation:

$$\begin{pmatrix} \dot{P}_u \\ \dot{P}_v \\ \dot{P}_w \end{pmatrix} = [T]\,[T_o]'\,[T_R] \begin{pmatrix} V_N \\ V_E \\ V_Z \end{pmatrix} \tag{4}$$

where P_u, P_v and P_w are relative positions in the u, v and w axes; ($\dot{P}_u$, $\dot{P}_v$, $\dot{P}_w$) and (V_N, V_E, V_Z) are velocity components; [T], [T_o] and [T_R] are transformation matrices:

[T] is the transformation from grid origin local vertical geographic to the relative grid coordinates

[T_o] is the transformation from local vertical geographic at the grid origin to the earth centered rotating coordinates

[T_R] is the transformation from local vertical geographic at the vehicle to the earth centered rotating coordinates

$$[T] = \begin{bmatrix} \cos\beta & -\sin\beta & 0 \\ \sin\beta & \cos\beta & 0 \\ 0 & 0 & 1 \end{bmatrix} \tag{5}$$

where β is the u-axis misalignment with respect to the direction of geographic north at the grid origin.

$$[T_o] = \begin{bmatrix} -\sin L_o \cos \lambda_o & -\sin \lambda_o & \cos L_o \cos \lambda_o \\ -\sin L_o \sin \lambda_o & \cos \lambda_o & \cos L_o \sin \lambda_o \\ \cos L_o & 0 & \sin L_o \end{bmatrix} \tag{6}$$

where L_o and λ_o are latitude and longitude of the grid origin. Let L_R and λ_R represent a user's position in the relative grid coordinates, then $[T_R]$ takes the same form as $[T_o]$, and

$$[T_o]'\,[T_R] = \begin{bmatrix} 1 & \sin(\lambda_R-\lambda_o)\sin L_o & \sin(L_R-L_o) \\ -\sin(\lambda_R-\lambda_o)\sin L_R & 1 & \sin(\lambda_R-\lambda_o)\cos L_R \\ -\sin(L_R-L_o) & \sin(\lambda_R-\lambda_o)\cos L_o & 1 \end{bmatrix} \tag{7}$$

For small values of $(\lambda_R - \lambda_o)$ and $(L_R - L_o)$, $[T_o]'[T_R] = [I]$, and equation (4) becomes

$$\begin{pmatrix} \dot{P}_u \\ \dot{P}_v \\ \dot{P}_w \end{pmatrix} \doteq \begin{bmatrix} \cos\beta & -\sin\beta & 0 \\ \sin\beta & \cos\beta & 0 \\ 0 & 0 & 1 \end{bmatrix} \begin{pmatrix} V_N \\ V_E \\ V_Z \end{pmatrix} \tag{8}$$

In other words, for short relative ranges, the major attitude rotational difference between vehicles geographic local vertical coordinates and the relative grid coordinates is an azimuth angle β about the Z-axis. When the above assumption holds, relationship among x-y-z, N-E-Z and u-v-w coordinates can be drawn. However, for visual clarity, only a two-dimensional figure is given;

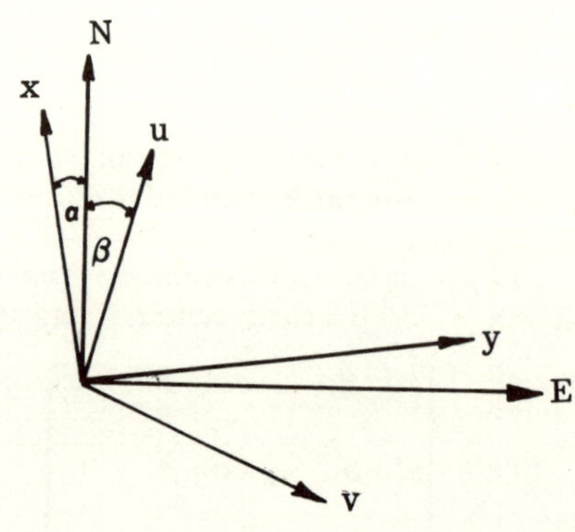

In relative navigation literature, β is referred to as the grid rotational error of the navigation controller. In general, due to errors in its dead reckoning equipment, the navigation controller also suffers from a grid translational error as illustrated below:

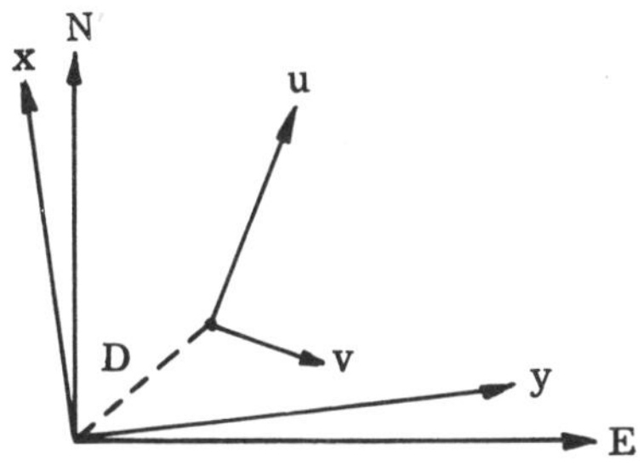

in which D is the distance that the origin of the relative grid has been drifted away. In order to eliminate β and D, ground (stationary) members must participate in the exercise, since by accurately observing geographic positions, the navigation controller can align its relative grid with respect to the N-E-Z coordinates, hence "stabilizing" the grid. Normally one ground station helps to eliminate the translational error; two or more ground stations are needed to eliminate the rotational error. However, when only one is available, rotational error can still be eliminated if the operation follows some specific rules. For example, consider a community consisting of a mobile navigation controller, an aircraft and a ground station. The relative grid, established by the navigation controller, will experience a rotation at a rate consistent with INS errors of the aircraft. If there is sufficient relative motion between the aircraft and the ground station (e.g., the aircraft circles around the station, then the maneuver would produce TOA data which appear to the relative navigation filter as multiple ground stations that are participating in the exercise, hence the rotational error would be eliminated. When operating in the open ocean where physical ground stations are not available, grid errors can also be eliminated if the navigation controller is able to obtain accurate geographic position reference data from outside of the community, since these data are equivalent to inputs (to the relative navigation filter) provided by ground stations. These operational procedures designed to eliminate grid translational and rotational errors are incorporated into the current JTIDS relative navigation algorithm, and various capabilities have been tested and verified to be satisfactory.

III. FUTURE R&D AREAS

Future research and development (R&D) areas in relative navigation are very broad since the development of this new technology is still in its infancy and the formulation of the theory is far from complete. It is seen that extensive analyses and simulations are needed to achieve refinements in the filtering algorithm, for example, the non-linearity effect; a study of effects due to different types of dead reckoner used within the community is required to obtain a better structure of the "mixed" community, especially when it is composed of a large number of users; an analysis of role assignment procedures based on accuracy quality levels is also needed to accomplish improvements in relative navigation operations. To this end, the following R&D topics, not all inclusive, are recommended for investigations.

a. Nonlinearity Effect

The current method of treating nonlinear effects in JTIDS relative navigation equations is essentially a first order linear approximation using the Taylor series expansion. It is found that under certain conditions this approach is unsatisfactory. For example, consider a special case of equation (2) in which the TOA range measurement is shown at the top of the next page, where R_A is the actual range, R_c is the measured range, which contains errors e_u, e_v and e_w, hence

$$R_A = \left[(R_c + e_u)^2 + e_v^2 + e_w^2\right]^{\frac{1}{2}} \tag{9}$$

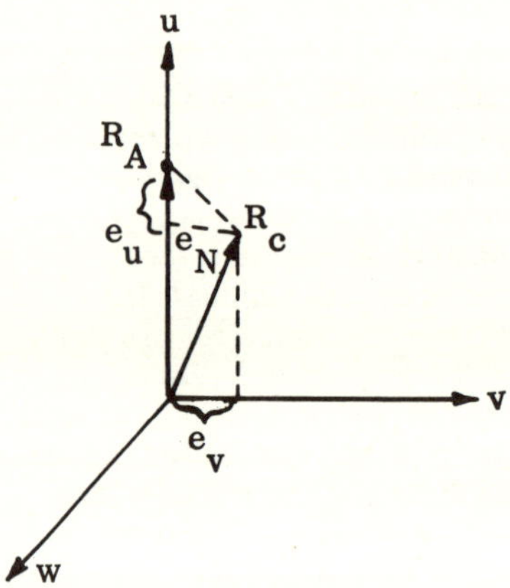

Expansion of equation (9) and retaining only the linear and quadratic terms:

$$R_A = R_c + e_u + \frac{1}{2R_c}\left(e_v^2 + e_w^2\right) \tag{10}$$

Let η be the additive noise of the TOA range measurement error, then in accordance with equation (2):

$$R_A - R_c = e_u + \frac{1}{2R_c}\left(e_v^2 + e_w^2\right) + \eta \tag{11}$$

For purposes of illustration, let $e_v = e_w = 2$ km and $R_c = 200$ km, then the quadratic term equals 20 meters. Therefore, when this term is dropped (linearization), the 20-meter measurement difference ($R_A - R_c$) is considered to be evidence of a 20-meter error in $|e_u|$(η is usually small compared to 20 meters). Such a gross approximation will produce large errors in the Extended Kalman filter estimates, and it should be anticipated that filter divergence will occur if frequent measurements are made within a range of 50 km. For this reason, it is recommended that detailed analysis and computer simulations be performed to refine the current relative navigation filter algorithm. For example, the Gaussian Second-order filter described by Widnall [3] and the Partitioned Estimation Algorithm given by Lainiotis [4] should be considered.

b. Mixed-Community Effects

The current filter design for JTIDS relative navigation assumes that each platform (user) is equipped with an INS. Furthermore, it is assumed that each INS is identically mechanized to the local vertical configuration, hence all dead reckoner error signatures are the same throughout the community. Under these assumptions, the relative navigation problem was solved by a decentralized estimator which consists of multiple Extended Kalman filters estimating subsets of the total system states. The filter is partitioned in such a way that each platform estimates its own errors and correlations among estimation errors are largely suppressed. As such the filter is subject to instabilities if the exchange of data is totally unrestricted. In practical operations, users will have different dead reckoners such as inertial and doppler systems whose error signatures are different from one another. Under this condition, the ability of a platform to accurately lock to the relative grid is questionable due to the fact that relative position errors are not properly accounted for in the filter. Research of this and other issues (for example, development of a generalized coordinate transformation algorithm independent of dead reckoner identities) have been initiated, and tentative results have been obtained. While verification of the study results are in progress, efforts to solve the mixed-community problem employing different methods are encouraged. For example, the theory of decentralized control discussed in reference [5] should be consulted.

c. Protocol Problems

The current JTIDS relative navigation community organization designates a navigation controller, which is responsible to establish a common grid for all users. Navigation data and associated accuracy, expressed in terms of "quality levels," are being exchanged through the JTIDS communication link. The navigation controller, by definition, has the highest position quality level relative to other members. Suppose due to system malfunction, a member transmits a message erroneously indicating its position quality level higher than that of the controller. When this event occurs, the current relative navigation algorithm assigns the malfunctioning member to take over the role of the legitimate controller. In order to reduce the effects of invalid quality level reports, it is recommended that either some kind of reasonableness test should be performed or a history of quality levels rather than individual data point quality level should be utilized in determining usefulness of data. For example, regression techniques discussed by Tukey [6] may be applicable.

Another interesting parameter, which may play an important role in determining the relative navigation hierarchy, is the pointing accuracy since the knowledge of the pointing error is most critical to weapon delivery. Therefore, consideration of reporting the pointing quality level within the community is recommended for future developments. Currently, a detail analysis of the relative navigation pointing error is needed. Also needed are improvements of the current source selection criteria. For example, usage of measurement data could be more dependent on geometry and be proportional to reported quality levels, instead of selecting the best observations over a short period of time (current implementation).

IV. SUMMARY

The definition of relative navigation and hardware/software elements required to implement its principles were introduced. Also the concept of relative grid and its errors affecting the system performance were given. This general introduction is followed by a discussion of the current JTIDS relative navigation functions, which encompass the dual-grid hybrid data processing method, community organization and hierarchy structures, the relative navigation filter, relationship between geographic grid and relative grid coordinates, and grid stabilization techniques. Study of the current implementation of these functions revealed that a number of improvements in the design of the relative navigation algorithm are needed. Specifically, it is indicated that issues involving nonlinearity, mixed dead reckoner community and protocol problems should be investigated in the near future.

REFERENCES

[1] Danik, B. "Hybrid-Inertial Navigation With Range Updates in a Relative Grid," AIAA Guidance and Control Confidential August 1973.

[2] Fried, W. R., "Principles and Simulation of JTIDS Relative Navigation," IEEE Trans. Vol. AES-14, No. 1, Jan. 1978.

[3] Widnall, W. S., "Enlarging the Region of Convergence of Kalman Filters Employing Range Measurements," AIAA Journal, Vol. 11, No. 3, March 1973, pp 283-287.

[4] Lainiotis, D. G.. "Partitioned Estimation Algorithms, I: Nonlinear Estimation" J. Inform. Sciences, Vol. 7, No. 3, pp 203-235,, 1974.

[5] IEEE Transaction on Automatic Control, Vol. AC-23, No. 2, April 1978.

[6] Tukey, "Exploratory Data Analysis," Addison-Wesley, 1977.

[1] Position Location Reporting and Control Tactical Air
[2] Integrated Tactical Navigation System

PARAMETER SENSITIVITY IN NONLINEAR DYNAMIC SYSTEMS

Bernt Jarmark Hans-Ove Jonsson*

SAAB-SCANIA AB Aerospace Div. S-58188 Linkoping Sweden
*also Linkoping Univ/Mathematics S-58183 Linkoping Sweden

ABSTRACT

When projecting a system used in an optimal or non-optimal way it is desirable to know the influence of the involved design parameters on the system's performance. The parameter sensitivity is studied for nonlinear dynamic systems with known parameters. This is accomplished by extending a Differential Dynamic Programming Algorithm introducing the parameters as additional states. In choosing the best parameter setting, the optimization of the parameters is achieved in parallel with the optimization of the trajectory. This gives a system which will have the best performance.

The discussed examples are within aircraft application, i.e., how an optimal trajectory depends on the individual parameters. It is shown in designing an aircraft how to choose a best parameter setting to meet a static constraint in the parameters and fulfil a given mission in an optimal way.

1. INTRODUCTION

In the design of a non-linear dynamic system it is often beneficial to know the parameters' influence on the system's performance. The mutual influence among the different parameters is usually not obvious. Consequently, it is interesting to know how the given parameters affect the dynamic response. With this knowledge, it will be possible to design an optimal system which is used in an optimal way.

D. G. Lainiotis and N. S. Tzannes (eds.), Applications of Information and Control Systems, 115-126.

The optimal control is determined by means of a first order Differential Dynamic Programming (DDP) algorithm [Refs. 1 and 2]. The extension of the DDP-method for parameters is wellknown from e.g. Refs. 3 and 1 (p.78). In the latter reference, the constant parameters are treated as null state described by $\dot{p}=0$. The convergence control in the iterative DDP-algorithm is accomplished by the technique described in Ref.2. Some concepts from Ref.4 have also been used in order to obtain a well behaved algorithm. Because it is an effective and easy-to-use algorithm, the parameter sensitivity technique described in this paper can be considered a new application field of the DDP method. The technique has turned out to be attractive for practical purposes, in this case problems associated with aircraft and missiles.

A short description of the optimal design procedure: Perform several iterations in order to optimize with respect to the control variables. Then make a discrete change in the parameters followed by further iterations in the control variables, etc. The result is a system which, for a given mission, is parameter-optimized for optimal control. Accordingly, the solution obtained is the superior solution to fulfil a given mission in a local domain. Naturally, this is a question of uniqueness concerning non-linear problems. However in practice, depending on the level of ambition, the algorithm can be started from different nominal control and parameter values to check for various optima.

The trajectory-parameter optimization technique has proven to be a powerful tool in, e.g., a composite system consisting of an aircraft launching a missile, [Ref.5]. In this problem, the resource allocation of each vehicle is determined in minimizing a cost function involving the parameters of the entire system.

In Ref.6 a non-linear control law is designed for a missile vs. a maneuvering target. The well known proportional navigation law is modified, and a couple of parameters associated with the control law are introduced. The parameter sensitivity analysis presented in this paper is then used in designing a good control law.

The purpose of this paper is to extend the DDP-algorithm to parameter sensitivity analysis including search procedures for optimal parameter settings. This is demonstrated in numerical case studies which are usually hard to solve, such as optimal climb path problems for aircraft. Two examples show the features of the parameter sensitivity analysis. In addition, a design problem with demanding reqiurements is solved.

2. PROBLEM STATEMENT

The continuous-time dynamic systems are assumed to be described by the following nonlinear vector state equation

$$\dot{x} = f(x,p,u;t) \qquad x(t_o) = x_o \tag{2.1}$$

$x(t)$ is an n-dimensional state vector, p is an l-dimensional constant parameter vector, $u(t)$ is an m-dimensional control vector and t is the time, $t \in [t_o, t_f]$.

The performance of the system is measured by the cost function

$$V(x_o,p;t_o) = \int_{t_o}^{t_f} L(x,p,u;t)\, dt + F(x(t_f),p;t_f) \tag{2.2}$$

where L and F are scalar nonlinear functions. The final time t_f may be given explicitly or implicitly.

The control vector u may be required to satisfy the state dependent constraints

$$g(x,p,u;t) \leq 0 \tag{2.3}$$

where g is an r-dimensional ($r \leq m$) nonlinear vector function.

Thus, the parameter vector p may be involved in the state equation, the cost function and the constraints. Furthermore, the control law may be parameterized via p, i.e., $u(x,p;t)$ as in Ref.6.

The object of the trajectory optimization problem is to choose $u(t)$ for all $t \in [t_o, t_f]$, such that the system equation (2.1) and the constraints (2.3) are satisfied and V, given by Eq.(2.2), is minimized. The parameter sensitivity analysis is made in accordance with the technique described in the next two sections including determination of an optimal parameter setting.

3. DDP ALGORITHM

To summarize the DDP algorithm [Refs 1 and 2] which usually is an iterative procedure, assume that we have a nominal control $\bar{u}(t)$ with the corresponding nominal state trajectory $\bar{x}(t)$ obtained from the state equation (2.1). Define the Hamiltonian function,

$$H(x,p,u,V_x;t) = V_x^T f(x,p,u;t) + L(x,p,u;t) \tag{3.1}$$

where V_x is a vector whose components are the partial derivatives of V with respect to the components of x. Expanding the Hamilton-Jacobi-Bellman(HJB) partial differential equation about the nominal trajectory $\bar{x}(t)$ to first order and neglecting higher order terms we have the additional differential equation including terminal conditions for the first order DDP algorithm as

$$-\dot{a} = H(\bar{x},p,u^*,V_x;t) - H(\bar{x},p,\bar{u},V_x;t), \quad a(t_f) = 0 \tag{3.2}$$

$$-\dot{V}_x = H_x(\bar{x},p,u^*,V_x;t) + [\mu^{\hat{r}}(\bar{x},p,u^*;t)^T g_x(\bar{x},p,u^*;t]^T,$$
$$V_x(\bar{x}(t_f),p;t_f) = F_x(\bar{x}(t_f),p;t_f) \tag{3.3}$$

$$u^* = \arg \min_{u \,|\, g(\bar{x},p,u;t) \leq 0} H(\bar{x},p,u,V_x;t) \tag{3.4}$$

The components of the vector multiplier function $\mu(x,p,u;t)$ are defined by

$$\begin{aligned} \mu_i &= 0 && \text{for all } g_i < 0 \\ \mu^{\hat{r}} &= -[\hat{g}_u \hat{g}_u^T]^{-1} \hat{g}_u H_u && \text{for } \hat{g} = 0 \end{aligned} \tag{3.5}$$

where it is assumed that $\hat{r}$ components of g are active ($\hat{r} \leq r$). Then $\hat{g}$ stands for these constraints and $\mu^{\hat{r}}$ is the corresponding vector multiplier ($\hat{r}$-dimensional).

The quantity $a(t_o)$ is the predicted change in cost $V(x_o,p;t_o)$ obtained using $u^*(t)$ compared with that using $\bar{u}(t)$. Due to the first order expansion of HJB-equation, the quantity $\Delta x(t)=x^*(t)-\bar{x}(t)$ must remain sufficiently small for the expansion of the HJB-equation to be valid. If $a(t_o)$ and $\Delta V = V(u^*) - V(\bar{u})$ are about equal, the expansion of the HJB-equation is valid. Otherwise, we use the Convergence Control Parameter technique (CCP) in accordance with Ref 2 to keep Δx within the proper range in every iteration.

The DDP-procedure can now be extended to include parameter sensitivity analysis, where the time-invariant parameter vector is incorporated in the problem formulation in accordance with section 2. Form a new state vector (x,p), where p satisfies the differential equation [Ref 1, p.78]:

$$p = 0 \qquad p(t_o) \text{ is given} \tag{3.6}$$

With this extension we get the additional equation:

$$-\dot{V}_p = H_p(\bar{x},p,u^*,V_x;t) + [\mu^r(\bar{x},p,u^*;t)^T g_p(\bar{x},p,u^*;t)]^T,$$
$$V_p(\bar{x}(t_f),p;t_f) = F_p(\bar{x}(t_f),p,t_f) \tag{3.7}$$

This concept is useful both in the case of optimal and nonoptimal trajectories. In the non-optimal case we calculate $V_x(t)$ and $V_p(t)$ for given $\bar{u}(t)$.

Note that g_x and g_p are (n x r) and (1 x r) matrices respectively, and V_x, V_p, etc., are considered as column vectors.

4. PARAMETER SENSITIVITY

The parameter sensitivity is studied with respect to the value of the cost function (2.2). Setting $p = \bar{p} + \delta p$ and expanding Eq. (2.2) about $\bar{p}$ to first order and neglecting higher order terms yields,

$$V(x_o,\bar{p}+ \delta p; t_o) = V(x_o,\bar{p};t_o) + V_p^T(x_o,\bar{p};t_o) \cdot \delta p \qquad (4.1)$$

where the parameter gradient, V_p, is determined by Eq.(3.7). Thus, the magnitudes of the components of V_p give a measure of the individual parameters' effect on the cost function. Furthermore, changes in p satisfying

$$V_p^T \cdot \delta p = 0 \qquad (4.2)$$

means that this particular change, δp, does not change the cost value.

In practice, inconsistencies can occur in V_p between iterations when close to an optimum and F is small in Eq.(2.2). When the F-value gets small we may obtain a variation, due to overshoots of the optimum, in the direction of V_x which gives various V_p's between iterations. Therefore, in this paper we will formulate a cost function such that the desired optimal problem corresponds to a higher F-value. For example, in time optimal problems put a penalty on t_f hard enough to get a higher value of F. Also, when designing a system for a best performance the inconsistencies in V_p may be avoided by replacing the left hand member of Eq.(4.1) by a given positive number, V_{ref}. Then the desired final state can be approximatively obtained by giving a reference state in Eq.(2.2) which is somewhat harder to reach and the cost value approaching V_{ref} corresponds to the desired state. The incremental improvement, δp, of the parameter vector will then be in accordance with

$$V_{ref} - V(x_o,\bar{p},t) = \Delta V = V_p^T \cdot \delta p \qquad (4.3)$$

The design procedure is usually subject to constraints in the parameters. Assume there are k conditions in p;

$$G(p) = 0 \qquad (4.4)$$

where G is a k-dimensional non-linear vector function. Expanding Eq.(4.4) to first order yields the proper change of p in satisfying Eq.(4.4),

$$G(\bar{p} + \delta p) = G(\bar{p}) + G_p(\bar{p}) \cdot \delta p \tag{4.5}$$

If $l = k + 1$ then δp might be solved from

$$\delta p = \varepsilon \begin{bmatrix} V_p^T \\ G_p \end{bmatrix}^{-1} \begin{bmatrix} \Delta V \\ -G(\bar{p}) \end{bmatrix} \quad , \qquad 0 < \varepsilon \leq 1 \tag{4.6}$$

If $l > k + 1$ then $l - k - 1$ parameters are free.

If $l \leq k$ then a compromise must be reached, as for example, by finding the least-squares solution.

In practice, when G and/or the problem is non-linear, δp from Eq.(4.6) has to be considered as an improvement in p and an iterative procedure has to be carried out as

$$p^i = p^{i-1} + \delta p \tag{4.7}$$

5. APPLICATIONS

For aircraft design purposes we select certain parameters which specify the quality of the vehicle considered. Their influence on the performance will be demonstrated and an optimal choice will be determined. The design parameters studied are the zero-lift drag coefficient, the wing area and the size of the engine. The engine is assumed to be unconstrained whereas, in practice, there are a finite number of engines to choose from. The aircraft is assumed to be used in an optimal way which is calculated by the DDP+CCP-algorithm discussed in Sec.3. The algorithm includes an interpolation routine described in Ref 7 to estimated numerical values and derivatives of components which are given by plots. In this study, a representative use of the aircraft is a movement in the altitude-velocity space, i.e., optimal transition between different energy states. Then the dynamic model, with the angle of attack assumed to be zero, will be

$$\dot{h} = v \cdot \sin\gamma/1000 \tag{5.1}$$

$$\dot{v} = [T(h,v,p) - D_o(h,v,p) - D_i(h,v,p) \cdot n^2]/m - g \cdot \sin\gamma \tag{5.2}$$

$$\dot{\gamma} = g \cdot (n - \cos\gamma)/v \tag{5.3}$$

$$\dot{m} = -b(h,v,p) \tag{5.4}$$

where h is the altitude in kilometers, v is the velocity in m/s, γ is the climb angle, m is the mass and g is the acceleration due to gravity. The load factor, n, is the control variable. The thrust $T(h,v,p)$ has been given the simplified form

$$T(h,v,p) = p_1 \cdot 70000 \cdot \rho(h) \cdot (1 + .8 \cdot M) \tag{5.5}$$

where M is the Mach number, and the density, $\rho(h)$, follows the ICAO's standard atmosphere. The drag forces D_o and D_i have the forms

$$D_o = C_{D_o} \cdot S \cdot q \tag{5.6}$$

$$D_i = (C_{D_i}/C_L^2) \cdot m^2 \cdot g^2/(S \cdot q) \tag{5.7}$$

where q is the dynamic pressure, S is the reference wing area and C_{D_o} is the zero drag coefficient which will be split into two parts, p_2 times the subsonic drag level and p_3 times the transonic drag rise respectively, see Fig. 5.1.

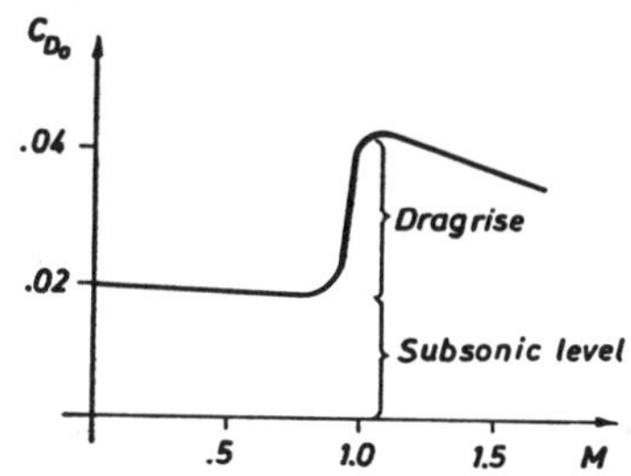

Figure 5.1: Zero drag coefficient vs. mach No.

The induced drag coefficient represented by C_{D_i}/C_L^2 is also taken from a plot corresponding to Fig. 5.1, and is considered fixed in the example below. The fourth parameter determines the wing area as

$$S = 40 \cdot p_4 \tag{5.8}$$

The mass dependence of the parameters is determined principally by the engine and the wing area:

$$m(p) = 1245 \cdot p_1 + 1440 \cdot p_4 + 1300 \cdot p_4^{.7} + m_1 \tag{5.9}$$

where m_1 is the remaining mass, including the fuel.

The fuel rate $b(h,v,p)$ is simplified as

$$b(h,v,p) = T(h,v,p) \cdot .0003 \tag{5.10}$$

The cost function (2.2) considered is

$$V(x_o,p:0) = k_1(h_{ref}-h(t_f))^2 + k_2(v_{ref} - v(t_f))^2+k_3t_f^2 \qquad (5.11)$$

where h_{ref} and v_{ref} are the desired terminal state and k_1,k_2 and k_3 are specified weighting factors.

Example 1

Suppose we want to find out the parameters influence on two slightly different missions. The missions are time optimal climbs from $\gamma = h = 0$, $v = 150$ m/s and $m_1 = 5181$ kg to $h_{ref} = 5$ km, $v_{ref} = 350$ m/s and 400 m/s respectively. The parameter vector is set to $p^T = (1,1,1,1)$, which gives the initial mass equal to 12000 kg. The weighting factors in the cost function (5.11) are set to $k_1 = 1000$, $k_2 = 1.$ and $k_3 = .02$. The parameter gradients obtained are

$$V_p^T = 245\ (-1.,.3,.33,.6) \qquad v_{ref} = 350 \text{ m/s} \qquad (5.12)$$

$$V_p^T = 950\ (-1.,.33,.49,.72) \qquad v_{ref} = 400 \text{ m/s} \qquad (5.13)$$

where the components of V_p are normalized to the engine (-1.). The factors 245 and 950 respectively are due to different optimal cost values.

From the V_p:s we conclude that supersonic flight accentuates the transonic drag rise. This has a 50 % greater effect on that mission, whereas the wing area has a 20 % greater effect and the subsonic draglevel has only a 10 % greater effect. All figures refer to the engine. Applying Eq.(4.2) on V_p we note e.g. that 1 % increase in wing area can be compensated by a .6 % increase in engine size, for case (5.12), to keep the same cost value. The trend in these results is not surprising whereas the degree of interdependence among the parameters may not be trivial.

Example 2

Numerical data describing the characteristics and performance of a realistic system are always in error. For example, in calculating optimal trajectories off-line, there will be problems of accuracy in the data for individual components of the aircraft. One example might be the importance of a theoretical correction term, $\Delta C_{D_o}(h,M)$, to be added to C_{D_o} in Fig. 5.1 compared to the accuracy of engine data.

$$C_{D_o}^{corr.} = C_{D_o} + \Delta C_{D_o} \qquad (5.14)$$

Multiply the engine data by p_1. All the data T, C_{D_o}, C_{D_i}/C_L^2, b and ΔC_{D_o} are taken from plots for a realistic aircraft. The weighting factors in Eq. (5.11) are $k_1 = 1000$, $k_2 = 1.$ and $k_3 = 0$. The desired final states are $v_{ref} = 390$ m/s and $h_{ref} = 6$ km at the fixed final time of 70 sec. Start conditions are $h = 0, \gamma = 25^o$, m = 15000 kg and v = 280 m/s. The parameters are set to 1. This gave a cost value of 290 (i.e. the desired final state is not fully reached) and the parameter gradient

$$V_p = \begin{pmatrix} -5305 \\ \vdots \\ 51.3 \end{pmatrix} \tag{5.15}$$

Using Eq. (4.2) the relation for unchanged cost will be

$$\delta p_1 = .00967 \cdot \delta p_5 \tag{5.16}$$

That is a 100% change in ΔC_{D0} can be compensated by a 1 % change in engine size, which might be within the tolerance of the engine data. Hence the correction term can be neglected for this kind of mission.

Example 3

Assume that we want to design an aircraft using the engine size and the wing area as design parameters. The mission is to make an optimal climb to 25 km altitude and have a velocity 590 m/s (M = 2) in 100 seconds. The initial conditions are $h = \gamma = 0$, $v = 150$ m/s and $m_1 = 3000$ kg. A ground roll distance (grd) of 150 m is also desired, representing a parameter condition. Then Eq.(4.4) will be

$$G(p) = 150 - \frac{m(p)^2}{T(p_1)\cdot S(p_4)\cdot 1.6} \cdot g \tag{5.17}$$

where T(1) has been given an average value of 101779 with respect to M. The weight factors in Eq.(5.11) are k_1=500, k_2 = .13 and k_3 = 0. The reference state is set to h_{ref}=25.3 km and v_{ref} = 609 m/s. This will give a final state close to the desired by setting V_{ref}=100 in Eq.(4.3). Substitute Eq.(4.6) in the search procedure (4.7), which is repeatedly applied after some improved DDP-iterations, starting with p_1 = .4, p_4 = .15 and $p_2 = p_3$ = 1. in all iterations.

The optimal trajectory for the optimal parameter setting is shown in Fig. 5.2, where curves for constant specific excess power (SEP) are also plotted. The optimal parameters are p_1^* = .92 and p_4^*=.28. This gives S= 11.2 m^2 and the takeoff mass = 5087.4 kg. Note that this example includes a difficult mission and a hard constraint to satisfy. The vehicle will therefore be small and have a good acceleration.

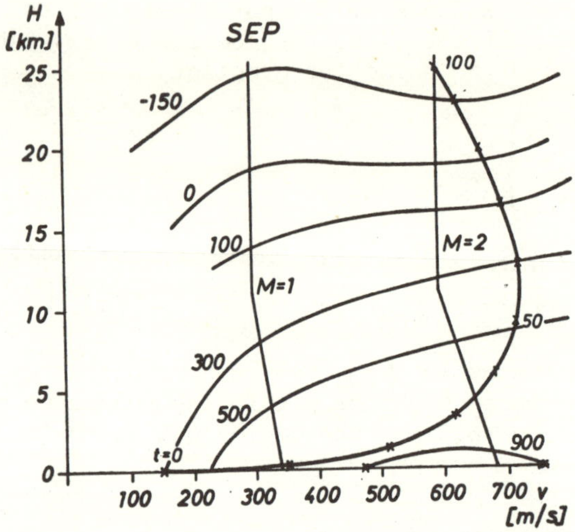

Figure 5.2: Optimal trajectory and SEP for the optimal vehicle obtained.

Table 5.1: Convergence scheme

I	p_1	p_4	ε	grd	$V(x_o,p;t_o)$
0	.4	.15	-	413.24	145.000
4	.4260	.1528	.1	388.55	87.000
6	.4391	.1604	.1	366.00	67.000
8	.4540	.1676	.1	345.52	53.000
10	.5195	.1942	.4	280.89	44.000
12	.7005	.2524	1.	190.57	31.000
14	.8468	.2816	1.	157.50	12.500
16	.9053	.2847	1.	150.49	3.200
18	.9194	.2825	1.	150.02	586
20	.9208	.2822	1.	150.00	128
21	.9208	.2822	-	150.00	92

Table 5.1 shows the convergence scheme, where I is the iteration of the trajectory optimization (DDP-iterations). The p-values are fixed during certain iterations. The $V(x_o,p;t_o)$ shown represent, in most cases, an acceptable "optimal" trajectory from where the changes in p are made.

It is necessary to choose a small ε in Eq.(4.6) in the first iterations due to the fact that the trajectory is far from optimum and the fact that V_p takes large values causing wild changes in the trajectories between the DDP-iterations. After some iteration it is possible to increase ε to 1 and finally the p-values approach the optimal ones very nicely. The parameter optimization procedure works very well, and the satisfaction of the parameter constraint is very good. Some further iterations would give $V(x_o,p;t_o)$ closer to V_{ref}. The precautional steps and the problem's sensitivity are due to the DDP-iterations. However, this example must be considered as a difficult task to solve.

It is worthwhile noting that the optimal parameters to some extent depend on their initial values, and, on the nominal initial trajectory which are needed to start the DDP-algorithm. For example a completely different start condition (p_1=2 and p_4=1) gives p_1^*=1.04 and p_4^*=.26 and a slightly different optimal trajectory. This is due to the fact that V_p gets different values in the two cases when approaching the optimum from different directions. For that reason a V_{ref} is needed and a better initial choice of V_{ref} could give a "more unique" p*. Also the first-order DDP method gives some contribution to this difference and furhter iterations would diminish it somewhat.

6. SUMMARY

A first order DDP-algorithm has been extended to include the possibility of studing the influence of known problem parameters with respect to the cost value. The mutual amount of the parameters' effect on the cost value has been demonstrated in two examples, where the parameter sensitivity analysis has served as an aid to form an opinion when designing a dynamic system. In addition, a design problem illustrate the technique considered.

The DDP-approach for the parameter sensitivity analysis is stable and behaves well. The first order parameter searching procedure works very well. The difficulties which appear belong to the trajectory optimization phase. This is well known from previous studies concerning optimal trajectories. However, those difficulties can be controlled by several tricks and improvements of the DDP-algorithm (see Ref.4). Consequently, the design procedure will be attractive for designing systems associated with aircraft.

All examples are within the field of aircraft application, where the choice of good parameters is not always obvious. Also, they are somewhat simplified but represent an attempt to model a realistic situation. However, the DDP-algorithm is not limited to simple examples. A fully realistic model with data from plots, etc., can be implemented in the DDP-algorithm. This has been partially illustrated in the examples considered.

REFERENCES

1 Jacobson, D.H. and Mayne, D.Q., Differential Dynamic Programming. New York: Elsevier 1970.

2 Järmark, B.S.A., On Convergence Control in Differential Dynamic Programming Applied to Realistic Aircraft and Differential Game Problems. Proceeding of the IEEE 1977 Conference on Decision and Control, New Orleans, December 7 - 9, 1977.

3 Mc Reynolds, S.R., J. Math. Anal. Appl. 19, 565, 1967.

4 Järmark, B.S.A., Some Aspects on Improving a First Order DDP-algorithm. SAAB TN73 Ae, Oct. 1978. SAAB-SCANIA AB, S-581 88 Linköping, Sweden.

5 Järmark, B.S.A., Preliminär utvärdering av avvägning plattform - vapen. Reg. No. FKU-2000-78.28, Sept. 1978. SAAB-SCANIA AB, S-581 88 Linköping, Sweden (in swedish).

6 Jonsson, H-O, Järmark, B.S.A., Evaluation of Nonlinear Control Laws - A Missile vs Aircraft Study. Reg.No. FKU-O-79.22, Sept 1979. SAAB-SCANIA AB, S-581 88 Linköping, Sweden.

7 Akima, H., A New Method of Interpolation and Smooth Curve Fitting Based on Local Procedures. Journal of the Association for Computing Machinery, Vol. 17, No. 4, Oct. 1970, pp 589 - 602.

ON THE IMPROVEMENT OF RADAR SIGNAL DETECTABILITY WITH COHERENT AND NONCOHERENT INTEGRATION

W. M. Waters

Search Radar Branch
Radar Division
Naval Research Laboratory
Washington, DC USA

Signal detection in the presence of receiver noise depends primarily upon the total echo energy returned to the radar from a target. In the simplest case the echo is completely known except for delay so that the well known matched filter and threshold detector maximize the likelihood of separating echo from noise. In actual practice, however, echoes must be detected in the presence of interference other than receiver noise such as environmental echoes (clutter) and jamming; also, unknown target motion further complicates the problem of choosing an optimum receiver which minimizes echo power needed for detection.

In this paper we choose the familiar radar receiver structures illustrated in Fig. 1 to show that processing combinations improve signal detectability more than processing all echo pulses noncoherently or even coherently. In both diagrams we assume that the RF and IF sections are matched to the pulse waveform and that a total of M pulses are available for coherent and noncoherent processing. We also assume that target radial velocity is known in the upper "reference" processor; velocity is assumed unknown but constant in the lower case which we call the peak-select/integrate (PSI) processor. In both cases we wish to find the value of N, for fixed M which minimizes $\mathcal{R}$, the SNR needed for detection.

Another important assumption is illustrated in Fig. 2 which shows the transmitted signal. Note that a total of M pulses is divided into blocks each containing N pulses. Echoes from pulses within one block are assumed to fluctuate from scan-to-scan (specifically, we use Swerling I fluctuations) whereas echoes

D. G. Lainiotis and N. S. Tzannes (eds.), Applications of Information and Control Systems, 127-134.

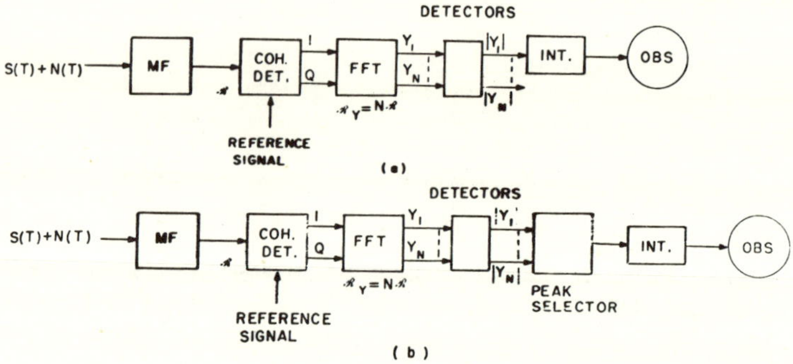

Fig. 1. Block-to-block radar signal processing systems.

from any one block are assumed independent of those from any other block. We further assume that block-to-block fluctuations are governed by Swerling's Case II distribution.

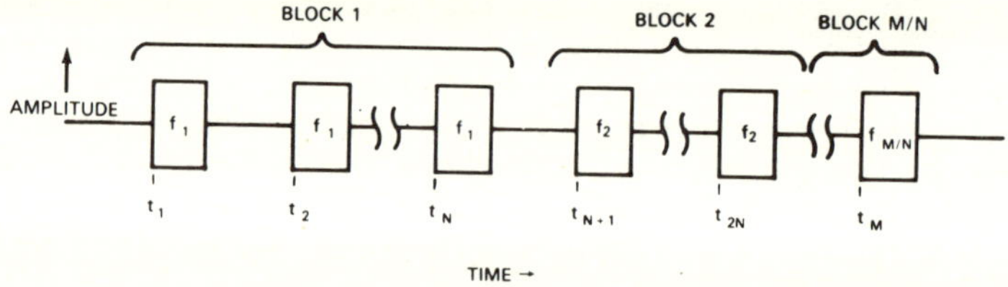

Fig. 2. Block-to-block frequency-agile radar signal (f_n = carrier frequency).

This model corresponds to a radar which changes frequency block-to-block and to a target producing scan-to-scan echoes fluctuating according to a "Swerling I" distribution. Frequency changes from block-to-block are assumed to decorrelate echoes from one block to the next, hence the assumption of a Swerling II distribution block-to-block. One could also conceive of an agile beam radar conforming to this model where, as in the frequency agile case, a fixed number of pulses are available for integration and where the beam may be scanned in other directions and returned according to an arbitrary schedule. Hence N pulses would be coherently integrated before the beam is moved; it would return M/N times with the coherent results integrated noncoherently. Figure 3 illustrates the effect of integrating all pulses coherently (or noncoherently) for non-fluctuating

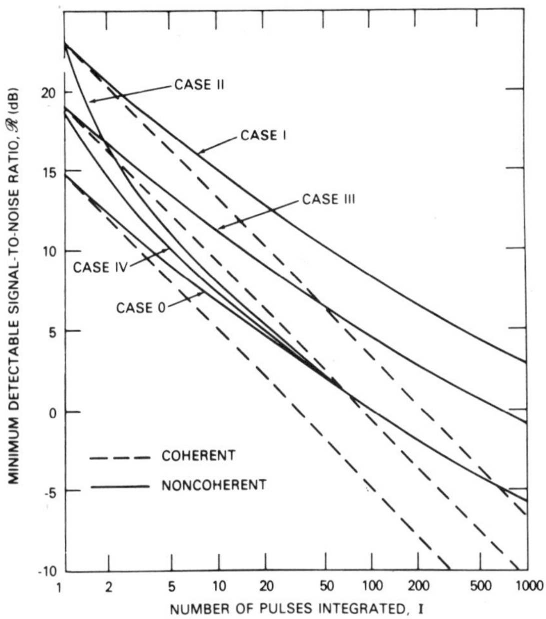

Fig. 3. SNR Improvement (Cases 0 through IV).

(Case 0) and fluctuating echoes (Cases I through IV). Cases I and III correspond to differently distributed scan-to-scan fluctuations; II and IV describe differently distributed pulse-to-pulse changes. Here we deal with an example involving Case I and II to describe scan-to-scan and block-to-block fluctuations respectively.

We will show that by an appropriate combination of both coherent and noncoherent processing, significantly less echo power is needed for detection than if processing is all non-coherent or all coherent. Making use of notation introduced in Fig. 1 we write expressions from which the SNR required for detection may be calculated for both Reference and PSI processors. For the Reference processor we have,

$$\mathcal{R} = \mathcal{R}_y(P_d, P_{fa}, M/N)/N \tag{1}$$

where $\mathcal{R}_y$ is the SNR required to achieve P_d and P_{fa} detection and false alarm probabilities (respectively) given M/N noncoherently integrated pulses (e.g., solid curve for Case II in Fig. 3).

For the PSI processor we have,

$$R = K_y(P_d, P_{fa}, M/N)NL_c(R,N) \tag{2}$$

where

$$L_c = \frac{NR(1-P_r)}{1+2NR}\left[1 + \frac{\sqrt{1+2NR}}{(1-P_r)NR}\right]$$

and $P_r(N,R)$ are tabulated error function integrals. Note that we have included a factor, L_c which expresses the degradation caused by unknown doppler frequency. Since multiple doppler channels are "collapsed" via the peak selector onto a single port, we refer to this as collapsing loss. L_c is a function of N (number of doppler channels), R and P_r, the probability that noise-only (from a doppler channel without signal) will exceed signal-plus-noise. This is well tabulated and is also the error probability associated with "M-ary" communications.

Detectable SNR, R is plotted in Fig. 4 for both processors using solid and dotted curves. As the latter case (unknown doppler) requires more complex, expensive calculations only two curves were computed for M = 100 and M = 200.

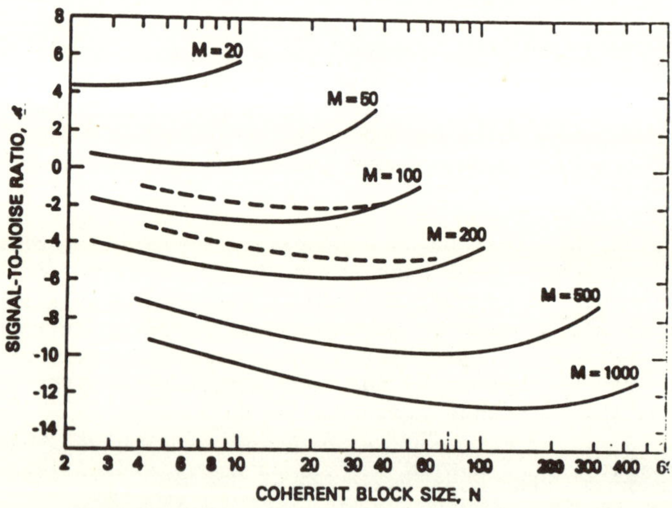

Fig. 4. Required SNR (for $P_d = 0.9$, $P_{fa} = 10^{-8}$), for fixed values of M (pulses per dwell time), vs coherent block size

In the reference case, note that an optimum split between coherent and noncoherent integration exists (minima at N_{opt}).

Note also that a slightly different value of N_{opt} exists for the PSI case. Moreover, $R(N_{opt})$ is substantially better than if processing were entirely coherent (N = M) or noncoherent (N = 1). These comparisons are made for the reference processor in Table I and for the PSI processor in Table II.

TABLE I - Comparison of Reference Processor ($N = N_{opt}$) With All Coherent and All Noncoherent Integration

M	N	R(dB)
100	100	3.0
100	1	0.1
100	$17 = N_{opt}$	-2.6
200	200	0
200	1	-2.0
200	$33 = N_{opt}$	-5.8

TABLE II - Comparison of PSI Processor ($N = N_{opt}$) With All Coherent and All Nocoherent Integration

M	N	R(dB)
100	100	3.0
100	1	0.1
100	$20 = N_{opt}$	-2.1
200	200	0.1
200	1	-2.0
200	$50 = N_{opt}$	-5.1

A Monte Carlo simulation of the reference and PSI processors was performed in which complex samples of noise were added to complex signal samples and processed by an 8 point DFT to simulate coherent integration of 8 pulses (N = 8). The zero-doppler channel was selected and a 10-pulse integration of its magnitude completed the simulation of the reference processor (M = 80). The PSI processor was simulated identically except the largest output from each DFT doppler channel was selected for noncoherent integration.

In each case, P_{fa} was plotted against the level of a threshold detector where due to time constraints only $P_{fa} > 10^{-4}$ were measured as shown in Fig. 5. Finally, the probability of detection was measured for different values of SNR using threshold levels corresponding to $P_{fa} = 10^{-2}$, 10^{-3} and 10^{-4} plotted in Fig. 6 where it should be noted that the "unknown doppler" case (PSI processor) requires about 2 dB more signal-to-noise ratio than the reference case in which doppler frequency is known.

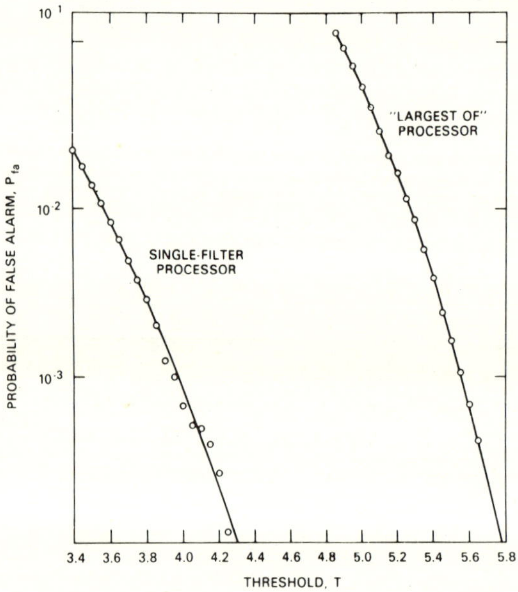

Fig. 5. Probability of false alarms as a function of threshold setting.

The equations shown earlier were used to compute values of $\mathcal{R}$ with N = 8 and M = 80, $P_d = 0.9$, and $P_{fa} = 10^{-4}$ for comparison with values from the simulation. Note that the simulation and theoretical results in Table III indicate about the same difference in the processors.

In summary, we have shown that by an appropriate combination of coherent and noncoherent processing (i.e., with $N = N_{opt}$) that substantially less signal-to-noise ratio is needed for detection than if either all coherent or all noncoherent integration is used. This is true whether doppler frequency is known or unknown.

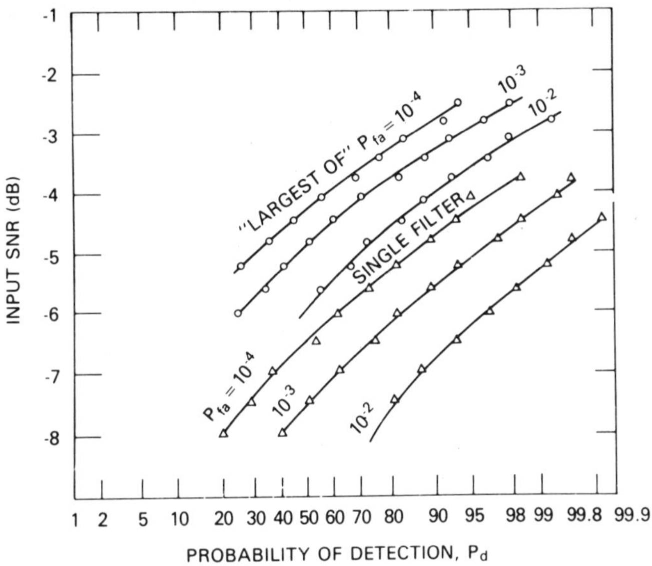

Fig. 6. Results of the simulation of both processors.

TABLE III - Comparison of Reference and PSI Processors Using Equations (1) and (2) and Simulation to obtain SNR* Values

Processor	Method	SNR (dB)
Reference	Eq. (1)	-4.9
Reference	Simulation	-4.7
Peak Select/Integrate	$L_c = 1 - P_r$	-3.3
Peak Select/Integrate	$L_c = \sqrt{1 - P_r}$	-3.9
Peak Select/Integrate	Simulation	-2.8

*$P_d = 0.9$, $P_{fa} = 10^{-4}$, $M = 80$, $N = 8$

We have also shown analytically and experimentally that the detection penalty due to the lack of doppler information is about 2 dB where $P_d = 0.9$, $P_{fa} = 10^{-4}$; for $P_{fa} \simeq 10^{-10}$ the difference is about 1 dB. Earlier work by Dow et al in connection with sonar showed that for much larger N and M (e.g., N = 400 and M = 12,000) the difference between knowing and not knowing doppler is about 2 dB. Overall, it appears that the penalty is greater for low threshold values than for high values. Thus it would seem of special importance to account for this effect in radar system designs involving further processing beyond the threshold as in track-while-scan systems.

Acknowledgement

This work was sponsored by the United States Navy and was performed at the Naval Research Laboratory. Special acknowledgement is due Mr. George J. Linde who performed the digital computer simulations.

TACTICAL RADAR: BRIGHTENING A GLOOMY FORECAST

Roy W. Jacobus and Charles A. Fowler

The MITRE Corporation
Bedford, Massachusetts, U.S.A.

Military radars for tactical surveillance must expect to confront physical assaults and electronic countermeasures from the enemy, and it appears that future radars will be forced to operate in exceedingly hostile environments. The advantage at present seems to lie with the enemy, and some observers have begun to mourn tactical radar as a dying art. This paper reviews the major offensive enemy measures which threaten to reduce the utility of tactical radar, and then examines a variety of design approaches that offer promise of improving radar's performance and survivability. The discussion supports our belief that modern radar is not doomed, but that designers must work very much harder to retain useful service under battlefield conditions. It is also clear that these radars and their supporting subsystems will be more elaborate and therefore more expensive than present-day versions, and that some performance limitations cannot be overcome despite the vigorous application of advanced techniques.

Recent technological advances have greatly extended the arsenal of modern weapons that can be brought to bear against our tactical surveillance radars. Furthermore, the enemy will presumably expend a significant fraction of his resources against our radars. These observations have led to the gloomy prediction that future radars will have very little utility in the midst of an active tactical battle. While we share these concerns, we note that a corresponding array of defensive techniques can be applied to the design of surveillance radars and their supporting systems.

D. G. Lainiotis and N. S. Tzannes (eds.), Applications of Information and Control Systems, 135-142.

Since this paper is unclassified, it is obvious that the discussions here are intended to deal only with basic threat/defense relationships, and do not pertain to the specific programs, plans, or systems of any country.

We shall focus our interest on the surveillance of aircraft in the tactical theater. We view the theater to be divided by an irregular time-varying battle line which separates friendly from enemy territory; under favorable conditions, our surveillance radars (assumed to be located on or over friendly territory) are required to detect and track all aircraft over friendly territory and 100 miles or so into enemy-held territory.

THREATS AGAINST TACTICAL RADARS

We shall confine our attention to the two major forms of attack against a surveillance radar: physical attack whose goal is to destroy the radar installation, and electronic jamming which attempts to blind the radar sensor. Both types of threat rely to some extent on the ability of the enemy to locate the radar site. Since the radar must emit a powerful electromagnetic signal to achieve useful detection range against small tactical aircraft, it also thereby provides the enemy with an intense beacon which he can utilize as an important part of his location strategy.

Can we, as radar designers, use advanced techniques (e.g., wideband spread-spectrum modulation) to prevent the detection of our radar signals by the enemy? The figure below illustrates, for a set of reasonably conservative assumptions, that we cannot escape detection. A simple incoherent receiver can achieve a

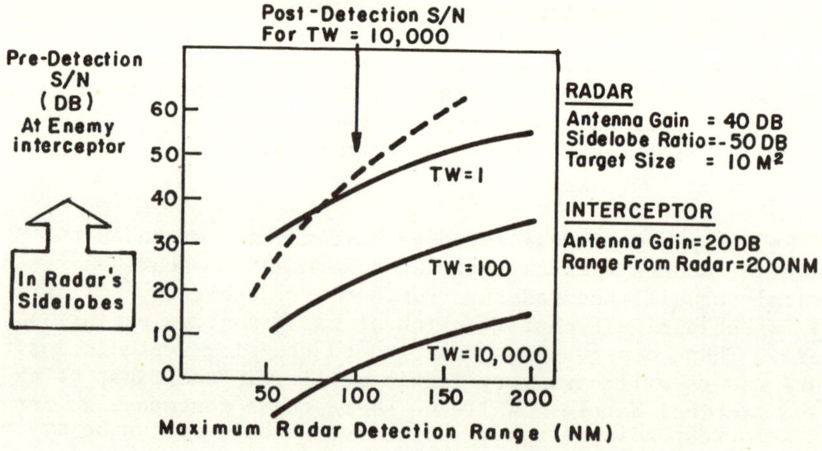

large signal-to-noise ratio even when the intercept site is 200 miles from the radar. No long-range radar can achieve "low probability of intercept" if the enemy chooses to implement the appropriate intercept receiver and antenna system.

Once the radar's location has been established, using direction-of-arrival or time-difference-of-arrival (TDOA) techniques, numerous weapon systems -- some conventional and others that could be developed in the future -- can be called upon to destroy the site. They include aircraft carrying bombs or short-range missiles, long-range supersonic missiles, subsonic cruise missiles and remotely-piloted vehicles, and artillery and rockets. Many methods for guiding the weapon warheads to the radar target are available, at least in concept. Terminal seekers can aim for the radar signal or infrared radiation from warm radar equipment. Active terminal-guidance radars can search for targets having large cross-section. If ultra-precise TDOA location information is available, the weapons can be directed to the radar from a distant command center; alternatively, on-board map-matching sensors can guide the weapon to a predetermined aim point.

Many forms of electronic countermeasures against surveillance radars are available today, including barrage jamming, spoofing, and chaff. By combining real-time computers with sophisticated signal processing, it is now possible to employ "smart" jamming that responsively matches the appropriate jamming to the radar signal environment. Jammers can take advantage of the beam

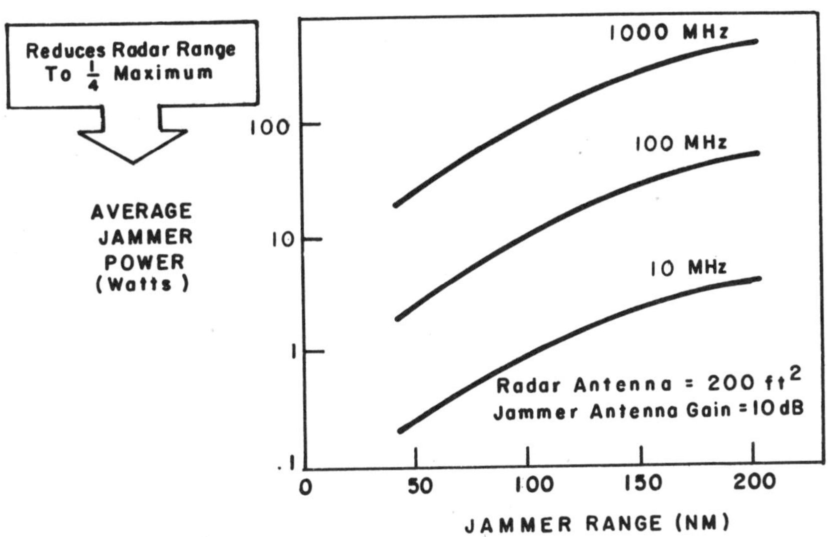

agility and transmit/receive independence offered by phased-array antennas. Disposable jammer packages can be deployed in large numbers over the battle area, so that the radar is always forced to look directly at one or more of them; they can be carried by RPV's or balloons, or dropped by parachute, or sewn on the ground by aircraft or artillery.

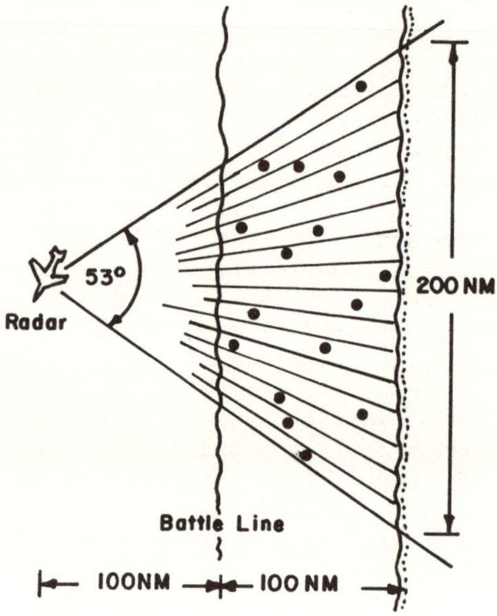

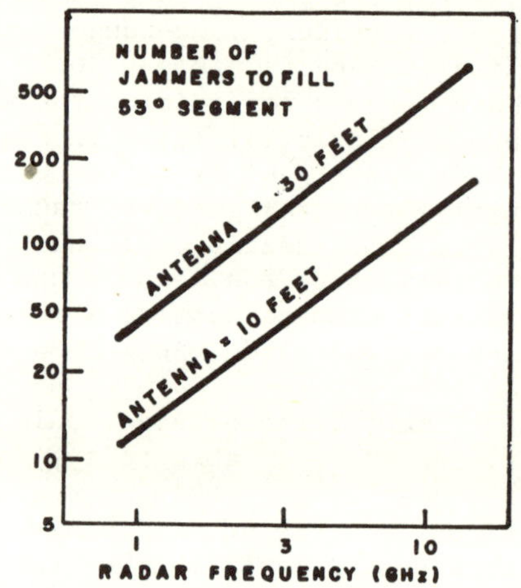

Jammer noise which is allowed to enter the radar's main beam is an especially difficult threat to counter. The graph on the preceding page shows that main-beam jammers need to emit very little power, and therefore could be modest in cost. The sketch above illustrates one deployment scenario where a radar surveillance sector of 53° is completely filled with enemy main-beam jammers (note that the radar will generally have poor resolution in elevation). Also shown is an estimate of the total number of enemy jammers needed to accomplish this mission, for two different dimensions of the radar antenna, and under the other assumed conditions.

This concludes our brief and necessarily incomplete review of threats to surveillance radars: location, destruction, and jamming. Now we shall consider the major design principles which can improve the radar's chances for survival and functional utility.

METHODS FOR ENHANCING PHYSICAL SURVIVAL OF TACTICAL RADARS

The radar survival techniques available to the design engineer are generally quite distinct from those used to improve the radar's detection performance; they occasionally conflict with detection requirements, and often require considerable help from systems external to the radar itself. Thus physical survival is costly, and demands a great deal of design attention and coordination, but it can have a high payoff in terms of increasing the cost and decreasing the effectiveness of the enemy's counter-radar actions.

The techniques may be grouped into five categories. The first is the *direct reduction of physical vulnerability*, and includes armor-plating, damage-tolerant design, and support from local point-defense weapons. Mobility is an essential element to survival, for we must give the enemy uncertainty about the radar's location. The figure below illustrates the influence of mobility on radar utility (as measured by percentage on-time) when constrained by the need to survive against radiation-seeking missiles. It is assumed that the radar periodically shuts off and moves one mile to avoid the incoming missiles.

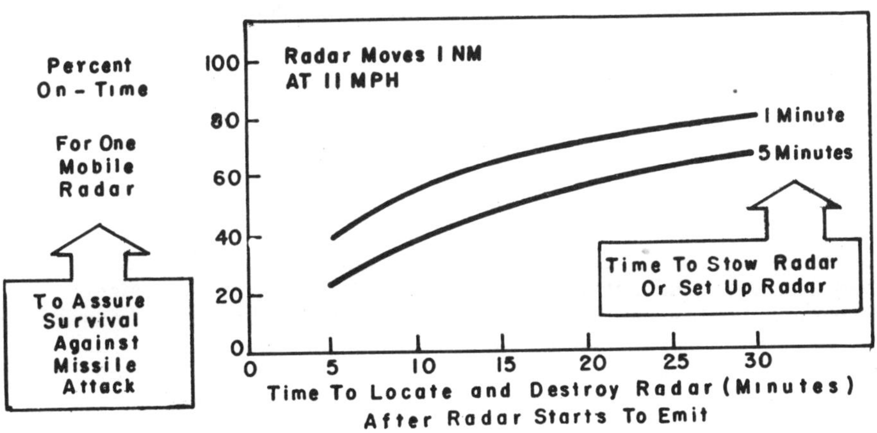

The second category is the *introduction of ambiguity*. This can be accomplished through the deployment of high-fidelity decoys; to increase the element of confusion, the decoys can cooperatively blink. Further difficulties with sorting and tagging are posed for the enemy if we randomly and abruptly vary the radar's signal parameters. In the third category we can place those techniques which reduce the accuracy of the enemy's location systems, i.e., the *introduction of measurement errors*. One example is the use of relatively long spread-spectrum coded waveforms with smoothly tapered envelopes (to avoid precision time-of-arrival measurements).

The fourth consists of methods for *confusing the terminal-guidance portions of the enemy's weapon systems*. This includes the deployment of nearby decoys, the coordinated blinking of the radar and its decoys, the provision of missile-detection sensors to allow "smart" responsive signal management, jamming of the weapons' electronics, the use of effective camouflage, and the careful suppression of infrared radiation from the radar site. In the final category we place *direct attacks on the enemy's command-and-control resources*: in particular, the airborne platforms used by his emission-location systems, and the various communication links associated with location and weapon guidance.

The numerous techniques described above, if used simultaneously and generously, can assure reasonable levels of survivability for tactical radars. However, any radar which dares to operate within range of the enemy's artillery can emit for only brief periods, and must be prepared to move almost continually.

Now we turn to the fundamental radar issue. Given that our radars survive, will their detection performance be adequate in the face of enemy jamming?

METHODS FOR IMPROVING PERFORMANCE AGAINST JAMMING

Let us first discuss the techniques available to the radar designer, and then consider the factors which affect the overall utility of a radar employing these countermeasures. The radar should strive for the narrowest possible azimuthal beamwidth, the largest time-bandwidth product for the transmitted waveform, and the widest band over which to frequency-hop; these steps minimize the problems from main-beam jammers or, alternatively, increase the enemy's cost to blanket a tactical area with effective main-beam jammers. Resistance to sidelobe jamming can be sought through antennas having low sidelobe gain, and from techniques to conceal the instantaneous transmitting frequency of the radar: frequency hopping combined with spoofing or jamming in the radar's sidelobes. A limited number of sidelobe jammers can be rejected by adaptive sidelobe-cancellation loops. Various multistatic system configurations and the use of cooperative radar networks offer some additional hope of improved performance against sidelobe jamming.

The antilocation techniques discussed earlier are also helpful in the jamming situation, because they can prevent the enemy from aiming high-gain jamming antennas at the radars. Since the jamming struggle can always be won by the enemy if he deploys a sufficiently large jammer, one objective of all these techniques is to drive the size, weight, and cost of the jammer to the point where it becomes an attractive target for physical attack by our weapon systems.

The outcome of the radar/jammer contest is a strong function of the tactical scenario. In particular, if the radar is required to detect aircraft which are flying over enemy territory, then the design of a jam-resistant radar is much more difficult. Part of the difficulty can be expressed in terms of the relative cost of the radar and the jammer. The figure below illustrates the problem

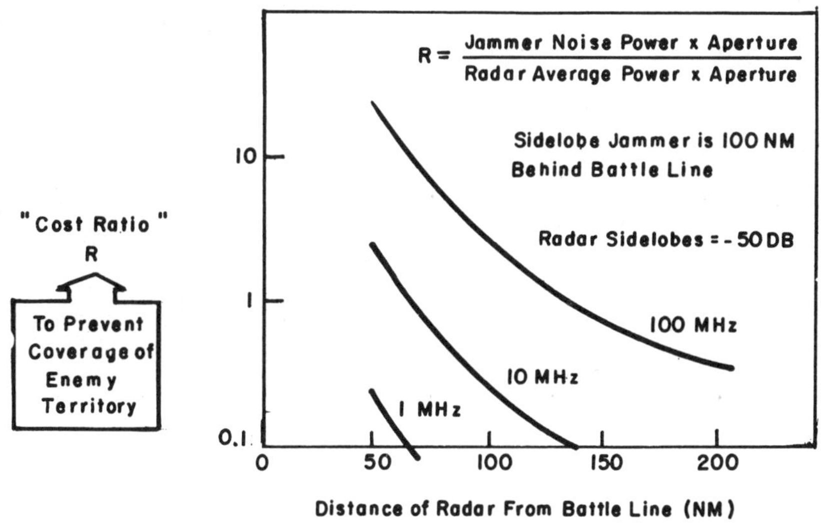

for one set of conservative conditions: The enemy sidelobe jammer is designed to produce noise at the radar which equals the received signal strength from an aircraft target of 10 square meters cross-section. The radar can detect aircraft up to the battle line, but cannot see beyond into enemy territory. The figure shows the "cost ratio", in terms of relative power-aperture products, as a function of radar bandwidth and distance from the battle line. Unless the

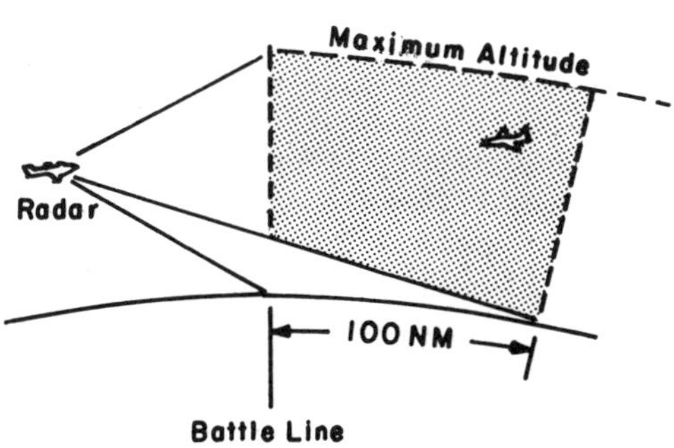

radar has been designed for wideband operation, the jammer will generally be smaller than the radar.

Some simple geometrical considerations are also important when we attempt to detect aircraft over enemy territory. As shown here, when our radar is airborne, it will be within

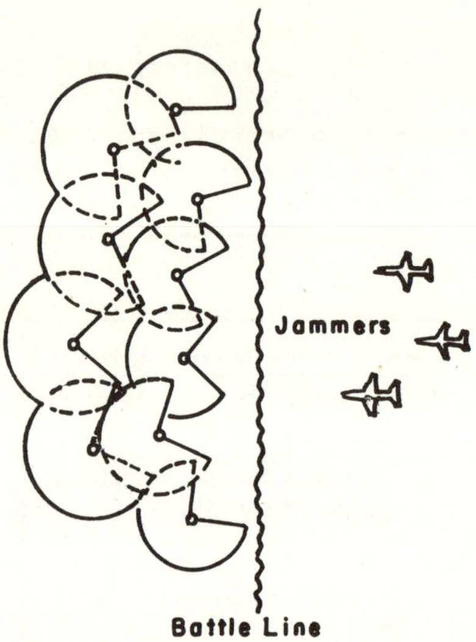

line-of-sight of enemy-held land. He can deploy large numbers of main-beam jammers or very powerful side-lobe jammers at relatively low cost because they can be ground-based. On the other hand, if we place our radar on the ground well behind the battle line, the enemy must now jam from airborne platforms which are expensive, limited in power, and vulnerable to attack; however, we will not have coverage of low-flying aircraft, and will thus have lost much of the value of surveillance.

If we can be satisfied with surveillance coverage primarily over our own territory, the struggle is much more favorable to the radar. In fact, we can guarantee limited coverage by netting multiple short-range radars as shown here.

CONCLUSIONS

Future tactical surveillance radars can expect to confront vigorous and sophisticated enemy countermeasures. If the radars and their supporting systems are designed from the outset with this hostile countermeasures environment in mind, and if the available techniques (as described earlier) are pursued with determination and imagination, then the physical survival of the radars seems likely, and the surveillance function should be performed satisfactorily in most cases. However, our ability to reliably detect aircraft flying over enemy-held territory remains in doubt. The surveillance function appears so valuable in a tactical battle that we can expect substantial expenditures by the enemy to destroy or disable our radars; correspondingly, we must plan to increase the complexity and cost of our radars and to make full use of decoys and deception techniques. The jammers are by nature in a reactive position, and if the radar designers are aggressive, they can force the enemy to make similar increases in cost, complexity, and vulnerability. To achieve these goals will require a fundamental change in our traditional approaches to the design and operation of tactical surveillance radars.

SECTION III

POWER SYSTEMS

HIERARCHICAL OPTIMIZATION OF LARGE-SCALE ELECTRIC POWER SYSTEMS

S. A. MANESIS and A. J. GRAMMATICOS

Dynamical Systems and Measurements Laboratory
University of Patras,
Patras, Greece

ABSTRACT: An hierarchical method is developed for solving the optimum economic operation problem of a large interconnected electric power system. The economic opetation control of the interconnected system is exerted via a two-level hierarchical structure. An "Area Economic Criterion" under certain conditions is introduced. First an optimum load dispatch between areas is achieved at the upper hierarchical level and subsequently the "optimum area load" is distributed optimally between the generating units of each area (lower hierarchical level).

The main charachteristics of the method are the successful decomposition of the large- scale problem, the fast response of the control system and the avoidance of coordination procedures. All calculations in this approach were performed off-line. However the method can be applied on-line.

INTRODUCTION: An electric power company is a complex enterprise whose basic goal is to supply electric energy as required by customers and for as long period of time as required. The electric power generation and transmission system is a typical example of an intergrated large-scale system. The operation of such a system is associated with the solution of several complex control problems. One of these is the optimum economic operation problem which has become very important over the recent years, especially in connection with the worldwide efforts to economize energy. The optimum state is selected subject to three basic conditions. a) The real power is generated from generating units whoose position and size are constant. b) The generated power at any time must be equal to the load plus the losses. c) The load is subject to wide but slow changes during the day. Hence-

D. G. Lainiotis and N. S. Tzannes (eds.), Applications of Information and Control Systems, 145-158.

forth, each change of the load must be followed by a new solution of the optimum economic operation problem.

For many years, additional generating units were connected to the existing units for obvious economic reasons. As a result, there are today national transmission networks with several hundred branches and nodes, as well as continental transmission networks with a few thousands branches and nodes. If one were to try to solve such a problem he would be faced with the task of having to solve thousands of nonlinear equations subject to additional stability constraints. Moreover there would be additional difficulties such as real time response and available computer memory. Therefore, the complexity of such a problem is great, especially when it must be solved several times each day and any attempt to obtain a centralized solution would be unrealistic (1).

The introduction of the "Area Economic Criterion" is the basic idea of the method which follows. If an area, which contains several generating units, could be substituted by a single "equivalent generating unit", then the manipulation of the variables corresponding to the generating units of an area could be substituted by fewer variables corresponding to the "equivalent generating unit", and a central computer would solve a reduced problem for a set of areas. Moreover, if further reduction in complexity is desired, it can be accomplished through integration of two or more areas in a larger "Equivalent Area". Such a reduction is possible if we apply the Toyoda, Chen, Hoover "equivalent generator" concept (2) to the optimum economic operation problem. The variables of an area must satisfy certain conditions in order to apply the above mentioned concept. First we apply this concept at a higher level for a set of areas without regard to specific conditions and then we secure satisfaction of the conditions at a lower level, for each area separately. Consequently a two-level hierarchical structure is introduced. The goal of each level is the optimal economic control.

DECOMPOSITION OF A POWER SYSTEM INTO AREAS

An interconnected electric power system may be decomposed into areas (Fig.1) according to various criteria such as the geographic location of the generating stations, the heavy load centers, the system interconnections, ownership or political boundaries, the system reliability etc. (3).

Every area of the system will contain a definite number of generating units and its structure could be considered unchanged

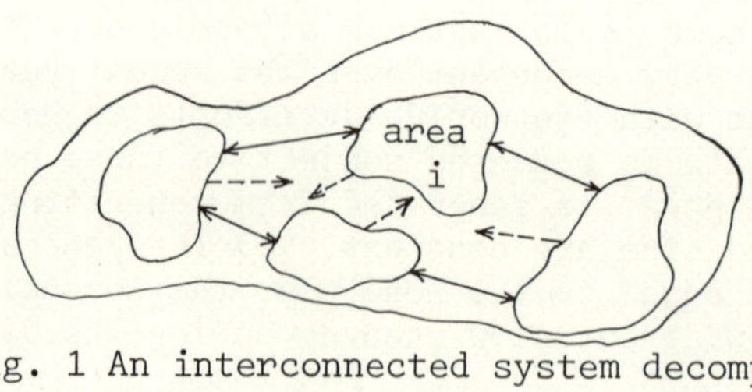

Fig. 1 An interconnected system decomposed into areas.

exept in special cases such as addition of new units, expansion of existing ones, a unit out of order etc. In formulating the problem we suppose that the system has been decomposed into areas by means of a decomposition method and that a general "Decomposition condition" which will secure the equivalence between the overall system and the decomposed system is satisfied.

Let S be a system with N generating units, (Fig.2) which has two inputs, the disturbance vector $\underset{\sim}{D}$ and the generating or control vector $\underset{\sim}{X}$ and one output, the cost vector $\underset{\sim}{C}$. Hence S is a mapping,

$$S : \underset{\sim}{D} \times \underset{\sim}{X} \longrightarrow \underset{\sim}{C}$$

We assume that the system S, is decomposed into M areas, where the i_{th} area is a mapping

$$R_i : \underline{D}_i \times \underline{X}_i \longrightarrow \underline{C}_i \qquad i = 1,\ldots,M$$

and $\underline{D}_i$, $\underline{X}_i$, $\underline{C}_i$ are components of vectors $\underset{\sim}{D}$, $\underset{\sim}{X}$, $\underset{\sim}{C}$.

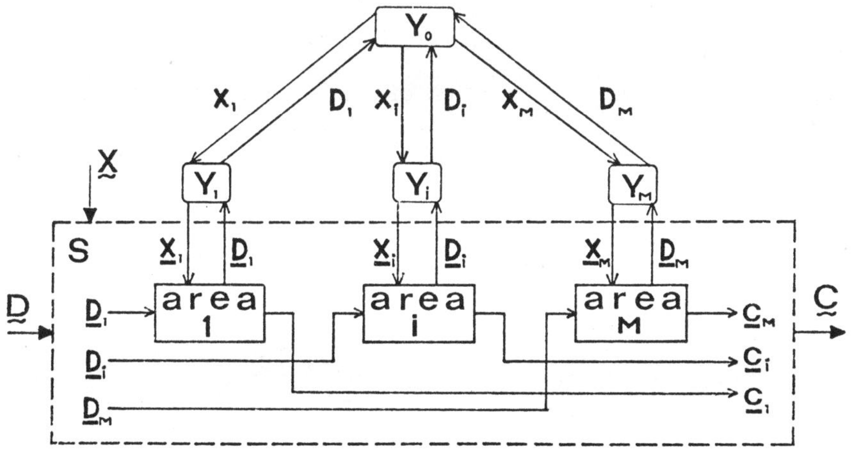

Fig. 2. Hierarchical economic control system.

We define a function $\bar{S}$

$$\bar{S}(\underset{\sim}{D},\underset{\sim}{X}) = \left[R_1(\underline{D}_1,\underline{X}_1),\ldots,R_M(\underline{D}_M,\underline{X}_M)\right].$$

Then the system S is equivalent to the M-area system if and only if

$$\underset{\sim}{C} = \bar{S}(\underset{\sim}{D},\underset{\sim}{X}) \Longleftrightarrow \underset{\sim}{C} = S(\underset{\sim}{D},\underset{\sim}{X}).$$

"AREA ECONOMIC CRITERION".

We consider an area i (Fig.3) which contains K generating units. We know the operation cost function for each unit, which may be a quadratic or a higher order function of the generated power X.

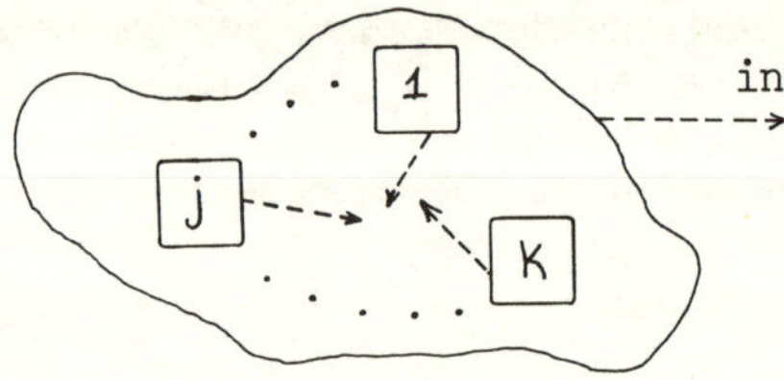

Fig.3 area i

We consider the following two conditions.

a) The transmission losses (L) are neglected.
b) The generating units operate optimally, i.e. at the same incremental cost df_{ij}/dX_{ij}

where X_{ij} (MW) the generated power and $f_{ij}(X_{ij})$ the operation cost function of unit $j(j=1,\dots,k)$.
Suppose $f_{ij}(X_{ij})$ is quadratic,

$$f_{ij}(X_{ij}) \triangleq a_{ij}(X_{ij})^2+b_{ij}(X_{ij})+c_{ij} \qquad \$/h.$$

Then the operation cost function of the area i is expressed by

$$F_i(X_i) = A_i(X_i)^2+B_i(X_i)+C_i \qquad \$/h \qquad (1)$$

where X_i is the total generated power of area i, $X_i = \sum_{j=1}^{K} X_{ij}$, and A_i, B_i, C_i are related to a_{ij}, b_{ij}, c_{ij} (2) as follows:

$$A_i = \frac{1}{\sum_{j=1}^{K} \frac{1}{a_{ij}}} \qquad B_i = \frac{\sum_{j=1}^{K} \frac{b_{ij}}{a_{ij}}}{\sum_{j=1}^{K} \frac{1}{a_{ij}}} \qquad (2)$$

$$C_i = \sum_{j=1}^{K} c_{ij}, + \frac{1}{4}\left[\frac{\left(\sum_{j=1}^{K} \frac{b_{ij}}{a_{ij}}\right)^2}{\sum_{j=1}^{K} \frac{1}{a_{ij}}} - \sum_{j=1}^{K} \frac{b_{ij}^2}{a_{ij}}\right]$$

If the operation cost functions f_{ij}'s are quadratic then an analytic determination of F_i is possible. However, if the f_{ij}'s are of higher order, then F_i exists under the same conditions but an analytic determination of it is difficult. In this case, we can determine F_i numerically as follows:

First we calculate the optimum generated power of each unit for several values of the load. Then we calculate the total optimum operation cost for each value of the load. The operation cost function $F_i(X_i)$ is then computed as a least square fit of the pairs (total optimum operation cost, total generated power). For a small number of generating units in an area a graphic solution of the optimum load dispatch is possible. Figures 4 and 5 show the incremental cost as a function of the generated power, for the cases of quadratic and cubic operation cost functions f_{ij}'s, of the area generating units and the "Equivalent Generating Unit".

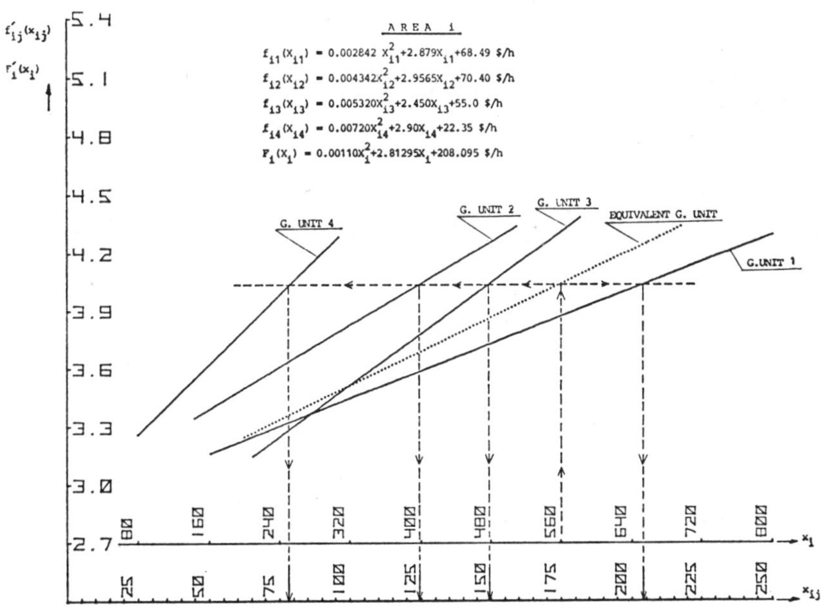

Fig.4

The intersection of the incremental cost curve of the "equivalent generating unit" and the curve of constant load defines the curve of constant incremental cost. This curve intersectes the incremental cost curves of generating units at points which have as abscissas the optimum generated power of each unit satisfied the area load.

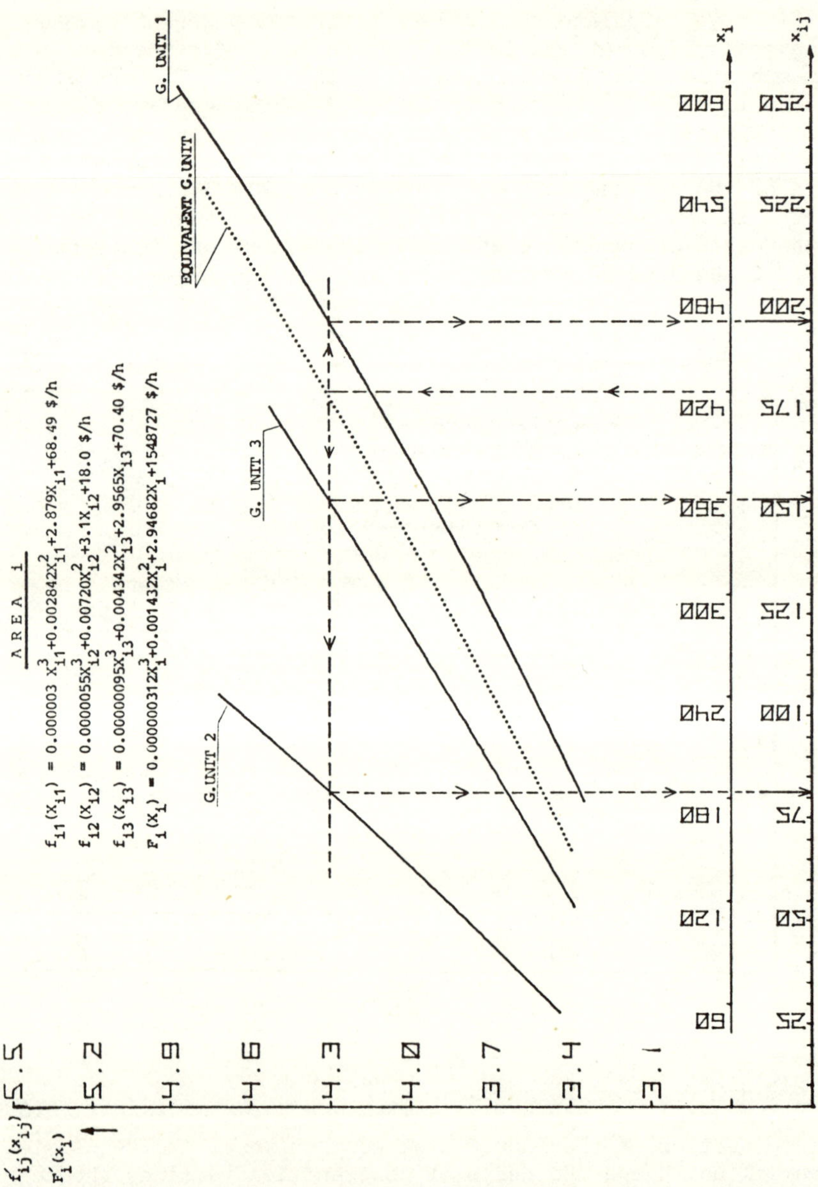
A R E A 1
f11(x11) = 0.000003 X11^3+0.002842X11^2+2.879X11+68.49 $/h
f12(x12) = 0.0000055X12^3+0.00720X12^2+3.1X12+18.0 $/h
f13(x13) = 0.00000095X13^3+0.004342X13^2+2.9565X13+70.40 $/h
F1(x1) = 0.000000312X1^3+0.001432X1^2+2.94682X1+1548727 $/h
G. UNIT 1
EQUIVALENT G.UNIT
G. UNIT 3
G.UNIT 2
f'ij(xij)
F'1(x1)
5.5
5.2
4.9
4.6
4.3
4.0
3.7
3.4
3.1
60
120
180
240
300
360
420
480
540
600
x1
25
50
75
100
125
150
175
200
225
250
xij

Fig. 5

PROBLEM FORMULATION

Let us assume that we have an interconnected electric power system which has been decomposed into M areas. Each area contains K generating units. The "overall problem" ($\mathcal{S}_G$) is now to distribute a load D to the KxM generating units optimally. The problem will be solved by a two steps procedure as follows:

First step:

Minimize $F_{total} = F_1(X_1)+\ldots+F_M(X_M)$

under constraining relation $\sum_{i=1}^{M} X_i = D$

we find the optimum generated power ($X_{i,optim.}$) of areas with the classical Lagrange multipliers method.

Second step:

Minimize $F_{i,total} = f_{i1}(X_{i1})+\ldots+f_{iK}(X_{iK})$

under constraining relation $\sum_{j=1}^{K} X_{ij} = X_{i,optim.}$

where $X_{i,optim.}$ has been computed at the first step. We find from the second step the optimum generated power ($X_{ij,optim.}$) of generating units for each area.

The first step is dedicated to a central computer (Y_o) and the second step, which is performed separately for each area, is dedicated to M local computers. So an hierarchical economic operation control system is formed as shown in Fig.2. The central computer (Y_o), upper hierarchical level, is interconnected with the M local computers. The i_{th} local computer transmits information about the load D_i from the area i to the central computer. The central computer receives the M information signals concerning the D_i's and procedes to the solution of the dedicated "upper level problem" ($\mathcal{S}_D$). The output vector $\underline{X}$ of Y_o has as components the optimum generating power ($X_{i,optim.}$) of each area which is transmitted as a command signal to the local computers. The local computers (Y_i), lower hierarchical level, solve the M independed problems which constitute the "lower level problem" ($\mathcal{S}_X$) . The output vector $\underline{X}_i$ of the i_{th} local computer has as components the optimum generated power ($X_{ij,optim.}$) of each generating unit. Furthermore, the i_{th} local computer receive information about the vector $\underline{D}_i$ from the area i, which has as components the loads D_{ij} of area i, and transmits upward an information signal about the component D_i of the vector $\underline{D}$ where $D_i = \sum_{j=1}^{K} D_{ij}$. We thus conclude that the action of the i_{th} local computer depend on all M areas but influence only the i_{th} area.

AUTO COORDINATION

In any hierarchical system coordination is required i.e. the "lower level problem" must be coordinable relative to the "upper level problem" and to the "overall problem". In the present case the proposed hierarchical system is coordinable automatically since each subsystem (central and local computers) performes the computation task in such a way that the coordination criteria (4) are always satisfied.

In fact the "lower level problem" is coordinable relative to the "upper level problem" since

$$(\forall \underline{x})(\forall(\underline{x}_1,\dots,\underline{x}_M))\left[P(\underline{x},\mathcal{S}_{\underline{D}})\wedge P((\underline{x}_1,\dots,\underline{x}_M),\mathcal{S}_{\underline{x}})\right]. \quad (3)$$

where generally $P(x,\mathcal{S}_Y)$ means that x is a solution of problem $\mathcal{S}_Y$ which is completely defined from Y. Also the "lower level problem" is coordinable relative to the "overall problem" since

$$(\forall(\underline{x}_1,\dots,\underline{x}_M))\left[P((\underline{x}_1,\dots,\underline{x}_M),\mathcal{S}_{\underline{x}})\Longleftrightarrow P((\underline{x}_1,\dots,\underline{x}_M),\mathcal{S}_G)\right] \quad (4)$$

The proof of condition (3) is obvious; the proof of condition (4) is given in Appendix A.

THE SYSTEM UNDER STUDY

We have applied the proposed method to a hypothetical electric power system consisting of 44 generators. The operation cost functions (Table 1.) of the generating units are based on real data. The electric power system has been decomposed into seven areas. The operation cost function of each area ("area economic criterion") is determined from Eqs (1), (2) and is shown in Table 2.

The 44 generating units of the system must supply optimally a load of 6,000 MW. The load is considered distributed to the 44 generating buses.

The incremental cost curves of generating units and "equivalent generating units" as a function of the generated power, for areas 5 and 7 are shown respectively in Figure 6 and 7.

The classical method for solving the optimum economic operation problem is the direct solution of the "overall problem" $\mathcal{S}_G$. The results of the solution to problems $\mathcal{S}_G$, $\mathcal{S}_{\underline{D}}$ and $\mathcal{S}_{\underline{x}}$ are shown in Table 3.

The same computer program was used to solve all problems so that a comparison of computation time be possible. The total computation time for solving problems $\mathcal{S}_D$ and $\mathcal{S}_x$ was about 25 seconds, while the solution of $\mathcal{S}_G$ took about 45 seconds. Note that the local computers operate simultaneously. No attempt was made to select the most effective computer program but the difference in computation time between the two methods is expected to be roughly proportional for any program. Such difference will be more

Table 1. DATA FOR THE SYSTEM UNDER STUDY

AREA NUMBER	GENERATOR NUMBER	GENERATOR COST EQUATION $f_{ij}(x_{ij})=a_{ij}(x_{ij})^2+b_{ij}(x_{ij})+c_{ij}$ a_{ij}	b_{ij}	c_{ij}	x_{ij}, MIN / x_{ij}, MAX
1	G 11	0.002842	2.8790	68.49	50/250
	G 12	0.003342	2.7700	96.40	60/220
	G 13	0.004290	2.8425	57.35	60/180
	G 14	0.00720	3.100	18.00	10/100
	G 15	0.0018	3.050	70.40	50/250
	G 16	0.0033	3.100	50.35	25/180
2	G 21	0.003342	2.8500	86.40	50/220
	G 22	0.003290	2.8425	57.35	60/220
	G 23	0.00386	2.6410	55.35	100/220
	G 24	0.00180	3.0500	15.00	50/250
	G 25	0.00180	3.0500	15.00	50/250
	G 26	0.00532	2.450	55.35	75/180
	G 27	0.002342	2.77	96.40	100/250
3	G 31	0.004290	2.8425	57.35	60/180
	G 32	0.007200	3.1000	18.00	10/100
	G 33	0.00400	2.950	60.40	50/180
	G 34	0.00532	2.450	28.00	50/180
	G 35	0.00380	2.850	65.40	75/180
	G 36	0.00360	2.800	90.30	60/200
4	G 41	0.00386	2.6410	55.35	100/220
	G 42	0.00532	2.450	55.00	50/180
	G 43	0.00520	2.900	50.35	50/150
	G 44	0.004342	2.9565	70.40	50/160
	G 45	0.003342	2.7700	96.40	60/220
	G 46	0.00400	3.000	60.40	50/180
	G 47	0.00330	3.100	18.00	25/180
5	G 51	0.00556	2.540	50.35	60/180
	G 52	0.00420	3.1425	48.00	10/150
	G 53	0.00720	2.900	22.35	25/100
	G 54	0.003342	2.770	96.40	60/220
	G 55	0.003642	2.870	98.00	60/200
	G 56	0.003342	3.150	66.49	10/200
	G 57	0.00250	3.200	69.99	10/250
6	G 61	0.00532	3.0025	55.00	25/150
	G 62	0.00250	3.2000	67.50	10/250
	G 63	0.002842	2.8790	97.40	50/250
	G 64	0.004360	2.8600	58.15	60/200
	G 65	0.00320	3.1500	18.95	25/220
	G 66	0.003448	2.9365	70.90	50/250
7	G 71	0.00386	2.6410	55.35	100/220
	G 72	0.00532	3.0025	55.00	25/150
	G 73	0.002842	2.970	95.40	60/250
	G 74	0.00320	3.150	26.95	25/220
	G 75	0.00486	2.742	53.00	50/200

significant if the program is supplied with additional requirements such as stability constrains and power flow computation at the various interconnections.

Table 2. AREA QUADRATIC COST EQUATIONS

AREA NUMBER	AREA COST EQUATION $F_i(x_i) = A_i(x_i)^2+B_i(x_i)+C_i$ A_i	B_i	C_i	x_i, MIN / x_i, MAX
1	0.00053144	2.959507	353.5774	255/1,180
2	0.00038635	2.87178	358.9402	458/1,590
3	0.00074025	2.82697	309.1270	305/1,020
4	0.00058075	2.84658	388.1480	385/1,290
5	0.00054649	2.97686	429.9317	235/1,300
6	0.00056442	3.01930	359.5628	220/1,320
7	0.00075921	2.917027	274.3685	260/1,040

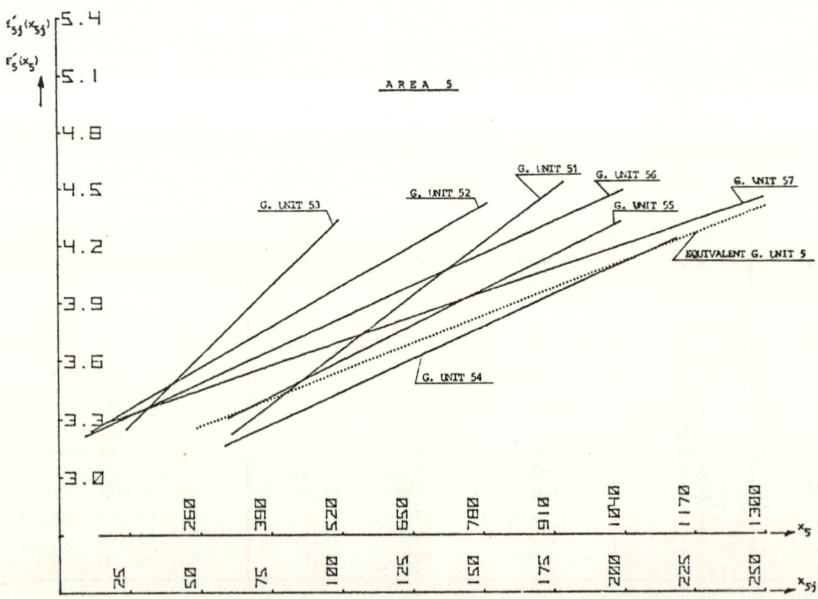

Fig. 6

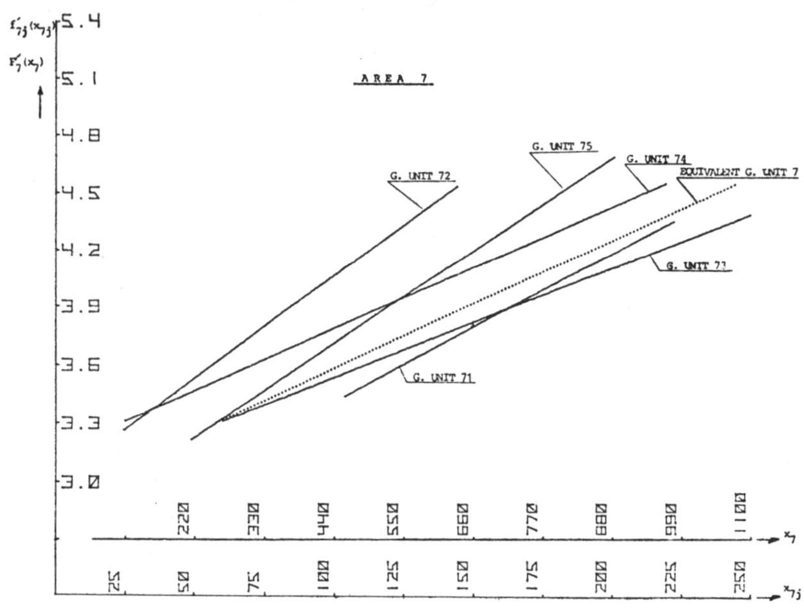

Fig. 7

CONCLUSIONS

First the optimum economic operation problem has been solved applying the proposed method. Consequently the same problem has been solved directly by the classical method. By inspection of Table 3, where the results for both cases are shown, the improved accuracy of the data obtained by the proposed method becomes obvious.

With the auto coordination of subsystems we avoid the coordination loops among them which are usually required by any coordination method (5). Hence the response of such a hierarchical optimization system is faster.

It must be noted that the method presented in this paper is computationaly more efficient since it avoids the interchange variable which have been used by other authors and therefore it leads to a faster response of the system (6).

Also the method can be applied to two or more large scale power systems with hierarchical economic operation control via a third level central computer.

Table 3. COMPUTER RESULTS

AREA NUMBER	GENERATOR NUMBER	PROBLEM $\mathcal{X}_G$ x_{ij}, optimum	PROBLEM OF UPPER LEVEL x_i, optimum	PROBLEM OF LOWER LEVEL x_{ij}, optimum
1	G 11	176.2588	866.8215	176.2615
	G 12	166.1961		166.1985
	G 13	121.0204		121.0222
	G 14	54.2260		54.2271
	G 15	230.7931		230.7974
	G 16	118.3114		118.3137
2	G 21	154.2273	1305.8811	154.2297
	G 22	157.8047		157.8072
	G 23	160.6030		160.6051
	G 24	230.7931		230.7977
	G 25	230.7931		230.7977
	G 26	134.4788		134.4804
	G 27	237.1595		237.1630
3	G 31	121.0204	711.8300	121.0186
	G 32	54.2260		54.2250
	G 33	116.3569		116.3550
	G 34	134.4788		134.4774
	G 35	135.6388		135.6368
	G 36	150.1187		150.1167
4	G 41	160.6030	890.4471	160.6020
	G 42	134.4788		134.4781
	G 43	94.3130		94.3123
	G 44	106.4434		106.4426
	G 45	166.1961		166.1950
	G 46	110.1069		110.1059
	G 47	118.3114		118.3102
5	G 51	120.5805	827.0730	120.5795
	G 52	87.8994		87.8981
	G 53	68.1149		68.1141
	G 54	166.1961		166.1945
	G 55	138.7775		138.7759
	G 56	109.3440		109.3423
	G 57	136.1710		136.1688
6	G 61	82.5521	763.2032	82.5534
	G 62	136.1710		136.1737
	G 63	176.2588		176.2612
	G 64	117.0705		117.0721
	G 65	114.1961		114.1982
	G 66	136.9424		136.9444
7	G 71	160.6030	634.7435	160.5986
	G 72	82.5521		82.5489
	G 73	160.2490		160.2430
	G 74	114.1961		114.1908
	G 75	117.1661		117.1626

λ optimum = 3,8808

APPENDIX A

The overall problem $\mathcal{S}_G$ is to,

$$\text{minimize } F_{total} = \sum_{i=1}^{M} \sum_{j=1}^{K} f_{ij}(X_{ij})$$

$$\text{under constraining relation } \sum_{i=1}^{M} \sum_{j=1}^{K} X_{ij} = D$$

Applying the Lagrange multipliers method we obtain the set of equations,

$$f'_{ij}(X_{ij})+\lambda_o = 0 \qquad (5)$$

$$\sum_{i=1}^{M} \sum_{j=1}^{K} X_{ij} = D \qquad (6)$$

Applying the Lagrange multipliers method to the problems $\mathcal{S}_{\underline{D}}$ and $\mathcal{S}_{\underline{X}}$ we obtain the set of equations,

$$F'_i(X_i)+\lambda = 0 \qquad (7)$$

$$f'_{ij}(X_{ij})+\lambda_i = 0 \qquad (8)$$

$$\sum_{i=1}^{M} X_i = D \qquad (9)$$

$$\sum_{j=1}^{K} X_{ij} = X_i \quad i=1,\ldots,M \qquad (10)$$

Differentiating Eqs. (1) results in

$$F'_i(X_i) = 2A_iX_i+B_i$$

Substituting Eqs. (7) and (2) into above equation yields

$$-\lambda = \frac{2X_i}{\sum_{j=1}^{K} \frac{1}{a_{ij}}} + \frac{\sum_{j=1}^{K} \frac{b_{ij}}{a_{ij}}}{\sum_{j=1}^{K} \frac{1}{a_{ij}}} \qquad (11)$$

Differentiating the generating unit cost function results in

$$f'_{ij}(X_{ij}) = 2a_{ij}X_{ij}+b_{ij}$$

Substituting Eq. (8) into above equation yields

$$X_{ij} = -\frac{\lambda_i+b_{ij}}{2a_{ij}}$$

By substituting this equation into Eq. (10) it becomes

$$X_i = -\sum_{j=1}^{K} \frac{\lambda_i+b_{ij}}{2a_{ij}} \qquad (12)$$

Substituting Eq. (12) into Eq. (11) we obtain

$$-\lambda = \frac{-2\sum_{j=1}^{K}\frac{\lambda_i+b_{ij}}{2a_{ij}}}{\sum_{j=1}^{K}\frac{1}{a_{ij}}} + \frac{\sum_{j=1}^{K}\frac{b_{ij}}{a_{ij}}}{\sum_{j=1}^{K}\frac{1}{a_{ij}}}$$

or $\lambda = \lambda_i$. This result signifies that the solutions of Eqs. (8), (9) and (10) are equivalent to the solutions of Eqs. (5) and (6).

REFERENCES

1. G. Quazza, "Large Scale Control Problems in Electric Power Systems". IFAC Symp. on "Large Scale Systems Theory and Applications", Udine, June 1976.
2. W. Hoover, J. Toyoda, M.S. Chen, "A new Operation Economics Approach Applied to the Realistic Scheduling of Large Thermal Machine Down Time". IEEE Conference Record of Control Power Systems, Oklahoma City, March 1976.
3. Thomas E. Dy-Liacco, "Control of Power Systems via the Multilevel Concept", Systems Research Center Cleveland, June 1968.
4. M.D. Mesarovic, D.Macko, and Y. Takahara, "Theory of Hierarchical, Multilevel, Systems". Academic Press, New York, 1970.
5. L. Hakkala and H. Blomberg, "On-Line Co-Ordination Under Uncertainty of Weakly Interacting Dynamical Systems", Automatica vol. 12, p.p.185-193, Pergamon Press, March 1976.
6. J.F. Aldrich, H.H.Happ, and J.F. Leuer, "Multi-Area Dispatch", IEEE Trans. Power App. Syst., vol. PAS-90, p.p. 2661-2670, Nov/Dec. 1971.

SCHEDULING OF ELECTRIC ARC FURNACES TO REDUCE PEAK POWER IN A STEEL PLANT

Hai Hoc HOANG and Michael P. POLIS

Department of Electrical Engineering
Ecole Polytechnique de Montréal
Montreal, Quebec, CANADA

ABSTRACT

We consider a steel mill undergoing a major expansion. The expansion calls for a reorganization of the mill, gives rise to the problem of scheduling the furnaces and permits the introduction of energy conservation considerations at the design stage of the mill. A conservation measure consists in constraining the maximum allowed level of power demand (contract demand) since electrical energy cost is proportional to this level. The constraint, however, must be set high enough so that production is not seriously affected. A power demand controller monitors and measures the power demand for the plant, and taps down or shuts off the furnaces, if necessary, to keep the power demand below the contract demand. The problem consists of scheduling the furnaces to maximize production while at the same time observing the power demand constraint. We will determine a mode of operation such that the schedule for either all or some of the furnaces is periodic. This permits the use of either a graphical approach or dynamic programming to determine the optimal furnace schedules over a given period.

1. INTRODUCTION

A steel mill melt shop consisting of two 110 ton electric-arc furnaces will be equipped with two additional 150 ton electric-arc furnaces. Two new overhead cranes (one for loading and one for tapping the 150 ton furnaces) as well as a continuous slab casting machine and an ingot caster will also be added to the mill. This expansion calls for a quasi-complete re-organization of the mill, and gives rise to serious problems of coordinating the operations of all the equipment. Moreover, faced with rising

D. G. Lainiotis and N. S. Tzannes (eds.), Applications of Information and Control Systems, 159-168.

energy costs it is desirable to introduce energy conservation considerations at the design stage of the mill expansion.

Since the total energy required to produce one ton of molten steel from a given mix of scrap and iron pellets is approximately constant, the electric energy costs are determined in large part by peaks in power demand. The contract with the utility company calls for the energy cost to be based on the maximum energy consumed during each 15 minute period as well as the total energy used per "month". Consequently, the power demand controller (PDC) which monitors and measures the power demand for the whole plant subdivides the time into 15 minute periods. Since the major portion of the consumed energy goes to the electric furnaces, the PDC taps down or shuts off the furnaces, if necessary, to keep the power demand below a required level during each 15 minute period. Two important questions must be answered:

1) What should be the required level below which the power demand must be kept?

2) What should be an efficient control to satisfy such a power demand constraint?

We have attempted to develop the necessary tools to answer these questions.

The first question may be answered by an economic/cost-benefit analysis since energy costs are directly related to the maximum allowed level of power consumption, and this level (within a plausible range) determines the attainable steel production.

Once the power demand constraint has been fixed, several approaches may be taken to answer the second question:

. It is possible to simulate the melt shop.
. A mode of operation can be choosen which renders all furnaces schedules periodic, thus allowing a graphical method which rapidly determines optimal furnace schedules.
. By choosing the operation of some of the furnaces to be periodic, dynamic programming may be used to determine the optimal schedules for the remaining furnaces.

The first approach is discussed in [1], and we will consider the second and third approaches in this paper.

In order to obtain an effective model in terms of computer memory and computing time requirements, we concentrated only on the electric furnaces and the material handling equipment necessary for charging and tapping them. The results provide the production manager with targets which can possibly be attained by coordinating efficiently the steel making activities.

In the next section of this paper, we give a brief description of the operating characteristics of the melt shop and the power demand controller (PDC). In section 3, we give the problem formulation. In section 4, we discuss two possible solution approaches and present some results.

2. DESCRIPTION OF THE MELT SHOP AND THE PDC

2.1 Plant Description

The plant considered here is shown schematically in Figure 1. It consists of two existing 110 ton electric-arc furnaces A and B, two new 150 ton electric-arc furnaces C and D, two loading cranes PR1 (serving furnaces A and B) and PR2 (serving furnaces C and D), two tapping cranes PR3 (serving furnaces A and B) and PR4 (serving furnaces C and D), two laddle stirring stations SB1 and SB2 used in conjunction with cranes PR3 and PR4 respectively, a cold scrap yard, two billet casters, one continuous slab caster and one ingot caster. In addition there is the PDC which is not shown in Figure 1.

For the purposes of this study we make the following assumptions: i) the cold scrap yard is operated in such a way that full scrap buckets are always available to be picked up by the loading cranes, ii) the casting machines are always ready to receive molten steel from the laddles which are transported to the casters by the tapping cranes. Thus it is assumed that no delays are introduced by equipment located at the exterior of the dotted lines shown in Figure 1.

2.2 Plant Operation

The operation of the melt shop consists of four concurrent furnace cycles and each furnace cycle is broken down into four phases. A brief description along with the operating characteristics of each phase is given below.

2.2.1 *Preparation:* At the beginning of this phase the operating personnel carry out various tasks to assure that the furnace is ready to be loaded. A short time before the furnace is ready, a loading crane picks up a loaded cold scrap bucket at the scrap yard. The crane then transports the bucket to the furnace, loads the furnace with the scrap and returns the empty bucket to the scrap yard. At the same time that the crane is returning the empty bucket the top of the furnace is closed and the electrodes are lowered.

2.2.2 *Melting:* At the start of this phase, power is applied and melting begins. After the scrap is almost completely melted, pig iron pellets are added to the molten scrap through a hole in the top of the furnace (this addition is accomplished by gravity feed through a flexible pipe). Heating then proceeds until the bath reaches a desired temperature.

2.2.3 *Refining:* During this phase the metal is refined by adding various chemicals to obtain the desired type of steel.

2.2.4 *Tapping:* The steps in this phase depend on the type of casting machine to which the molten steel is to be directed, as well as the type of furnace which is being tapped. For steel which is destined for either the billet or slab casters, the contents of the furnace are emptied into a laddle and, after certain characteristics of the steel in the laddle are measured, chemicals may be added to insure that the steel meets specifications. A tapping crane then transports the laddle to the laddle stirring station. When the stirring is completed the laddle is transported by the tapping crane to the billet or slab caster, positioned, and emptied into the caster. For laddles destined for the ingot caster, the stirring operation is not necessary.

Apart from the PDC, the melt shop may be considered from an operational viewpoint as two separate plants: one above and one below the double horizontal line shown in Figure 1. This is due in large measure to structural considerations which make it undesirable for the new cranes to service the existing 110 ton electric-arc furnaces.

2.3 The Power Demand Controller (PDC)

It is necessary to determine the contract demand such that the melt shop is operated with a load factor as close as possible to unity. Once this contract demand is specified, the function of the PDC is to assure that the total energy consumed by the melt shop during each 15 minute demand period does not exceed the prespecified limit. The mode of operation of the PDC is detailed in [1].

3. PROBLEM FORMULATION

To simplify the problem, the following assumptions are made: there is perfect mixing in the bath, there are no heat losses, only one typical type of steel is produced, the mix cold scrap-pig iron pellets is fixed. Under these assumptions the energy required for the melting and refining phases is fixed for each type of furnace. It is further assumed that the time required for the preparation and tapping phases is fixed equal to 15 minutes.

If there were no constraint on power demand, the energy vs time profiles of the two furnace types would be as shown in Figure 2, with the values of the parameters given in Table 1. If a given furnace follows the profile shown in Figure 2, it will produce a batch of steel in minimum time. In reality $\Delta = 0$ since the furnaces do not require energy during the tapping and preparation phases. However, to assure that the state of the furnaces determines uniquely the time we have artifically introduced a value of $\Delta = 0.15$ MWH. The power delivered to each type of furnace to produce a batch of liquid steel in minimum time is thus shown in Figure 3.

The problem may be formulated as follows:

Let $E_i(t)$ be the total energy received by the i th furnace during the interval (t_o, t). Thus

$$E_i(t+1) = E_i(t) + \Delta t \, P_i(t) \tag{1}$$

and

$$E_i(t_o) = \text{given}, \; i = 1, \ldots, 4 \tag{2}$$

The control $P_i(t)$ is the power delivered to the i th furnace in the interval $(t, t+1)$. The controls are constrained as follows:

$$\begin{aligned} &\text{For } 0 \le [E_i \text{ modulo } E_i^f] \le E_i^m && P_i \le P_{i_{max}}^m \\ &\text{For } E_i^m < [E_i \text{ modulo } E_i^f] \le E_i^r && P_i \le P_{i_{max}}^r \\ &\text{For } E_i^r < [E_i \text{ modulo } E_i^f] \le E_i^f && P_i = \frac{\Delta}{.25 \text{ Hrs}} \end{aligned} \tag{3}$$

The furnaces are coupled through the constraint that the sum of the energies received by the four furnaces during each 15 minute period is less than or equal to the contract demand, i.e.

$$\sum_{i=1}^{4} [E_i (\tfrac{15n}{\Delta t}) - E_i (\tfrac{15(n-1)}{\Delta t})] \le E_{cd} \tag{4}$$

$n = 1,2,3,\ldots$

where E_{cd} is the contract demand.

The performance criterion is

$$J = \sum_{i=1}^{4} E_i(t_f) \tag{5}$$

where t_f is the time period over which the optimization is performed.

We want to find $\underset{P_i}{\text{Max}}\,(J)$ which is equivalent to maximizing the production over the time period t_f (under the assumption of no heat loss).

Rather than keeping the total energy consumed over each 15 minute period $\leq E_{cd}$ we may require

$$\sum_{i=1}^{4} P_i(t) \leq P_{max} \qquad \forall t \qquad (6)$$

Note that this constraint is more severe than (4). That part of the available energy not used may serve as a performance measure, i.e. in discrete time

$$J_1 = P_{max} \cdot t_f - \sum_{i=1}^{4} \sum_{j=1}^{M} P_i(t_j) \Delta t_j \qquad (7)$$

where $\sum_{j=1}^{M} \Delta t_j = t_f$ and $\sum_{i=1}^{4} \sum_{j=1}^{M} P_i(t_j) \Delta t_j$ is the energy used during the time period t_f.

4. SOLUTION METHODS AND SOME RESULTS

4.1 General Considerations

There are a number of difficulties associated with the problem formulation presented in Section 3. First, it is not possible to write the necessary conditions of optimality since the constraints, as shown in Figure 3, are not differentiable. Second, a range of values of P_{max} should be investigated. P_{max} must be sufficiently large to allow several furnaces to operate simultaneously in the melting phase, otherwise production will be seriously affected. If all furnaces were simultaneously in the melting phase, power requires would be 182 MW. Thus, P_{max} should not exceed 182 MW. Moreover a furnace in the tapping and preparation phases consumes no energy. It is thus possible to reduce P_{max} by avoiding simultaneous tapping and preparation of the furnaces. The same consideration holds for the refining phase since the power required is much less than that required for melting. Thirdly, since an optimization on $(0, t_f)$ yields no information for $t > t_f$, the final states must be taken into account. It is possible that the final states would be such that some furnaces must be tapped simultaneously just after t_f. To avoid this certain final states should be avoided, for example, by penalizing them. Lastly, the problem considered is of high dimensionality. Using dynamic programming and N values of E_i, for example, the computational complexity is N^4 per iteration. Fixing the operation of some of the furnaces reduces the dimensionality of the problem. Hence both solution approaches discussed below are based on the idea of forcing at least some of the furnaces to produce liquid steel in such a way that their operation becomes periodic.

4.2 A Graphical Approach

An optimal solution which satisfies (1)-(3), (6) and minimizes J_1 would at best use all the available energie, i.e. $J_1=0$.

The graphical approach attempts to use as much of the available energy as possible and yields both the furnace schedules as well as the level of P_{max}. To maximize production at least some of the furnaces could be operated so that they produce in minimum time. The minimum times are 130 min. and 120 min. for the 110 ton and 150 ton furnaces respectively. Thus for $t_f \geqslant 130$ min. each furnace may produce a batch of steel in t_f minutes (assuming P_{max} sufficiently large and a proper choice of initial conditions). For $t_f = 130$ min. the time when each furnace enters the tapping phase is fixed so that only one furnace at a time is in a refining, tapping or preparation phase. The operation of each of the 110 ton furnaces is then fixed a priori so that they produce in minimum time. Properly controlling the power delivered to the 150 ton furnaces during melting allows them to produce a batch in t_f minutes and renders approximately constant the sum of the powers delivered to the four furnaces. For $t_f > 130$ min. the power supplied to the 110 ton furnaces during melting is reduced so that they each produce in t_f minutes.

The graphical method thus allows the rapid calculation of P_{max} and forces the operation of the furnaces to be periodic with period t_f. A typical result, for $t_f = 140$ min. is shown in Figure 4.

4.3 Dynamic Programming

In order to reduce the dimensionality of the problem, the operation of the 150 ton furnaces is fixed a priori so that they produce a batch of liquid steel in minimum time. For a given level of P_{max} dynamic programming is used to optimally allocate the remaining power to the 110 ton furnaces. Several remarks are necessary. First, the discretization of the state variables E_i should be sufficiently refined to reflect possible controls achieved by tapping up/down the furnaces. Although there are about half a dozen possible tapping levels, the control may be considered continuous since it is possible to combine tapping levels over time. Secondly, the choice of t_f and weights on the final states should be determined. Practically, the period t_f should be sufficiently long to allow all furnaces to produce at least one batch of liquid steel, but not so long that random interruptions, e.g. replacement of broken electrodes, would almost certainly occur. Based on operational data, we chose $t_f = 4$ hours. Weights on the final states are much more difficult to fix. We proceeded by extending the optimization horizon to 8 or 12 hours, for example, and retaining only the first four hour period for scheduling purposes. The trajectories obtained are used cyclically in the sense that the final states for one period become the initial states for the next. A typical result is shown in Figure 5.

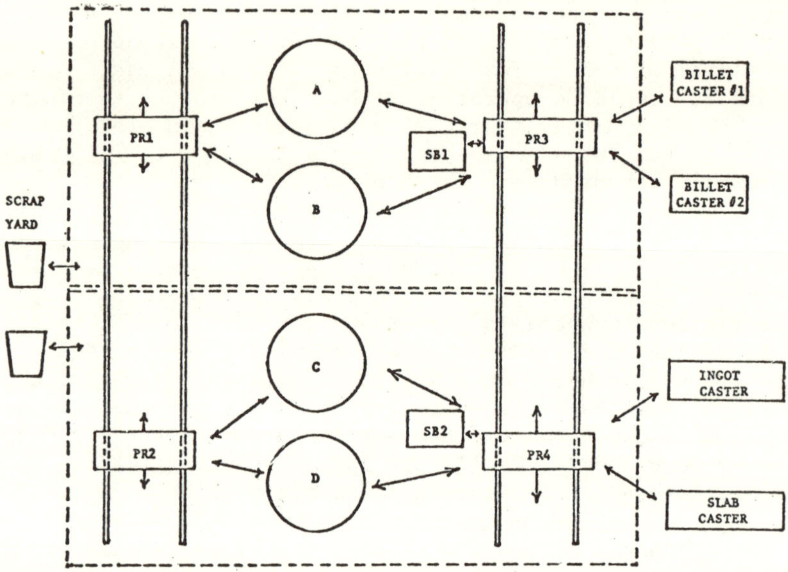

Figure 1 A schematic of the plant.

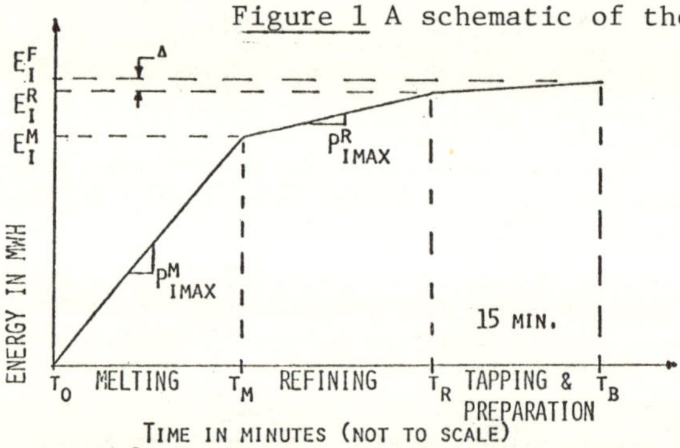

Figure 2 Energy vs time profiles for the two furnace types.

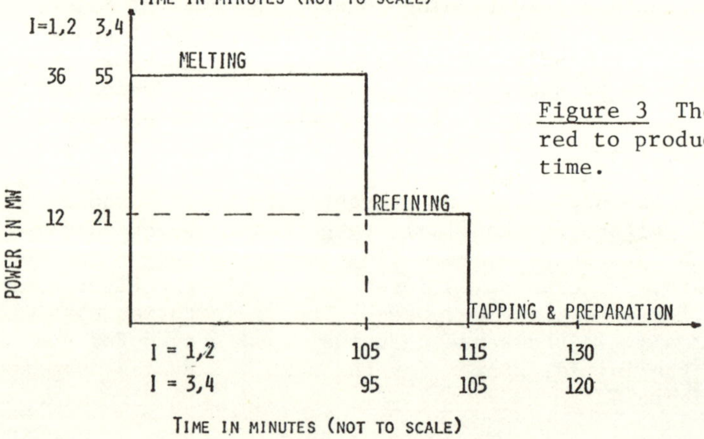

Figure 3 The power required to produce in minimum time.

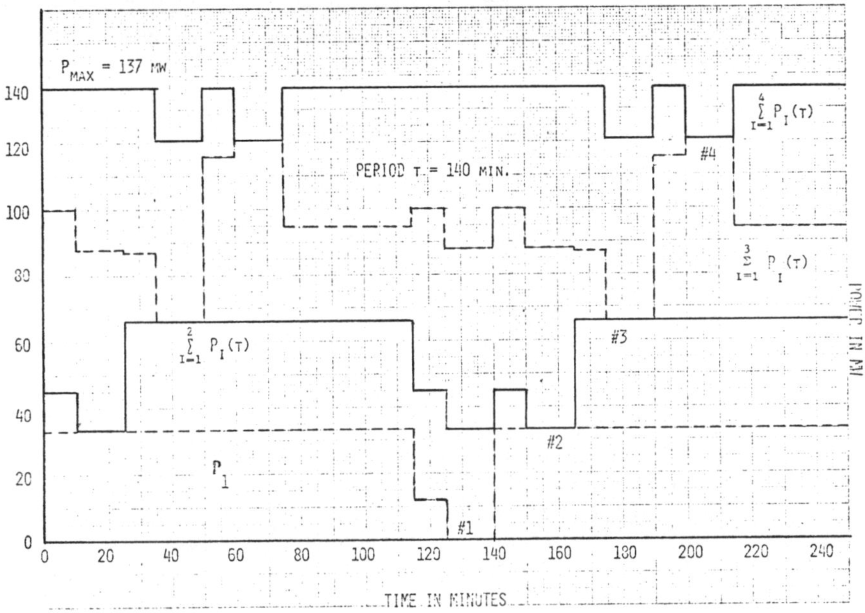

Figure 4 A typical result by the graphical method.

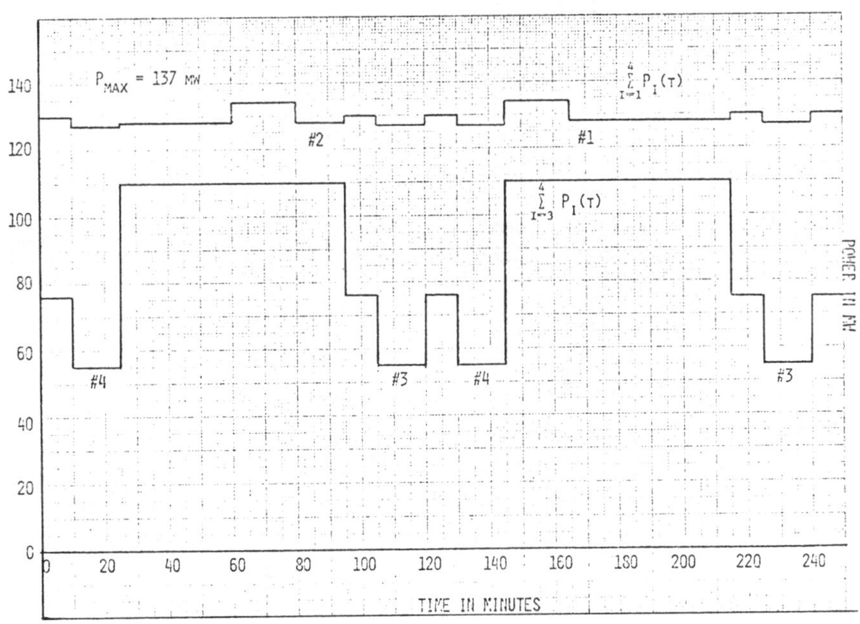

Figure 5 A typical result using dynamic programming.

Table 1 Parameter values for the two furnace types

If $T_0 = 0$

VARIABLE UNITS	I -	E_I^M MWH	E_I^R MWH	T_M MIN.	T_R MIN.	T_B MIN.	P_{IMAX}^M MW	P_{IMAX}^R MW
110 TON FURNACES	1,2	63	65	105	115	130	36	12
150 TON FURNACES	3,4	87	90.5	95	105	120	55	21

$\Delta = \begin{cases} 0 \\ 0{,}15 \end{cases}$ MWH (CHOOSEN $\neq 0$ SO THAT DURING TAPPING AND PREPARATION THE STATE OF THE FURNACE DETERMINES UNIQUELY THE TIME).

The following summary of results shows how effective the two proposed solution approaches are. For $P_{max} = 150$ MW an energy utilisation factor of .956 (.936) and hourly production of 240 (225) tonsare obtained by the graphical method (dynamic programming). For $P_{max} = 137$ MW the corresponding figures are .975 (.935) and 223 (210) tons,respectively.

5. CONCLUSIONS

To conclude we make the following remarks.

- Using a 4 hour period for scheduling is reasonable since there are unpredictable delays such as broken electrodes, mechanical breakdowns, etc. which preclude scheduling over a longer time.
- The techniques proposed in this paper allow the rapid estimation of the level of contract demand which should be negociated with the utility company.
- Scheduling the furnaces using either the graphical approach or the dynamic programming approach for P_{max} not exceeding 150 MW shows that it would be possible to use almost all of the available energy.
- An extension of this work would be to incorporate a model of the casting machines to permit an evaluation of the delays they may introduce.

ACKNOWLEDGEMENTS

This work was supported in part by Natural Sciences and Engineering Research Council Canada Grants Nos. A-8670 and A-8816.

REFERENCE

(1) P. Mérette, H.H. Hoang & M.P. Polis, "Efficient Use of Energy in Steel Making", Proceedings of 1978 Summer Computer Simulation Conference, Newport Beach, California, July 24-26, 1978, pp. 772-777.

STEADY STATE NETWORK EQUIVALENTS FOR POWER SYSTEMS

E. C. Housos, G. Irisarri,
R. M. Porter, A. M. Sasson

American Electric Power Service Corporation
2 Broadway, New York, New York 10004

ABSTRACT

Steady State equivalents are of importance for the study of the static characteristics of a large power system when either the computer facilities are limited or the solution time must be decreased. Several equivalents are examined and compared. Emphasis is placed on the REI (Radial Equivalent Independent) equivalencing technique and some practical considerations with regard to its implementation will be discussed. Test load flow results will be presented in order to assess the accuracy of the examined equivalents. The major conclusion of the paper is that, as long as there exists some reactive power support from the equivalent, it will perform satisfactorily.

INTRODUCTION

Steady state equivalents are of great importance when there is an interest in studying a small part of a large system. In such a case it would be desirable to equivalize the system around the area of interest and experiment with the reduced system instead of the large system. Another reason for using equivalents exists when either the computer facilities are limited or the solution time must be decreased (e.g. interactive load flow, real-time contingency analysis).

Several types of equivalencing techniques have been considered to date. Since all equivalents are calculated from a solved base case load flow they are exact only for the base case. In addition, there are other characteristics of equiva-

D. G. Lainiotis and N. S. Tzannes (eds.), Applications of Information and Control Systems, 169-181.

lents such as, reactive power support, accuracy, sparsity considerations, conceptual simplicity, etc., which characterize some of them as better than others.

One of the most widely used equivalencing methods is the one due to Ward (1). The major problem with the Ward equivalent is that it does not allow for reactive power support in the equivalized area. This can be explained by noticing that although the real power is always specified for every bus but the stack bus, the reactive power is not specified for every bus and may vary. If the equivalent assumes that the reactive power at the regulated buses will remain at its base case value, even under outages the results are usually unacceptable. For the Ward equivalent once the equivalent has been formed there is no way of producing more or less reactive power in the equivalized area if the outaged case so requires. Therefore it is expected that the behavior of an equivalent network with reactive power support capabilities would be closer to the behaviour of the unreduced network. Several equivalencing techniques such as REI (5). Ward equivalent with buffer (2) and Extended Ward equivalent (3) satisfy the above requirement.

Experiments were conducted with the REI equivalent and the Ward equivalent with a buffer zone and although the REI type equivalent performed satisfactorily, the final conclusion was that the Ward Equivalent with a carefully selected buffer zone is a very good equivalencing method. This equivalent is the easiest to implement and understand and thus it has been recommended for future usage at AEP.

GENERAL PROBLEM

The general problem can be formally stated as follows. Given a solved load flow model of an interconnected power system PS and an area of interest AI within PS find a new equivalent load flow model PE that has a smaller number of buses and branches than PS. PE should be such that for changes of the operating conditions within AI the results of PE are close to the results of PS. If $AE = \{PS\} - \{AI\}$ then schematically the steady state equivalent problem is as shown in figure 1.

If PS were a linear system then the problem posed would be easily and exactly solved. This can be seen from the following matrix manipulations. Since it was assumed that PS can be described by a set of linear equations and without loss of generality these could be nodal equations then

$$\begin{array}{c} \text{AE} \\ \text{AI} \end{array} \begin{bmatrix} Y_{11} & Y_{12} \\ Y_{21} & Y_{22} \end{bmatrix} \begin{bmatrix} \underline{v}_1 \\ \underline{v}_2 \end{bmatrix} = \begin{bmatrix} \underline{I}_1 \\ \underline{I}_2 \end{bmatrix} \qquad (1)$$

solving for $\underline{v}_1$

$$\underline{v}_1 = Y_{11}^{-1} * Y_{12} * \underline{v}_2 + Y_{11}^{-1} * \underline{I}_1 \qquad (2)$$

substituting into (1)

$$Y_{22\ EQ} * \underline{v}_2 = \underline{I}_{2\ EQ} \qquad (3)$$

where

$$Y_{22\ EQ} = Y_{22} - Y_{21} * Y_{11}^{-1} * Y_{12}$$

$$\underline{I}_{2\ EQ} = \underline{I}_2 - Y_{21} * Y_{11}^{-1} * \underline{I}_1$$

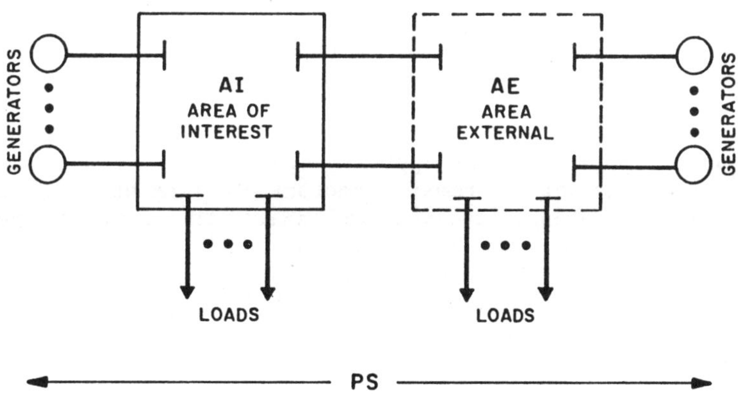

FIGURE I

According to the previous definition of PE the set of equations (3) is a description of the equivalized system and it contains the same number of buses as AI. If any elements of AI are altered the solution of PE is exactly the same with the solution of PS because AE does not change whenever AI changes. This is very critical because as it will be seen later in the paper this does not hold for non-linear systems.

For power systems the equations that govern the steady state behaviour are non-linear. At first sight it appears that there is no way of getting a smaller system PE to perform as the original system PS when AI changes. Indeed there is no analytic way that would produce a system PE which produces the same answer with PS for any conceivable change in AI. However, there exist many methods capable of producing a PE which gives answers satisfactorily close to PS for a set of reasonable changes (line outages, generator outages, etc.) in AI. The next few sections attempt to describe and qualify some of the known equivalencing techniques.

WARD EQUIVALENT WITHOUT BUFFER (1)

This is the most used equivalencing technique. Using the voltage solution of PS, the power injections are transformed either to impedances or to currents. Then the system is treated as a linear system and an equivalent system PE is constructed which consists only of the buses belonging to AI. The main disadvantage of this equivalent is the fact that there is no reactive power support from the external system. In order to construct the equivalent the base case injection conditions were used. The real and reactive injections as it is well known from load flow theory are fixed for load buses but the reactive power injection of a generator is allowed to vary within limits in order to keep the voltage at that bus constant. As it will be shown, this is the main source of error for all load flow equivalents. That is, the approximation involved in assuming that the voltages of AE remain constant is more or less valid but the approximation involved in fixing the reactive power generation of a generator bus to its base case generation value is not always valid. It is often the case that large amounts of reactive power must be imported into AI from AE as it will be shown from the experimental results.

WARD EQUIVALENT WITH BUFFER (2)

As mentioned in the previous section, the main problem with Ward equivalent without a buffer zone is that no reactive power can be imported from AE if the outaged case so requires. If instead of eliminating all buses of AE a few generators which are capable of producing large amounts of reactive power whenever needed were kept the quality of the equivalent should improve. The problem now is transformed into one of finding a sufficient number of generators to be retained in order to be able to provide enough reactive power support whenever needed. From the experiments performed it seems that a small number of generators with large reactive power capability suffices to achieve an accurate solution.

Some criteria to be used in deciding which of the generators of AE should be kept are the following:

1) Keep the generators with maximum reactive power capability. That is, keep the generators for which the difference between the base case reactive power production and the reactive limits is the largest.

2) Eliminate all the generators which do not produce enough real or reactive power to cover their own local requirements.

3) The number of generators to be kept should be a small percentage of the total number of buses in AE.

EXTENDED WARD (3)

In this equivalencing technique (originally developed for real-time applications (3,4)) a different way of providing reactive power support is considered. It involves the creation of a number of new generators equal to the number of boundary load buses. That is, one new fictitious generator is connected to each boundary load bus in such a way as to not affect the base case solution of the reduced system. The new generators do not produce any real power but are capable of producing reactive power whenever the internal system so requires. For multisystem planning applications the number of additional generator buses would in general be large. For this reason this scheme was not considered for the simulations presented in this paper.

REI EQUIVALENT (5,6,7)

Another equivalent that has been suggested is the REI (Radial Equivalent Independent) equivalencing technique. In this case, AE is transformed into a passive network by aggregating all the generation of AE in one new bus and all the load of AE in another new bus so that the rest of AE becomes passive. Schematically the REI equivalent inserts a lossless network N_{REI} between the eliminated buses and their aggregated injections as shown in figure 2, where NL is the number of load buses and NG is the number of generator buses. The REI equivalent (N_{REI}) is formed as a function of the injected power of the buses included in the equivalent and the base voltage of these buses. The parameters involved are shown in figure 3.

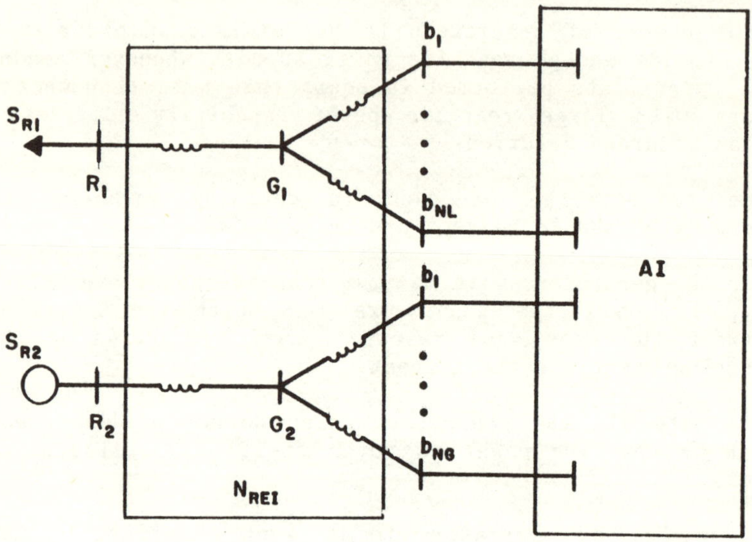

FIGURE 2

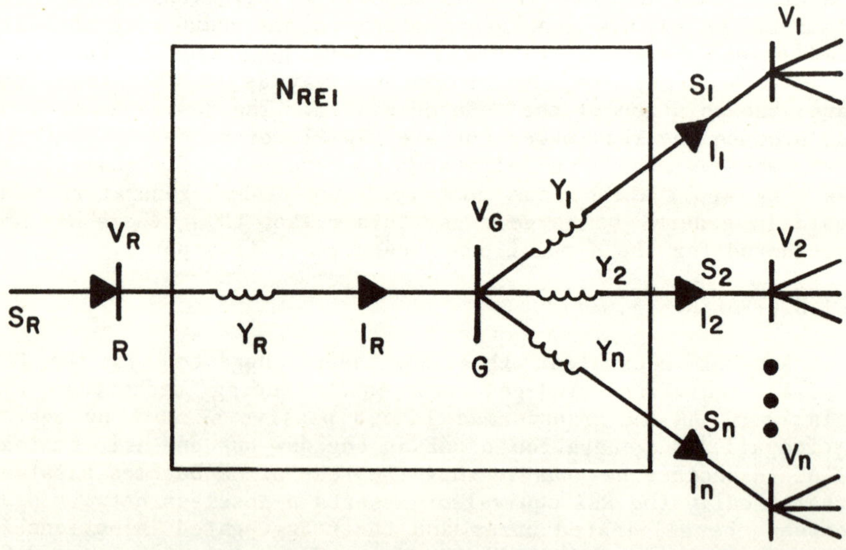

FIGURE 3

From figure 3 and letting $V_G=0$, then

$$S_R=\sum_i S_i \qquad I_R=\sum_i I_i \qquad v_R=\frac{S_R}{I_R^*}$$

$$Y_R=\frac{S_R^*}{|v_R^2|} \qquad Y_i=\frac{-S_i^*}{|v_i^2|}$$

Notice that the network N_{REI} is a lossless network for the base case conditions. In fact, by construction the power injected into R is $S_R = \sum_i S_i$ and the power injection into the ith bus of the REI equivalent is equal to S_i. This can be seen also from the previous figure. Thus all the power that comes into N_{REI} comes out of N_{REI} as well. Hence, N_{REI} is a lossless network.

Improvements of the REI Equivalent

The REI equivalent as suggested by P. Dimo and his colleagues (5) and exemplified by W. Tinney (7) did not give satisfacory results when implemented at AEP. Several problems occurred but the most important was the one concerning the voltage magnitude of the newly created REI buses. As is known it is essential to have the voltage magnitudes of all the buses and in particular the voltage controlled buses close to 1 p.u. to achieve load flow convergence. As proposed the equivalent did not make any provisions to avoid this low voltage condition at the newly created buses. Even if all the generators are grouped together and all the load buses are grouped together a low voltage condition may still exist at the REI buses. The above considerations will be further illustrated with the use of an example.

Example 1

Given $v_i = 1.0$ for all i

Then $v_R = 1.0$ for all values of the S_i's

Example 2

Given $v_1 = 0.95 \qquad v_2 = 1.0 \qquad v_3 = 1.0$

$S_1 = 5.0 \qquad S_2 = -2.0 \qquad S_3 = -1.0$

Then $v_R = 0.88$ low!

Example 3

Given $v_1 = 0.95$ $v_2 = 1.0$ $v_3 = 1.0$

$S_1 = 50.0$ $S_2 = -40.0$ $S_3 = -5.0$

Then $v_R = 0.655$ very low!

From the above examples it can be seen that care should be exercised in forming the equivalent. It should be avoided incorporating in the same equivalent both negative and positive values of the net injected power. This is in part avoided if two REI equivalents are formed for each equivalized area, one for the PV and one for PQ buses. However, it was found experimentally that this does not necessarily guarantee that the REI voltages will not be low. Of course, restrictions of the kind referred to above do not necessarily conform with possible sparsity considerations that could be used in forming the equivalent (8).

To resolve the above problem various possibilities of node groupings were considered. All these schemes will be described next and the better ones will be highlighted.

1) REI1
Group all the PV buses together in the PV-REI bus and group all the PQ buses together in the PQ-REI bus.

2) REI2
Same as REI1 but instead of using the base case voltages use the "flat start" voltages for all buses.

3) REI3
Same as REI1 but if the real net power of a PV bus is less than zero put the net real power of that bus in the PQ-REI bus.

4) REI4
Same as REI3 but instead of putting the net real power in the PQ-REI bus put the real load in the PQ-REI bus and the real generation in the PV-REI bus.

5) REI5
Same as REI4 but put the real generation of a PV bus satisfying all the requirements of REI4 in the PQ-REI bus if it is negative.

6) REI6
Same as REI5 but if a PV bus has reached one of its reactive power limits, treat is as a PQ bus.

7) REI7
Same as REI3 but if a PV bus has reached one of its reactive power limits treat it as a PQ bus and thus assign it to the PQ-REI bus.

For the experiments performed it was clear that REI6 and REI7 were better. The voltage of the REI buses was always close to 1 p.u. for both REI6 and REI7.

Another parameter that entered into consideration at this point was the total real and reactive power losses encountered in the reduced system compared to the unreduced system. Since the total injected power is less for REI7 than for REI6 the total losses encountered when REI7 is used should be less than the total losses of REI6. This was also confirmed experimentally.

Another justification for using REI7 could be given intuitively as follows. Since the line impedances around the REI buses have abnormal values it would seem reasonable to assume that the smaller the flow on these lines the more accurate the results would be. This is precisely the difference between REI6 and REI7 and for that reason REI7 has been chosen as the best procedure for forming an REI equivalent of a given area.

While this paper was being written reference (9) was brought to the authors' attention by Dr. P. Dimo. In (11) it is shown that a criterion for grouping buses so as to guarantee adequate voltage values for the REI nodes is to only aggregate generator buses with current angle of a similar value. It was suggested there that only those generator buses whose current angle is within 10° of each other be grouped together. It can be shown that the criteria used in both REI6 and REI7 approximately satisfy the above requirement.

The above considerations can also be interpreted as a relaxing of the colinearity condition given by F.F. Wu and N. Narasimhamurthi in (8).

Limits on the PV-REI Buses

The limits on the reactive power produced by the PV-REI buses were set equal to the sum of the limits of the PV buses incorporated in that REI equivalent. This is a conservative limit and it means that the REI bus will hit one of its reactive power limits only if all the generators of the unreduced system hit their limits. This means that using the equivalent it is more difficult to simulate the precise behaviour of the generators once these are lumped into one generator. The experimental results show that although the individual generators lose their identity the REI bus is a source for reactive power support and the results were acceptable. Since the main problem with Ward equivalents without a buffer zone is the fact that there is no reactive power support the REI equivalent overcomes this problem.

Since the PV-REI bus provides the reactive power needed there is no need for using a PQ-REI bus as well. This means that the load buses and all generators on their limits could be eliminated. This was done in the reported experiment.

COMPUTATIONAL RESULTS

In order to test the equivalents under realistic conditions a very large load flow case of the Eastern United States interconnected system was used. The case originally had 3561 buses and 6003 lines. About 2500 buses were eliminated using the different equivalencing techniques. Two critical transmission lines were outaged in the internal system and the limits of the generators in the internal system were reduced in order to force the equivalized part of the system to produce all the reactive power support needed. For the test system, there was sufficient internal reactive support considering normal reactive limits. Under these conditions the Ward equivalent performs well as the outages require no external reaction. Therefore, it was not enough to just outage two lines connected to a boundary bus in order to force a large import of reactive power but it was also necessary to reduce the limits of all generators in the internal system. This is a very important point because from the simulations it was concluded that most equivalents behave satisfactorily if the outages are not such as to require a significant external reaction.

It is common practice when testing equivalencing methods to outage tie lines between AE and AI; however, since the worst test for an equivalent is when it is forced to import large amounts of reactive power from the equivalized area, it is not at all clear that outaging a tie line will necessarily affect the equivalent significantly. It may happen that by taking a tie-line out of service the internal system is forced to satisfy all its needs with its own generators.

Table I defines the different cases tested and gives the number of buses and lines of every case. The results of all the previously described experiments are given in Table II.

The main conclusion that can be drawn is that any equivalent which can provide some reactive power support will perform satisfactorily. If we take into consideration that conceptual difficulty of the equivalent, its computational economy and sparsity, it seems that a Ward equivalent with an appropriate buffer zone will perform satisfactorily even under very stringent conditions and thus it is the suggested equivalent to use.

TABLE 1

	BUSES	LINES	DESCRIPTION
UNREDUCED	3561	6003	INITIAL BASE CASE.
WARD	1013	2464	WARD EQUIVALENT WITHOUT BUFFER.
WARD 1	1178	2882	WARD EQUIVALENT WITH BUFFER CONTAINING ALL THE GENERATORS OF ALL NEIGHBORING SYSTEMS.
WARD 2	1112	2679	SAME AS WARD 1 BUT ALL GENERATORS THAT HAVE REACHED ONE OF THEIR REACTIVE LIMITS ARE ELIMINATED.
WARD 3	1056	2474	SAME AS WARD 2 BUT EVEN MORE GENERATORS ARE ELIMINATED.
WARD 4	1028	2372	SAME AS WARD 1 BUT ONLY ONE GENERATOR PER NEIGHBORING SYSTEM IS RETAINED.
WARD 2/REI	1058	2520	SAME AS WARD 2 BUT THE GENERATORS OF EACH NEIGHBORING SYSTEM ARE AGGREGATED INTO A REI GENERATOR.

CONCLUSION

The most important conclusion is that all equivalents that can provide reactive power support perform satisfactorily. In particular the Ward equivalent with a buffer zone is the easiest to implement and understand and it is suggested that this equivalencing technique be used in the future.

More work and experiments on automatic ways of selecting generators to be retained would be of importance. Although there seem to exist several degrees of freedom in choosing these generators it would be beneficial to find the relative importance of the trade-offs involved.

TABLE II LINE FLOWS IN THE VICINITY OF THE OUTAGES

LINE FLOWS		UNREDUCED		WARD		WARD 1		WARD 2		WARD 3		WARD 4		WARD 2/REI	
FROM	TO	MW	MVAR	MW	MVAR	MW	MVAR	MW	MVAR	MW	MVAR	MW	MVAR	MW	MVAR
1523	1522	-724.5	-4.2	-724.6	28.7	-727.6	-3.5	-728.2	-2.7	728.6	-2.6	-728.1	-7.4	-730.0	-1.0
	1528	84.4	-246.9	81.9	-255.4	83.2	-249.0	83.3	-248.4	83.4	-248.5	84.1	-252.7	82.7	-246.4
	1580	0.0	0.0	0.0	0.0	0.0	0.0	0.0	0.0	0.0	0.0	0.0	0.0	0.0	0.0
	1605	0.0	0.0	0.0	0.0	0.0	0.0	0.0	0.0	0.0	0.0	0.0	0.0	0.0	0.0
	3237	640.2	-452.1	642.7	-405.2	644.4	-450.4	644.8	-451.7	645.2	-451.6	644.0	-440.2	647.3	-455.8
1522	1472	240.8	31.7	240.0	26.0	242.1	30.8	241.5	31.7	241.5	31.6	241.3	30.6	241.4	32.0
	1523	726.3	-289.4	726.4	-258.7	729.4	-289.6	730.0	-290.2	730.4	-290.2	729.9	-284.4	731.8	-291.9
	1600	-967.1	-18.7	-966.3	-39.3	-971.5	-17.5	-971.4	-17.7	-971.9	-17.5	-971.2	-21.6	-973.2	-16.4
1512	3239	336.2	-20.0	338.3	-3.7	336.8	-20.9	336.8	-20.3	336.9	-20.4	337.3	-17.3	336.4	-20.9
	3256	205.6	-43.4	207.0	-30.4	206.0	-44.0	206.0	-43.9	206.0	-43.8	206.3	-40.9	205.7	-44.6

* TIE LINE
\$ OUTAGED LINE

REFERENCES

(1) Ward J. B., "Equivalent Circuits for Power Flow Studies", AIEE trans., Vol. 68, pp. 373-382, 1949.

(2) H. Duran and N. Arvanitidis, "Simplification for Area Security Analysis: A New Look at Equivalencing", IEEE TPAS, Vol. PAS-91, pp. 670-679, 1972.

(3) Monticelli A., Deckman S., Garcia A., and Stott B., "Real-Time External Equivalents for Static Security Analysis," IEEE TPAS, Vol. PAS-98, pp. 498-508.

(4) Dopazo J.F., Irisarri G., Sasson A.M., "Real-Time External System Equivalent for On-Line Contingency Analysis," F78 674-4 presented at the 1978 IEEE PES Summer Meeting Los Angeles, July 1978.

(5) Dimo P., Groza L., Ionescu S., Ungureanu B., and Petcu I., "Research Concerning the Generalized Utilization of a Single Equivalent to the Power System State Analysis," Computer Methods of Power System Analysis and Control, Joint Research Seminar between Romania and USA, Bucharest, Romania, June 1974.

(6) Dimo P., Nodal Analysis of Power Systems, Abacus Press, England, 1975.

(7) Tinney, W. F. and Powell, W. L., "The REI Approach to Power Network Equivalents," PICA Conference, Toronto, May 1977.

(8) F. F. Wu and N. Narasimhamurthi, "Necessary Conditions for REI Reduction to be exact", A79 065-4, IEEE PES Winter Meeting, New York, Jan. 1979.

(9) P. Dimo, "La Structure Hierarchique des Modeles REI pour les Reseaux Vastes. Criteres pour leur Construction", Rev. Poum. Sci. Techn.-Electrotechn. et Energ. 23, 4, p. 511-521, Bucarest, 1978.

NUMERICAL DETERMINATION OF SYNCHRONOUS MACHINE TRANSIENT ERRORS WITH REPRESENTATIVE EXCITATION

Demetrios P. Papadopoulos

Public Power Corporation of Greece

ABSTRACT

This paper presents a numerical program which determines for a synchronous machine supplying power to a large system via a step-up transformer and a transmission line: the operating point, the linearized model parameters (K_1 to K_6) and the transient errors of the system variables by solving the state-equations subject to input perturbations for constant excitation and IEEE standardized Type-1 exciter with rate feedback and power system stabilizer. The program can handle other types of exciters and permits parameter sensitivity analysis to seek better system transient error performance. It is applied to a 160 MVA machine operating in the above mode.

LIST OF SYMBOLS

E_{FD}	generator field voltage
H	inertia constant
K_A, τ_A	regulator gain and time constant
K_E, τ_E, S_E	self-excited field constant, time constant and saturation function of exciter
K_F, τ_F	regulator rate feedback gain and time constant
K_R, τ_R	input transducer and filter constant and time constant
K_o	speed gain
K_1 to K_6	parameters of linearized machine model
P, Q	real and reactive power
R_e, X_e	equivalent resistance and reactance of external system
T_a, T_e, T_m	accelerating, electrical and mechanical torque
V_e, V_R	regulator error and output voltage

D. G. Lainiotis and N. S. Tzannes (eds.), Applications of Information and Control Systems, 182-189.

$V_{REF.}$	reference voltage
V_t	machine terminal voltage
δ	angle between quadrature axis and infinite bus
τ'_{do}	d-axis transient open circuit time constant
τ_o	reset time lag constant of PSS
$\tau_1,\tau_3;\tau_2,\tau_4$	lead and lag constants of PSS
ω	per unit speed deviation from synchronous ($\dot{\delta}$)
ω_B	base system radian frequency
o subscript:	steady-state or operating point value
Δ subscript:	small change of state variable
prefix :	perturbation of state-equation inputs

1. INTRODUCTION

The continuous developments and improvements with respect to size and technology of synchronous machines have analogous impacts on their excitation systems (1). It is a known fact that the excitation systems of machines and especially those of modern technology (2,3,4) play an important role in the safe, stable and economic operation of power systems as far as normal operating conditions (i.e. small system disturbances or impacts) pertinent to dynamic stability, and abnormal operating conditions (5) (i.e. large system disturbances or impacts) pertinent to transient stability are concerned.

To check or improve various designs of such systems through simulation and analysis, one needs appropriate model representation (6,7)and also analog and/or numerical tools to carry out the desired analysis. Under normal operating conditions a power system experiences almost continuously small disturbances due to modetate load variations and minor scheduled or not system configuration changes. These disturbances set-up transient error oscillations about the respective operating point, which if not damped fast enough can either lead to instability or cause over long periods of time fatigue on system components and thus reduce substantially their useful life and also the reliability and availability of the system. The relevant technical literature shows that for many years this area has been explored extensively. Recent works (8,9,10) have developed insights of various controls with compensating effects to offset reductions in stability margins and damp effectively the oscillations resulting from small perturbations.

The present work offers a computing capability for the determination of the system transient errors caused by input perturbations. Such analytical computations are important to determine the effects of excitation control on the damping of oscillations, which characterize the dynamic stability. The program computes: the operating point, the linearized machine model parameters (K_1 to K_6) and their dependence on machine loading (P and Q) and on external system equivalent parameters R_e and X_e, the eigenvalues of the system for stability reasons, and the system transient errors by solving the linearized system state-equation. It is possible to incorporate in the program other types of excitation systems and investigate their suitability in terms of overall system transient error performance. Finally, the program offers the possibility to conduct parametric sensitivity analysis on a simulated system, to test and improve existing excitation designs or develop new ones with more desirable features. The program is applied to a 160 MVA machine with constant excitation and with Type-1 exciter having rate feedback and power system stabilizer signals.

2. PHYSICAL SYSTEM AND RELATED MATHEMATICAL MODEL

The physical system consists of a machine with excitation system delivering power to a large power network via a step-up transformer and a transmission line. The linearized version of the system in block diagram form is shown in Fig. 1. The basic assumptions necessary for the development of the linearized model are discussed in References 7,8,9. The linearized model can predict the behaviour of the system about its operating point for small variations of the input variables, which correspond to small impacts of the power system. It should be noted that when switches SW_1, SW_2/SW_2^- and SW_3/SW_3^- are open Fig. 1 depicts the linearized machine model with constant excitation. On the other hand, when the above switches are closed Fig. 1 depicts the linearized machine model with representative excitation system having rate feedback and power system stabilizer signals. These cases are studied seperately and the transient errors of the system variables are analyzed.

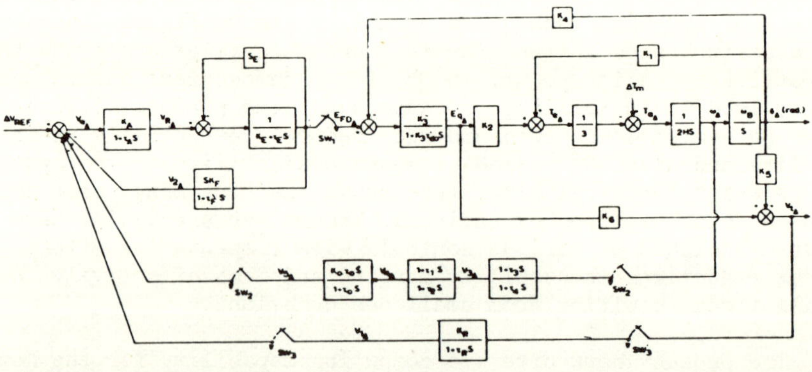

Fig. 1 Block diagram representation of system

The state-equation model that describes the system of Fig. 1 is of the general form $\dot{\underline{X}}_\Delta=\underline{A}\underline{X}_\Delta+\underline{B}\Delta\underline{u}$; for the case of linearized machine with representative excitation system it is given by state-equation

$$\begin{bmatrix} \dot{E}'_{q\Delta} \\ \dot{\omega}_\Delta \\ \dot{\delta}_\Delta \\ \dot{V}_{1\Delta} \\ \dot{V}_{2\Delta} \\ \dot{V}_{3\Delta} \\ \dot{V}_{4\Delta} \\ \dot{V}_{5\Delta} \\ \dot{V}_{R\Delta} \\ \dot{E}_{FD\Delta} \end{bmatrix} = \begin{bmatrix} -\frac{1}{K_3\tau'_{do}} & 0 & -\frac{K_4}{\tau'_{do}} & 0 & 0 & 0 & 0 & 0 & 0 & \frac{1}{\tau'_{do}} \\ -\frac{K_2}{6H} & 0 & -\frac{K_1}{6H} & 0 & 0 & 0 & 0 & 0 & 0 & 0 \\ 0 & \omega_B & 0 & 0 & 0 & 0 & 0 & 0 & 0 & 0 \\ \frac{K_6K_R}{\tau_R} & 0 & \frac{K_5K_R}{\tau_R} & -\frac{1}{\tau_R} & 0 & 0 & 0 & 0 & 0 & 0 \\ 0 & 0 & 0 & 0 & -\frac{1}{\tau_F} & 0 & 0 & 0 & \frac{K_F}{\tau_F\tau_E} & -\frac{K_F(S_E+K_E)}{\tau_F\tau_E} \\ -\frac{K_2\tau_3}{6H\tau_4} & \frac{1}{\tau_4} & -\frac{K_1\tau_3}{6H\tau_4} & 0 & 0 & -\frac{1}{\tau_4} & 0 & 0 & 0 & 0 \\ -\frac{K_2\tau_1\tau_3}{6H\tau_2\tau_4} & \frac{\tau_1}{\tau_2\tau_4} & -\frac{K_1\tau_1\tau_3}{6H\tau_2\tau_4} & 0 & 0 & \frac{\tau_4-\tau_1}{\tau_2\tau_4} & -\frac{1}{\tau_2} & 0 & 0 & 0 \\ -\frac{K_0K_2\tau_1\tau_3}{6H\tau_2\tau_4} & \frac{K_0\tau_1}{\tau_2\tau_4} & -\frac{K_0K_1\tau_1\tau_3}{6H\tau_2\tau_4} & 0 & 0 & \frac{K_0(\tau_4-\tau_1)}{\tau_2\tau_4} & -\frac{K_0}{\tau_2} & -\frac{1}{\tau_0} & 0 & 0 \\ 0 & 0 & 0 & -\frac{K_A}{\tau_A} & -\frac{K_A}{\tau_A} & 0 & 0 & \frac{K_A}{\tau_A} & -\frac{1}{\tau_A} & 0 \\ 0 & 0 & 0 & 0 & 0 & 0 & 0 & 0 & \frac{1}{\tau_E} & -\frac{(S_E+K_E)}{\tau_E} \end{bmatrix} \begin{bmatrix} E'_{q\Delta} \\ \omega_\Delta \\ \delta_\Delta \\ V_{1\Delta} \\ V_{2\Delta} \\ V_{3\Delta} \\ V_{4\Delta} \\ V_{5\Delta} \\ V_{R\Delta} \\ E_{FD\Delta} \end{bmatrix} + \begin{bmatrix} 0 & 0 \\ 0 & \frac{1}{2H} \\ 0 & 0 \\ 0 & 0 \\ 0 & 0 \\ 0 & \frac{\tau_3}{2H\tau_4} \\ 0 & \frac{\tau_1\tau_3}{2H\tau_2\tau_4} \\ 0 & \frac{K_0\tau_1\tau_3}{2H\tau_2\tau_4} \\ \frac{K_A}{\tau_A} & 0 \\ 0 & 0 \end{bmatrix} \begin{bmatrix} \Delta V_{REF} \\ \Delta T_M \end{bmatrix} \qquad (1)$$

system (1). The analytical effort aims to study the behaviour of the transient errors of the system variables caused by input perturbations for the two excitation cases.

3. COMPUTER PROGRAM

The developed computer program is written in Fortran IV language for the H-6050 Honeywell computer of the Greek Public Power Corporation. A simplified version of the program flow diagram is shown in Fig. 2.

In the following an attempt is made to give a qualitative description of the functional operation of the program. In block No 1 the program reads all required input data, i.e. of machine, excitation system, step-up transformer and transmission line. In block No 2 the program calculates, through a subroutine called SMLIM, the linearized machine model parameters K_1 to K_6 for a specified machine loading condition, and the corresponding operating point using the mathematical expressions given in Reference 8 or 9. In block No 3 the program forms the matrices $\underline{A}$ and $\underline{B}$ of the state-equation system, whereas in block No 4 a subroutine called EIGEN is used to determine the eigenvalues of matrix $\underline{A}$ by solving the system $(\underline{A}-\lambda\underline{I})=\underline{0}$. Block No 5 tests exhaustively the real part of all eigenvalues to see if it is negative. If at least one eigenvalue is found to have positive real part control passes to block No 6 which adjusts selectively certain system parameters to alleviate this undesirable condition by forcing the program to make a new complete cycle (blocks 2 through 5). If all eigenvalues are found to have negative real part program control goes to block No 7. This block uses the subroutine called RKAM to solve for specified initial conditions the system state-equation $\underline{\dot{X}}_\Delta=\underline{A}\underline{X}_\Delta+\underline{B}\Delta\underline{u}$. The subroutine RKAM requires subroutine DERFUN which defines the set of the "n" simultaneous first-order system differential equations to be solved. Finally in block No 8 the output variables of the program are given.

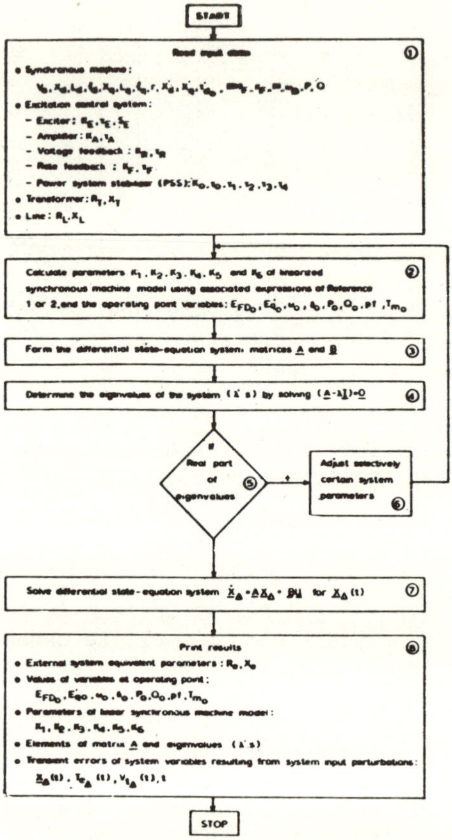

Fig. 2 Flow diagram of computer program

4. APPLICATION OF COMPUTER PROGRAM AND DISCUSSION OF RESULTS

The program was used to study two cases shown in Fig. 1, i.e. (a) the linearized machine model with constant excitation, and (b) the linearized machine model with Type-1 excitation system along with rate feedback and power system stabilizer. The numerical values of the fixed parameters used in the execution of the two cases were those of a typical 160 MVA machine, i.e. V_a=1.0, X_d=1.7, X_q=1.6, X'_d=0.245, r=0.001096, r_F=0.000742, P=1.0, Q=0.6, R_T=0.0134, X_T=0.364, R_L=0.02, X_L=0.036 p.u., τ'_{do}=5.9, H=2.37, τ_E=0.95, τ_A=0.05, τ_R=0.05, τ_F=1.0, τ_o=10, $\tau_1=\tau_3$=0.4401, $\tau_2=\tau_4$=0.0294 sec., ω_B=377 rad/sec., K_E=-0.17, S_E=0.95, K_A=50, K_R=1.0, K_F=0.04, K_o=1.0. The operating point and the parameters of the linearized machine model for a 100% loading were: E_{FDo}=2.6405, E'_{qo}=1.0408, P=1.0, Q=0.6, T_{mo}=1.0 p.u., δ=1.1732 rad., and K_1=1.0889, K_2=1.2692, K_3=0.3072, K_4=1.7301,

K_5=-0.0469, K_6=0.4941. The eigenvalues for case (a) were: λ_1=-0.2104, λ_2=-0.1706+j5.3644, $\lambda_3=\lambda_2^*$, whereas those of case (b) were: λ_1= -0.9385+j1.3423, $\lambda_2=\lambda_1^*$, λ_3=-2.0362, λ_4=-0.1178, λ_5=-34.2599+j2.4041, $\lambda_6=\lambda_5^*$, λ_7=-18.7091+j1.8140, $\lambda_8=\lambda_7^*$, λ_9=-0.2654+j5.3846, $\lambda_{10}=\lambda_9^*$.

Taking advandage of the program capabilities several important aspects of the problem at hand were investigated and various graphical results were constructed. The dependence of the parameters K_1 to K_6 on machine loading was examined. A typical result regarding the parameter K_5, which is the only one that takes on positive and negative values whereas the others are only positive, is shown in Fig. 3. The importance of these six parameters on the dynamic stability of the system and their dependence on machine loading is discussed in Ref. 9. The dependence of the same six parameters on variations of R_e and X_e for fixed Q and variable P machine loading was analyzed. Typical results for parameter K_5 are shown in Fig. 4.

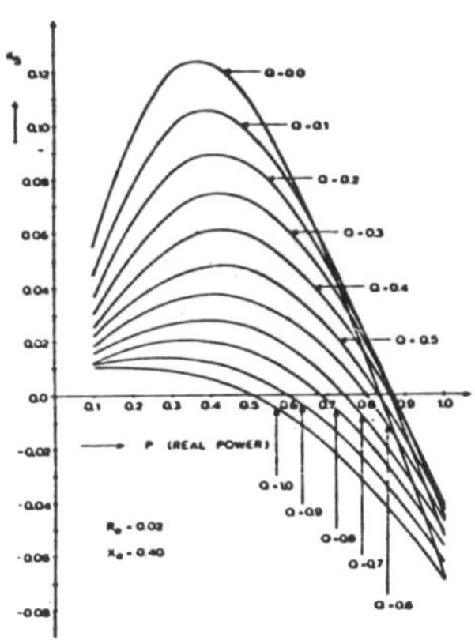

Fig. 3 K_5 vs P and Q

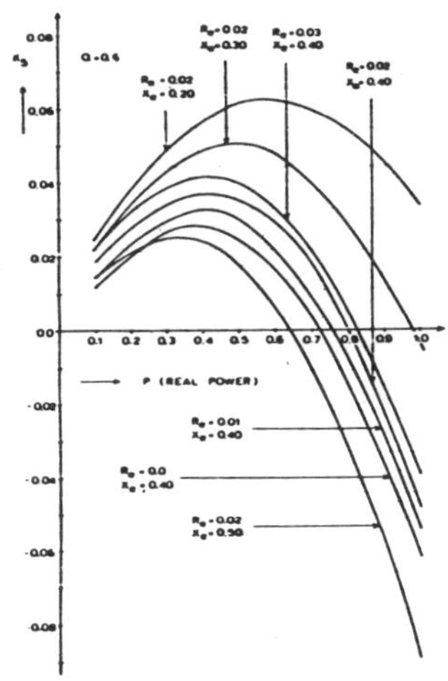

Fig. 4 K_5 vs P, R_e and X_e

The transient errors about the operating point of some system variables subject to the input perturbations ΔE_{FD}=0, ΔT_m=10%; ΔV_{REF}= 0, ΔT_m=10% and ΔV_{REF}=5%, ΔT_m=0, for the two excitation cases are shown in Figs. 5 through 8. The superiority of the dynamic performance of the system with Type-1 exciter is evident.

5. CONCLUSIONS

A computer program has been presented for the determination of transient errors of the system variables about an operating point subject to input perturbations. Through the use of the program the following were determined: the operating point of the machine; the

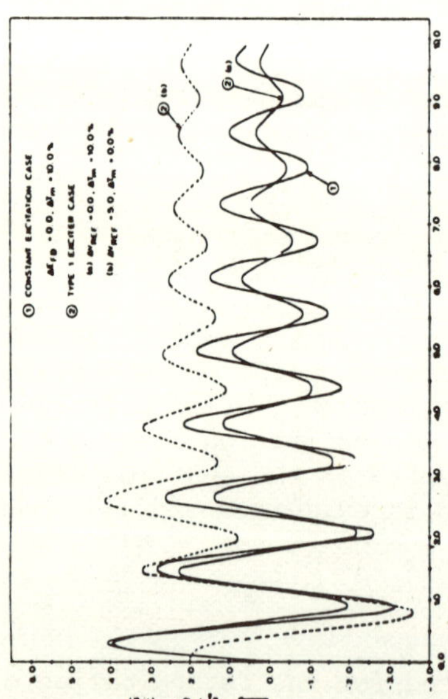

Fig. 5 Transient error of ω_Δ

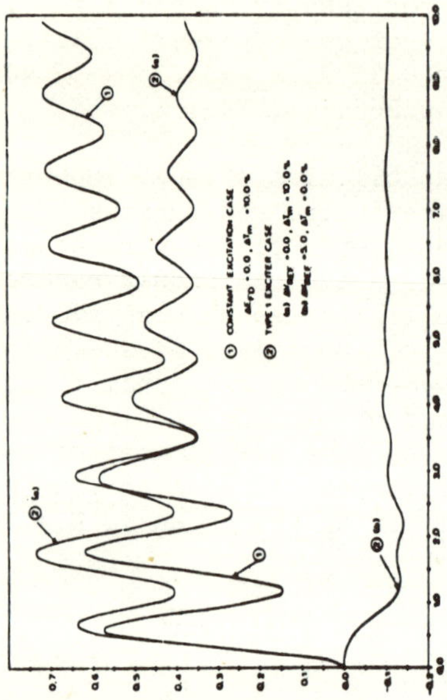

Fig. 6 Transient error of δ_Δ

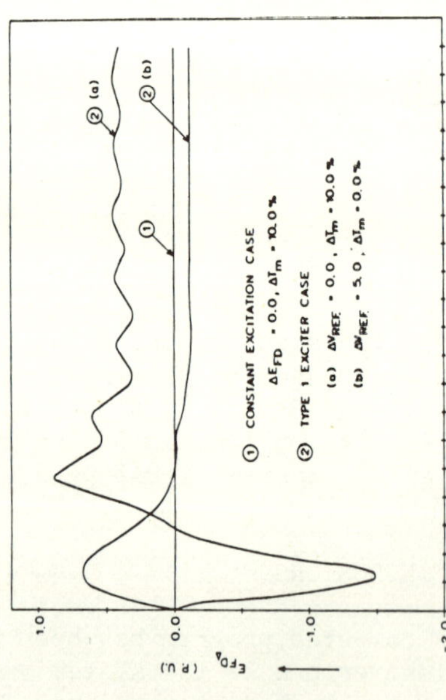

Fig. 7 Transient error of E_{FD_Δ}

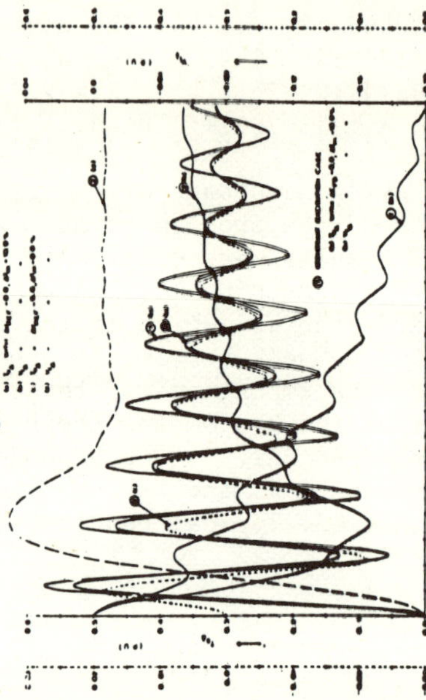

Fig. 8 Transient error of T_{e_Δ} and V_{t_Δ}

linearized parameters (K_1 to K_6) of the machine at the specified operating point and their dependence on various machine loading conditions (P and Q) and on variations of the equivalent external system parameters R_e and X_e; and the system eigenvalues.

This program can also simulate other types of excitation systems and through parameter sensitivity analysis can improve small signal dynamic system performance. In other words, it gives the possibility to analyze and improve certain designs based on criteria of effective damping and minimum steady-state error.

REFERENCES

1 "Review of trends in excitation systems and possible future developments", Fenwick, D.R., Wright, W.F.: 1976, Proc. IEE 123 (5), pp. 413-420.

2 "Improved stability with low time constant rotating exciter", Perry, H.R., Luini, J.F., and Coulter, J.C.: 1971, IEEE Trans. PAS-90, pp. 2084-2089.

3 "A high initial response brushless excitation system", Dillman, T.L., Skooglund, J.W., Keay, F.W., and South, W.H.: 1971, IEEE Trans. PAS-90, pp. 2089-2094.

4 "Design of a power system stabilizer sensing frequency deviation", Keay, F.W., and South, W.H.: 1971, IEEE Trans. PAS-90, pp. 707-713.

5 "Excitation system performance during abnormal conditions", Baldwin, M.S., and Wagner, C.L.: June 1974, Power Engineering, pp. 48-51.

6 "Computer representation of excitation systems", IEEE Committee Report: 1968, IEEE Trans. PAS-87, pp. 1460-1464.

7 "Power excitation control and stability", Anderson, P.M., and Fouad, A.A.: 1977, Iowa State University Press, pp. 208-366.

8 "Concepts of synchronous machine stability as affected by excitation control", Demello, F.P., and Concordia, C.: 1969, IEEE Trans. PAS-88, pp. 316-329.

9 "Dynamic system stability Part I - Investigation of the effect of different loading and excitation systems", El-Sherbiny, M.K., and Mehta, D.M.: 1973, IEEE Trans. PAS-92, pp. 1538-1546.

10 "Concepts of power system dynamic stability", Demello, F.P., and Laskowski, T.F.: 1975, IEEE Trans. PAS-94, pp. 827-833.

FLAT TIE-LINE CONTROL FOR THE GREEK POWER SYSTEM

N.C. Voulgaris and P.D. Charamoglis

Greek Public Power Corporation

An electronic control system has been developed and installed to maintain at a predetermined level the power flow on the 150 KV, 100-MW tie line between the Yugoslav and Greek power systems. This level is maintained by controlling two 100-MW generating units of a hydroelectric power station located about 400 km away from tie line.
The power flow is measured by an analog power transducer, is converted into a digital coded word and is fed serially into a power line carrier channel through an FSK modem. At the hydroelectric station, the digital word is decoded and converted back into an analog signal which is applied to a controller for processing. The analog signal is then compared with front-panel settings of the desired exchange program and "raise-lower" commands are issued to the generating units. In order to insure safe operation of the generating units, their instantaneous power outputs are compared with upper and lower limits set through front-panel switches of the controller unit.
The maximum hourly exchange error achieved is ± 2 MWH.

1. INTRODUCTION

Continuity of service and economy are the main reasons that power generating systems around the world operate interconnected. Power companies in neighboring countries find it useful and desirable to share generation and reserves under normal and emergency conditions. Interchanges of power are made according to a program which is implemented with automatic control equipment (1 - 5).

D. G. Lainiotis and N. S. Tzannes (eds.), Applications of Information and Control Systems, 190-198.

This paper describes the development of an electronic control system which is used to maintain at a programmed level the power flow on the 150-KV, 100-MW tie line between the Yugoslav and Greek power systems. More specifically, the paper describes the development of a) terminal units - a transmitter and a receiver - used for data transmission from the interconnection point to a hydroelectric power control station some 400 km away and b) a flat tie - line controller, located at the hydroelectric station.

The electronic control system described here was installed in the summer of 1977 and has been working ever since. Results taken from actual operating conditions of the system show that the maximum average exchange error achieved is of the order of ± 2 MWH.

2. PRELIMINARY CONSIDERATIONS

When an electric power network operates independently, all load variations and power generation changes influence the system frequency. The governors of the generating units which follow closely the system frequency increase or decrease the generated power output in such a way that the frequency deviation is corrected. As a result of this automatic primary frequency control, the system frequency is maintained within a deviation which is around ± 100 mHZ for the Greek power system.
If such an electric power network is interconnected with a much larger power network, whose frequency is maintained within a much smaller deviation, e.g. ± 10 mHZ from 50 HZ the frequencies of the interconnected power networks are equalized. Consequently,any load variations and power generation changes which occur within the smaller power network do not affect significantly the system frequency and thus, the governors of the generating units will not respond. All power variations, therefore, will be reflected to the tie line which will be acting as a frequency regulator for the smaller power network. Such an operation, however, may not be acceptable by the larger power company and the small power company will be required to provide means to maintain at a programmed level the power flow on the tie line.

The electronic control system described here serves this purpose for the Greek power network in a 150-KV, 100-MW tie line with the Yugoslavian power network.

3. GENERAL SYSTEM DESCRIPTION

A block diagram of the tie line control system is shown in Fig. 1. The actual power flow on the tie line is measured at the Ptolemais substation by a power-to-voltage converter(unit 1)

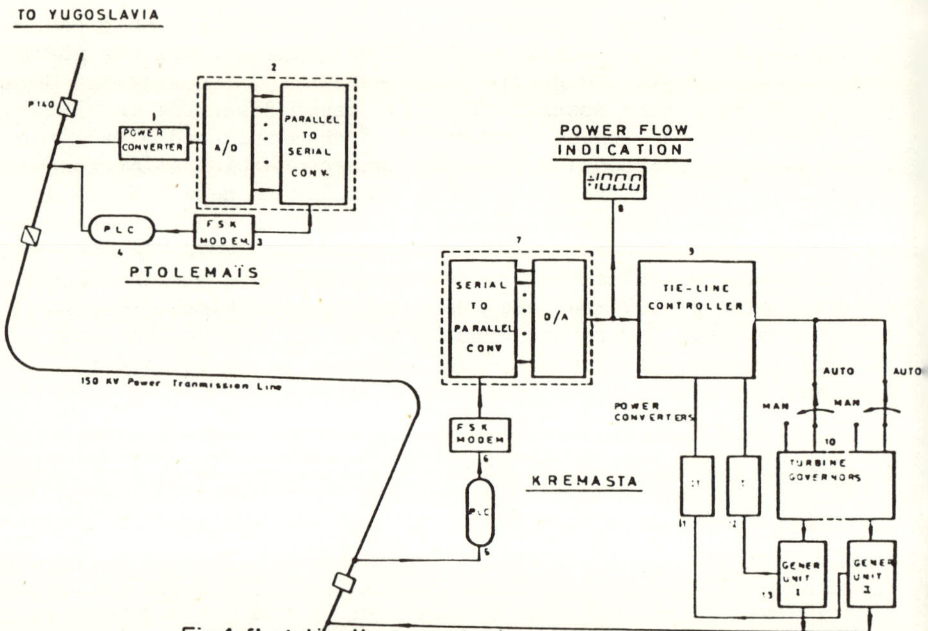

Fig 1. flat tie line control system block diagram

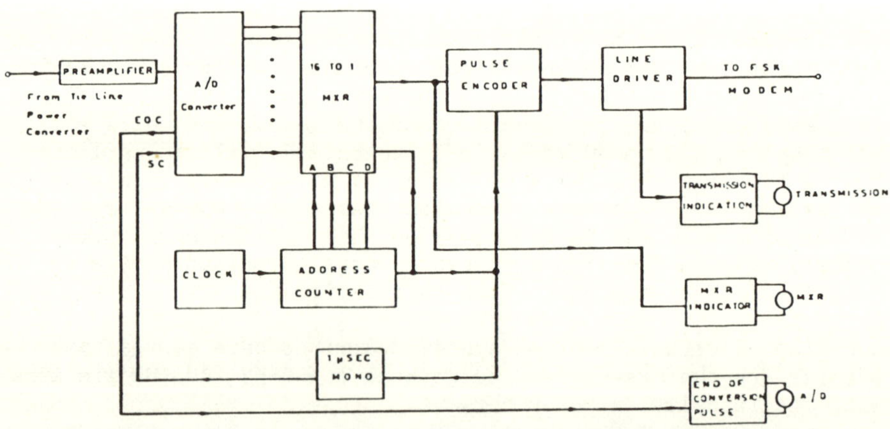

Fig. 2: transmitter block diagram

which gives an output of ± 10V for a ± 100 MW power flow. This analog voltage is converted into a 12-bit digital signal which is applied to a parallel-to-serial converter (unit 2) and then to a 50-baud FSK modem (unit 3). The modem output is fed into a power line carrier (PLC) channel (unit 4).

At the Kremasta hydroelectric station the PLC (unit 5) selects the appropriate signal which is converted back into a digital signal by FSK modem (unit 6). The digital signal is then converted in parallel form and finally into an analog signal (unit 7) which is in the range of ± 10V and is proportional to the actual power flow on the tie line. The analog signal is applied to a flat tie-line controller (unit 9) for further processing and control.

4. DETAILED DESCRIPTION OF SYSTEM UNITS

4.1. Transmitter

A block diagram of the transmitter is shown in Fig. 2. The output of the tie-line power converter is preconditioned in a Preamplifier and is then applied to a 12-bit A/D Converter. In this design a bit indicating the state of the automatic tie-line switch is also transmitted. A 16-to-1 Multiplexer is used as a sync-bit inserter and as a parallel-to-serial converter. The serial pulse waveform is then encoded with clock pulse information in the Pulse Encoder so that the clock pulses can be regenerated in the receiver (6). The Line Driver provides the pulse amplitude required by the FSK Modem used in the PLC channel.

Start-convert pulses for the A/D converter are provided by a 1-μsec monostable multi triggered by the Address Counter. LED indications are provided to monitor the correct operation of several transmitter subsystems, such as End-of Conversion Pulses, Multiplexer and Line Driver outputs.

4.2. Receiver

Figure 3 shows a block diagram of the receiver. The FSK modem output is a digital signal which is applied successively to a Pulse Shaper, a Limiter, two Integrators and Comparators and steering Logic which produces a pulse series identical to that obtained at the multiplexer output of the transmitter. The clock pulses needed for the receiver operation are generated from the incoming digital signal by a Pulse Edge Detector and a Monostable multi, while the Sync Pattern Recognition circuit produces reset pulses to synchronize the Address Counter. The

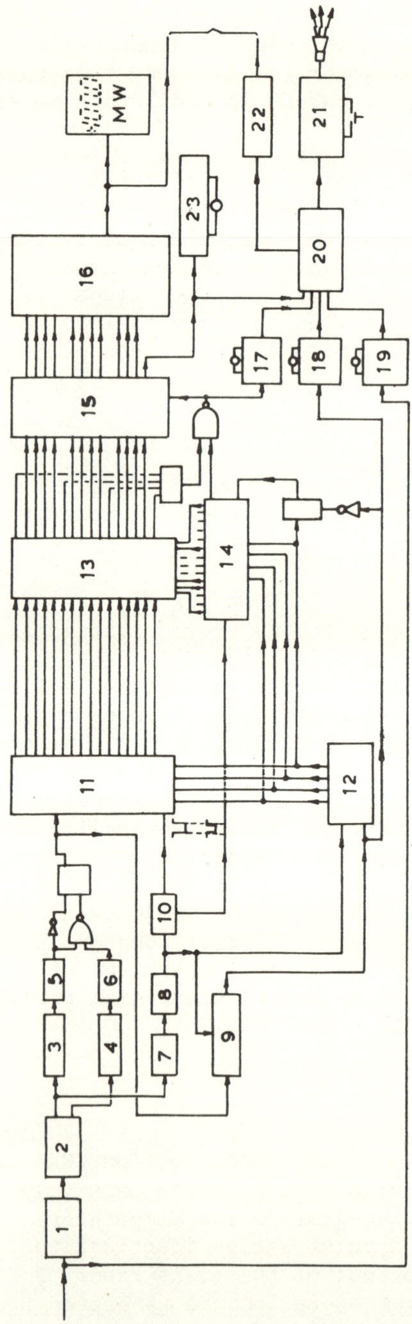

1 : PULSE SHAPER
2 : LIMITER
3 : + INTEGRATOR
4 : - INTEGRATOR
5 : + COMP
5 : - COMP
7 : PULSE EDGE DET.
8 : 30 MSEC MONO

9 : SYNC PATTERN RECOGNITION
10 : STROBE GENERATOR
11 : 1 TO 16 DMXR
12 : ADDRESS COUNTER
13 : ADDRESSABLE LATCH
14 : LATCH CLOCK
15 : MEMORY
16 : D/A CONVERTER

17 : LATCH CLOCK STATUS
18 : RESET PULSE STATUS
19 : INPUT DATA STATUS
20 : ERROR DETECTOR
21 : ALARM CONTROL
22 : DISABLE COMMAND
23 : TIE - LINE SWITCH STATUS

Fig 3 receiver block diagram.

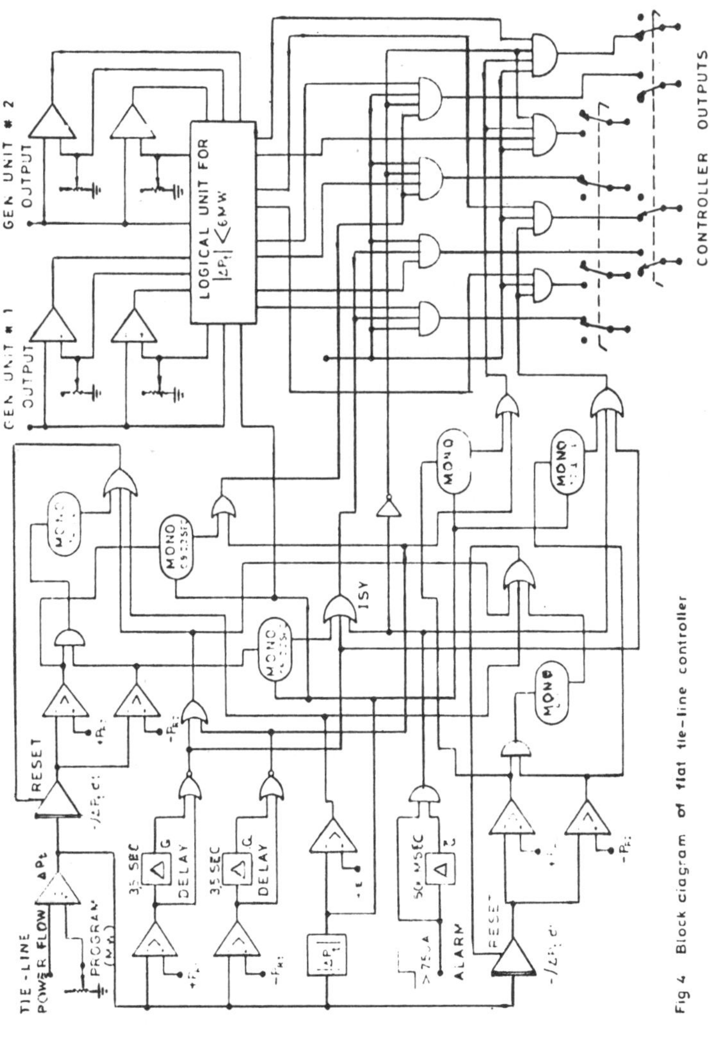

Fig 4 Block diagram of flat tie-line controller

1-to-16 Demultiplexer is used as a serial-to-parallel converter and the information bits are selected at the output of the Address--able Latches and are stored in Memory while the sync bits are subjected to a check. If this check is satisfactory and if the Latch Clock pulses operate normally, the Memory is enabled to feed a D/A Converter which provides an analog signal proportional to the power converter output at the tie-line point.

Error detection circuits are included in the receiver to warn for mulfunctions such as absence of a) sync bits or latch clock pulses b) address counter reset pulses c) input data and d) tie-line switch status. A detection of any of these errors will generate a disable signal to the controller and will activate a sound alarm. LED indications for these four cases are provided to monitor the signal presence.

4.3. The Flat Tie-Line Controller

A block diagram of the controller is shown on Fig. 4. An OP AMP forms the area control error ΔP_t which is the difference between the programmed power interchange and the actual power flow on the tie-line. The aim of the controller is to reduce ΔP_t to zero by issuing appropriate raise/lower commands to either the synchronize input or the speed adjusting input of the turbine governor. These inputs differ in their speed of response. The time duration of raise/lower commands is varied according to the absolute value of ΔP_t. Thus, for 6 MW $< |\Delta P_t| <$ 30 MW the raise/ /lower commands to the synchronize input last between 0.9 and 2.2 sec, while for the speed adjusting input, they last between 1.2 and 3.2 sec. The commands are issued at fixed time intrervals which are 14 sec for the synchronize input and 9 sec for the speed adjusting input. The selection of the above numbers was based on experimental results on the system to give a near - optimum response.

For $|\Delta P_t| >$ 30 MW, continuous raise/lower commands are issued, after an initial delay of 3.5 sec, until the error is reduced below 30 MW.
In both cases above the controller issues commands to one or two generators simultaneously.

For $\varepsilon < |\Delta P_t| <$ 6 MW, where ε is a minimum error value, the raise/lower commands are issued only to the generating unit that has the greater error correction capability.
The power output of each generating unit is monitored and no command that would tend to drive a unit off its generation limits is issued. The controller also has the capability to accept a digital input from the tie-line in order to issue raise generation commands when the tie-line current is greater than a preset value.

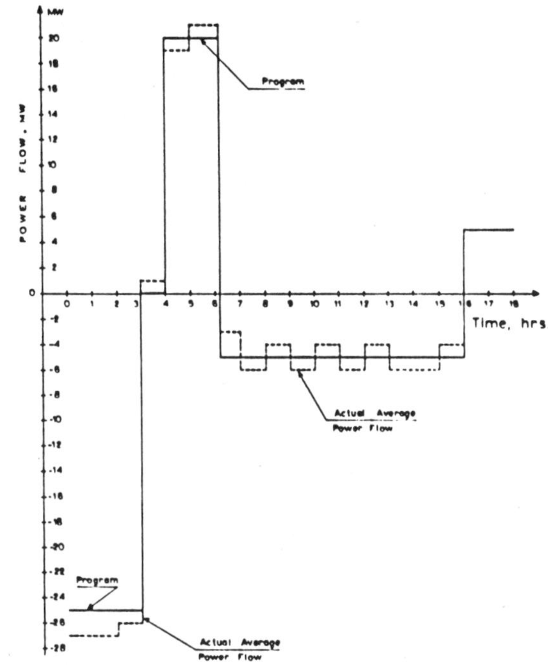

Fig. 5 Programmed and Actual Average Hourly Power Flow VS. Time

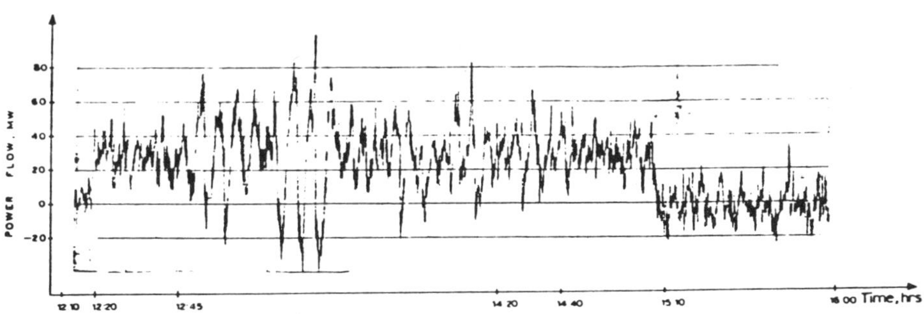

Fig. 6 Instantaneous Actual tie-line Power Flow for several programs
12:10 – 12:20 P = 0 MW 12:20 – 15:10 P = +20 MW
15:10 – 16:10 P = -5 MW

5. EXPERIMENTAL RESULTS

Fig. 5 shows programmed and actual power flows on the tie--line as computed from hourly energy recorders. Average hourly errors of ± 1 MWH and a maximum of + 2 MWH are observed on these and other similar curves of the system.
Fig. 6 shows the instantaneous actual power flow on the tie-line as obtained on a recorder. Several interchange program levels can be easily seen on this plot. The variations shown on this figure are attributed to load changes of the interconnected power networks.

6. CONCLUSIONS

The development of an electronic control system designed to maintain at a predetermined level the power flow on a 150 KV, 100 MW tie-line between the Yugoslav and Greek power systems has been described.

Results gathered from a two-year actual operation prove the reliability and satisfactory operation of the system. A maximum average hourly error of ± 2 MWH is achieved and is acceptable. This error could be further minimized at the expense of more frequent operation of the generating units. Although a close loop feedback system, no oscillations have been observed at any power interchange program, because command time durations and command delays were selected experimentally to satisfy a near - optimum condition.

REFERENCES

1. "Automatic Generation Control of Interconnected Power Systems". A Rigos, G. Monastiriotis and D. Beis. Greek Public Power Corp. Internal Report, July 1975.
2. "Power Flow Control-Basic Concepts for Interconnected Systems". Nathan Cohn. Preceedings of the Midwest Power Conference, Chicago, Illinois pp 159-175, 1950.
3. "Fundamentals for control of Generation and Power Flow on Interconnected Systems". Nathan Cohn. Presented at the Seminar on Control of Interconnected Electric Power Systems. Lima, Peru, November 1970.
4. "Techniques for Improving the Control of Bulk Power Transfer on Interconnected Systems". Nathan Cohn. IEEE Annual Winter Power Meeting, January 1971.
5. "Automatic Economic Loading Practices of Interconnected Power Systems in U.S.A. ". S.B. Morehouse. CIGRE paper No 315, Paris 1966.
6. "Terminal Units for Continuously Monitoring the State of a Group of Switches. "N.C. Voulgaris. Greek Public Power Corporation Internal Report, July 1974.

DYNAMIC SIMULATION OF POWER SYSTEMS INTERCONNECTED VIA PARALLEL AC AND DC LINES

N.A.Vovos G.D.Galanos

Power Systems Laboratory
School of Engineering
University of Patras,Patras Greece

Abstract - This paper describes a mathematical model for the dynamic simulation of power systems interconnected with parallel ac and dc lines. The two basic characteristics of the proposed mathematical model are:a)The detailed representation of the power system components, using their exact differential equations.b)The introduction of tensor techniques for the simulation of line and bus faults which results in increased flexibility of the computer program.

Although the system modelled in the paper consists of only two buses, the formulation of the equations used permits the extension of the simulation to any size system without altering the basic structure of the model.

INTRODUCTION

One of the most promising applications of HVDC transmission is the improvement of the transient stability of the interconnected ac power systems, utilising the fast electronic power flow control capability of the dc links (1),(2).

A number of dc power flow control strategies have been proposed and tested using simulation techniques (3),(4),(5),(6),(7),(8). All these methods, although they can provide a good indication of the response of an integrated ac-dc system subjected to a particular control strategy, they can never provide the exact performance of the system due to either the simplicity of the analogue models or the approximations in the mathematical formulation of the digital models.

The aim of this paper is the development of a digital simulation method for the detailed and exact representation of rea-

D. G. Lainiotis and N. S. Tzannes (eds.), Applications of Information and Control Systems, 199-206.

listic configurations of ac-dc systems, suitable not only for transient stability studies but also for the study of injected harmonics and any other application requiring the calculation of the instantaneous values of the system's state variables.

DESCRIPTION OF THE SYSTEM MODELLED

A power system consisting of two finite control areas interconnected via two ac tie lines and a dc link is modelled. Each control area is represented by a turbine-generator set equipped with power frequency and voltage-reactive power control systems. To each generator bus are also connected all the essential components used in real ac-dc schemes sush as:

a) AC filters for the 5th,7th,11th,13th harmonics and a high pass filter.
b) Switchable capacitor banks for reactive power compensation.
c) Static loads of variable magnitude and power factor.
d) Generator, line and converter transformers.
e) Protective relays and autoreclosing circuit breakers.
f) A communication link for the sychronization of the converters

The analysis and mathematical modelling of all the above subsystems and components are described in the following sections, together with the development of a flexible method for simulation bus faults, load switching, unbalanced loading conditions and all types of line faults at any desired point of the system.

An important advantage of the model is that any type of control system can be easily simulated and tested using the model.

REPRESENTATION OF THE SYNCHRONOUS MACHINES

The synchronous generators in the system are represented using the three phase terminal equations.The equations describing the machine including the effect of the step-up transformer are:

$$\frac{d}{dt}(i) = (L)^{-1}\left[-(V)-(R)(i) - \frac{d}{dt}(L_s)(i)\right] \tag{1}$$

The elements of the inductance matrices are functions of the rotor angle ∂ and $d\partial/dt$. These variables are calculated at each step from the solution of the swing equation of the machine.

$$P_{Ti}-P_{Gi} = \frac{H_i}{\pi f_0} \frac{d^2\delta_i}{dt} + D_i \frac{d\delta_i}{dt} \quad \text{PU MW} \tag{2}$$

where: $D_i(d\delta_i/dt)$ is the per unit power in the damper windings and H_i is the per unit inertial constant of the turbine-generator set.

The per unit electrical output of the generator is:

$$P_{Gi} = \left[(V_{si}) + (L_t^i)\frac{d}{dt}(i_s^i)\right]^t (i_s^i) \quad \text{pu MW} \tag{3}$$

THE POWER AND VOLTAGE CONTROL SYSTEMS

The block diagram of the power-frequency and voltage-reactive power control system is shown in Fig.1. From the block diagram the differential equations of the control loops are:

$$\frac{d}{dt} P_{ci} = -K_{Ii}\Delta f_i \tag{4}$$

$$\frac{d}{dt} Y_i = \frac{1}{T_{Gi}} \left(P_{ci} - \frac{1}{R_i}\Delta f_i - Y_i\right) \tag{5}$$

$$\frac{d}{dt} P_{Ti} = \frac{1}{T_{Ti}} (Y_i - P_{Ti}) \tag{6}$$

$$\frac{d}{dt} V_{ri} = \frac{1}{T_{Ei}} (K_{Ei}V_{2i} - V_{ri}) \tag{7}$$

$$\frac{d}{dt} V_{2i} = \frac{1}{T_{Ai}} \left[K_{Ai}((V)_{ref} - (V) - V_{si}) - V_{2i}\right] \tag{8}$$

$$\frac{d}{dt} V_{si} = \frac{1}{T_{Ei}T_{si}}(K_{si}K_{Ei}V_{2i} - K_{si}V_{ri} - T_{Ei}V_{si}) \tag{9}$$

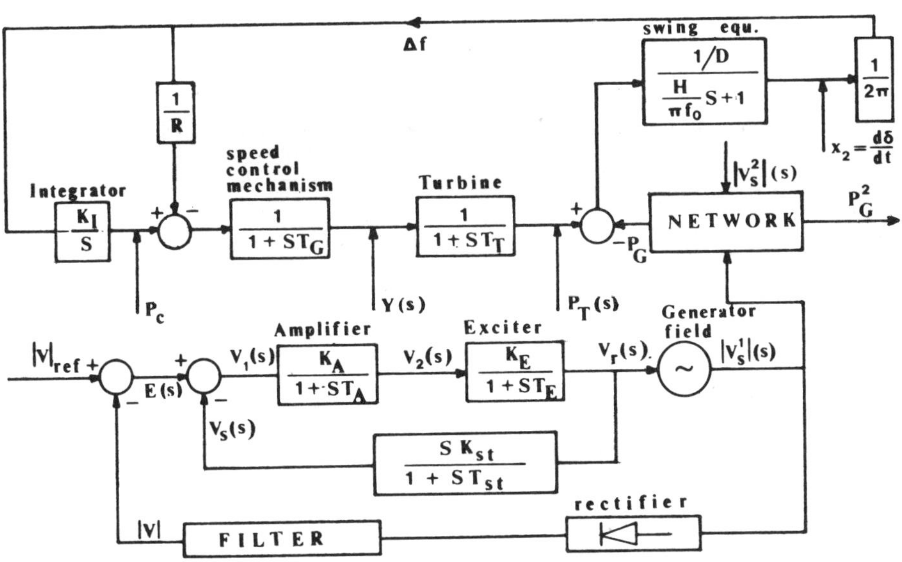

Fig.1 The block diagram of the P-f and Q-V control systems.

REPRESENTATION OF THE TRANSMISSION SYSTEM

The various types of line and bus-bar faults can be simulated by switching in and out branches of the general network in Fig.2. Mathematically, this can be achieved by transforming the loop impedance matrix of the general network using incidence matrices. The elements of these matrices are logical functions of the state of the system's switches.

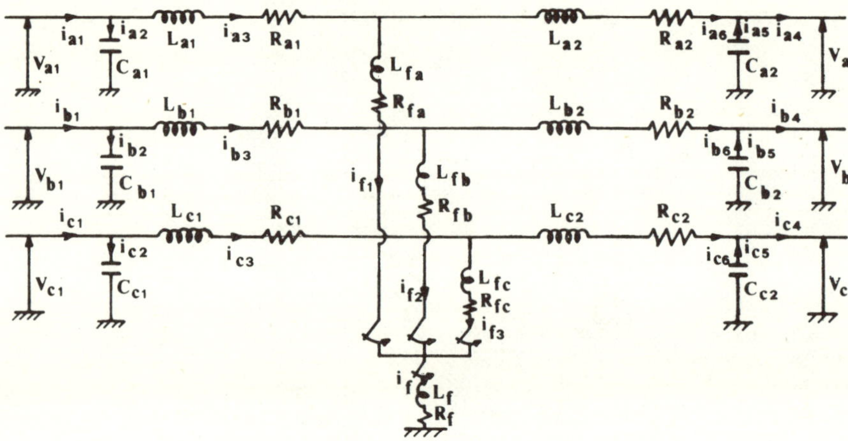

Fig.2 The generalized transmission system.

The loop equations of the generalized transmission system are:

$$(V_i) = (Z_i)(I_i) \tag{10}$$

The equation (10) is valid only when all switches are closed. For any other state of these switches the independent currents of the transmission system are given by:

$$(I_i) = (C_p)(I_p) \tag{11}$$

where C_p is an incidence matrix and its elements depend on the particular conduction state of the switches. By using the special form of the matrices and partitioning them as follows:

$$(V_i)=\begin{bmatrix} V_{s1} \\ \hline V_{s2} \\ \hline V_2 \end{bmatrix} \quad (I_i)=\begin{bmatrix} I_I \\ \hline I_{II} \\ \hline I_2 \end{bmatrix} \quad (Z_i)=\left[\begin{array}{cc|c} Z_{c1} & 0 & \\ 0 & Z_{c2} & Z_2 \\ \hline \multicolumn{2}{c|}{Z_3} & Z_4 \end{array}\right] \quad (C_p)=\left[\begin{array}{c|cc} C_1 & \multicolumn{2}{c}{C_2} \\ \hline & C_{41} & C_{42} \\ C_3 & C_{43} & C_{44} \end{array}\right]$$

where: $(Z_{ci}) = \text{diagonal } (\frac{1}{C_{ai}}, \frac{1}{C_{bi}}, \frac{1}{C_{ci}}) \quad i = 1,2$

(Z_{44}) is a 3xm (m=1,2,3) matrix, the elements of which take values 1,-1,0 depending on the topology of the network.

The eq.(10) using eq.(11) can now be written as follows:

$$\frac{d}{dt}(V_{s1}) = (Z_{c1})(I_I) \tag{12}$$

$$\frac{d}{dt}(V_{s2}) = (Z_{c2})(I_{II}) \tag{13}$$

$$\frac{d}{dt}(I_2) = (L_m)^{-1}\left[(V_m)-(R_m)(I_2)\right] \tag{14}$$

Where:

$$(Z_4) = (R_4)+s(L_4) \tag{15}$$

$$(V_m) = (C_4)^t(V_2) \tag{16}$$

$$(R_m) = (C_4)^t(R_4)(C_4) \tag{17}$$

$$(L_m) = (C_4)^t(L_4)(C_4) \tag{18}$$

These equations represent the single line transmission system for any fault condition provided that the incidence matrix is known.

REPRESENTATION OF TWO PARALLEL LINES

The simulation method developped in the previous section can be easily extended for the case of two parallel ac lines. In this case we obtain:

$$\frac{d}{dt}(V_s^1) = (Z_{c1}^1)(I_I^1) \tag{19}$$

$$\frac{d}{dt}(V_s^2) = -(Z_{c2}^1)(I_{II}^1) \tag{20}$$

$$\frac{d}{dt}(I_2^1) = (L_m^1)^{-1}\left[(V_m^1)-(R_m^1)(I_2^1)\right] \tag{21}$$

$$\frac{d}{dt}(I_2^2) = (L_m^2)^{-1}\left[(V_m^2)-(R_m^2)(I_2^2)\right] \tag{22}$$

$$(I_I^2) = (Z_{c1}^2)^{-1}(Z_{c1}^1)(I_I^1) \tag{23}$$

$$(I_{II}^2) = (Z_{c2}^2)^{-1}(Z_{c2}^1)(I_{II}^1) \tag{24}$$

where the upper index 1 or 2 refers to the line 1 or 2 respectively. The currents I^1_I and I^1_{II} are used to define the bus voltage and it is necessary to express them in terms of the other state variables of the system. Applying kirchhoff's current law to each bus-bar we obtain:

$$(I^1_I)=\left[(u)+(Z^2_{c1})^{-1}(Z^1_{c1})\right]^{-1}\left[(I^1_s)-(B)(C^1_{44})(I^1_2)-(B)(C^2_{44})(I^2_2)-\right.$$
$$\left.-(I^1_L)-(I^1_c)-(I^1_F)\right] \qquad (25)$$

$$(I^1_{II})=\left[(u)+(Z^2_{c2})^{-1}(Z^1_{c2})\right]^{-1}\left[(I^2_s)+(I^1_{III})+(I^2_{III})-\right.$$
$$\left.-(I^2_L)-(I^2_c)-(I^2_F)\right] \qquad (26)$$

where: $(B)=\boxed{u\,|\,u}$ and (u) is a 3x3 unit matrix

$(I^j_L),(I^j_c),(I^j_F)$ the current vectors of the static loads, the ac-dc converters and the filters respectively.

The way the equations (19) to (26) are formulated permits the addition of any number of lines and busses to the model and thus makes the dynamic simulation program very flexible.

THE CONSTRUCTION OF THE MATRIX C_p

The matrix C_p defined in the previous section can be constructed if the elements of the submatrix C_{44} are calculated. The matrix C_{44} relates the currents i_{f1}, i_{f2}, i_{f3} to the independent loop currents corresponding to the particular topology of the fault simulating network. The dimensions of the matrix C_{44} are 3xl where l is the number of the links of this network, thus l is equal to the number of the elements switched in minus one.

If a state variable X_i for i=1,2,3,4 is defined 1 if the branche of the currents i_{f1}, i_{f2}, i_{f3}, i_f respectively exist and 0 if doesn't exist, then the existing non zero elements of the matrix C_{44} are given below.

$$C_{11} = 1 \text{ if } X_1 = 1 \qquad (27)$$

$$C_{21} = \begin{cases} 1 \text{ if } \bar{X}_1 X_2 = 1 & (28) \\ -1 \text{ if } X_1 \bar{X}_3 X_4 = 1 & (29) \end{cases}$$

$$C_{31} = \begin{cases} 1 \text{ if } \bar{X}_1 \bar{X}_2 X_4 = 1 & (30) \\ -1 \text{ if } X_3 \bar{X}_4 = 1 & (31) \end{cases}$$

$$C_{22} = 1 \text{ if } X_1X_2(X_3+X_4) = 1 \tag{32}$$

$$C_{23} = \begin{cases} 1 \text{ if } X_3X_4(\bar{X}_1X_2+X_1\bar{X}_2) = 1 & (33) \\ -1 \text{ if } X_1X_2X_3\bar{X}_4 = 1 & (34) \end{cases}$$

$$C_{33} = 1 \text{ if } X_1X_2X_3X_4 = 1 \tag{35}$$

REPRESENTATION OF THE DC LINK

The differential equations describing the dc link are automatically assembed using tensor techniques to deal with the time varying topology of the network caused by the switching action of the converter valves. This method have been described in details in references (7),(8).

REPRESENTATION OF THE FILTERS STATIC LOADS AND CIRCUIT BREAKERS

The harmonic filters are of the series RLC type tuned to the 5th, 7th, 11th and 13th harmonics. The set of the differential equations of the four harmonic filters is:

$$\frac{d}{dt}(I_F^i) = (L^i)^{-1}\left[(V_s^i)-(V_c^i)-(R^i)(I_F^i)\right] \tag{36}$$

The high pass filter consists of a capacitance in series with a parallel RL network. The differential equations are:

$$\frac{di}{dt} = \frac{1}{L}(V_s-V_c) \tag{37}$$

$$\frac{dV_c}{dt} = \frac{1}{C}\left[i+\frac{1}{R}(V_s-V_c)\right] \tag{38}$$

The static loads are represented by series RL networks. The differential equations for the loads are:

$$\frac{d}{dt}(I_L^i) = (L_L^i)^{-1}\left[(V_s^i)-(R_L^i)(I_L^i)\right] \tag{39}$$

The index i=1,2 for the buses 1 and 2 respectively.

The switching action of the circuit breakers is represented by a resistance which increases with time. From experimental resistance-time curves we obtain the analytical function using polynomial approximation.

CONCLUSIONS

This paper presents for the first time an exact and detailed dynamic simulation of a realistic integrated ac-dc system. Although the model used consists of only two buses the formulation and the simulation techniques used can be directly applied to systems with any number of buses and lines.

The system modelled includes all the essential components of the real ac-dc schemes sush as ac filters, capacitor banks for reactive power compensation and autoreclosing circuit breakers.

An important characteristic of the mathematical model developped is the flexibility with which all types of line and bus faults, load changes, line switching and autoreclosing can be programmed.

Although a three phase representation of the system is used, by applying sparsity techniques, the computer storage requirements are minimized.

A digital computer program based on the mathematical model developped in this paper was succesfully used for the study of the transient stability of ac-dc systems (2). The results of this study shown the effectiveness of the mathematical model.

REFERENCES

(1) N. A. Vovos and G. D. Galanos, "Damping of power swings in ac tie lines using a parallel dc link operating at constant reactive power control", IEEE Transactions on Power Apparatus and Systems, Vol. PAS-98, pp. 416-425, March/April 1979.

(2) N. A. Vovos and G. D. Galanos, "Transient Stability of AC-DC Systems", IEEE Transactions on Power Apparatus and Systems, Vol. PAS-98, pp. 1375-1383, July/August 1979.

(3) H. A. Peterson, P. C. Krause, J. F. Luini and C. H. Thomas"An analog computer study of a parallel AC and DC power systems", IEEE Transactions on Power Apparatus and Systems, Vol. PAS-85 pp. 191-209, March 1966.

(4) E. Uhlmann, "Stabilization of an AC link by a parallel DC link", Direct Current, pp. 89-94, August 1964.

(5) H. A. Peterson and P. C. Krause, "Damping of power swings in a parallel AC and DC system", IEEE Transactions on Power Apparatus and Systems, Vol. PAS-85, pp. 1231-1239 , December 1966.

(6) C. N. Weygandt and R. F. Chu, "Resent results on the stabilizing effect of a DC transmission link", Presented at the 1977 Winter Meeting, New York, NY, Jan. 30-Feb. 4, Paper No. F 77-061-5.

(7) J. Milias-Argitis and G. Galanos, "Dynamic simulation of HVDC transmission Systems", IEEE Transactions on Power Apparatus and Systems, Vol. PAS-92, May/June 1976.

(8) J. Milias-Argitis, G. Giannakopoulos and G. Galanos, "Dynamic simulation for multiterminal HVDC Systems", IEEE Transactions on Power Apparatus and Systems, Vol. PAS-97, pp. 587-593, March/April 1978.

SECTION IV

BIOENGINEERING

RESIDUAL FUNCTION AFTER THE LESION IN FORMAL NEURAL NETS

A. Delgado and J. Mira

Fac. of Sciences. Univ. of Granada, SPAIN

ABSTRACT.- If we represent the activity of cortical neurons at physical level (spike generation), we can consider the neuron as a finite deterministic automaton with many boolean inputs and one output, result of a measure on the internal state.
A standard lesion eliminate 80.000 units, more or less, which means to eliminate $2^{80.000}$ states and, nevertheless, does not interrupt the overall function. Obviously, the functions must be cooperative and asociated at macroscopical levels of cerebral dynamics. Otherwise, the effect of lesion should be catastrophic.
In this paper we propose a logical model of cooperativity and a calculus of the residual function after the lesion in formal neural nets. The functional deficit depends on the degree of conectivity, and the measure used to define macro-states.

1. THE LOGIC LEVEL ON CEREBRAL DYNAMIC

Our work is an essay of operational formulation of some aspects of cerebral dynamic. We are searching for the theoretical bases that may explain the high operative abilities that remain on the nervous tissue after traumatic and localized lesions (1,2,3). This - trend have directed us to the study of the cooperative processes, namely conectivity degree and language used by cooperants modules.
To speak of brain language is equivalent to speak of neuronals codes (4). This codes are at least, analo-

D. G. Lainiotis and N. S. Tzannes (eds.), Applications of Information and Control Systems, 209-216.

gics, logics and symbolics. The analogical level is the more studied and the symbolic the less because of its codes are difficult to be identified. It is needed to identify the neurophysiological symbols and "words" following a physico-conductal isomorphisme. Between these two level we find the logic one, that possibly works us a control structure of the others.

We are going to analyze in this paper the concepts of lesion, functional deficit and residual function at logical level.

In order to build the model we have used the Moreno Diaz's (5,6,7) theory of formal nets, considering that the neurone can be described as a finite and deterministic automaton with M booleans inputs ($x_1 \ldots x_M$) and a only output, y, that results from the probabilistic or deterministic measure on the N variables of state ($s_1 \ldots s_N$). This variables specify the microscopic description of the neurones by means of logical functions of the inputs and state vectors:

$$s_k(t) = f_k\left[\bar{S}(t-1); \bar{X}(t-1)\right], \quad k = 1,2,\ldots N \tag{1}$$

$$y(t) = F\left[\bar{S}(t-1)\right] \tag{2}$$

From this microscopic description we can get the macroscopic one, constructing the set of inputs $\{X_m\}$, the macrostates $\{S_i\}$ and the transition matrices in which are projected the functional matrix $\mathcal{M}_{ij}(X_m)$.

$$\mathcal{M}_{ij}(X_m) = \sum_{m=1}^{2^M} T_{ij}^m \cdot X_m = \sum_{m=1}^{2^M} T_{ij}^m \cdot x_1^\alpha \cdot x_2^\beta \cdots x_M^\mu, \quad X_m = (\alpha, \beta, \ldots, \mu), \; m = 1,2,\ldots,2^M \tag{3}$$

The jump from the macroscopic description to the microscopic one is a synthesis problem. The inverse jump is a problem of analysis and its correlated algoritms are

$$s_k(t) = \sum_{i=1}^{2^N} \sum_j \mathcal{M}_{ij}(X_m) \cdot s_1^a \cdot s_2^b \cdots s_N^f, \quad \forall (a,b,\ldots,f) \mid s_k = 1 \tag{4}$$

$$m_{ij}(X_m) = \prod_{k=1}^{2^N} \left[f_k(X_m, S_i)\right]^{s_k^j} \tag{5}$$

The outputs of the neuron define its observable macroscopic state that follows the all or none law. Thus, each output function F(S) introduce a partition of the set of microstates in two equivalence clases $\{0\}$ and $\{1\}$, in such a way that there are only changes of macroscopic states when the microscopical one link microstate of different class.

This two levels description process (micro and macroscopic) is iterative. Thus, a net of Q formal neurones can be represented by means of the conection of Q·finites automatas, in such a way that each of these automatas may receive external inputs from the same set (x ...x) and feedbacks of outputs functions of all the automatas (y ...y), itself included. The neuronal macrostates are now the state variables describing the microscopical state of the neuronal net.

2. LESION

At a logical level the cooperativity develop because of all the neurones (Q) take part in all the net microstates. Each neurone bring a state variable and the - global function correlated to the net is the automaton. Thus all neuronal net are cooperative if the correlated automaton can be separate into its constituent parts.
After lesion, there is a deep alteration in the microscopical behaviour of the net. Each neurone eliminated divide by two the number of microstates present, so that there is an exponential decrease (2^{-L} for L neurons) in the number of automaton states. If the global function is present yet, although depressed, is due to the fact that it is associated to macroscopical states, in such a way that the residual function mainly depends of the measure criterium employed to define the equivalence among microstates.
The relevant partitions are those that have the property of sustitution. So that all the microstates of a subset have the same behaviour for any input in relation with some element of the same or other subset (8). Opperationally this means that

$$\sum_i m_{ij}(X_m) = cte., \ \forall i,j \in \{k\} \tag{6}$$

being $\{k\}$ the subset of microstates whose measure is equal to k. The functional matrix for macrostates is obtained as logical sum of the elements that link microstates that belong to each one of the macroscopic state. This functional matrix is constitued by submatrices whose rows have the same "minterms" in equal or different order. An example should be

$$\mathcal{M}_{ij}(X_m) = \begin{vmatrix} x & 0 & \bar{x} & 0 \\ 0 & x & 0 & \bar{x} \\ 0 & 1 & 0 & 0 \\ 1 & 0 & 0 & 0 \end{vmatrix} \longrightarrow \mathcal{M}^*(X_m) = \begin{vmatrix} x & \bar{x} \\ 1 & 0 \end{vmatrix} \tag{7}$$

Let us see another example for a net with (N=3,M=1) that uses a counter as measure function. Thus, all the microstates with the same number of fired neurons are equivalent. There are then, 4 macrostates

$$S_0=\{000\},\ S_1=\{001, 010, 100\},\ S_2=\{011\ 101\ 110\},\ S_3=\{111\}$$

Intended for arbitrary partition we use the probabilistic formulation. The firing probabilities of formal - neurones being:

$$\begin{array}{c|cccccccc|cccccccc|cccccccc|} & \multicolumn{8}{c|}{p(y_3=1)} & \multicolumn{8}{c|}{p(y_2=1)} & \multicolumn{8}{c|}{p(y_1=1)} \\ x=0 & 0{,}5 & 0 & 0 & 0 & 0 & 0{,}8 & 0 & 1 & 0 & 0 & 0 & 0{,}5 & 0{,}8 & 1 & 0 & 0{,}5 & 0 & 1 & 1 & 0 & 0 & 0 & 0 & 0 \\ x=1 & 0{,}1 & 0 & 0{,}8 & 0 & 0 & 0 & 0 & 0{,}9 & 0 & 0 & 0 & 0{,}5 & 0 & 0 & 0 & 0{,}5 & 0 & 0 & 0 & 0 & 0{,}9 & 0 & 0 & 0{,}1 \end{array}$$

Applying the analysis algorithm we obtain the functional matrix for micro ($\mathcal{M}$) and macrostates ($\mathcal{M}^*$). In order to normaliza, we have divided each elements of $\mathcal{M}^*$ by the number of rows added to each macrostate.
The elimination of a formal neuron, let it be y_2 , exclude all the probabilities of neuronal shuting correlated to minterms that have such a neurone fired. That is to say the microfunction of y_1 and y_3 stand, only that many input configurations will never be present. However, others will be present more frequently and because so the mean of the counter response, in the time, will be modified. The reduced functional matrix are:

$$\mathcal{M}^*\left|\begin{matrix}\text{before}\\ \text{lesion}\end{matrix}\right| = \begin{pmatrix} 0{,}5\bar{x}+0{,}9x & 0{,}5\bar{x}+0{,}1x & 0 & 0 \\ 0{,}2\bar{x}/3+1{,}3x/3 & 2{,}8\bar{x}/3+1{,}7x/3 & 0 & 0 \\ 1{,}5\bar{x}/3+2{,}5x/3 & 0{,}7\bar{x}/3+0{,}5x/3 & 0{,}8\bar{x}/3 & 0 \\ 0{,}045x & 0{,}5\bar{x}+0{,}455x & 0{,}5\bar{x}+0{,}455x & 0{,}045x \end{pmatrix} \quad (8)$$

$$\mathcal{M}^*\left|\begin{matrix}\text{after}\\ \text{lesion}\end{matrix}\right| = \begin{pmatrix} 0{,}5\bar{x}+0{,}9x & 0{,}5\bar{x}+0{,}1x & 0 & 0 \\ \bar{x}/2+1{,}1x/2 & \bar{x}/2+0{,}9x/2 & 0 & 0 \\ 0{,}2\bar{x}+x & 0{,}8\bar{x} & 0 & 0 \\ 0 & 0 & 0 & 0 \end{pmatrix} \quad (9)$$

At a microscopical level the product of the lesion is a disaster. Nevertheless over the reduced functional matrix for macrostates, the effect of the lesion is much more limited. The functional organization stands, although depressed, according to experimental results. An algorithm that enable us obtain directly the functional matrix correlated to the injured neuronal net from that of the normal net is the following. Being $\{y_j\}$, $j \in \{k\}$ the set of neurones that eliminates the lesion, since the reduction process is iterative, it will be sufficient to analyze the effect of just one neurone elimination. Let it be y_j.

1. This neurone define a partition of the microstates set: $\{S^1_j\}$ and $\{S^0_j\}$.
2. The set $\{S^1_j\}$ can not be present. Thus we will eliminate all the rows of the functional matrix, whose subindex, in binary, includes the bit j at high logical state. That is to say all the row starting by initial state that belong to the set $\{S^1_j\}$.
3. We add couples of colums. Any given colum whose subindex, in binary, belong two $\{S^0_j\}$, is changed by the addition of itself plus the one that has all the other bits equal, and the j at a high logical state.
4. We eliminate all the rows whose bit j is at a high logical state.
5. We repeat the process as many time as eliminated neurones.

3. RESIDUAL FUNCTION

After lesion a residual function remins, that must be substracted from the function of the normal net, in order to evaluate the functional lost. To carry out this, we introduce the concept of quantity of function using the following opperational definition.

1. We obtain the sequence of values $\{y(S_0, X_m)\}$ that the neuronal outputs take from the initial state S_0 and under the constant input X_m.
2. We define a lineal measure over the microstates

$$F[y(t)] = \sum_{k=1}^{N} \alpha_k \cdot y_k(t), \quad y(t) \in (01), \quad 0 \leq \alpha_k \leq \alpha_{max} \tag{10}$$

3. We specify a set of intervals $(I_1, \ldots, I_F)$ in such a way that we do them equivalents all those microstates whose measure belong to the same intervel.
4. The quantity of function is the average of these measures. Being p_i the probability (relative frequency) of the macrostate whose measure is I_i, we have:

$$QF = \sum_{i=1}^{F} p_i \; I_i \tag{11}$$

To the digital simulation of this algorithm we use iterative nets (N=10, M=2) working under constant input and with a variable degree of conectivity. The figure 1 shows the schedule of the calculus, in where the conectivity is controled by mean of distributors. The lesion is simulated by mean of a mask. The last step is the totalizer that include a measure function and the needed counters to evaluate the probability of macrostates.

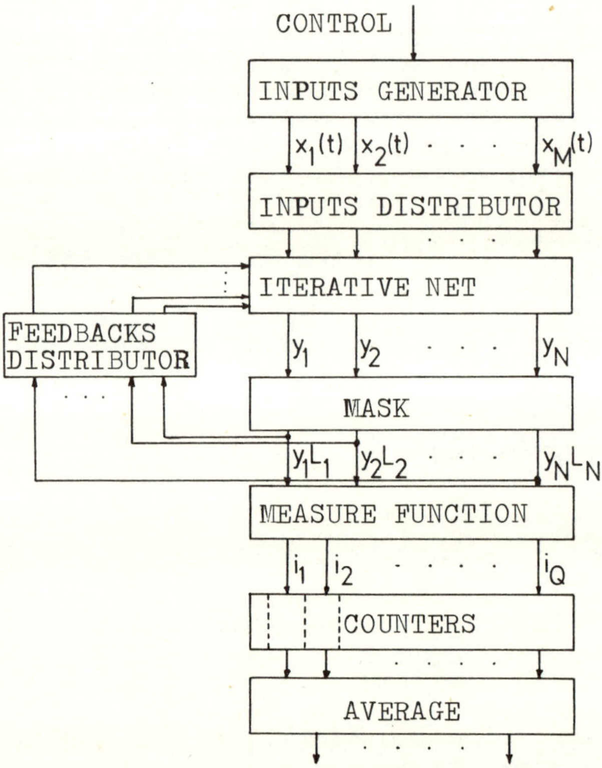

Figure 1. Calculus of residual function

We obtain thus the residual function curves that represent the change of the quantity of function with the span of the lesion. We have selected the results of the simulation according to the curves length and the initial states, which enhances the comparative process between residual functions associated to nets

with different degrees of conectivity.
The figure 2 shows the results of OR nets with conectivity 2 and 5. Open circles represents the quantity of function and the stars its normalized value. Normalizing result we eliminate the dependence of the quantity of function with the length of a transitory string or a cicle of oscillation.
The figure 3 shows the results of residual function of AND nets, and figure 4 those of XOR function.

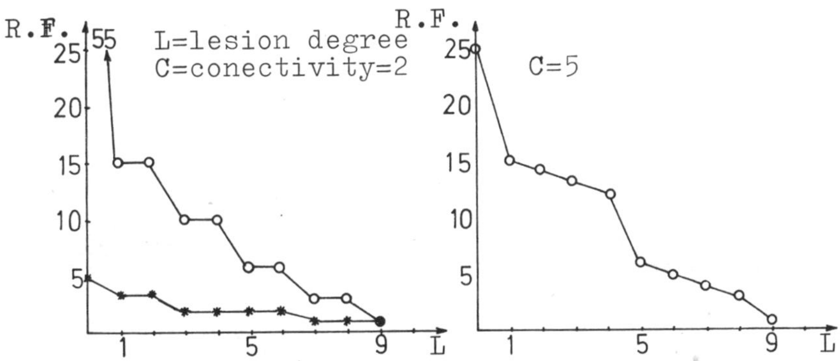

Figure 2. Residual Function for OR nets

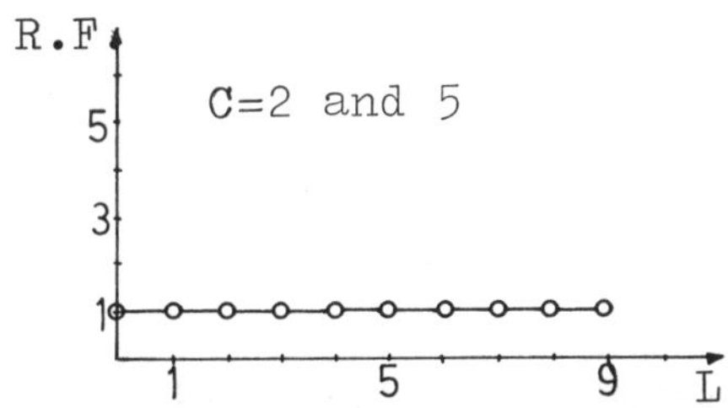

Figure 3. Residual Function for AND nets

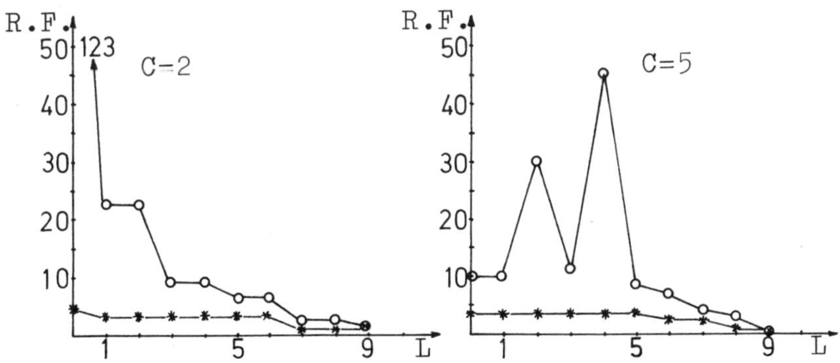

Figure 4. Residual Function for XOR nets.

When conectivity increase the residual function curves are continues and decrease in a progressive way following the Lashley's Law of Mass Action. The functional deficit is proportional to the eliminated mass tissue.

4. CONCLUSIONS

Cooperative process at a logit level "explain" some of Lashley's results and this is so not because of the goodness of our model but because of the weakness of Lashley's experimental data.

If the logic level is the correct to describe the cerebral dynamics, at least as a structure of control, the anatomo-functional redundance must be spectacular. We need to know more potents neuronal codes than the biophysical and logic ones. This codes are not sufficient for to explain complex cooperation. The dynamic of the brain is mainly bassed upon neurophysiological symbols. All the facts exposed direct us to a social theory of neural nets.

REFERENCES

(1) Delgado, A.: 1978, "Modelos Neurocibernéticos de Dinámica Cerebral". Tesis doctoral. E.T.S.I. Telecomunicación. Univ. Politécnica, Madrid.

(2) Delgado, A., Mira, J. and Moreno, R.: 1979, "Un Modelo Conceptual de Dinámica Cerebral". Rev. de Automática, nº 39, Madrid.

(3) Mira, J., Delgado, A. and Moreno, R.: 1979 "Cooperative Processes in Cerebral Dynamic". In these Proceeding, pp. 273.

(4) Pribram, K.H.: 1971, "Languages of the Brain". Prentice-Hall, inc. New Jersey.

(5) Moreno-Diaz, R. and McCulloch, W.S.:1969, "Circularities in nets and the Concept of Functional Matrices". Biocib. of the C. N. S.,L. Brown, Boston

(6) Moreno-Diaz, R.: 1971, "Deterministic and Probabilistic Neural Nets with Loops". Mathematical Biosciences, 11, pp 129-136.

(7) Santesmases, J.G. and Col.: 1977, "Formal Neuronal Nets". Proc. of the First Int. Conf. on Inf. Sc. and Syst., vol. 2, D.G. Lainiotis and N.S. Tzannes editors. Hemisphere Publishing Corporation, London.

(8) Booth, T.L.:1967, "Sequential Machines and Automata Theory". J. Wiley, N. York.

GRAMMARS FOR EFFICIENT NON-CONTEXT-FREE PARSING

Michael Kunze

Technische Hochschule Darmstadt

INTRODUCTION

From the point of view of applications such as to programming languages or computational complexity the theory of formal languages seems to be rather disappointing: Their algebraic theory is well developped by now and has become part of mathematics - regardless of its external motivation - as far as rational languages [3] are concerned and also for context-free languages to a certain extend. But even for the latter the algebraic theory [2] and the algorithmic theory of parsing [1] are remarkably separated. Clearly, for the whole of the context-sensitive languages we cannot expect a uniform and complete theory. Therefore we have to restrict our investigation to a subclass. Nevertheless it may be surprising that it is possible to keep the following features close together despite their divergence in common experience:

(A) to deal with classes containing languages interesting with respect to applications (as $\{\#w\#w \mid w \in \Sigma^*\}$, $\{\#w\#v\#u \mid w = u \cdot v$ as integers at base $|\Sigma|\}$, or programming languages)

(B) to have a deterministic device for efficient parsing (with respect to generating grammars of suitable type)

(C) to treat (B) in an algebraically reasonable manner (only using such algebraic structures that are already studied in mathematics)

D. G. Lainiotis and N. S. Tzannes (eds.), Applications of Information and Control Systems, 217-225.

1. LP - MACHINES

As usual, formal grammars are given by (Σ, V, P, S) where Σ is the terminal alphabet, $V \setminus \Sigma$ is the set of nonterminals, $P \subseteq V^+ \times V^*$ is the set of productions and $S \in V \setminus \Sigma$ is the start symbol. Beyond their use for generating languages it is possible to interpret some special types of grammars as automata:

Definition: An LP-machine (linear parallel processing machine) is a grammar (Σ, V, P, q) such that $P \subseteq \Sigma(V \setminus \Sigma) \times V^*$.

1.1. Interpretation as Automata

In order to explain how an LP-machine $\mathcal{A} = (\Sigma, V, P, q)$ is working, we understand

Σ as the union of input symbols, output symbols (if any), and of symbols for internal use such as marked symbols or punctuation marks

$V \setminus \Sigma$ as the set of operator types

P as the set of processing rules, a rule $aA \longrightarrow B_1 \ldots B_n b_1 \ldots b_m$ saying that the operator A reading the symbol a has the possibility of changing 'a' into '$b_1 \ldots b_m$' and simultaneously generating the sequence $B_1 \ldots B_n$ of operators for further computation to the left

q as the initial operator which has to start the computation on the input word written left of q

V^* as the set of all configurations of the LP-machine $\mathcal{A}$ so that $\mathcal{A}$ may be thought of as working with a single, but extensible tape unit (therefore the name 'linear' p.p. machine) allowing to erase and to print without shifting of the neighboring symbols

We consider $\overset{*}{\underset{P}{\Longrightarrow}}$ - derivations as computations of $\mathcal{A}$. Thus, $\mathcal{A}$ has a variable number of active control units which is regulated by the input but has no fixed upper bound in general. All operators have a unique processing direction from right to left determined by the form of the productions in P. As a characteristic of LP-machines, we point out that they don't have a master control which might be referred to as state in the usual sense.

It turns out to be convenient to define LP-machines by their processing tables. In fact very often this is much more convenient than the same task for Turing machines where such tables grow tediously long.

$\mathcal{A}_=$	initial operator S	A_0	A_1	B_0	B_1	E	operators
terminals: 0	A_0S	$A_0 0$	$A_1 0$	ε	/	/	
I	A_1S	$A_0 I$	$A_1 I$	/	ε	/	
#	E	$B_0\#$	$B_1\#$	/	/	ε	

$\mathcal{A}_2$	E	I	Q
a	IE	Qa	I
#	ε	#	/

ε denotes the empty word.

A processing rule $aA \longrightarrow \alpha$ is represented by an entry α in row a and column A.

<u>Definition</u>: Let be $\Gamma \subseteq \Sigma$ a subalphabet of Σ. A formal language $L \subseteq \Gamma^*$ <u>is accepted</u> by an LP-machine $\mathcal{A} = (\Sigma, V, P, q)$, if

$$L = \{w \in \Gamma^* \mid wq \overset{*}{\underset{P}{\Longrightarrow}} \varepsilon\} .$$

$\mathcal{A}_=$ is accepting the language $\{\#w\#w \mid w \in \{0,I\}^*\}$, $\mathcal{A}_2$ accepts $\{\#a^{2^n-1} \mid n \in \mathbb{N}_0\}$. We do not define how to compute functions by LP-machines because we don't study computational complexity of functions in this paper. We only mention that LP-machines are equivalent to Turing machines. Special programs like one for the multiplication of binary numbers can be done very fast by parallelism. On the other hand universal programs are slow, because they require a lot of copying enforced by the right-to-left processing. For LP-machines the natural notion of determinism is very smooth:

<u>Definition</u>: An LP-machine $\mathcal{A} = (\Sigma, V, P, q)$ is d e t e r m i n i s t i c, if $P: \Sigma(V \setminus \Sigma) \longrightarrow V^*$ is a partial function.

As a justification of this definition we notice that given any configuration α of $\mathcal{A}$ subconfigurations to be processed cannot overlap because of the type of rules in P. As a consequence one can prove by induction:

<u>Lemma</u>: Let be $\alpha \overset{*}{\underset{P}{\Longrightarrow}} \beta$ and $\alpha \overset{*}{\underset{P}{\Longrightarrow}} \gamma$ computations of a deterministic LP-machine $\mathcal{A} = (\Sigma, V, P, q)$. If γ is a terminal configuration of $\mathcal{A}$ (i.e. no further processing is possible), then $\beta \overset{*}{\underset{P}{\Longrightarrow}} \gamma$.

Therefore, the result of deterministic LP-computations is unique whenever existing.

1.2. Classical Types of Automata

Restrictions of the processing rules of LP-machines (Σ, V, P, q) lead to special types of machines, some of which are well studied in classical theory of automata.

a) $P \subseteq \Sigma(V\setminus\Sigma) \times (V\setminus\Sigma)^*\Sigma^*$: LP-machines in normal form

This type of processing rules doesn't yield a proper restriction with respect to their computational power which is equivalent to that of Turing machines: Deterministic LP-machines in normal form accept precisely the recursively enumerable languages.

b) $P \subseteq \Sigma(V\setminus\Sigma) \times [(V\setminus\Sigma)^* \cup (V\setminus\Sigma)^*\Sigma]$: bounded LP-machines

These automata are one of our main subjects of interest. In the example above $A_=$ and A_2 are deterministic bounded LP-machines. A language $L \subseteq \Gamma^*$ is called a bounded LP-language, if it is accepted by a bounded LP-machine (Σ, V, P, q) where $\Gamma \subseteq \Sigma$. Considering leftmost computations, i.e. leftmost derivations according the grammar (Σ, V, P, q), we obtain a linear space bound (cf. Lemma). Therefore we know

Proposition 1: Each bounded LP-language is context-sensitive.

On the other hand considering rightmost derivations, bounded LP-machines are easily seen to be generalizations of pushdown automata with a single state Thus every context-free language is a bounded LP-language.

c) $P \subseteq \Sigma(V\setminus\Sigma) \times (V\setminus\Sigma)^*$: homogeneous pushdown automata

These LP-machines are equivalent to pushdown automata with a single state which therefore may be dropped for convenient notation. $V\setminus\Sigma$ is used as pushdown alphabet, and the top symbol of the pushdown store being the leftmost one the machine operates on the input word reading it from right to left. Allowing nondeterministic accepting devices all context-free languages are accepted by such automata, but concerning deterministic automata the situation is less trivial:

d) $P \subseteq \Sigma(V\setminus\Sigma) \times [(V\setminus\Sigma)^* \cup \Sigma]$: state-free pushdown automata

In order to simulate a pushdown automaton in the usual sense by an LP-machine called state-free pushdown automaton, we have to encode the states and put them at the border between input word and pushdown store, a state being an operator if the next input symbol should be read, being a terminal if an ε-move is proposed, and changing its nature time by time like an amphibious state. As a result, all deterministic context-free languages (with respect to reversed direction of reading) with added endmarker are accepted by deterministic bounded LP - machines.

e) $P \subseteq \Sigma(V\setminus\Sigma) \times [(V\setminus\Sigma) \cup \{\varepsilon\}]$: finite automaton in LP-notation,
$P \subseteq \Sigma(V\setminus\Sigma) \times [(V\setminus\Sigma) \cup \{\varepsilon\}]\Sigma^*$: generalized sequential machine in LP-notation. These types of automata are useful as subroutines of more complex LP - machines.

1.3. Algebraic Interpretation

From the algebraic point of view we study structures (M,Λ) consisting of a monoid M and a unary relation Λ, i.e. a distinguished subset of M [4,5], with the essential feature that an effective representation is given for the set M, its monoid operation $\cdot$, and mostly also for the relation Λ. For the important case where the monoid in question is a free monoid Σ^*, the interest is concentrated on representations of the subset. This is often achieved by a grammar (Σ, V, P, S) generating the language $L \subseteq \Sigma^*$. Actually this concept of formal grammars is nothing but that of embedding, that is embedding the structure (Σ^*, L) into a monoid (M,Λ) with distinguished subset Λ as a substructure such that (M,Λ) has an internal effective description. In our case this is $M := V^*$ and $\Lambda := \{\alpha \in V^* \mid S \overset{*}{\underset{P}{\Rightarrow}} \alpha\}$ where $\overset{*}{\underset{P}{\Rightarrow}}$ is the transitive and compatible hull of the finite relation $P \subseteq V^* \times V^*$. Then we have $(\Sigma^*, L) = (\Sigma^*, \Lambda \cap \Sigma^*) \subseteq (V^*, \Lambda)$. Clearly, $\overset{*}{\underset{P}{\Rightarrow}}$ being a finitely generated compatible preorder relation is a very natural relation in algebraic terms, compared for example to the idea of regulated rewriting systems [6].

While one fixes the multiplication in the extension structure (M,Λ) to be the concatenation for describing the generation of languages, the algebraic interpretation of acceptors for languages uses a fixed definition of the relation Λ but varying definitions of the multiplication in the extension monoid. To this end consider LP - machines algebraically : Let be $\mathcal{A} = (\Sigma, V, P, q)$ a deterministic LP - machine in normal form, and define an operation $\cdot$ for elements of Σ and $\Gamma := V\setminus\Sigma$ by

$$a \cdot A := \begin{cases} B_1 \ldots B_n b_1 \ldots b_m & \text{, if } aA \longrightarrow B_1 \ldots B_n b_1 \ldots b_m \text{ in } P \\ 0 & \text{otherwise, where } 0 \notin V \text{ is a new element} \end{cases}$$

$$a \cdot b := ab, \qquad A \cdot B := AB, \qquad A \cdot a := Aa$$

This operation defines a multiplication in $M := \Gamma^* \Sigma^* \cup \{0\}$:

$$(\ast) \qquad \alpha\cdot\beta = \begin{cases} \gamma & \text{, if } \alpha\beta \overset{*}{\underset{P}{\Rightarrow}} \gamma \in \Gamma^*\Sigma^* \\ 0 & \text{otherwise} \end{cases}$$

Then the language $L \subseteq \Sigma^*$ accepted by $\mathcal{A}$ is the set of left inverses of q in M:

$$L = \Lambda = \{ w \in M \mid w\cdot q = 1 \}.$$

Therefore the processing table for a deterministic LP-machine may be read as multiplication table of an effectively represented monoid containing the free monoid Σ^* of terminal words and the free monoid Γ^* of operators as submonoids. Associativity is guaranteed by the Lemma. The following result is stated without explicit reference to the notion of LP-machines.

Theorem 1: Let Σ, Γ be finite alphabets and M,$\cdot$ a monoid satisfying

(i) Σ^+, Γ^+, $\Gamma^+\cdot\Sigma^+ \cup \{0\}$ are subsemigroups of M,$\cdot$ where Σ^+, Γ^+ denote the corresponding free semigroups and 0 is a zero element of M,$\cdot$

(ii) $M = \Sigma^+ \cup \Gamma^+ \cup (\Gamma^+\cdot\Sigma^+ \cup \{0\}) \cup \{1\}$

(iii) $\Sigma^+ \cap \Gamma^+ = \Sigma^+ \cap (\Gamma^+\cdot\Sigma^+ \cup \{0\}) = \Gamma^+ \cap (\Gamma^+\cdot\Sigma^+ \cup \{0\}) = \emptyset$

(iv) $\cdot : \Gamma^+ \times \Sigma^+ \rightarrow \Gamma^+\cdot\Sigma^+$ is injective

Let be $q \in \Gamma$ and $L_M^{inv}(q) := \{ w \in M \mid w\cdot q = 1 \}$. Then $L_M^{inv}(q) \subseteq \Sigma^+$, and the language $(\Sigma^*, L_M^{inv}(q))$ is

(1) rational,
if $\Sigma\cdot\Gamma \subseteq \Gamma \cup \{0,1\}$ (1a)

(2) accepted by a deterministic homogeneous pushdown automaton,
if $\Sigma\cdot\Gamma \subseteq \Gamma^* \cup \{0\}$ (2a)

(3) (reversal of) a deterministic context-free language,
if $\Sigma\cdot\Gamma \subseteq \Gamma^* \cup \Sigma \cup \{0\}$ (3a)

(4) a deterministic bounded LP-language,
if $\Sigma\cdot\Gamma \subseteq \Gamma^* \cup \Gamma^*\Sigma \cup \{0\}$ (4a)

(5) recursive,
if all the computations of products $\alpha\cdot\beta \in \Sigma^*\cdot\Gamma^*$ by $\alpha\beta \overset{*}{\underset{\cdot|\Sigma\times\Gamma}{\Rightarrow}} \ldots$ terminate (5a)

(6) recursively enumerable,
if $\cdot$ satisfies $(\ast)$ where $P = \cdot|_{\Sigma\times\Gamma}$ (6a)

Conversely, for each language $(\bar{\Sigma}^*, L)$ of type (k), $k \in \{1,2,3,4,5,6\}$, there exists a monoid M,$\cdot$ satisfying (i) – (iv), (ka) with appropriately chosen alphabets $\Sigma \supseteq \bar{\Sigma} \dot{\cup} \{\#\}$, Γ such that $\#L = L_M^{inv}(q)$ for an element $q \in M$.

2. CONTEXT-SENSITIVE LP-LANGUAGES

Deterministic LP-machines are supposed to yield the decision about acceptance of the input word during the computation, that is, without being allowed to look for the end of input and to find out whether there is still another symbol in the input or not. Clearly, only those languages which are prefixes can be accepted in such a way deterministically. Therefore it is convenient to add an endmarker to the input, if deterministic acceptance is required in a more general setting.

Definition: A language $L \subseteq \Sigma^*$ is an unpointed bounded LP-language, if its pointed version $\#L$ is accepted by a bounded LP-machine $\mathcal{A} = (\Sigma', V, P, q)$ where $\Sigma \dot{\cup} \{\#\} \subseteq \Sigma'$. L is said to be deterministic, if $\mathcal{A}$ may be chosen deterministic.

For natural languages like Hebrew punctuation sometimes is less trivial than adding just a dummy symbol at the end of words.

2.1. Parsing

Theorem 2: Let be $G = (\Sigma, V, R, S)$ and $\mathcal{A} = (\Sigma, V, P, S)$ grammars satisfying

$$R \subseteq (\Sigma \cup \{\varepsilon\})(V\setminus\Sigma) \times (V\setminus\Sigma)^*\Sigma,$$

$$P \subseteq \Sigma \quad (V\setminus\Sigma) \times (V\setminus\Sigma)^*(\Sigma \cup \{\varepsilon\}),$$

such that the productions in R and P correspond to each other by

$$A \rightarrow B_1 \ldots B_n a \text{ in } R \iff aA \rightarrow B_n \ldots B_1 \text{ in } P,$$

$$bA \rightarrow B_1 \ldots B_n a \text{ in } R \iff aA \rightarrow B_n \ldots B_1 b \text{ in } P.$$

Then L(G) is accepted by $\mathcal{A}$.

Thus generating grammars for bounded LP-languages may be easily given in generalized Greibach normal form, allowing one-sided interchanging rules for terminals and variables. For these grammars the concept of rightbound derivations is the natural refinement of rightmost derivations for context-free grammars, i.e., to apply productions as far right as possible without disturbing the intended result. To visualize the interchanging rules within derivation trees we use the natural order relation for nodes at the same level and define crossings of edges between succeeding levels. As an example of how to use a bounded LP-machine for parsing we reconstruct the derivation tree with crossings from the LP-computation of $\mathcal{A}_=$ for input #10#10:

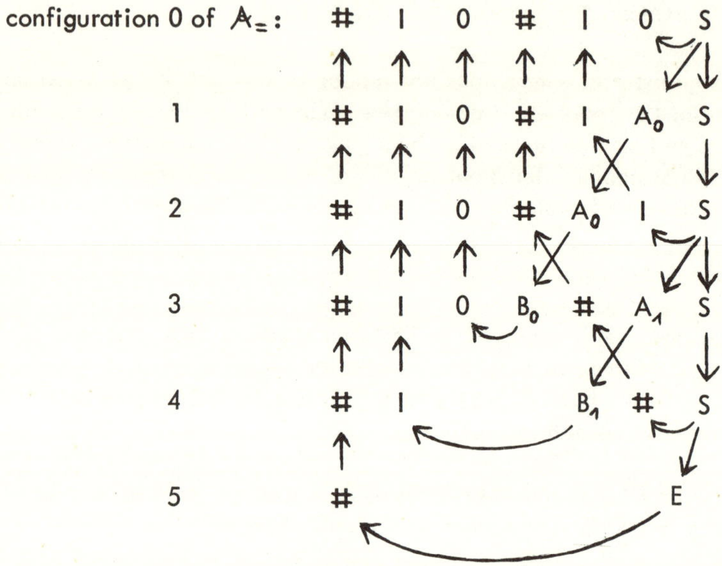

Result: a derivation tree according the grammar which corresponds to $A_=$ by Theorem 2

Of course, this construction has to be done much more carefully in the proof of Theorem 2 by an appropriate induction.

2.2. Unpointed Languages

We list some properties of bounded LP-languages.

Proposition 2: Given a deterministic bounded LP-machine A, it is undecidable whether the language accepted by A is empty or not.

This fact is reduced to Post's correspondence problem by the bounded LP-language of word-by-word translations $L_D = \{ \# * w_n * \dots * w_1 \# * u_n * \dots * u_1 \mid (w_i, u_i) \in D \}$ where $D \subseteq \Sigma^+ \times \Sigma^+$ is a finite dictionary. As a bounded LP-machine admits bounded LP-machines as subautomata, one proves

Proposition 3: The class of unpointed bounded LP-languages is closed under ε-free substitution. Especially it is closed under ε-free homomorphisms, union, concatenation, and catenation closure.

Of course, this composition of LP-machines cannot preserve determinism in general. As usual by compression of symbols and computation steps we obtain

Proposition 4: The class of (deterministic) unpointed bounded LP-languages is closed under inverse homomorphisms, left quotients, and right quotients with respect to a given word $s \in \Sigma^+$.

Since the class of deterministic context-free languages is neither closed under union nor intersection, the following fact is quite interesting:

Proposition 5: The class of deterministic unpointed bounded LP-languages contains the deterministic context-free languages and is closed under Boolean operations.

Moreover, from the point of view of applications, many useful languages are accepted deterministically by bounded LP-machines : for example languages like the graphs of addition and multiplication of binary numbers, and translations of various notations for arithmetic expressions. Finally, LP-machines provide an efficient accepting device also for some context-free languages which are not deterministic (as context-free languages) but are recognizable by deterministic bounded LP-machines.

References

[1] Aho, A./Ullman, J.: The theory of parsing, translation, and compiling. Vol. I, II. Englewood Cliffs, N.J. 1972/73.

[2] Berstel, J.: Transductions and context-free languages. Stuttgart 1979.

[3] Eilenberg, S.: Automata, languages, and machines. Vol. A, B. New York, London 1974/76.

[4] Kunze, M.: 'On syntactic monoids with distinguished subsets'. Proc. of the 2nd workshop-meeting on Categorical and Algebraical Methods in Computer Science and System Theory, Haus Ahlenberg, Herdecke 1978. Forschungsbericht Nr. 74, Abt. Informatik, Univ. Dortmund 1978. pp. 54 - 62.

[5] Sakarovitch, J.: 'Monoides pointes'. Semigroup Forum 18 (1979) 235-264.

[6] Salomaa, A.: Formal languages. New York, London 1973.

THE FIRST PASSAGE TIME PROBLEM WITH APPLICATIONS TO NEURONAL MODELING

L.M. Ricciardi and L.Sacerdote

Istituto di Scienze dell'Informazione, Università di Salerno, Salerno, Italy

ABSTRACT

A problem arising in a variety of subjects consists of evaluating first passage distributions for random functions defined by stochastic differential equations. This paper provides closed form expressions for mean and variance of the first passage time through a constant boundary for the Ornstein-Uhlenbeck (O.U.) process. Because of the complexity of the closed form expressions, asymptotic behaviors are studied. Furthermore, the statistics of the first passage time is investigated by the aid of a computer to shed some light on the input-output behavior of a formal neuron whose dynamics is modelled by a diffusion process of the O.U. type.

1. A DIFFUSION MODEL FOR A FORMAL NEURON

Consider a neuron characterized by $p+q$ dendrites, firing threshold S and zero resting potential. In virtue of the Kintchine theorem we may assume that each dendrite carries point-like excitatory or inhibitory signals that may be considered to be roughly Poisson distributed. This is a reasonable assumption when the frequency of post synaptic potentials is large enough. Let α_k $(k=1,2,\dots,p)$ be the rate of the excitatory pulses, of magnitude e_k, $(e_k>0)$ arriving along the k-th pathway and β_j $(j=1,2,\dots,q)$ the rate of the inhibitory pulses, of magnitude i_j $(i_j<0)$ traveling along the j-th pathway. If x is the membrane potential at a certain time, the arrival at that time of an excitatory (inhibitory) input along the k-th excita-

D. G. Lainiotis and N. S. Tzannes (eds.), Applications of Information and Control Systems, 226-236.

tory (j-th) inhibitory) pathway produces to the instantaneous transition

$$x \to x + e_k \qquad k = 1,2,\dots,p \qquad (x \to x + i_j \quad j = 1,2,\dots,q) \tag{1.1}$$

These assumptions imply that the membrane potential undergoes a random walk in continuous time. Furthermore, it is a time homogeneous Markov process so that its transition probably density function (p.d.f.) satisfies the Chapman-Kolmogorov equation:

$$f(x, t+\Delta t \mid x_0, t_0) = \int_{-\infty}^{\infty} dz\, f(x, \Delta t \mid z, t) f(z, t \mid x_0, t_0) \,. \tag{1.2}$$

Here the transition p.d.f. is defined as:

$$f = f(x, t \mid y, \tau) = \frac{\partial}{\partial x} \Pr\{ X(t) = x \mid X(\tau) = y \} . \tag{1.3}$$

If we assume that in the absence of external inputs the membrane potential exponentially decays toward its resting value with a time constant θ , recalling (1.2) we can write:

$$f(x, \Delta t \mid z) = \Big\{ 1 - \Delta t \Big[\sum_{1}^{p}{}_{k} \alpha_k + \sum_{1}^{q}{}_{j} \beta_j \Big] \Big\} \delta [x - (z - z\Delta t/\theta)] + \\ + \Delta t \sum_{1}^{p}{}_{k} \alpha_k \delta [x - (z - z\Delta t/\theta + e_k)] + \Delta t \sum_{1}^{q}{}_{j} \beta_j \delta [x - (z - z\Delta t/\theta + i_j)] . \tag{1.4}$$

Substituting (1.4) in (1.2) and making use of the approximation $(1 - \Delta t/\theta)^{-1} \approx 1 + \Delta t/\theta$ one thus obtains:

$$\frac{\partial f}{\partial t} = \frac{\partial}{\partial x}\Big(\frac{x}{\theta} f\Big) + \sum_{1}^{p}{}_{k} \alpha_k [f(x - e_k, t \mid x_0) - f(x, t \mid x_0)] + \\ + \sum_{1}^{q}{}_{j} \beta_j [f(x - i_j, t \mid x_0) - f(x, t \mid x_0)] . \tag{1.5}$$

Expanding the functions $f(x - e_k, t \mid x_0)$ and $f(x - i_j, t \mid x_0)$ as Taylor series about x Eq. (1.5) can be written as

$$\frac{\partial f}{\partial t} = \sum_{1}^{\infty}{}_{m} \frac{(-1)^m}{m!} \frac{\partial^m}{\partial x^m} (A_m f) , \tag{1.6}$$

Where we have set:

$$A_1 = -\frac{x}{\theta} + \eta_1$$

$$A_r = \eta_r \qquad (r = 2, 3, 4, \dots) \tag{1.7}$$

$$\eta_m = \sum_{1}^{p}{}_{k} \alpha_k e_k^m + \sum_{1}^{q}{}_{j} \beta_j i_j^m .$$

We now perform a diffusion approximation and set

$$\eta_r = 0 \qquad (r = 3, 4, \ldots) \tag{1.8}$$

While keeping $|\eta_1| < \infty$ and $0 < \eta_2 < \infty$. Expression (1.8) may be rigorously obtained by suitably letting all α_k's and β_k's diverge while simultaneously making all e_k's and i_k's infinitely small (1). Due to (1.8) from (1.6) we obtain

$$\frac{\partial f}{\partial t} = -\frac{\partial}{\partial x}\left[-\left(\frac{x}{\theta} + \eta\right) f\right] + \frac{\sigma^2}{2}\frac{\partial^2 f}{\partial x^2} \tag{1.9}$$

Where $\eta = \eta_1$ indicates the net mean input and $\sigma^2 = \eta_2$ indicates the magnitude of input fluctuations. Let us now remark that the change of variables

$$\begin{aligned} x - \eta\theta &\to x \\ x_0 - \eta\theta &\to x_0 \end{aligned} \tag{1.10}$$

throws Eq. (1.9) into:

$$\frac{\partial f}{\partial t} = \frac{\partial}{\partial x}\left(\frac{x}{\theta} f\right) + \frac{\sigma^2}{2}\frac{\partial^2 f}{\partial x^2} \tag{1.11}$$

We have thus restrict our analysis to the O.U. process described by Eq. (1.11) but the extension of our results to the case of Eq. (1.9) is straightforward through (1.10). In particular, the first passage time p.d.f. through S $g(t) = g(S, t|x_0)$ for the O.U. process provides us, on account of (1.10), with the neuron's firing probability density function.

2. MEAN AND VARIANCE OF THE FIRST PASSAGE TIME

The moments of the first passage time through a constant threshold for the O.U. processes may be determined by a straightforward differentiation of the Laplace transform of the first passage time p.d.f. Indeed, let

$$g_\lambda = g_\lambda(S|x_0) = \int_0^\infty e^{-\lambda t} g(S, t|x_0)\, dt \tag{2.1}$$

be the Laplace transform of $g(S, t|x_0)$ and let

$$t_n(S|x_0) = \int_0^\infty dt\, t^n g(S, t|x_0) \tag{2.2}$$

be the n-th order moment of the random variable T, defined by

$$T = T(S, x_0) = \inf\{t : x(t) > S \mid x(0) = x_0\}. \qquad (2.3)$$

One then has:

$$t_m(S \mid x_0) = (-1)^m \left(\frac{d^m g_\lambda}{d\lambda^m}\right)_{\lambda=0} \qquad (m = 1, 2, 3, \ldots) \qquad (2.4)$$

In particular, for the mean M and the variance V of the first passage time we have

$$M = M(S \mid x_0) = -\left(\frac{dg_\lambda}{d\lambda}\right)_{\lambda=0} \qquad (2.5)$$

$$V = V(S \mid x_0) = \left(\frac{d^2 g_\lambda}{d\lambda^2}\right)_{\lambda=0} - \left|\left(\frac{dg_\lambda}{d\lambda}\right)_{\lambda=0}\right|^2 .$$

In order to determine mean and variance via (2.5) the Laplace transform of the first passage time p.d.f. has to be known. Siegert (2) has shown that for the Ornstein-Uhlenbeck process g_λ is the solution of the ordinary differential equation

$$\frac{\sigma^2}{2}\frac{d^2 g_\lambda}{d\lambda^2} + \frac{x_0}{\theta}\frac{dg_\lambda}{d\lambda} - \lambda g_\lambda = 0 \qquad (2.6)$$

satisfying

$$\lim_{x_0 \to S} g_\lambda(S \mid x_0) = 1$$
$$\lim_{x_0 \to -\infty} g_\lambda(S \mid x_0) = 0 \qquad (2.7)$$

Making use of (2.6) Capocelli and Ricciardi (3) found that the Laplace transform of the firing probability density function is

$$g_\lambda(S \mid x_0) = \exp\left\{\frac{(x_0 - \eta\theta)^2 - (S - \eta\theta)^2}{2\sigma^2\theta}\right\} \frac{D_{-\lambda\theta}\left[\left(\frac{2}{\sigma^2\theta}\right)^{1/2}(\eta\theta - x_0)\right]}{D_{-\lambda\theta}\left[\left(\frac{2}{\sigma^2\theta}\right)^{1/2}(\eta\theta - S)\right]}. \qquad (2.8)$$

Where $D_\nu(z)$ denotes the Parabolic Cylinder Function. After conveniently expressing $D_\nu(z)$ in terms of the Kummer function $\phi(a,c;z)$, $M(S|x_0)$ and $t_2(S|x_0)$ can be calculated via (2.4). Thus doing, the mean firing time is found to be

$$M = M(S|x_0) = \theta\left\{\frac{1}{2}\left[\sum_{m=1}^{\infty}\frac{y^{2m}}{m(2m-1)!!} - \sum_{m=1}^{\infty}\frac{x^{2m}}{m(2m-1)!!}\right] + \left(\frac{\pi}{2}\right)^{1/2}\left[x\,\phi\left(\frac{1}{2},\frac{3}{2};\frac{x^2}{2}\right) - y\,\phi\left(\frac{1}{2},\frac{3}{2};\frac{y^2}{2}\right)\right]\right\} \qquad (2.9)$$

whereas the second order moment is given by

$$\begin{aligned} t_2(S|x_0) = \theta^2\Bigg\{&\left[\sum_{m=1}^{\infty}\frac{x^{2m}}{m(2m-1)!!} - \sum_{m=1}^{\infty}\frac{y^{2m}}{m(2m-1)!!}\right]\left[\left(\frac{\pi}{2}\right)^{1/2} y\,\phi\left(\frac{1}{2},\frac{3}{2};\frac{y^2}{2}\right)\right. \\ &\left.-\frac{1}{2}\sum_{m=1}^{\infty}\frac{y^{2m}}{m(2m-1)!!}\right] + \left[x\,\phi\left(\frac{1}{2},\frac{3}{2};\frac{x^2}{2}\right) - y\,\phi\left(\frac{1}{2},\frac{3}{2};\frac{y^2}{2}\right)\right]\Bigg[(2\pi) \\ &-\left(\frac{\pi}{2}\right)^{1/2}\sum_{m=1}^{\infty}\frac{1}{m(2m+1)} + \left(\frac{\pi}{2}\right)^{1/2}\sum_{m=1}^{\infty}\frac{y^{2m}}{m(2m-1)!!} - \pi\,y\,\phi\left(\frac{1}{2},\frac{3}{2};\frac{y^2}{2}\right)\Bigg] + \\ &+\frac{1}{2}\left[\sum_{m=2}^{\infty}\frac{x^{2m}}{m(2m-1)!!}\sum_{k=1}^{m-1}\frac{1}{k} - \sum_{m=2}^{\infty}\frac{y^{2m}}{m(2m-1)!!}\sum_{k=1}^{m-1}\frac{1}{k}\right] + \\ &+(2\pi)^{1/2}\left[\sum_{m=1}^{\infty}\frac{y^{2m+1}}{m!\,2^m(2m+1)}\sum_{k=1}^{m}\frac{1}{2k-1} - \sum_{m=1}^{\infty}\frac{x^{2m+1}}{m!\,2^m(2m+1)}\sum_{k=1}^{m}\frac{1}{2k-1}\right] \end{aligned} \qquad (2.10)$$

Here we have set

$$x = \left(\frac{2}{\sigma^2\theta}\right)^{1/2}(\eta\theta - x_0)$$
$$y = \left(\frac{2}{\sigma^2\theta}\right)^{1/2}(\eta\theta - S) \qquad (2.11)$$

From (2.11) and (2.9) the variance can be immediatly written down. A neurobiologically interesting choices of the values for the parameters has been made to compute mean and variance. It is important to remark that the only approximation made in order to compute the above statistics arises due to the finiteness of the number of terms taken into account to estimate infinite series. For our choices of values of the parameters excellent results may be obtained with speady computation because only the first few terms in the series contribute appreciably.

A first analysis of (2.10) and (2.9) has been performed under the assumption that no excess of input exists, that is $\eta = 0$ Furthermore it is not restictive the assumption $X_0 = 0$, being the process linearly position dependent.

Figure 1 shows the behavior of V as a function of M for the following choice of the parameters $\theta = 5$ msec, $S = 30$ mV. It is interesting to note that for $\sigma^2 > 5\ 10^4\ (mV)^2/ms$ this function is well fitted (in the least square sense) by the straigh line $y = 7.31x - 0.05$. This may be understood by analysing the asymptotic behavior of M and V for $\sigma^2 >> S^2/\theta$. Setting $X_0 = 0$ from (2.3) and (2.10) we obtain

$$M \sim \theta \sqrt{\pi} \frac{S}{\sigma\sqrt{\theta}} \qquad (2.12)$$

$$V \sim 2\theta^2 \sqrt{\pi} \left(1 - \sum_1^\infty {}_m \frac{1}{2m(2m+1)}\right) \frac{S}{\sigma\sqrt{\theta}} = \theta^2 \sqrt{\pi} \ln 4 \frac{S}{\sigma\sqrt{\theta}}.$$

Thus, whenever the threshold is negligible with respect to the magnitude of input fluctuations, the neuron's output is approximatively Poisson distributed if time is expressed in units of θ. This feature is of great importance in the analysis of the validity of the considered model because Poisson distributed neuron's outputs have been experimentally observed. It is to be remarked that another linear fitting holds for a second range of σ^2. However, the reason for such linear dependence is not to be searched into the asymptotic approximation (2.12). Figure 2 shows the behavior of $M(3010)$ as a function of the magnitude of the input fluctuations. We remark that M is larger than 10 msec (a neurobiologically standard situation) only for small values of σ^2 ($\sigma^2 < 400\ mV^2/msec$).Furthermore, as shown by Fig.3, $V(3010)$ rapidly increases as σ^2 decreases. Thus, in this situation, the mean firing time is not a good statistical parameter, due to the spread of the range of the most likely firing times. All this suggests that the neuronal model considered may be improved by considering both an absolute and a relative after firing refractoriness. This will be the object of a paper in preparation.

Figures 4 and 5 shed some light on the role of the neuronal threshold. This is of great importance in our study because a

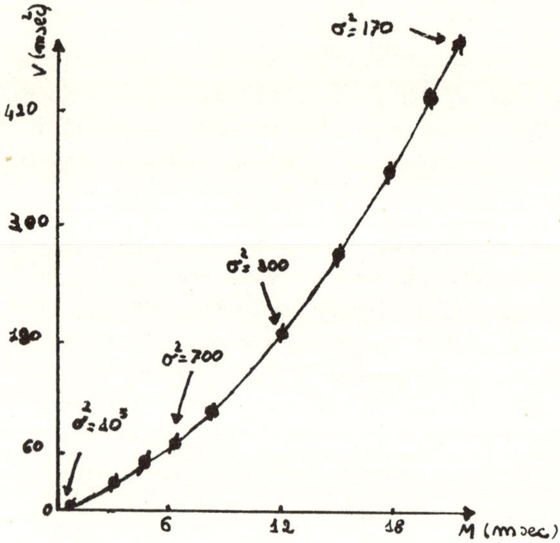

Fig.1 For $\eta = 0$ and $\theta = 5$ ms the variance $V = V$ (30|0) is plotted as a function of the mean firing time $M = M$ (30|0) for different values of magnitude σ^2 of the input fluctuations. Starting from the bottom σ^2 varies between 10^5 and 170 in units of $(mV)^2/msec$.

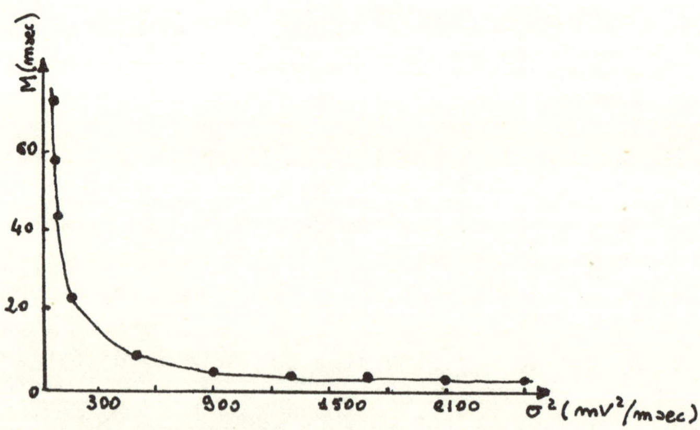

Fig.2 The decreasing of the mean firing time as a function of σ^2 is indicated for the same choice of η and θ as in the previous figures.

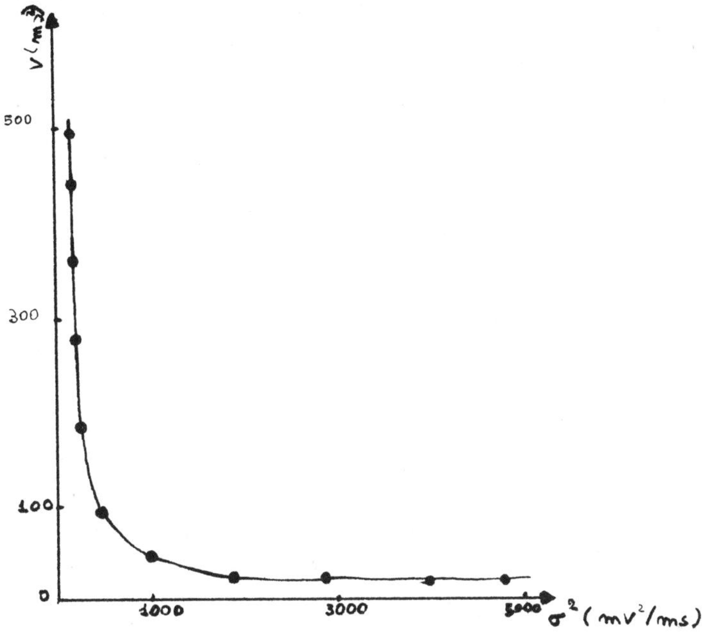

Fig.3 For $\eta = 0$ and $\theta = 5$ ms the variance of the firing time is plotted versus σ^2

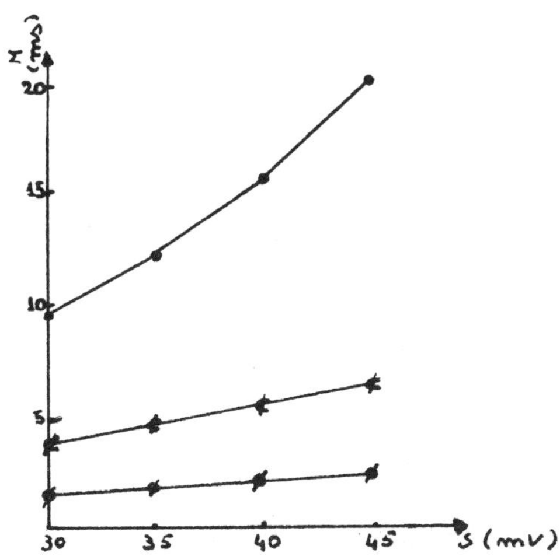

Fig.4 Mean firing time as a function of S for $\eta = 0$ and $\theta = 5$ msec Here σ^2, in units of (mV²/msec), takes the values 400 (•), 1500 (¢) and 6700 (ϕ).

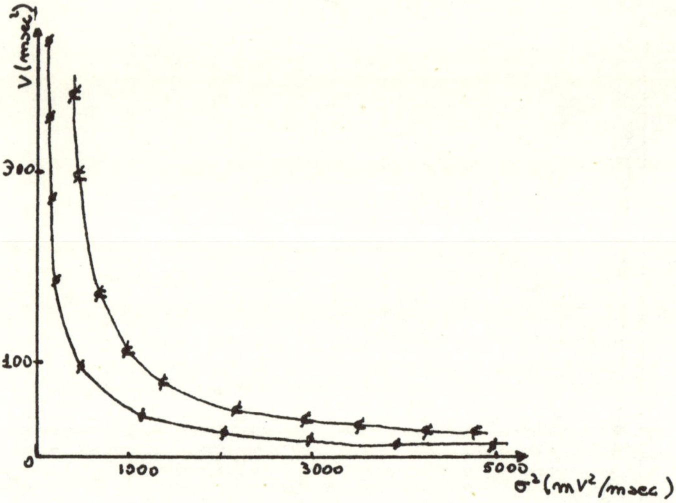

Fig.5 Variance of the firing time versus σ^2 for $\eta = 0$, $\theta = 5$ms and thresholds (in mV): $S = 45$ (ɸ) and 30 (ɸ)

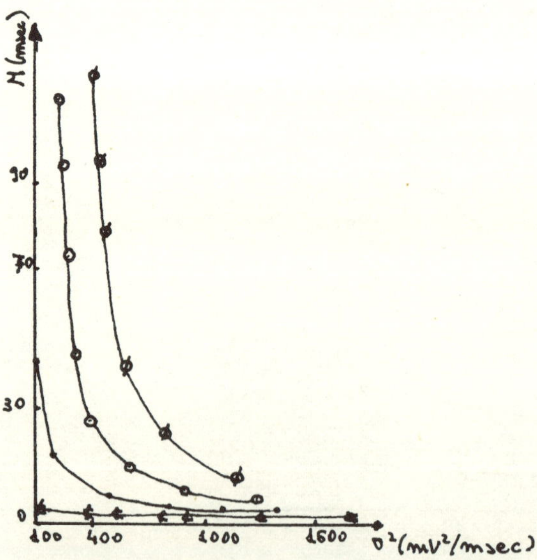

Fig.6 For $\theta = 5$ms and $\eta = -10$mV(ɸ), -5mV(o), 0 (•) and 10 mV (ɸ) the mean firing time $M = M\,(30|0)$ is plotted as a function of σ^2.

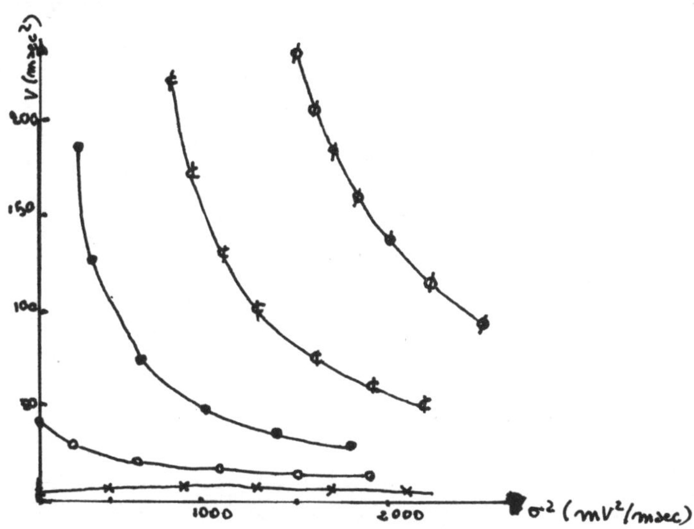

Fig. 7 For θ = 5ms and η= -10mV (ϕ) -5mV (¢), 0 (•) 5mV (o) and 10mV(X) the variance V(30/0) of the firing time is plotted as a function of σ^2

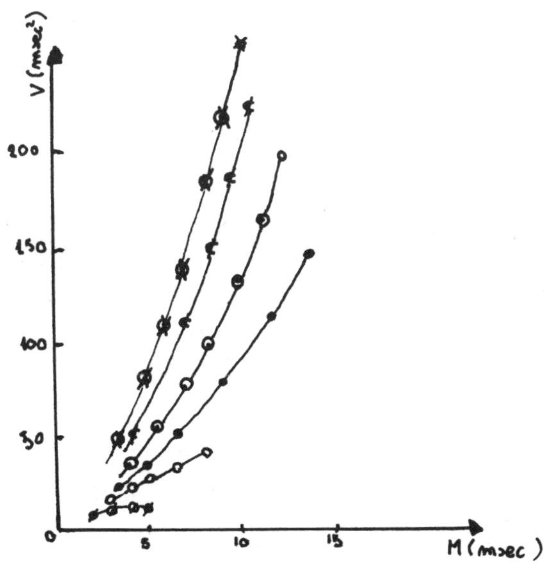

Fig.8 Variance V=V(30|0) versus M=M(30|0) for different values of η and σ^2.Starting from the left,the plotted curves correspond to η -10mV, -5mV, 0,3 mV,5mV and 7mV and 10 respectively.

possible way to account for neuron's firing refractoriness is to use time dependent thresholds. From figure 4 and 5 we note that for small values of σ^2 relevantly different results arise as the threshold is made to change. Figures 6,7 e 8 illustrate finally the role of the net mean input η . Fig. 6 shows that a larger mean firing time arises in the presence of a non-zero net inhibition while the opposite occurs for non-zero net excitation. Fig.8 shows the behavior of V as a function of M for various values of η . As before, as σ^2 increases a linear fitting because possible, the range of this fitting depending now upon the value of η . A more comprehensive study may be found elsewhere (4).

CONCLUSION

Mean and a variance of the first passage time prob.p.f. have been evaluated for a O.U. process through a constant threshold attempting to shed some light on the validity of a model of a neuron. However our results may be useful in other fields. Our discussion provides an indirect confirmation of the validity of the considered model and suggests some possible ways to improve the model. The method used to evaluate mean and variance seems speedier than other methods known in the literature (5).

ACKNOWLEDGMENTS

This work has been supported in part by the Group for Mathematical Information Sciences (GNIM) of the National Council for Research.

REFERENCES

(1) Ricciardi,L.M.: 1976, Biol.Cybernetics 24,pp.237-240

(2) Siegert,A.J.F.: 1951, Phys.Rev.81,pp.617-623

(3) Capocelli,R., Ricciardi,L.M.: 1971, Kybernetik 8,pp.214-223

(4) Ricciardi,L.M., Sacerdote,L.: 1979,Biol.Cybernetics

(5) Keilson,J., Ross,H.: 1975, Sel.Tables Math.Stat.3,pp.233-328

DIGITAL FREQUENCY ANALYSIS OF THE ALCOHOL CHANGED ELECTRORETINOGRAM

J. Levett, R. Jaeger and K. Ladien

Depts. of Physiology and Ophthalmology
Rush Presbyterian-St. Lukes Medical Center
1753 West Congress Parkway
Chicago, Illinois 60612 U.S.A.

The electrical response of the retina to a bright flash of light is termed the electroretinogram (ERG). The ERG waveform may be altered by changing the intensity of the background illumination or by changing the color of the flash. Using a white flash at a 10 foot lambert level of background illumination, we have studied the effects of ethyl alcohol (ETOH) on the ERG. The ERG waveform may be divided into components, the best known being the b-wave. Our studies have centered around a secondary oscillation of the ERG which we have termed the b'-wave. The b'wave has been found to be selectively and reversibly attenuated by both acute and chronic ETOH treatment. So far most of our work has centered around frogs (1), although we have also observed similar ERG changes with ETOH in humans and rabbits. Our data collection procedures have been summarized elsewhere (1); briefly ERGs are obtained with corneal or saline moistened electrodes, cotton wicks, coupled to silver-silver chloride electrodes. ERGs are amplified (1000x) and filtered for oscilloscope display, processing on-line by an averaging computer, and stored on tape by a digital computer for later analysis.

Typical frog ERG data under normal and alcohol conditions are shown in Fig. 1A. Past analyses have focused on the amplitude and latency measurements of the peaks in the ERG. These have typically been the amplitudes of the b- and b'-waves and their latencies to peak from delivery of stimuli. The ratios of amplitudes and latencies have also been used. In the present study ERGs were collected and stored on a digital computer. The ERGs were bandpass filtered (0.25-50 Hz) prior to digitizing at 1000 samples per sec. for 256 msec.

D. G. Lainiotis and N. S. Tzannes (eds.), Applications of Information and Control Systems, 237-239.

The data were plotted in the time domain to study the ERG waveform. The frequency analysis consisted of multiplication by a cosine taper time window followed by fast fourier transformation (FFT). Only the magnitude of the FFT is included in the present study. The frequency characteristics shown in Fig. 1C are for the ERGs of Fig. 1A. Note that the loss of the b'-wave corresponds to a reduction in the frequency components around 15 Hz.

To explain the oscillatory potentials we make the assumption that the light adapted oscillations subsequent to the b-wave are generated by the response of the Muller glial cell to potassium. Glial cells are known to behave as K+ electrodes (2), but oscillatory potentials have not been observed in their electrical response (2,3). We postulate the reason for this is cell damage due to necessary impalement by the electrode. From an engineering point of view we can model the system as a complex unitary process.

We have so far modeled the generation of the ERG in response to light by a parallel RLC circuit driven by a K+ current source as shown in Fig. 2.Inductive elements are disturbing in membrane charcteristics since the Hodgkin-Huxley formulation (4) utilized only resistance and capacitance. Cole approached the same problem in an analysis of the squid axon (5) and found the presence of an inductive reactance. Teorell (6) demonstrated oscillation of trans-membrane potential and modeled its properties with RLC elements. More recently Deutsch (7) modeled sensory processes with RLC elements.

Our parallel RLC model (8) was simulated on a digital computer in an interactive mode with a CRT display of the ERG waveform that could be modified by changing R, L and C. A large number of simulations were run to give close matches to real data. (Simulated responses in Fig. 1B). An identical frequency analysis of the simulated responses gave results consistent with the actual data as shown in Fig. 1D. A parameter sensitivity analysis of the model give in Fig. 3 shows that increasing membrane resistance gives a change in the amplitude ratio of the b-wave/b'-wave ratio that is similar to what is observed in this ratio in real ERG data following ETOH.

The ERG is a sensitive indicator of alcohol. The light-adapted ERG offers a routine test for use with alcoholic populations utilizing the b'-wave.

Selective and reversible changes in both temporal and frequency characteristics of the ERG are seen with alcohol. One mechanism to explain these findings is that alcohol increases membrane resistance. A delay of the b'-wave subsequent to amplitude changes can be explained by the inductive term.

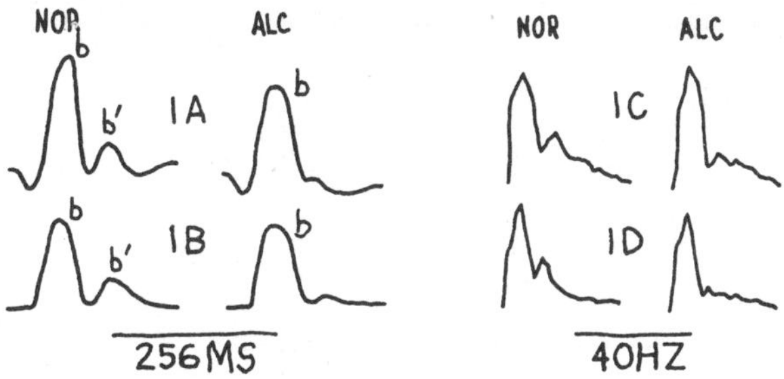

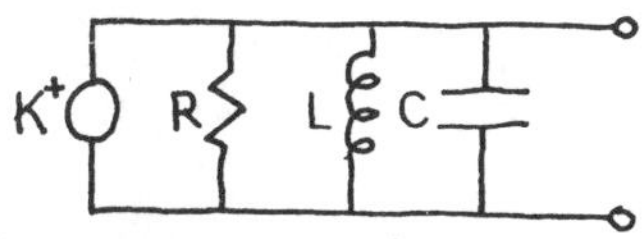

2

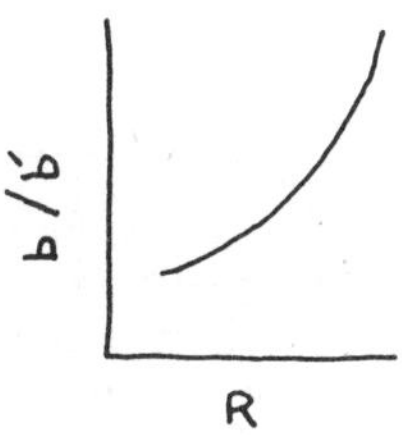

3

The use of frequency domain analysis permits the development of a dose-response characteristic in spectral terms for drug studies. Our current work involves extending the frequency domain analysis to permit transfer function determination.

Acknowledgement: Supported by the Helen Regenstein Foundation, Department of Ophthalmology, Rush Presbyterian - St. Lukes Medical Center

References:

(1) Levett, J. et al., Currents in Alcoholism 3: 47, 1978.
(2) Miller, R. & Dowling, J., J. Neurophysiol. 33: 323, 1970.
(3) Oakley, B. & Green, B., J. Neurophysiol. 39: 1117, 1976.
(4) Hodgkin, A. & Huxley, A., J. Physiol. 117: 500, 1952.
(5) Cole, K., J. Gen. Physiol. 25: 29, 1941.
(6) Teorell, T., J. Gen. Physiol. 42: 8, 1951.
(7) Deutsch, S., TIT J. Life Sciences, 29, 1971.
(8) Levett, J. and McAvinn, J.D., T.I.T.J. Life Sciences, 1979

A SCHEDULING ALGORITHM FOR A RADIOLOGICAL SERVICE

E. Burattini°, G. Della Monica°, M. Furnari*

° Laboratorio di Cibernetica, Arco Felice, Naples, Italy

* Istituto di Radiologia, II^ Facolta' di Medicina e Chirurgia, Via Pansini , Naples, Italy

ABSTRACT

Using a heuristic scheduling algorithm we have designed a simple computerized procedure by means of which it was possible to rationalize a X-rays service. In such a way we attempted to optimize the waiting times for radiological sections. A mathematical formalization of the problem, the complexity analysis of the algorithm and the experimental results showing the improvement obtained, are presented.

INTRODUCTION

In recent years, there has been increasing concern expressed regarding the efficiency of diagnostic radiology personnel as well as the utilization of radiographic equipments. The utilization of radiographic examination rooms or diagnostic rooms, is low and a typical total time spent by a single patient in the X-ray departement is more than 120 min. An analytical approach tò improving the staff and the equipment utilization, as well as decreasing patient waiting time, is impracticable because of the complexity of the operation of diagnostic radiologic departments [6]. Trial and error experimentation could be used, but the cost of such method is high in terms of disruption of the operation of the department and in terms of expenditures for equipment and personnel, especially if there is no assurance that the changes will be beneficial. The diagnostic radiology department is a service facility that provides health care primarily to patients in the Hospital. The problem areas in diagnostic radiology include utilization of facilities, technicians and radiologists. The managerial problems encountered

D. G. Lainiotis and N. S. Tzannes (eds.), Applications of Information and Control Systems, 240-248.

in the administration of the department are very similar to those encountered in any corporation management.

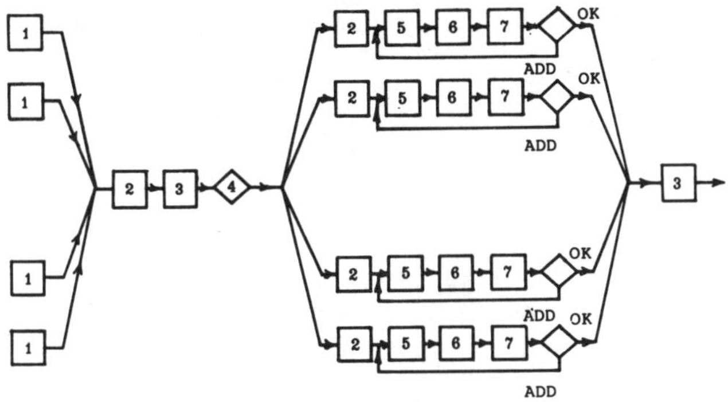

fig.1

Therefore we have sketched out (fig.1) a model for patient flow in the radiologic department of the Hospital of the II Faculty of Medicine and Surgery of the University of Naples. These clinics are spread over a large area and are connected by bus services. The daily patient flow is about 120, there are 9 diagnostic rooms and a radiologic staff of 100 persons. The model of fig.1 demonstrates the bottleneck in our diagnostic radiologic department for the resources utilization. Next we have sought a strategy to plan an efficient patient room assessment, based on the following assumptions: a) it must be close to the present organizational pattern, that is easily accepted by the radiological staff of the department; b) it must reduce the patient waiting time and increase the resource utilization avoiding at the same time the possibility of accepting patients who must not receive radiation due to preceding treatment; c) it must classify the applications according to the geographical position of the clinics.

COMPUTATIONAL COMPLEXITY

At a first approximation level the problem of patients scheduling for a radiological department consists in assigning to m diagnostic rooms the applications of the n patients in such a way that the final time in which every application is satisfied will be as minimal as possible. The scheduling must take into account that for any patient the related application either can or must be executed in different rooms and the duration of every requested examination is different. This problem can be formalized as a well known problem of operational research: the job-shop problem.
The general job-shop problem can be formulated as follows. There are n jobs $J_1,\ldots,J_n$ that have to be processed on m machines $M_1,\ldots$

..,M_m. Job J_i (i=1,...,n) consists of a sequence of j tasks (j=1,. ..n_i) which must be performed each on a single machine K in a time $t_{i,j,k}$, and must be completed before task j+1 can start. Then we must minimize the finish time FT(S) of the schedule S, that is minimize the time at which all tasks of all jobs have been completed. Clearly in our case any job represents a patient, the tasks of every job are the radiological examinations relative to each patient, the machines are the diagnostic rooms. Thus, algorithms solving the job-shop problem, can be used to schedule the patients of a radiological department. Analyzing the job-shop problem it is clear that there is an infinite number of schedules for any instance of it. The number of possible schedules is so large that solutions by exhaustive enumeration and comparative evaluation are not feasible even for relatively small problems and large computers. It is even difficult to give a good upper bound on the number of the busy schedules (i.e. schedules in which at all times from start to finish at least one processor is processing a task). In the general case the number of operations on a given machine could be greater than n, and the number of busy schedules far greater than $(n!)^m$. It should be sufficient to note that for n=6 and m=5,$(6!)^5$ is approximately 1.93×10^{14}, which is more than the number of micro seconds in six years [2]. It hardly seems likely that adequate computer time would ever be made available to solve a problem of this size by exhaustive enumeràtion, or that computer capacity will ever exist that could solve problems of realistic dimensions by this approach.
In terms of computational complexity we have that the job-shop problem is NP-complete [3], that is, it belongs to a set of combinatorial problems which include the travelling salesman, finding the chromatic number of a graph, etc. It is known that all the NP-complete problems are equivalent in the following intuitive sense. If one can find a polynomial time-algorithm (i.e. an algorithm running in a time polynomial in the size of the input) one can find a polynomial-time algorithm for every other problem in the class, according to a procedure that normally varies from one problem to another. Thus either there exist polynomial-time algorithms for all the NP-complete problems, or none of them has a polynomial-time algorithm; we do not known which of these two assertions is the correct one, however there is strong evidence to suggest that all NP-complete problems are inherently intractable. For these reasons we have to resort to some heuristic procedure which, running in a polynomial-time, gives results not too far from the optimal ones.
The use of heuristic leads obviously to the need for some "relative measure" of how good it is. Usually relative measure are defined as ratios between absolute measures and the measure of something sufficiently important to be taken as reference. In the case of the approximation of optimization problems, many authors identify this reference with the optimal solution; so that the relative measures usual in the litterature [5] are:

$$r'=q_A/q^* \quad (1) \qquad \text{and} \qquad r''=(q_A-q^*)/q^* \quad (2)$$

where q_A is the value of the optimizing function corresponding to the solution given by the algorithm A; q^* is the value of the optimizing function corresponding to the optimal solution.
In a previous paper [1] we have introduced and discussed a different "relative measure": $r''' = (q_A - q^*)/(\hat{q} - q^*)$ (3)
where $\hat{q}$ is the value of the optimizing function corresponding to the worst solution. For a more detailed discussion on this subject we refer the reader to [1]. In the present paper we will use the functions (1) and (3).
Let us now turn our attention to the finish time properties of busy schedules. In [4] it was demonstrated that:
Lemma 1. Let S^* be an Optimal Finish Time schedule for an m>2 processor n-job job-shop problem. Let S be any busy schedule for this problem. Then $FT(S)/FT(S^*) \leq m$.
This isn't a good bound since we cannot demonstrate that the ratio $FT(S)/FT(S^*) < m$. Moreover if we use function (3) we can show that $(FT(S)-FT(S^*))/(FT(\hat{S})-FT(S^*)) \leq 1$; i.e. it is always possible to have some input for which any "reasonable" algorithm gives the worst solution. As the algorithm we present in next section belongs to the previous class of "reasonable" ones, then also in this case we could have some instance for which the solution obtained is the worst possible. Analyzing the set of possible instances, that is of instances which really can occur utilising our procedure, it apperas, but has not so far been proved, that we will never have a worst case behaviour.

SCHEDULING ALGORITHM

After a careful examination of the present radiological service management, we have proposed a new procedure which is schematically shown in fig.2. From this scheme it appears that the applications

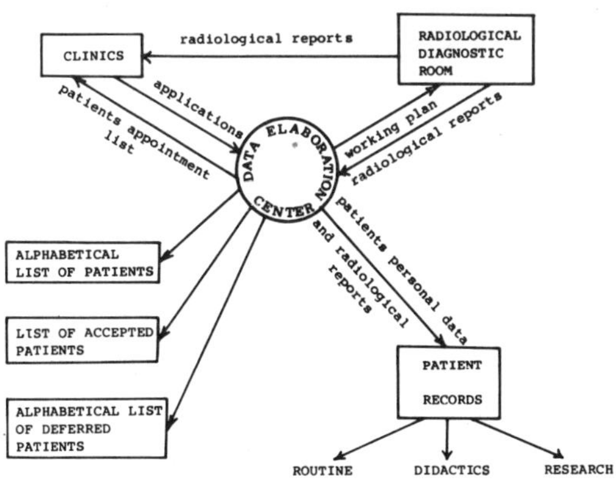

fig. 2

to radiology service must be applied by single clinics using suitable forms. All these forms will be the input for the scheduling algorithm. The output of the procedure will be: a) the work planning for the diagnostic rooms; b) the patient list for each clinic; c) a series of patient lists in lexicographic order, etc.
The purpose of our algorithm was principally to minimize the inconveniences for the patients, reducing their waiting times in the diagnostic rooms. This aim is reached by means of: 1) giving the right schedule for each patient; 2) organizing sequentially possible multiple radiological examination for a single patient; 3) minimizing the inactivity times for the diagnostic rooms. A pidgin Algol description of the scheduling algorithm is the following:
Let n be the number of patients for each of which the N_c clinics have requested j ($j=1,...,n_i$) radiological examinations for the m diagnostic rooms. Let N be the number of total examinations.

ALG. SCHEDULE

1. Sort for priority. For the same priority group together all the patients coming from the same clinic.
2. Initialize: r=0;
3. _if_ r=N _go to_ 7;
   else r=r+1, let h be the patient number and i the relative radiological examination number;
4. _if_ h≠h-1, i.e. the request is not relative to the preceding patient, _then go to_ 5;
   else if the request can be carried out in the same diagnostic room as the preceding ones
   then schedule the request; _else_
5. search for the least busy diagnostic room and schedule the request;
6. _if_ i=1 _then go to_ 3;
   else memorize the relative diagnostic rooms, _go to_ 3;
7. _if_ for all h i=1 then _go to_ 9;
8. _else_ examine again the scheduling exploiting to the utmost the inactivity times;
9. _write_ all the requested lists.

The relative flow-chart is shown in fig.3.

SOME EXPERIMENTAL RESULTS

From the theoretical results of second section it appears that a complete analysis of the behaviour of our algorithm in the worst case is very difficult to carry out. Moreover the order in which the list of applicant patients is formed and is presented to the algorithm depends strictly on some priorities relative to disease, waiting list, clinic of origin, etc. making, if possible, the analysis even harder. For these reasons, to give an idea of how our algorithm works, we report here the experimental results of its application to more than one hundred cases.

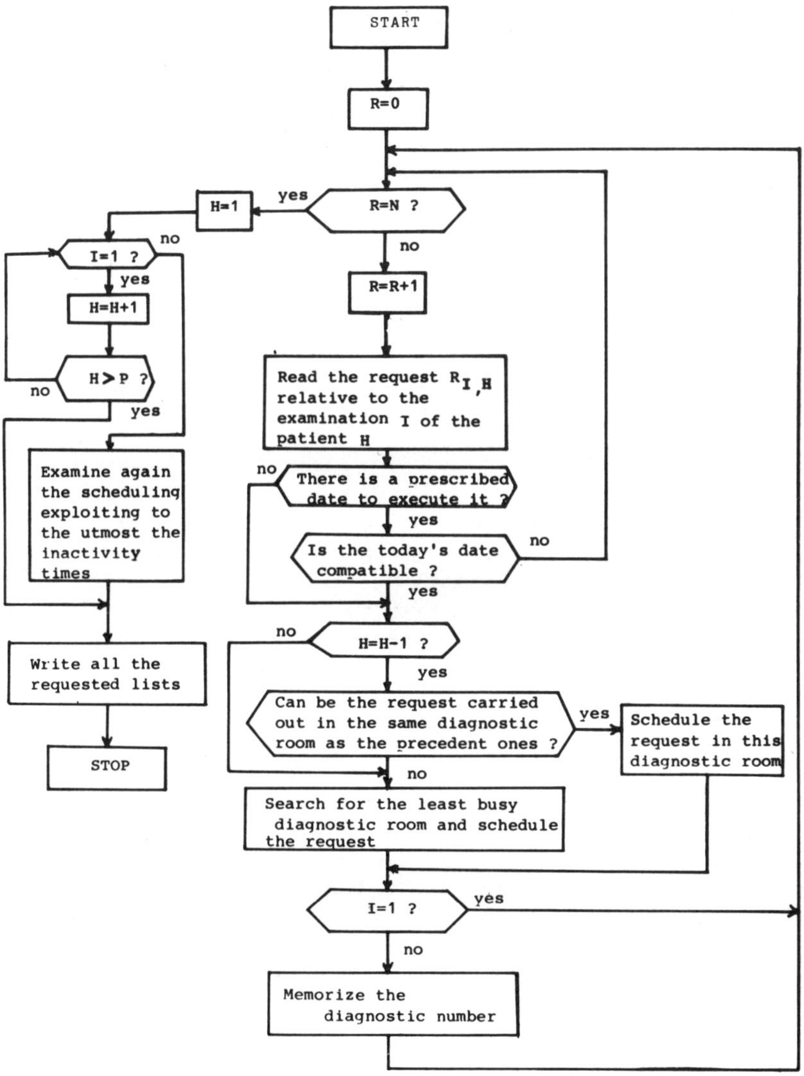
START
R=0
R=N ?
yes
H=1
no
I=1 ?
yes
H=H+1
H > P ?
R=R+1
Read the request $R_{I,H}$ relative to the examination I of the patient H
There is a prescribed date to execute it ?
Is the today's date compatible ?
Examine again the scheduling exploiting to the utmost the inactivity times
Write all the requested lists
STOP
H=H-1 ?
Can be the request carried out in the same diagnostic room as the precedent ones ?
Schedule the request in this diagnostic room
Search for the least busy diagnostic room and schedule the request
I=1 ?
Memorize the diagnostic number

Fig. 3

To measure the performance of the algorithm we report in fig.4 the total time (i.e. CPU time + I/O time) versus the input size ($|S| = m x n x N$). The results obtained show a quasi-linear dependence. This means that for actual inputs we have that the running time should be quite proportional to the size of the input. In Tab.1 we also report the running times relative to a little set of experiments. In average, we have no more than 30 sec to run our algorithm.

Tab. 1

m	n	N	S size	T sec	t min
7	52	64	23300	13.65	9.09
8	57	72	32800	14.59	19.60
8	62	81	40176	14.87	8.43
8	79	104	65728	19.72	7.06
7	96	119	79900	20.66	8.55
7	100	123	86100	21.41	6.36
8	108	134	115775	23.11	19.60
8	132	172	181600	26.16	21.00
8	141	174	196272	28.95	18.90

To study the heuristic behaviour of our procedure we thought it convenient to measure, using the evaluation function (3), the utilization of diagnostic rooms in terms of minimum finishing time.

To analize the minimum finishing time of the algorithm, since it was impossible for us to find an analitical expression for the value of the optimizing function corresponding to the optimal solution and that corresponding to the worst ones, we take as $FT(S^*)$ the value T/m (T=sum of task times for all jobs) and as $FT(\hat{S})$ the the value T, which can be considered the ideal values for the optimal and worst case. In fact, it is obvious that $FT(S^*) \geq T/m$ and $FT(\hat{S}) \leq T$ whatever can be the input.

Drawing the function $r_{FT} = (FT(S) - T/m)/(T - T/m)$ versus the size of input $|S|$ we observe (see fig.5) that for small inputs the value of the function is high and it decreases when the input size increases, becoming quite constant for inputs larger than $30 x 10^4$.

This behaviour can be explained by considering that when it is a large input we have many patients and the percentage of single examinations is higher. Consequently the possibility for the algorithm to cover the dead time of the diagnostic rooms is greater. Viceversa, with small inputs there are few possibilities of single examinations and therefore to cover the dead times of diagnostic rooms. However we believe that the value of 0.20 for the evaluation function in the majority of cases is a good value. In fact, it means that in average the distance of the solution, found by the algorithm, from the optimal one is of the order of 20% of the distance between the worst solution and the optimal one. Since in the actual situation the worst case is bounded (at most 12 hours) the maximum allowed error is also bounded (at most of the order of 2 h and 20 min).

The most evident result we obtained, using this procedure for historical data, was a drastic reduction of the dead times. In average comparing the past hand-made scheduling of patient applications with the scheduling carried out with our procedure, we have a reduction of finishing time of about 4 hours.

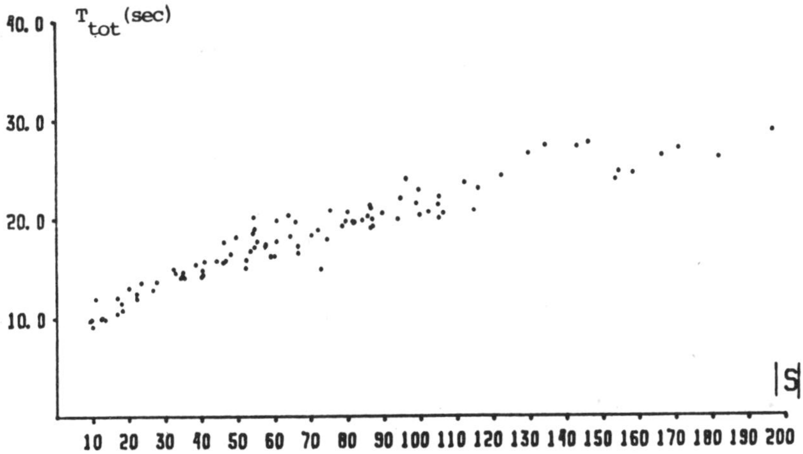

FIG. 4

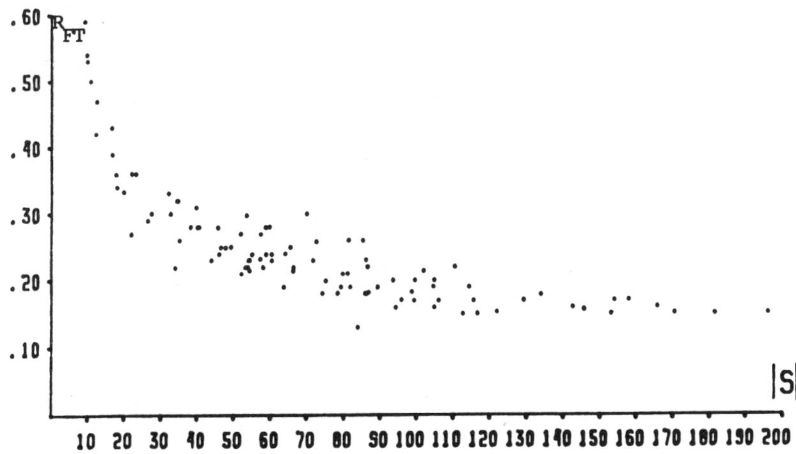

FIG. 5

From Tab.1 we deduct that the average waiting time for each patient is never greater than 30 min, that is, we reduce with the application of this algorithm tha waiting time of a factor 4.
In conclusion we can state that the waiting time is very well optimized even if the finishing time is not, having thus achieved our purpose.

The authors are very grateful to Prof. E.Porta and Prof. A.Sodano of the Institute of Radiology of the II Faculty of Medicine and Surgery of the University of Naples for their suggestions and support during the preparation of this work.

REFERENCES

1) Aiello,A., Burattini, E., Massarotti, A., Ventriglia, F., 1977 A new evaluation function for approximation algorithms, Proc. of Informatica 77, Bled, Yugoslavia.
2) Conway, R.W., Maxwell, W.L., Miller, L.W., 1967, Theory of scheduling, Addison Wesley Pub. Co., Massachusetts.
3) Garey, M.R., Johnson, D.S., Sethi, R., 1975, The complexity of Flowshop and Jobshop Scheduling, Tech. Rep. n.168, Comp. Science Dept., The Pennsylvania State University.
4) Gonzalez, T., Sahni, S., Flowshop and Jobshop Schedules: Complexity and approximation, Operation Research vol.26,n.1, 1978.
5) Johnson D.S., Approximation algorithms for combinatorial problems, J. Comp. Syst. Sc., 9, 256-278.
6) Lev, B., Caltagirone, R., Shea, F.J., 1972, Investigations in health care delivery. Patient flow and utilization of resources in a diagnostic radiology department, Investigative Radiology, Nov.Dic., vol.7, pag.517-525.

FUZZINESS IN THE MODELLING OF NEURAL SYSTEMS

Settimo Termini and Francesco Ventriglia

Laboratorio di Cibernetica , C.N.R.
Via Toiano, 2 - Arco Felice (Naples) Italy

Abstract. In this communication we are dealt with the uncertainty in neuronic systems. We propose a treatment of this phenomenon in the setting of the theory of fuzzy sets.

1. INTRODUCTION

The main aim of the present communication is to present a very rapid survey of some neurophysiological results which show, in a non ambiguous way, the existence of uncertainty phenomena in the nervous system. We shall indicate some of the approaches that have been made in order to overcome this problem.

This problem has been approached by making use of stochastic models and probabilistic assumptions. This approach has shown to be useful also for allowing a greater manageability of neural systems whose enormous complexity is by itself a deep obstacle to a deterministic treatment.

However some objections can be raised, at an interpretative level, against some of these stochastic assumptions.

It would then be very useful to admit the presence of this uncertainty without making the assumptions implied by the standard use of probability theory. All these requirements fit well with the aims and proposals of the theory of fuzzy sets. This last theory, however, is not so developed from the point of view of mathematical technique as to allow the formulations of clearly alternative mathematical models which can be tested experimentally and compared with the corresponding probabilistic or deterministic ones.

One can only slowly try to construct pieces of formalisms dressed upon the various problems which starting from the prop-

D. G. Lainiotis and N. S. Tzannes (eds.), Applications of Information and Control Systems, 249-257.

erties of the theory of fuzzy sets already established show how one can go along with uncertainty in a formal and quantitative way without probabilistic or statistical commitments.

We shall then show two items in which this can be done but shall in particular warn against wrong roads.

For which regards the references let us note that we have not aimed neither at completeness nor at historical precision for what regard the attribution of results and considerations discussed here. The references have then to be considered as exemplifications being in each case, the more known paper, the most clear example or the more available reference to us. In order not to make dull the paper with a list of references out of proportions to the length of the paper we refer the reader for the citations in the text which are not reported in our bibliography to the most exhaustive references of one of the following books [B≡ Burns, 1968; SC≡ Scott, 1977; PE≡ Popper-Eccles, 1977] .

2. UNCERTAINTY AT A NEURONIC LEVEL

In the present paragraph we shall collect and report some neurophysiological results leading, in our opinion, to the necessity of the presence of (the notion of) uncertainty in neuronic systems and we shall do some considerations tending to the clarification of way in which it is present and the reasons of this presence.

A first phenomenon is the fluctuation of the excitability of neurons. It essentially consists in the fact that iterated applications of a rectangular electrical stimulus produce an action potential in a fraction of all the trials. This fraction depends on the intensity of the stimulus.

Many investigations have been devoted to this phenomenon, we limit ourselves here to remembering [Hagiwara 1954,(B), Fatt and Katz 1952,(B), Verveen and Dirksen 1965].

A second related phenomenon is the latency fluctuations: the time interval between the application of the stimulus and the beginning of the action potential varies in a random way. The same spontaneous activity, registered with microelectrodes from the brain stem and the cerebral cortex has a time-variation which is so irregular as to support, in some way, the hypothesis of stochasticity of their behaviour in this phase [Poggio and Viernstein, 1964 (B), Bishop et al.,1964; Pfeiffer and Kiang, 1965].

One can then look, as has been proposed, at the fluctuations of the excitability of neurons as a neuronic noise and ascribe this second phenomenon to the same reason. Many neuronic models have been in fact, studied whose membrane potential is represented as a markov stochastic process [Gerstein and Mandelbrot 1964; Geisler and Goldberg 1966; Roy and Smith 1969].

Some authors have tried to elaborate theories or at least to provide (more or less quantitative) descriptions of neuronic systems in which this stochastic behaviour is taken into account superseding in this way one of the points of the original formulation of McCulloch and Pitts (1943) (SC). Among these we remember Burns (1968) who based himself on the two phenomena previously discussed and Taylor (1972) relying on another phenomenon, i.e. the (random) emission of packets of trasmitting substances to the presynaptic membrane [Katz 1966, Aidley 1971].

Statistical methods in the study of neurodynamics have been introduced by other authors on quite different bases and motivations namely the extreme technical difficulty of a deterministic treatment of systems so complex as the neuronical one [Rapoport 1950, Beurle 1956, Smith and Davidson 1962 (SC), Griffith 1963, 1971 (SC), Wilson and Cowan 1973 (SC), Ventriglia 1974 (SC), 1979].

Notwithstanding this graceful and appreciable agreement between the technical needs for manageable tools such as the statistical ones and the experimental evidence for non-deterministic behaviour as the ones described before the use of stochastic models presents some difficult problems. It is not clear, in fact, whether it is possible to provide always satisfactory interpretations of the reasons and origins of the stochasticity and moreover some authors [Chung et al. 1970 (SC)] warn against considering as noise the "spontaneous" activity and the bursty discharge of the neurons of the central nervous system. In fact "it is not obvious what any neuron is trying to say, and given the possibility that burstiness may itself be meaningful we have no basis a priori to decide what is noise and what is message".

All this neurophysiological evidence points towards the needs, for a theoretical understanding of this problem, of having from one side formal techniques powerful enough as to model these kind of systems in all their complexity and which are such, on the other side, as to allow plausible interpretations.

Let us remember again that probabilistic assumptions even if very useful both for mapping an evident non-crispness of the behaviour of some parameters and for reducing the enorm complexity of an hypothetical corresponding deterministic crisp model do not appear completely reasonable expecially at some levels of analysis as is the one, considered here, of the functioning of the net in which is taken into account the independent behaviour of the neurons.

Let us briefly discuss this point here which is applicable also at a higher level of organization which will be discussed in the following. We limit ourselves to noting that the previous difficulty arises from the fact that the most natural interpretation of statistical modelling is a frequentistic ones. In the case in which (as is the case considered here of a neuronic net) such a model is not convincingly interpretable (or representable) in frequentistic terms we are faced with a deep problem since the other classical interpretation of probability such as Carnap's logical approach or De Finetti's subjectivistic one are not adequate or natural for description of the natural sciences at a low level of abstraction.

Let us note, *en passant*, that his problem is methodologically identical to the one faced by microphysics and solved ingeniously (even if not satisfactorily according to many critics) by the classical interpretation of Quantum Mechanics. Let us note that objections against Carnap's or De Finetti's interpretations are less stronger (and in many cases disappears) for highly structured theories at a noteworth level of abstraction (in the Carnap's case) or for Social Sciences (in De Finetti's case), since in these cases one can organize the things as to refer probabilistic concepts at levels in which these two interpretations are reasonable. Of course this is not the case for modelling at low levels of abstraction of natural systems, such as in our case. We have then in order to overcome the previous difficulties to look for a more embracing alternative scheme allowing a more natural formalization of the uncertainty and indeterminateness presented by neuronic systems which admits an interpretation different from both a deterministic and a statistico-stochastic ones.

This scheme in our opinion can be successfully envisaged in the conceptual setting of the theory of fuzzy sets and, hopefully, formally developed with the aid of also the future achievements of the mathematical sides of this theory. An approach of this kind seems particularly promising also for the reason that in these last years one of the most promising lines of development of this theory has been the elaboration of "fuzzy logics", that is uncertainty from the point of view of logical-like laws and of approximate reasoning. Now it is well known that approximate reasoning is just one typical feature of natural neural systems and the one which more sharply differenziate natural from (presently known) artificial systems. In doing this however we have to be very careful and take into account the fact that such alternative approach, represented mainly - as we have already said - by the logical developments of the theory of fuzzy sets is "still in its infancy" as stressed by Bellman and Zadeh (1975). It comes natural to ask at this point whether it is possible (and worthwhile) to try to reconstruct the classical model of McCulloch and Pitts on the base of the previous considerations and by substituting the language of classical propositional logic with the language of one of the "deviant" logics of uncertainty and vagueness that has been developed in this last decade and, in particular with fuzzy logics as developed by Goguen 1969, Zadeh [1975] which all share the common property of being more akin to some macroscopic aspects of human reasoning. We shall enter here neither into the most elementary details of such a program which besides its technical difficulties needs, by itself, many compelling preliminary methodological and conceptual clarifications among which we list the following:

1. Is the one of the Kleene an adequate translation of the theory of McCulloch and Pitts, or, if not, what more apt interpretation of it can be done?
2. Does a Kleene-line (limitative) characterization theorem holds also for this hypothetical new interpretation?
3. The use of the calculus of propositional classical logic implies the description of neural nets in terms of the more fundamental logic (the LOGIC) or is only a useful formal tool and technique? This leads us to the other problem of
4. The plurality or uniqueness of logic.

Notwithstanding these difficulties a program of this kind has to be pursued at least for the fact that whatever the conclusions, these could throw light on some essential limitations existing on this kind of modelling and, indirectly, afford some indications on the points in which "a Science of the Mind" (if any) will differ from the "Sciences" to which we are presently accostumed.

3. UNCERTAINTY AT A CELL-ASSEMBLY LEVEL

In this section we shall first collect some neurophysiological proposals and results pointing toward the existence of a higher level of organization of the (neural net of the) brain; secondly we shall briefly report and comment some problems raised by the same existence of a multi-level (hierarchical) approach both from an epistemological point of view and a relationship with interpretation at a psychological level. Finally we shall comment on the usefulness of a presence of a formal explicatum of uncertainty also at this level and briefly indicate some points where it is possible to focus and find interesting problems to be tested.
In order to understand or only to speak of them we shall describe a hierarchical scheme for the analysis of neural processes which is widely accepted (Scott 1977). The most interesting points of that scheme are for our study the following: 3. Neuron behaviour from membrane dynamics and electromagnetic theory; 4. Neural net dynamics from neuron behaviour; 5. Cell assemblies from net dynamics; 6. Psychology from cell assemblies dynamics.

The most interesting proposal for treating the level fifth of the previous hierarchical is the one made in 1949 by Hebb (SC) which postulated the existence of a new entity, the cell-assembly, for organizing on new bases the knowledge of the inferior neurophysiological level. This notation is based on the assumption that every particular stimulus, frequently repeated, will lead to the slow development of a cell-assembly, a system of neurons able to work as a closed system, to provide facilitation to other systems and having, usually, a specific motor facilitation. The plausibility of this notion has been reinforced, even if with some modifications, by the physiological investigations of Mountcastle (1957) (PE) on the somatic sensory cortex and of Hubel and Wiesel (1962, 1974) (PE) on the visual cortex.

These studies lead to the individuation, in these particular cortical zones of neuronical organizations of cylindrical shape perpendicular to the cortical surface. Subsequent investigations of Szentagothai (1969) (PE), Szentagothai and Arbib (1975) (PE) on the microstructure of the cortex have lead to postulating that, both a structurally and functionally, every area of cerebral cortex is organized in (functional) units of about 10.000 neurons.

On the basis of this studies Eccles (1977) thinks that the way of working of the single column is such as to extend its own level of activity and to inhibit the neurons of the ones immediately surrounding it. A different behaviour in the laminae more superficial of the columns is superimposed to it. In these laminae there exist a finer level of inhibition and a weaker and dispersed level of excitation coming even from very distant points of the cortex; in such a way in this laminae is present an influence at the same time more subtle, gentler and more integrated. On the base of what has been said before about the impredictability - within some limits - of the behaviour of the single neuron even a cell-assembly is, in some way, smooth for what regards the predictability of its behaviour.

If one associates to the cell-assemblies (or to systems of cell-assemblies) the single perception-acts this would explain both the different reaction to identical stimuli and the possibility of precise descriptions of ambiguous situations (for instance, the identical classification of a same letter written in non identical ways). Also in higher levels of neuronic aggregations (as the aggregation of cell-assemblies in macro-cell assemblies) to which one could ascribe the higher activities of human brain, as, for instance, the production of language and its understanding, the previous non-rigidity of behaviour had to be present.

Under this hypothesis also the mistakes in the production of the discourse, as, for instance, those reported by Lasley in his contribution to the Hixon Symposium 1948 concerning the anticipation or posposition of letters or words, would find their explication un this margin of uncertainty present in the behaviour of the macro-cell-assembly responsible of that particular production.

By denying the presence of uncertainty and flexibility as essential ingredients of the behaviour it would be very hard to envisage also the possibility of a model mapping the reliability of human brain as, for instance, the capability of expressing one and the same idea in many different linguistic forms.

We think that a comparison between models of linguistic behaviour and neurophysiological facts done by using the same language of hierarchy of fuzzy sets and extended entropy measures could provide useful, at least, for testing the flexibility of this new language.

We shall briefly suggest here how one can take into account the uncertainty in firing of the single neuron and the consequence of it in the behaviour of the cell-assembly by using fuzzy sets [Zadeh, 1965]and their entropy measures [De Luca and Termini, 1972]. For a complete survey of the literature on fuzzy sets we refer to Gaines and Kohout (1977) and for recent surveys and developments of the entropy measures to the two complementary papers[Riera, Sanchis and Trillas, 1978; De Luca and Termini, 1979].

Fuzzy sets concepts can be associated to the neuron in the following way. Relying on the phenomenon of the fluctuation of the excitability previously discussed let us consider the class of the stimuli of the neurons. We have already observed that some stimuli always cause the firing of the neuron (those with I>102%T), (stimuli of type B) some never cause the firing (those with I<98%T) (stimuli of type A) and some that sometimes cause the firing and sometimes do not (those with $98\%T \leq I \leq 102\%T$) (stimuli of type C).

Let X be the class of all the stimuli and x the generic element. Let N be a neuron. We now consider the fuzzy set f_N defined on X in the following way.

$$f_N(x) = 0 \quad \text{for any} \quad x \in A$$

$$f_N(x) = 1 \quad \text{for any} \quad x \in B$$

$$f_N(x) = r \quad \text{for any} \quad x \in C, \quad 0 < r < 1$$

the particular value of r being determined according to the neighbourhood of I to T.

We note that no probabilistic (frequentistic) notions have been till now involved since the value r associated to an ambiguous stimulus has to be determined (on the base of rules to be eventually discovered) on the basis of the value of I and T only. To each neuron, which is in this scheme represented by a fuzzy set, one could associate an entropy measure measuring its fuzziness, i.e. measuring the degree of ambiguity in its behaviour.

These entropy measures could also serve to classify the various neurons for what regards the magnitude of the fluctuation of excitability independently of the particular shape of the function f_N. By suitably combining those contributions one could also try to associate quantitative evaluations of the ambiguity in the response of a cell-assembly trying to relate it to ambiguities present in the physiological level. This fuzzy behavior had to be, of course, more evident in the case in which the major part of the stimuli are of type C.

The phenomena of perceptual instability (i.e. those phenomena in which one has apparent inversions or fluctuations of the perceived properties of objects as Necker's romboid, Schröder stair,

Mach's book and so on) could be perhaps referred in this setting to a situation in which slight variations in the values of ambiguous stimuli (of class C) would produce variations in their contribution to the associated entropy measures as to produce changes of decisions. The relationship between the level of the neuron described by a fuzzy set and its entropy measure and the one of a cell assembly has not been till now clarified. We think that two proposals can be suggested at this very rough preliminary level. One is to work, at the level of the cell-assembly, by referring to the fuzzy set description of the single neurons and then using entropy and energy measures of them suitably combined. The other is to define the whole cell-assembly as a new fuzzy set and consider the entropy measures of the whole cell-assembly. This kind of "fuzzy-fuzzy sets" has already been proposed by J.A. Goguen (1974) for analyzing natural languages in his paper "Concept Representation in Natural and Artificial Languages".

A notion of entropy measure for fuzzy-fuzzy set has not yet been developed.

References. Aidley, D.J.: 1971, The Physiology of Excitable Cells. Cambridge University Press. Cambridge, England.

Bellman, R., and Zadeh, L.A.: 1975, Modern Uses of Multiple Valued Logic J.M. Dunn and G. Epstein (eds). D. Reidel (1977), 105-165.

Bishop, P.O., Levick, W.R., and Williams, W.O.: 1964, J. Physiol. 170, 598-612.

Burns, B.D.: 1968, The uncertain nervous system. E. ARNOLD (Publishers) L.t.D. London.

De Luca, A., and Termini, S.: 1972, Inform. and Control, 20, 301-312.

De Luca, A., and Termini, S.: 1979, in M.M. Gupta, R.K. Ragade and R.R. Yager (eds.) Advances in Fuzzy Sets Theory and Applications. North-Holland (to appear).

Eccles, J.C.: 1977, in Popper/Eccles: The self and its Brain. Springer International.

Gaines, B.R., and Kohout, L.J.: 1977, International J. Man-Machine Studies, 9, 1-68.

Geisler, C.D., and Goldberg, J.M.: 1966, Biophysical J. 6, 53-69.

Gerstein, G.L., and Mandelbrot, B.: 1964, Biophysical J. 4, 41-68.

Goguen, J.A.: 1969, Synthese, 19, 325-373.

Goguen, J.A.: 1974, Intern. J. Man-Machine Studies, 6, 513-561.

Katz, B.: 1966, Nerve, Muscle and Synapse. McGraw Hill Co. New York.

Pfeiffer, R.R., and Kiang, N.Y.S.: 1965, Biophysical J. 5, 301-316.

Rapoport, A.: 1950, Bull. Math. Biophysics, 12, 109-121.

Riera, T., Sanchis, C., and Trillas, E.: 1978, Preprint of the Department of Mathematics of the Polithechnic University of Barcelona (Spain).

Roy, B.K., and Smith, D.R.: 1969, Bull. Math. Biophysics 31, 341-357.

Scott, A.C.: 1977, Neurophysics. J. Wiley&Sons, N.Y. London Sidney Toronto.

Taylor, J.G.: 1972, Spontaneous Behaviour in Neural networks. J. Theor. Biol., 36, 513-528.
Ventriglia, F.: 1979, In these proceedings, pp.258.
Verveen, A.A., and Derksen, M.E.: 1965, Kybernetik Band 2 Heft 4, 152-160.
Zadeh, L.A.: 1965, Information and Control, 8, 338-353.
Zadeh, L.A.: 1975, Synthese 30, 407-428.

KINETIC THEORY OF NEURAL SYSTEMS: SIMULATION OF PROPAGATED WAVES

F. Ventriglia

Laboratorio di Cibernetica, C.N.R.
Via Toiano, 2 - Arco Felice (Na) - ITALY

ABSTRACT. - In this communication we examine the ability of a kinetic model of neural systems to propagate informational waves.

1. INTRODUCTION

The mathematical theory of neural systems at which we have been working at for some years [Ventriglia, 1973, 1974, 1978, 1979] belongs to the great stream of researchs which attempt to deduce the laws of the superior order neural processes, as the behavioral or psychological ones, from the laws which regulate the dynamic behaviour of the neuronic systems. On the ground of the extreme difficulty of a deterministic treatment of complex systems as the neural ones, which induced A. Rapoport in 1950 to use a statistical method for studying neurodynamical processes, also our theory is a statistical one. Its main difference from other statistical theories is a strict analogy with theories of physics. In fact it is grounded over the following basilar model of neural systems. It consists of a system of "particles" - neurons and impulses - which occupy a certain volume V of the three dimensional space and interact among them. The neuronic "particles" occupy fixed positions in space, while the impulses form a surrounding gas. The impulses are non-auto-reacting and their total number change after collisions impulse-neuron. They transport a quantum α of excitation-positive or negative. A parameter E of excitation is associated to neural particles, which varies as the time goes on.

D. G. Lainiotis and N. S. Tzannes (eds.), Applications of Information and Control Systems, 258-264.

The variations of E are due both to an internal drift (which is related to the exponential decay of the membrane's potential with constant μ), and to the absorption and to the emission of impulses. The parameter E is normalized. The statistical description of the system is obtained by means of three distribution functions: $\phi(x,e,t)$, distribution of the neuronic excitation; $f(x,v,t)$, distribution of the impulses velocities; $\phi_d(x,t)$, distributions of neurons which are in the period of absolute refractoriness. In the construction of the kinetic equations describing the time evolution of the distributions, we take into account almost all the significant characteristics of the biological neurons: the synaptic delay, t_o; the time summation with constant of time μ; the period of absolute refractoriness τ; the excitability with threshold; the hyperpolarization. In this paper we report the study of the ability of the model to propagate informational waves. The study is accomplished by means of computations on a computer of the solutions of the kinetic equations in the case of a one-dimensional neural system.

2. ONE-DIMENSIONAL KINETIC EQUATIONS

The kinetic equations for an unlimited one-dimensional neural system with the features reported in Introduction can be easily derived following the scheme of a previous work [Ventriglia, 1974]. The equations are:

$$\left[\frac{\partial}{\partial t} + v\frac{\partial}{\partial x}\right]f = -\sigma\psi|v|f + \lambda\sigma a f^{*}\int_{1-\alpha}^{1} de\,\phi\int dv|v|f(.,.,t-t_o) \quad (1)$$

$$\left[\frac{\partial}{\partial t}+\mu(e_r-e)\frac{\partial}{\partial e}\right]\phi = \sigma\int dv|v|f(.,.,t-t_o)[a\phi(.,e-\alpha,.)+ (1-a)\phi(.,e+\alpha,.)-\phi]+\delta(e-e_r)\phi_d(.,t-\tau) \quad (2)$$

$$\frac{\partial\phi_d}{\partial t} = -\phi_d(.,t-\tau) + \sigma a\int dv|v|f(.,.,t-t_o)\int_{1-\alpha}^{1} de\phi \quad (3)$$

with the following condition of normalization over ϕ :

$$\int_0^1 de\phi = \psi-\phi_d \quad \text{(where } \psi \text{ is the neurons' density).} \quad (4)$$

In these equations, the absorption coefficient σ, the coefficient of emission λ, the percentage of excitatory neurons a, the distribution of the velocities of the impulses emitted by firing neurons $f^*(v)$ are the statistical parameters describing the net of connections of the biological neural systems. The other symbols have the following explanation: $|v|$ is the absolute value of v, δ is the Dirac's function, e_r corresponds to the level of rest of the membrane's potential. By imposing to the system (1), (2), (3), (4) the following initial conditions:

$$\begin{aligned} f(x,v,0) &= F(x,v) \\ \phi(x,e,0) &= \Phi(x,e) \\ \phi_d(x,0) &= 0 \end{aligned} \qquad (5)$$

and by solving it we can study the ability of the model to transmit information. For semplifying such study we compute by starting from the distribution f and ϕ the two following statistical parameters:

$$\rho(x,t) = \int dv\ f(x,v,t) \qquad (6)$$

$$E(x,t) = \frac{1}{\psi} \int_0^1 de\ (x,e,t)\ e \qquad (7)$$

which denote, respectively, the local density of impulses and the mean local neural excitation. The solution of the system (1), (2), (3), (4) with the condition (5) is accomplished by means of a numerical technique of approximate solution, and the functions $\rho(x,t)$ and $E(x,t)$ are computed from f and ϕ by means of numerical integration formulae. A complete description of the method is reported in Ventriglia, 1979. Computational exigences impose that the initial conditions F and Φ be periodic with respect to space, x, so that also the distributions f and ϕ are periodic. The technique of solution has been implemented on a computer and the results of several elaborations are reported in the following section.

3. RESULTS

The initial conditions F and Φ can be considered as the waves of information which strike the neural system at time t=0. In this setting, the function f and ϕ can be considered as the reaction of the system when the time goes on. An initial neuronic excitation uniformly distributed from zero to one has been considered, while the function F simulates an excitatory wave. The characteristic behaviour of the two macroscopic parameters $\rho(x,t)$ and $E(x,t)$ is reported in fig. 1 for a particular choice of the parameter ψ, λ, a, σ .

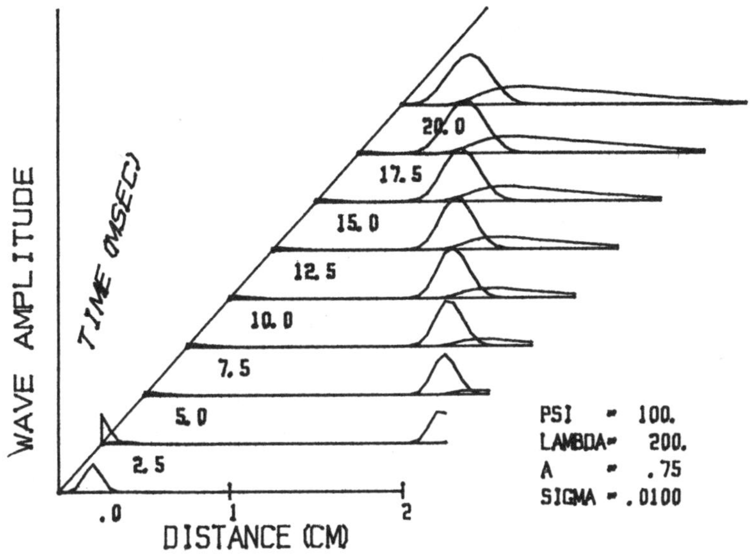

Fig. 1

The main aspect of this behaviour is the propagation of an information wave in consequence of the initial stimulus. The propagation of information is one of the basilar aspects of the neural activity; the fact that our model retain this characteristic is a very significant one for its validity. Furthermore, we can observe that the propagation is a dispersive one. Also the difference we note between the parameters ρ and E is reminescent of the biological reality. In fact, the more smooth and more lasting form of $E(x,t)$ can be easily associated to the slow variations of membrane's potential for subthreshold impulses to which some authors attribute to genesis of EEG [Elul, 1972].
The behaviour of the statistical parameter $\rho(x,t)$, when we simulate a lesion in a portion of the neural system, is reported in fig. 2. The "lesion" occupies the space from 0.95 to 1.25 cm. The simulation of this phenomenon has been obtained by putting $\sigma = 0.2$ and $f^*(v) \simeq 0$ in the above-mentioned interval.

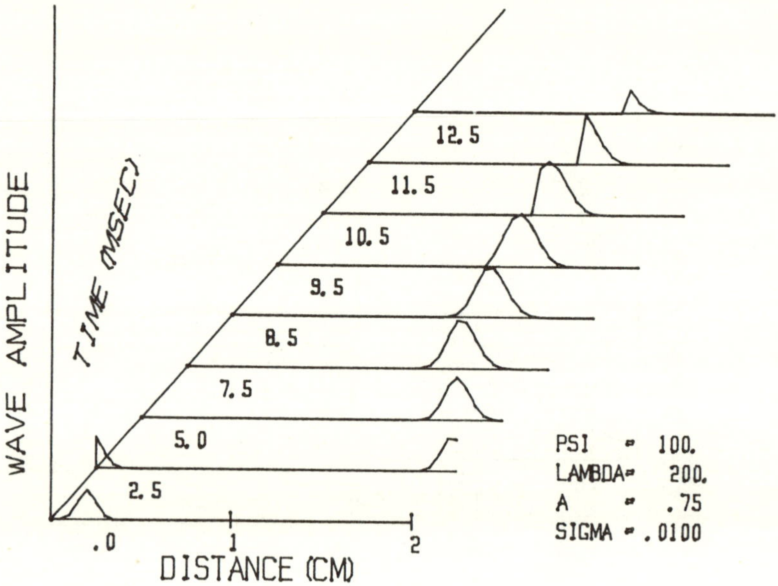

Fig. 2

From that figure we can note that the "lesions" acts as an absorbent boundary. In fact the informational wave disappears when it enters in that space. Once more the adherence of the model to the biological facts is shown.

If we assume the parameter a (percentage of excitatory neurons) changes with the position, we simulate a neural system with a variable number of excitatory neurons. We assume a = 0.40 in the intervals 0.0÷0.4 cm. 1,5÷2.0 cm. and a = 0.80 elsewhere. The correspondent behaviour of the density of impulses ρ is reported in fig. 3. This figure shows the difference from the case quoted in fig. 1 clearly.

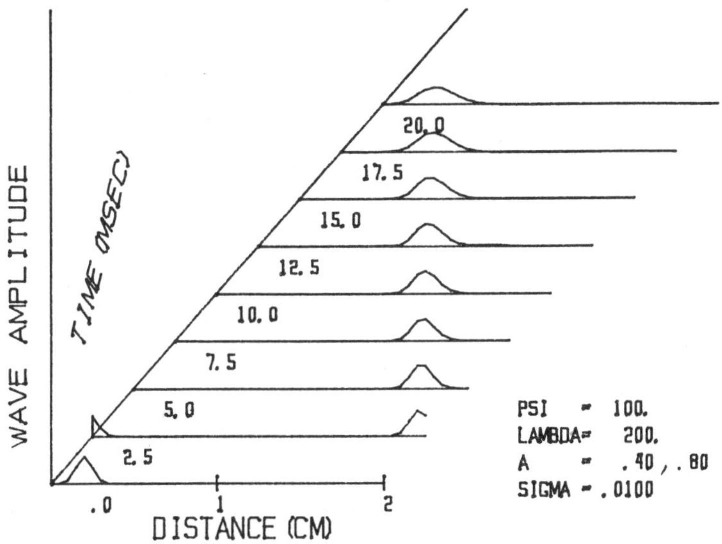

Fig. 3

4. COMMENTS AND CONCLUSIONS

The results of the previous section, besides confirming the validity of the model of neural system we proposed (at least for what concerns the simulation of some gross aspects of the activity of the biological neural systems), indicate some lines of development for future works. In fact, we can apply the method of approximate numerical solution to the solution of the kinetic equations for a model of a two-dimensional neural system. In this case, if we take care that the parameters ψ, λ, a, σ, f* reflect more closely the structure of some well known neural system (for example, a portion of cerebral cortex) and if we choose suitably the initial and boundary conditions, we could lay the foundations of a quantitative study of the dynamics of the cell-assemblies [Hebb, 1949; Eccles, 1977] , the neurophysiological entities which, presumably constitute the basilar elements for a neurophysiological theory of the psycology [Scott, 1977].

REFERENCES

Eccles, J.C.:1977, in Popper/Eccles:The self and its Brain. Springer Int.

Elul, R.:1972, Int. Rev. Neurobiol. 15, 227-272.

Hebb, D.O.:1949, Organization of Behavior. J. Wiley & Sons, N.Y. - London.

Rapoport, A.:1950, Bull. Math. Biophys. 12, 109-121.

Scott, A.C.:1977, Neurophysics. J. Wiley & Sons. N.Y.

Ventriglia, F.:1973, Intern. J. Neuroscience 6, 29-30.

Ventriglia, F.:1974, Bull. Math. Biol. 36, 535-544.

Ventriglia, F.:1978, Biol. Cybernetics 30, 75-79.

Ventriglia, F.:1979, Biol Cybernetics (to appear).

THE VAN DER POL OSCILLATOR AS AN ELEMENT IN AN HIERARCHICALLY STRUCTURED RESPIRATORY CONTROLLER

C. L. Nikias

Dept. of Electr. Eng.
SUNYAB
Amherst, NY 14260

W. F. Fincham

Dept. of Electr. Eng.
QMC, Univ. of London
London, E14NS

A proposal regarding the basis of the autonomous frequency of breathing in the human is examined. The time aspects in terms of experimental evidence which demonstrates that the phenomenon of frequency capture is possible when small periodic pressures are applied externally to the rib cage are also considered. It is suggested that a van der Pol type of oscillator is an appropriate model component, in which its autonomous frequency is parametrically defined in terms of the dissipation of energy in the respiratory mechanics.

1. INTRODUCTION

The motivation for developing a system description using a VDP element relates to experimental evidence which suggests that the respiratory system may be frequency captured when appropriately stimulated or forced [1]. It may well be that considerably more sophisticated adaptive and hierarchically structured models might exhibit the capture phenomena but they are unlikely to do so as rapidly (within one breath) as the experimental evidence just cited indicates and at the speeds of frequency transition possible in the van der Pol oscillator.

Essentially the respiratory airflow pattern as measured at the mouth is strongly cyclic, the period having a stochastic component [2]. The time for the inspiration of air is usually longer than the time for expiration which in turn, may be followed by a 'rest' period of zero airflow before the next inspiration. In ref. [3], the breathing pattern optimizer used

D. G. Lainiotis and N. S. Tzannes (eds.), Applications of Information and Control Systems, 265-272.

different performance functionals for inspiration and expiration. The model generated responses convincingly similar to those observed in practice under conditions of spontaneous breathing, the authors cautiously concluding that the performance criteria were used, in part, to minimize the energy cost of breathing. In this paper we present a scheme incorporating similar features with an additional element having the functional description of a van der Pol type oscillator (VDP).

2. MODEL FORMULATION

In order to develop a coordinated scheme containing an optimizer and a VDP element, a convenient link is provided by the model of chest and lung dynamics developed by Jodat, Horgan and Lange [4]. For convenience the modified form used in ref. [4] is reproduced in Fig. 1 and has the desirable attribute that the chest and abdomen appear separately and so permit an independent forcing function to be easily added. The figure shows this acting as an equivalent force on the chest wall musculature. The normal drive to the muscles is the single driving force P_M and is shown as the output of a parametrically driven VDP oscillator. Fincham and Nikias have shown [5] that by making the parameters of the VDP equation independent functions of time, the double cosinusoidal-pause waveshape proposed in [4] can be easily generated.

3. ENERGY FUNCTIONS

The work of breathing was estimated from:

$$W = - \int_o^T P_{p\ell} \dot{V}_\ell \, dt \tag{3.1}$$

where $P_{p\ell}$ is the pleural pressure and $\dot{V}_\ell$ the rate of change of lung volume. Integration is over one complete cycle of breathing, a time interval which is easily defined under steady-state conditions. This formula gives relative values nearest to the experimental results of McIlroy et. al. [6] and Milic-Emili and Petit [7] who used the lung volume and the intraesophageal pressure to form the integrand of Eq. (3.1).

3.1 Energy Functions Without Stimulus

This is the condition of spontaneous or autonomous breathing and Eq. (3.1) can be used directly because T is the well defined breath period. The drive function $P_M(t)$ was time apportioned as for a normal breath suggested by Agostoni [8], viz:

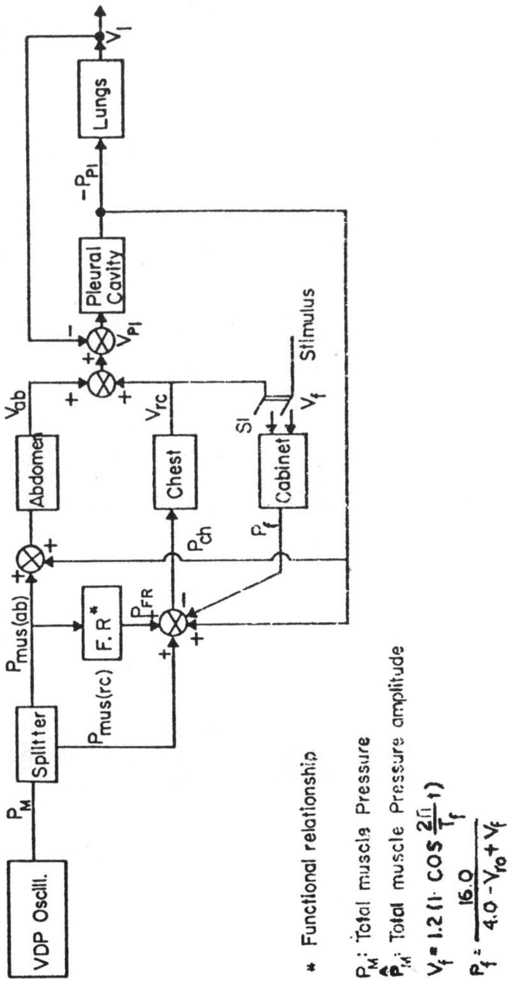

Fig. 1. Model of the respiratory mechanics

i) inspiratory duration $T_1 = 0.5T$

ii) expiratory duration $(T_2-T_1) = 0.4T_1$

iii) rest period $(T-T_2)$

The tidal volume V_T is the maximum change in lung volume V_ℓ from a datum value and the alveolar ventilation is the average gas flow rate per breath. i.e., $\dot{V}_A = V_T/T$.

The basic van der Pol equation is:

$$\ddot{x} + \varepsilon(x^2-a)\,\dot{x} + \sigma x = 0 \tag{3.2}$$

and using the method given in ref. [5], the cosinusoidal wave-shaping parameters to match the above breath specification for T = 4 sec are

(i) $0 \leq t \leq T_1$: $\varepsilon = 0.1$, $\sigma = 2.4$

(ii) $T_1 \leq t \leq T_2$: $\varepsilon = 0.1$, $\sigma = 89$

(iii) $T_2 \leq t \leq T$: $\varepsilon = 10.0$, $\sigma = 0$

Parameter (a) defines the waveform amplitude and x(t) is equivalent to $P_M(t)$.

The system shown in Fig. 1 was simulated using the ISLII digital simulation language, a derivative of CSMP, with model parameters obtained from reference [4] and which corresponded to the breathing of a healthy person. By choosing various combinations of values of V_T and T, a family of curves was plotted to show the variation of the rate of change of the work of breathing W/T as a function of 1/T at different values of $\dot{V}_A$. Typical results are the solid curves of Fig. 2 which have a close qualitative agreement with the data given in ref. [6] and [7].

These results imply that the model structure and signal drive format used are highly eligible companions to describe some aspects of the mechanism of energy dissipation in the respiratory musculature.

3.2 Energy Functions and Stimulus

With the ganged switch S1 closed an external stimulus can be input to the model. The scheme shown in Fig. 1 is believed to represent the conditions existing in the experiments of Fincham and Liassides [1]. The chest musculature is shown subjected to a total pressure variation P_{ch} made up of the autonomous drive and a component P_f generated by enclosing the chest in a rigid cabinet driven by a piston of independently controlled swept volume V_f. In the simulation a sinusoidal waveshape of period T_f for V_f was used with an amplitude of 1.2 units, this value being considered appropriate as compared to the value of $\hat{P}_M$ = 4.8 units. In the experimental situation [1] the cabinet used, contained an inherent leak which could be simulated.

Examination of Fig. 1 clearly shows that the waveshapes of $P_{p\ell}$ and V_ℓ will be those of the spontaneous breathing case modulated by the stimulus signal. To estimate the work of breathing for one 'average' breath, it was necessary to perform the integration of Eq. (3.1) over an integral number of breaths n_2 selected so that at least one complete cycle of the $P_{p\ell}$ variation was experienced. So with the stimulus applied,

$$\bar{\dot{W}}_s = \frac{-1}{T_t} \int_0^{T_t} P_{p\ell} \, \dot{V}_\ell \, dt \qquad (3.3)$$

where $T_t = n_1 T_f = n_2 T$ n_1, n_2 integers. The average value of ventilation in T_t

$$\bar{\dot{V}}_A = \frac{\sum_{i=1}^{n_2} V_T(i)}{n_2 T} \qquad (3.4)$$

where V_T (i) is the tidal volume for breath i. Selection of a range of values for T and two values of T_f provided sufficient results to plot the dashed waves of Fig. 2. As a word of interpretation each curve is for a constant value of $\dot{V}_A$, the dashed curves being valid for the stated constant stimulus frequency $1/T_f$.

4. IMPLICATIONS OF THE RESULTS

The results plotted in Fig. 2 have the interesting feature that the intersections of each pair at a specific value of ventilation $\dot{V}_A$ are at either the fundamental frequency of the stimulus signal or at a sub-or multi- harmonic. Because these frequencies do not coincide with f_{opt}, the corresponding work values are higher than W_{opt}. Fig. 2 shows that for a

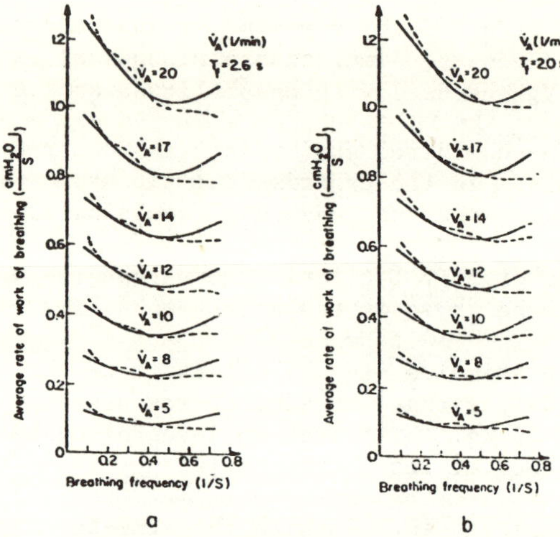

Fig. 2. Average rate of work of breathing V breathing frequency

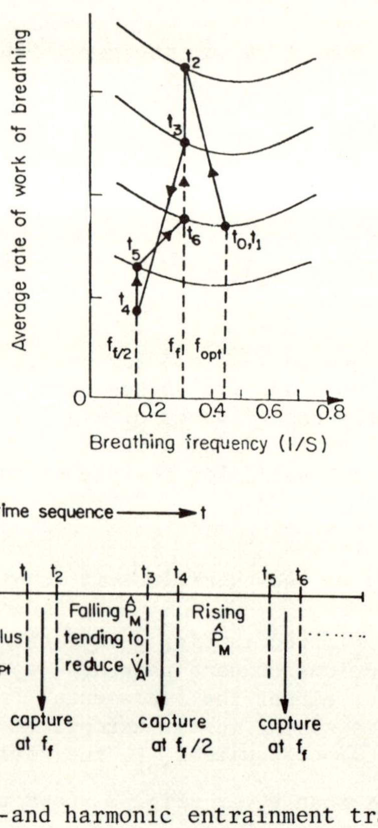

Fig. 3. Sub-and harmonic entrainment trajectories

$\dot{V}_A = 8\ \ell \cdot \min^{-1}$, the normal optimal frequency of breathing $f_{opt} = 0.42$. For $T_f = 2.6$ secs ($f_f = 0.38$) i.e.,

$$\bar{\dot{W}}(f_f = 0.38) < \bar{\dot{W}}(f_f/2)$$

In keeping with a minimum energy criterion, the controller should operate to select $T = T_f$ and not $2T_f$ it is is for 'some reason' prevented from selecting f_{opt}. The 'some reason' implies that a capture condition is in operation. A possible sequence of events in a stimulus type experiment would typically run as follows:-

i) Before application of the stimulus $f = f_{opt}$ and $\dot{W}$ is at a minimum compatible with a level of $\dot{V}_A$ required to maintain the chemical status of blood at an optimum value [2].

ii) Application of a stimulus of adequate amplitude causes capture within a breath [1]. This is only likely to result if quick acting neuro-muscular pathway are involved. It will be unlikely that $\dot{V}_A$ remains at its previous value. With synchronization of breathing present, its value will increase with a sequent loss of carbon dioxide.

iii) The shifted chemical status of the blood will attempt to force the controller to lower the alveolar ventilation. This could be achieved in a number of ways:

a) Maintain in-phase synchronization but at a lower value of $\hat{P}_M$.
b) Switch to out-of-phase synchronization.
c) Fall out of synchronism and go to an appropriate f_{opt} at a lower $\dot{V}_A$.
d) Maintain synchronism but at a sub- or multi-harmonic of the fundamental stimulus frequency.

Proposal 'b' can be deleted because it has not yet been observed in practice and 'c' only infrequently. Fig. 3 summarises a likely behaviour for the closed-loop system including the blood circulation elements. A similar explanation is possible for multi-harmonic entrainment.

5. FUTURE WORK

The model proposed does include features which are highly relevant when attempting to explain experimental results, and the next step must be a simulation exercise to determine the stochastic closed-loop behaviour of the system, this being the job of the filters and predictors[14] available in the control engineering area. The subtlety required of the experimentalist is enormous and realistic progress in the development of the model and will proceed only if there is cross fertilization of ideas between the experimentalist and the modeller-ideally perhaps one and the same person.

REFERENCES

[1] Fincham W. F., Liassides C., "The role of the van der Pol oscillation in the control of breathing in the human": 1978, Proc. of UKSC Conference, IPC Science and Techn. Press, pp 59-70.

[2] Priban I. P., Fincham W. F, "Self adaptive control and the respiratory system": 1965, Nature 208, No. 50008, pp 339-343.

[3] Hamalainen R. P., Viljanen A. A., "Modelling the respiratory airflow pattern by optimization criteria": 1978 Biol. Cybern. 29, pp 143-149.

[4] Jodat R. W., Horgan J., Lange R., "Simulation of respiratory mechanics": 1966, Bioph. Jour. 6, pp 773-785.

[5] Fincham W. F., Nikias, C. L., "Further periodic solutions of the van der Pol equation and its application to generalized waveform generation": 1979, submitted for publication.

[6] Mcllroy M. B., Marshall R., Christie, R. V., "Work of breathing in normal subjects": 1954, Clin. Sci. 13, pp 127-136.

[7] Milic-Emili J., Petit, J. M., "Mechanical efficiency of breathing": 1960, J. Appl. Phys. 15, pp 359-362.

[8] Agostoni E., "Diaphragm activity and thoracoabdominal mechanics during positive pressure breathing": 1962, J. Appl. Phys. 17(2), pp 215-220.

[9] Lainiotis, D. G., "Partitioning Filter": 1979, Inform. Scien. 17, pp 177-193.

COOPERATIVE PROCESSES IN CEREBRAL DYNAMIC

J. Mira, A. Delgado and R. Moreno

Universities of: Granada and La Laguna,Spain

ABSTRACT.- At length, we are interested in a conceptual model of cerebral dynamic which could be used to represent global properties of nervous system concerning with their functional stability after lesion. The study of traumatic and surgical lesions in animals and men, have sugested that cooperativity, functional multiplicity of the same anatomic area, and gradual distribution of functions are genuine properties of the information processing in the cortical tissue. A central problem is the study of the cooperation between neurons or patterns of neurons with the operational definition of density of function, functional deficit and residual function after the lesion. We propose a conceptual model based on this concept that "explains" the "Law of mass action" of Lashley and "Law of facilitation" of J. Gonzalo, pointing out the difficulties of dealing with data obtained from experimental input-output relations.

1. INTRODUCTION

Historically, there has been two opposed tendencies about functional organization of the brain: The localizationist (Broca, Wernicke) and the antilocalizationist (Jackson and Lashley, i.e.) with a progresive tendency towards integration (J. Gonzalo) in a theory of gradual distribution of functions with functional multiplicity of the same anatomic area.
The study of traumatic and surgical lesions in animals

D. G. Lainiotis and N. S. Tzannes (eds.), Applications of Information and Control Systems, 273-280.

Lashley (1) and men, Luria (2,3), J. Gonzalo (4,5), have sugested that the main relevant property of the information procesing by cortical tissue is cooperativity. Thus, we scan the data and hypothesis more relevants concerning to cooperative processes.
Lashley main points are: Equipotentiality of the tissue (1926), "Law of mass action" (the functional deficit is proportional to the mass of tissue resected, 1929), fields theory (1937) and Cooperativity with diffuse memory (1950).
Luria uses the concept of functional systems with the cooperation of many anatomical areas. Thus, the function can be depressed by lesions in several areas and consecuently, the same lesion can produce deficits in diverse functions.
The sensorial disturbances produced by traumatic lesions (scotomas like) are well described by the theories of J. Gonzalo (4,5,7,8) which include Lashley's theories and are more physical than that of Luria.
J. Gonzalo propose functional continuity and gradual intensity. So, the more peripherical is a cortical lesion, the more unimodal and contralateral is the functional deficit associated. If, the lesion comes to the centre, the functional deficit is plurisensorial and it is necessary to eliminate greater amount of tissue mass in order to produce the same deficit.

2. COOPERATIVE PROCESSES

From the bibliographic information is apparent that two of the relevant properties of cerebral dynamics are:

a. Cooperativity.

b. Functional multiplicity of the anatomical substratum

We carried out this paper in order to explore the posibilities of modeling these cooperative processes pointing out the difficulties of dealing with data obtained from experimental input-ouput relations. We start, from a structural point of view of an anatomical substratum, that support a certain distribution on functional modules (neurons or pools of neurons), polimodals and with a high degree of conectivity (see figure 1). If these modules are polymodal, on the same anatomical area can coexist diverse modalities (for example, the sensorial modalities, acoustic, optic and somesthetic). The outward information entering the brain structures is diffused among the modules (input distributor) correlated to their functional specificity. The modules converse with languages that specifie the level of cooperation and produces a global function by means of an outputs totalizer. We introduce in the cooperation schema a mask intended for simulate the lesions.

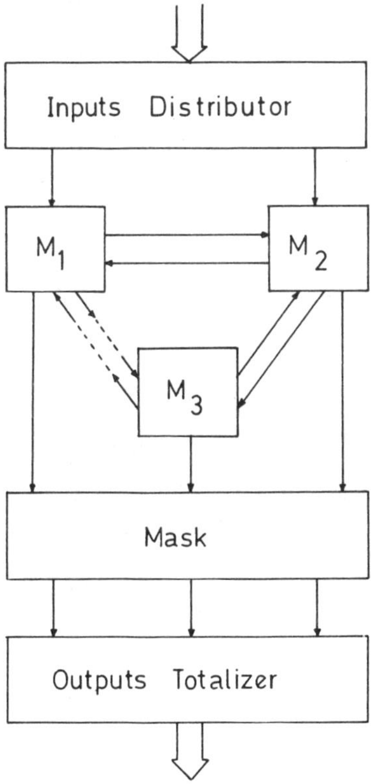

Figure 1. Cooperation schema

The neuronal cooperation can be analized according to the schema of figure 1 if we specifie the operational meaning of the following parameters:

1. Function.
2. Density of function.
3. Lesion and residual function.
4. Interaction (conectivity) and neural codes (analogic, logic or symbolic).
5. Neuropsicological experiments for evaluation of the cooperation schema.

3. NEURAL FUNCTIONS

According to Luria (3) a basic problem in cerebral dynamic is to specify the functional level. We can speak of the function of a neuron, a tissue or a pool of brain structures. At biophysical level we describe the neuroelectric state of the tissue (spikes or slow potentials) as physical magnitudes. The cooperative processes are linked to the integro-differential equation correlated with the spatio-temporal evolutions of this state.

Keeping us in the field of sensorial functions, we will consider a region of cortex organized as a "layered computer" (see figure 2). The layers corresponding to horizontal, strellate and pyramidal cells.

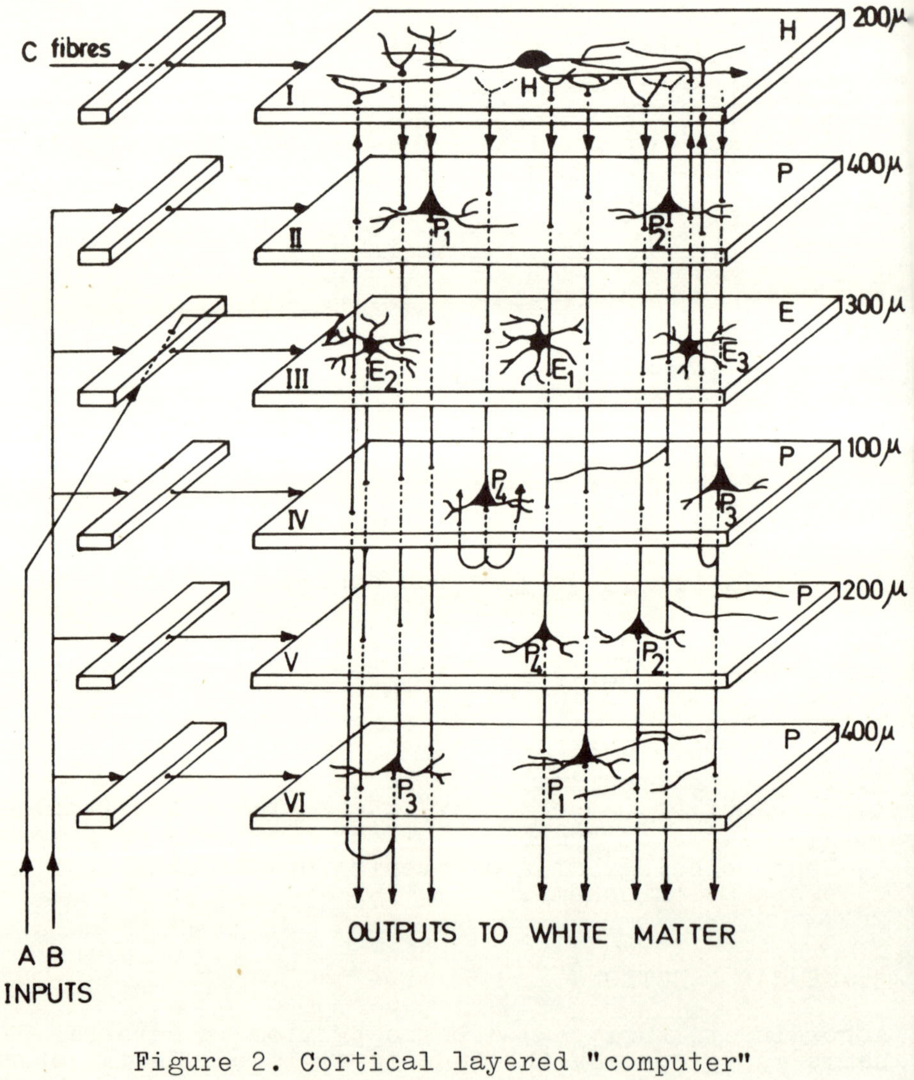

Figure 2. Cortical layered "computer"

The system receives specific and inspecific inputs (fibers A, B and C) and generate outputs to white matter. We assume the following hypothesis:

1. The layers are anatomically invariants and homogeneous.
2. In each layer there is a dominant cell whose anatomo-functional properties are extended to the layer as a whole.
3. For connection among layer, we use a generalization of the arch of Cajal.

The function performed by each layer can be represented by spatio-temporal filters (linear or nonlinear analogic process), followed by threshold (see figure 3). If f(x,y,t) and f(α, β, t) are the inputs in t at points (x,y), (α, β) the output at point (x,y) in (t+Δt) is:

if $g(x,y;t) \geq \Theta(t)$, $\Phi(x,y;t+\Delta t) = g(x,y;t)$ if not $\Phi(x,y;t+\Delta t)=0$

with $g(x,y;t) = \int_\Sigma f(\alpha,\beta;t)\, K(x,\alpha;y,\beta;t)\, d\alpha \cdot d\beta + \int_\Sigma g(\alpha,\beta;t-\ \ t)\, K^*(x,\alpha;y,\beta;t)\, d\alpha \cdot d\beta$

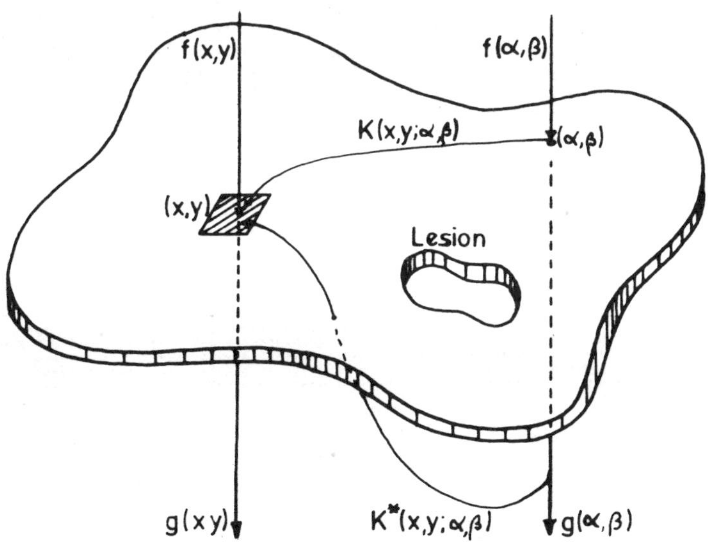

figure 3. Spatial Filter

where K and K are the kernels or weighting functions associated to the dendritic and axonix fields and is the threshold. If we consider stationary, time independent processes, the model can operate postulating the adequate kernels (7,8).
In digital simulation we use, as inputs, spatial pulses or stairs which stands during the running of the

program. The output is observed in the different layers at successive time intervals. Later, we will introduce a mask that simulates the lesion and analyzes its spatio-temporal evolution. The feedback among layer extend and make fuzzy this lesion (see figure 4). The main problem is the relation between the lesion and the functional deficit associated and this depends of the dendritic field and the feedback among layers. We have seen sensorial functions at analogic level. Here, the elimination of a module do not affect the action of the others. Physically, corresponds to weak interaction fenomena in which we can specify the function through the definition of an average potential. This simple cooperation is present in anatomic regions where the sensorial projections are "point to point" and the lesion generate deficits scotoma like. Increasing interaction means complex cooperation.

figure 4. Digital Simulation of the lesion

4. COOPERATIVE PARAMETERS

We can make a difference between anatomical and experimental parameters. Anatomically the cooperation depend of the degree of conectivity and interdependence between individual neurons.Over a surface of a layer, S, there are n·m neurons. Each one cooperates with a quantity of function, D, proportional to their area of calculus (dendritic field), from which we can obtain:

1. Redundancy,R. Measures the overlap of dendritic fields.

$$R=\sum_{i,j=1}^{n,m} \mu_{ij}\cdot D_{ij}/\bar{\mu}\cdot S\,, \qquad \mu=\text{density of function}$$

2. Conectivity, C. Measures how many neurons receive information from a specific one.

$$C=(A+D)/s_0\,,\quad A=\text{axonic field},\ s_0=\text{elemental area}$$

3. Functional Deficit, FD. Measures the quantity of function missing by the lesion L.

$$FD=R\cdot L$$

4. Residual Function, RF. Adding the functional deficit to the residual function we obtain the Redundancy.

$$RF=R\cdot(S-L)$$

A lesion, L, assumes the physical elimination of an area of calculus and the inputs to other layers for which this area was compelled patway. So, the more is the redundancy and the conectivity of a layer, the more is their reliability.

Experimentally, the cooperative process in cerebral dynamic are explored by subtracting the evaluations of a task before and after certain portions of the brain are removed. In our schema, some modules are masked to obtain the residual function.

The objective is just to make a map of the density of function of an area of nervous tissue, $\mu(\vec{x})$. In order to correlate this density, bivalent experiments (Luria) or polivalents ones, parametric type, (Lashley, J. Gonzalo) are usually used.

We can use as experimental dates the value of the stimulus, $I(\vec{x})$, needs to exceed a threshold Θ when the lesion is localized in $\vec{x}$ and has a span Δ. Then

$$I(\vec{x})\cdot\left[\int_{(X)}\mu(\vec{x})\cdot d\vec{x}-\int_{\Delta}\mu(\vec{x})\cdot d\vec{x}\right]=cte=1 \qquad \text{and} \qquad \int_{(X)}\mu(\vec{x})\cdot d\vec{x}\equiv F$$

where F is a measure of the whole function correlated

to the area X when no lesion is present. Supposing tha $\mu(\vec{x})$, changes softly in the region Δ and being V the volumen of this region, we have:

$$\int_{\Delta} \mu(\vec{x})\, d\vec{x} = V \cdot \mu(\vec{x}) \qquad \text{and} \qquad \mu(\vec{x}) = \frac{1}{V}\left[F - 1/I(\vec{x})\right]$$

Since $I(\vec{x})$ is the threshold stimulus when there is a lesion in $\vec{x}$, its inverse $1/I(\vec{x})$ give us a measure of the residual function of the resting tissue. As F is a measure of the global function, the density $\mu(\vec{x})$ can be obtained substracting the residual function $1/I(\vec{x})$. Please, notice that $I(\vec{x})$ is the experimental curve and then if the cooperative processes is complex it can be practically imposible to obtain $\mu(\vec{x})$ from $I(\vec{x})$.

REFERENCES

(1) Lashley, K.S.: 1929, "Brain mechanisms and intelligence". Chicago Univ. Press.
(2) Luria, A.R.: 1974, "Cerebro y Lenguaje". Ed. Fontanella, Barcelona.
(3) Luria, A.R.: 1974, "El cerebro en acción". Ed. Fontanella, Barcelona.
(4) Gonzalo, J.: 1952, "Las funciones cerebrales humanas según nuevos datos y bases fisiológicas". C.S.I.C. Vol. XLIV, Madrid.
(5) Gonzalo, J.: 1976-77, Personal comunication.
(6) Lashley, K.S.:1950, "In search of the engram". Soc. of Exp. Biol. Symposium. Cambridge Univ. Press.
(7) Moreno, R., Rubio, F. and Mira, J.: 1969. "Aplicacion de las transformaciones integrales al proceso de datos en la retina". Rev. de Automática, nº5, Madrid.
(8) Delgado, A.: 1978, "Modelos Neurocibernéticos de Dinámica Cerebral". Tesis doctoral. E.T.S.I.Telecomunicación. Univ. Politécnica. Madrid.

APPLICABILLITY OF COMPUTER SIMULATION IN BIODYNAMICAL SYSTEMS MODELLING

D.Matko, F.Bremšak, R.Karba, M.Porenta*, B.Zupančič

University of Ljubljana, Faculty of Electrical Engineering and *Medical Faculty, Department of Nuclear Medicine, Ljubljana, Yugoslavia

The paper deals with the computer simulation of the biodynamical systems and its applications. First the compartment model method and its applicability to identification of the biodynamical system parameters is described. Special attention is given to curve fitting procedure using an adaptive model and to identificabillity of parameters. In the second part of paper various dynamical models of radiocardiographic curves are reviewed The results of threecompartment model using variable gain, which is used at the Clinic for Nuclear Medicine in Ljubljana in the routine praxis, are shown.

1. INTRODUCTION

In the past few years the computer simulation is more and more applicated to the modelling of biodynamical systems. Most of the mathematical models used are so called "compartment" models, being obtained by the decomposition of the biodynamical system in his elementary parts. The function of each elementary part is then described by the first order differential equation.

One question appearing at this point is, if the first order differential equation really describes the function of particular part. In most cases the first order differential equation can be obtained under some assumptions and neglections using physical laws. Practical experiments usually verify the theoretical investigations.

The complete model is obtained by aggregation of the compartments into the biodynamical system which is described by a set of

D. G. Lainiotis and N. S. Tzannes (eds.), Applications of Information and Control Systems, 281-285.

first order differential equations. In the mathematical meaning the output of each compartment represents the "state" of the biodinamical system.

2. PURPOSES OF SIMULATION

Having the mathematical model another question appears: "What we shall need it for?" The answer to this question depends on the particulary application of biodynamical system modelling. In some cases the dynamic model is used for diagnostical purposess only. In this case we try to match the output of the model to the output of the measured biodynamical system. This is so called "curve fitting" procedure. In the mathematical language this procedure is called identification using the adaptive model. In order to describe the difference between the model and biodynamical system we introduce the quantity called index of performance of the identification procedure. This index can be the correlation quotient or the sum of squares of errors between the measured output of the biodynamical system and model response.

In order to identify properly the parameters of the biodynamical system, they have to be identifyable. This means that no two parameters should have exactly the same influence on the output of the biodynamical system. Due to the noise, beeing always present in the measured biodynamical systems output, the words "exactly the same" in the previous sentence should be replaced by the word "similar".

3. IDENTIFICATION

In the compartment model normally the time constants of each compartment, i.e. the time constant of the first order differential equation are to be determinated. If the compartment are aggregated in the cascade and only the outp ut of the last compartment is measureable, we should not be able to identify the time constants of all compartments. To do this, we would need more information about the biodynamical system, we would need more outputs. In the most suitable case, the whole state of the biodynamical system, i.e. the outputs of all compartments, could be measured. But this is in the praxis nearly unattainable. Normally the combination of the state variables can be measured.

If the model is identifyable, the equality of the output implicates the equality of the model and biodynamical system parameters what is the main point of the diagnosis procedure.

4. CONTROL ACTION

Having the model and its parameters the control action of the biodynamical system can be investigated. This means that if the biodynamical system is excitated, the influence of this excitation can be studied.

5. APPLICATIONS

One of possible applications of the biodynamical system modelling is the identification of the heart parameters by means of the radiocardiographic (RCG) curves. The RCG curves are obtained by injections of the radioactivity traser into patient´s vena and by measuring of the heart radioactivity.

The simple model of this biodynamical system, shown in Figure 1, consists of three compartments: right half of the heart, lungs and left half of the heart. The output of the biodynamical system is the sum of the right and left half of the heart. This simple model is described in (1).

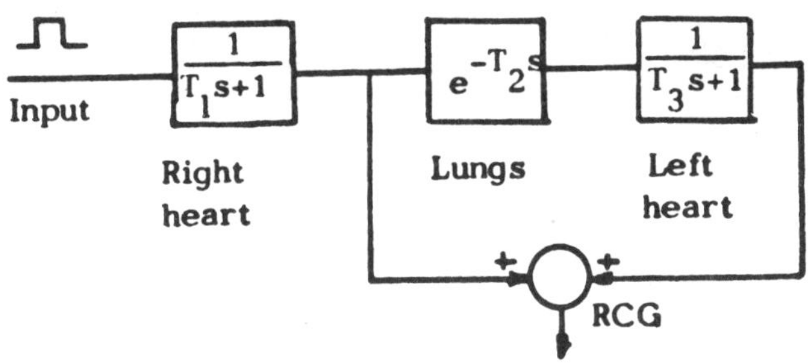

Figure 1. Simple RCG model

Firs extension of this model, we have done, is the variable gain of right and left heart chamber to the cardiac output, due to the possibility of skewed position of the scintillatic colimator. This model, given very good results, is now used at the Clinic for Nuclear medicine in Ljubljana in the routine praxis. Some results of the investigations are shown in Figure 2.

The next extension, beeing made is the decomposition of the circulatory system into four chambers of the heart, lungs consisting of the transport delay and lag and recirculation through the body. Corresponding model is shown in Figure 3. A similar model was described in (2). This model would enable simulation and identification of the intercardiac shunts, but the RCG curve of the whole heart radioactivity includes too little information, that it would be possible to distinguish the influence of different parameters. This would be possible by recording the output of each compartment, what can be done by the scintillation camera. The investigations in this direction are in progress.

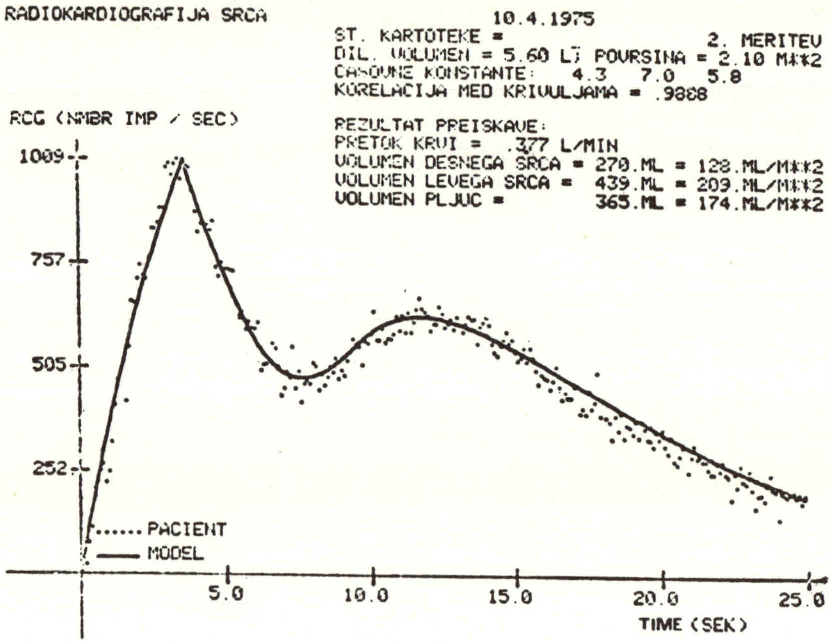

Figure 2. Results of investigation

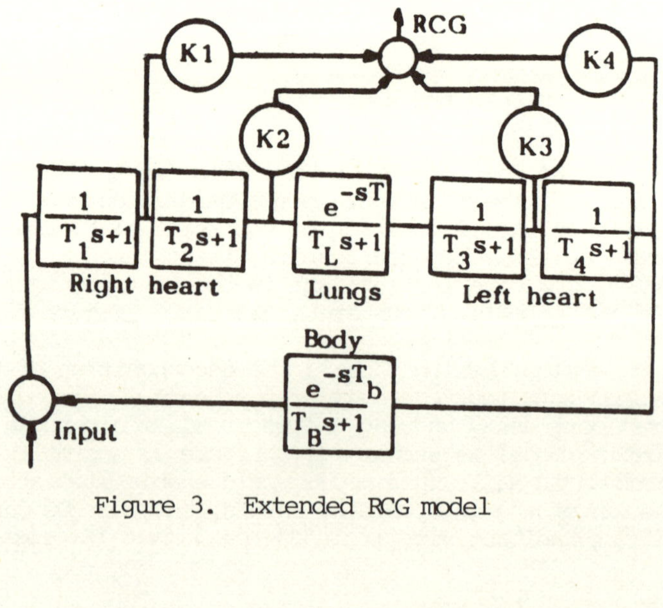

Figure 3. Extended RCG model

6. CONCLUSION

Special simulation program has given clinically applicable results on fractional parameters of the left and right heart hambers and the blood circulation in the lungs in patients with diffuse lesion of the miocardium, of various etiology and with valvular disfunctions. Therefore this method of analysis has already become part of the rutine procedure at the clinic for the Nuclear Medicine in Ljubljana.

REFERENCES

(1) Matko, D., Bremšak, F., Porenta, M., Šega, P., Karba, R.: 1976, Proceedings of the First International Conference on Information Sciences and Systems, Patras, Greece.

(2) Minato, K., M. Kuwahara, M., Yonekura, Y., Hirakawa, A.: 1978, Preprints of the Seventh Triennial World Congress IFAC, 1, Helsinki, Finland, pp. 533-540.

A SIMPLE MODEL FOR TREATMENT INDUCED RESISTANCE IN PEST POPULATIONS AND IN METASTASIZED TUMOR SYSTEMS*

James R. Thompson

Rice University
Houston, Texas 77001

Introduction, The nonimpact of control theory on cancer therapy should be a source of frustration for quantitatively oriented scientists in the field of oncology. Similarly, pest control has largely escaped serious attention by the applied control theorist. At first glance the argument for the use of control theory in both areas appears overwhelming. There are target populations which are to be destroyed without exceeding threshhold decimation levels of nontarget populations.

Closer examination reveals some substantial obstacles to the control theoretic approach. For example, there is the difficulty of formulating a natural criterion functional. In the case of radiotherapy, where cures are frequent, we might simply use survival fractions of treated cells. In the case of chemotherapy, cures are rare. Nevertheless, one of the most popular strategies is "Kill, Kill, Kill!" in which the cancer population is massively reduced near the time of treatment commencement.

If we do not like this criterion, which shall we pick? Shall we maximize the time to death of the patient? Shall we maximize the time to the end of "reasonable quality of life"? The list of possibilities is endless.

Still another stumbling block for the creation of useful oncological control theory is our lack of a good understanding of the kinetics of neoplasia. In the absence of a supportable mathematical model, there is a temptation to work with irrelevant but mathematically interesting models. Or, opting for empirical stability, one might look to simple models with varying degrees of plausibility (see, e.g., [1], [2], [3], [4]).

*This research supported in part by the United States Army Research Office under DAAG29-78-G-0817 at Rice University.

D. G. Lainiotis and N. S. Tzannes (eds.), Applications of Information and Control Systems, 286-293.

The problem here may be a tendency to confuse "validity" with plausibility and become excessively enchanted with a three to six parameter model because one can use it to obtain a reasonable fit of one time variable consisting of two monotone pieces.

A third difficulty with oncological and ecological control theory to be mentioned here is the manner in which data is selected and collected. I am not talking about a sloppy or cavalier attitude toward data collection. This is, of course, a problem. Rather, I am speaking of the fact that, in the absence of guidance from some speculative models, many of the variables selected for collection may be effectively irrelevant. Exploratory data analysis [6], in the absence of a model may enable one to extract much of the relevant information from a useful data set (although I would argue that in any EDA the brain is at least subconsciously sifting through its catalogue of models). But how much good can EDA be on an information set which excludes all really important variables from observation in the first place? And how can we have any notion as to which variable should be recorded without some preliminary speculative modelling?

There are very few tumor systems for which we have data sets showing the degree of reduction caused by a mode of treatment. As an example, one of our most reliable measures of survival fraction is that obtainable with multiple myeloma. For that system we can speak with some confidence that the minimum tumor mass during the course of therapy was at most 10^{-1} of the original mass. But we are effectively powerless to distinguish a minimum fraction of 10^{-2} from one of 10^{-4} . Given that for "cure" we need at least to bring the mass down by a factor of 10^{-5}, our difficulty is apparent. Most biostatisticians appear content to work with a weak correlate of tumor growth--namely tables of patient survival, and little pressure has been put on the medical community to obtain the kind of quantitative information really needed to design control strategies.

The result of near nonexistence of used control theoretic models in chemotherapy and pesticide management has been chaos in both fields. For example, in American agriculture, farmers are effectively allowed to use any of the currently approved agents in any proportion they wish. The result has frequently been the emergence of a target population resistant to large doses of everything legally available.

In chemotherapy resistance is even more of a problem and unlike the situation in pesticide management, may be irreversible. Yet treatment strategies are planned with a lack of careful design which most of us would find quite unacceptable, in, say, cake baking.

Discussion. In this paper I propose to carry out some simple preliminary speculative modelling to gain some insights into the problem of resistance in relation both to chemotherapy and to pesticide management. The models are certainly wrong but possibly they are useful.

Let us first consider a case where we have two cell populations--a normal and a neoplastic, both competing for the same life support. Let us suppose that in the untreated mode, growth proceeds according to

$$(1)\qquad \begin{cases} \dfrac{ds}{dt} = s(a - bg(s+n)) \\[2ex] \dfrac{dn}{dt} = -\dfrac{ds}{dt} \\[2ex] \dfrac{dp}{dt} = p(\alpha - \beta f(p)) \end{cases}$$

where s is the mass (i.e., number appropriately normalized) of cancer cells

n is the mass of competing normal cells

p is the mass of noncompeting normal cells (e.g., bone marrow, gut) most likely to be brought by the chemotherapeutic agent below a critical level

a, b, α, and β are positive constants

g is a monotone increasing function.

In the treated mode growth proceeds according to

$$(2)\qquad \begin{cases} \dfrac{ds}{dt} = s(a - a^*(t) - bg(s+n)) \\[2ex] \dfrac{dn}{dt} = -\dfrac{ds}{dt} \\[2ex] \dfrac{dp}{dt} = p(\alpha - ka^*(t) - \beta f(p)) \end{cases}$$

where $a^*(t)$ is a continuous (in time) non-negative control

k is a positive constant generally less than unity.

We note that in this model, $g(s+n)$ is a constant throughout the treatment; thus, the untreated cancer population is growing exponentially. Accordingly, we may indeed be well advised in this case to follow the "Kill, Kill, Kill!" strategy and utilize the maximal allowable (determined by constraints on p) dosage as soon as possible. If this model were applicable, our ability to effect cure would be limited only by the ability of the critical nonneoplastic population p to survive a killing dose of the target population. I fear this model is unrealistically optimistic.

Next, let us examine the situation where we have a second neoplastic population r which is totally resistant to the chemotherapeutic agent. So that in the untreated mode we have

$$(3) \quad \begin{cases} \dfrac{dr}{dt} = \lambda r[a - bg(r+s+n)] \\ \dfrac{ds}{dt} = s[a - bg(r+s+n)] \\ \dfrac{dn}{dt} = -\dfrac{dr}{dt} - \dfrac{ds}{dt} \\ \dfrac{dp}{dt} = p(\alpha - \beta f(p)) \end{cases}$$

where λ is a positive constant.

In the treated mode we shall assume

$$(4) \quad \begin{cases} \dfrac{ds}{dt} = s(a - a^*(t) - bg(s+r+n)) \\ \dfrac{dr}{dt} = \lambda r(a - bg(s+r+n)) \\ \dfrac{dn}{dt} = -\dfrac{ds}{dt} - \dfrac{dr}{dt} \\ \dfrac{dp}{dt} = p(\alpha - ka^*(t) - \beta f(p)) \; . \end{cases}$$

This model is closer to reality. Whereas it is probably inaccurate to consider a neoplastic subpopulation as being totally resistant to an agent, the usual failures of chemotherapeutic regimes following the lapse of initial remission appears to indicate the existence of neoplastic subpopulations more resistant to the agent than the critical tissue population p .

Again, there seems to be an advantage to massive treatment early on. Restraint in treating the susceptible population will not, under the conditions given, retard the growth of the resistant population. "Kill, Kill, Kill!" is optimal.

The models above consider a situation in which the life support system of the cancer and the normal tissue in the same region is constant. In many cases, however, this is unrealistic. The cancer cells seem to generate their own vascularization. Accordingly, let us consider the model

$$(5) \quad \begin{cases} \dfrac{ds}{dt} = s(a - a^*(t) - bg(s+r)) \\ \dfrac{dr}{dt} = \lambda r(a - bg(s+r)) \; . \end{cases}$$

Here competition for life support is between the susceptible and resistant cancer cells. Since we assume, as in the preceding model, that the resistant population is invulnerable to the treatment, there is no hope of cure. However, in this case, "Kill,

Kill, Kill!" is not optimal. The presence of the susceptible population inhibits the growth of the resistant population. Thus, the susceptible population acts as a control on the resistant population.

Problem 1. If $r(0)+s(0)=z$, where z is fixed, find a strategy which maximizes the time T for which $r(T+)+s(T+)$ is first greater than z.

Definition. Strategy A is that which maintains $a^*(t)$ so that $s(t)=z-r(t)$ until s is totally depleted.

Theorem 1. Strategy A solves Problem 1.

Proof. Consider a strategy, say B, which is different from A over the interval $(0,\Delta t]$ satisfying the condition

$$r(t)+s(t)<z\ . \tag{6}$$

Then

$$r_B(\Delta t)>r_A(\Delta t) \tag{7}$$

$$s_B(\Delta t)<s_A(\Delta t)\ . \tag{8}$$

But then any strategy starting at time Δt will have a class of control options which are a proper subset of those had we used Strategy A (since s is the only control on r).

Moreover, r has grown to a level greater than that had we used Strategy A.

Thus, Strategy A is optimal over $(0,\Delta t]$. But then $r(\Delta t)+s(\Delta t)=z$. So Strategy A is optimal globally.

Problem 2. If $r(0)+s(0)<z$, where z is fixed, find a strategy which maximizes the time T for which $r(T+)+s(T+)$ is first greater than z.

Definition. Strategy A^* is that which maintains $a^*(t)=0$ until $r+s=z$, then reduces s continuously to maintain $r+s=z$, then reduces s continuously to maintain $r+s=z$ as long as s is present (thus uses $a^*(t)=[a-bg(z)][s+\lambda r]/s$.

Theorem 2. Strategy A^* solves Problem 2.

Proof. Follows that for Theorem 1.

In Table 1 we show the ratios of time until total tumor load pass z. using Strategy A divided by the comparable time using "Kill, Kill, Kill!", in which the susceptible population is completely destroyed at time zero. All tumor masses are given in terms of fractions of the asymptotic mass $g^{-1}(a/b)$. The two g functions considered are the logistic $(g(s+r)=s+r)$ and the Gompertz $(g(s+r)=\log(s+r))$. We assume that at time zero $r(0)+s(0)=z$.

z	r_0	$\left(\frac{T_A}{T_{KKK}}\right)_{Logistic}$	$\left(\frac{T_A}{T_{KKK}}\right)_{Gompertz}$
.1	.01	1.067	1.443
	.001	1.086	1.820
	.0001	1.094	2.164
	.00001	1.099	2.485
.3	.1	1.163	1.407
	.01	1.296	2.106
	.001	1.345	2.712
	.0001	1.368	3.268
	.00001	1.381	3.792
.5	.1	1.465	1.934
	.01	1.703	2.980
	.001	1.800	3.900
	.0001	1.850	4.750
	.00001	1.880	5.555

Although Gompertzian models for neoplastic growth are currently in vogue (see, e.g., [5]) it is necessary to mention here that we really have virtually no reliable evidence to favor the logistic, the Gompertzian or some other model of tumor growth. However, Theorem 1 asserts that Strategy A is preferred for a large class of g functions.

Naturally, as a medical procedure, long term continuous therapy is not practical. However, it is clear that a discrete "pill-a-day" approximation is easily implemented. Also as a practical matter we should note that

$$a^*(t) = [a - bg(z)][s + \lambda r]/s \quad , \tag{9}$$

the control for Strategy A, is nearly constant over a long period of time, since typically $s(0) \gg r(0)$. When $a^*(t)$ becomes impractically large, the therapy will be discontinued. Even so, it is generally the case that the time until discontinuation of therapy can be very close to that obtainable with Strategy A--namely

$$T_A = \frac{\log \frac{z}{r(0)}}{\lambda[a - bg(z)]} \quad . \tag{10}$$

We have sidestepped the constraint that the total amount of agent given must not be in excess of some amount beyond which excessive damage to critical nonneoplastic tissue occurs. In the scenario considered here, the critical issue is the fact that the only inhibitor we have on the growth of the resistant neoplastic population is the presence of the susceptible neoplastic population. There comes a point beyond which further therapy simply has marginal effect on the reduction of tumor size. Typically, this point occurs before we pass the admissible level of destruction of critical neoplastic tissue. However, should we wish to add the constraint that

$$(11) \qquad \int_0^{T^*} a^*(t)dt \leq C$$

in Problem 1, it is clear that we achieve maximum T^* by using

$$a^*(t) = [a - bg(z)][s + \lambda r]/s$$

until

$$\int_0^{T^*} a^*(t)dt = C \ .$$

Let us turn now to an examination of the effect of resistance in insect populations. If we use here the same model as in (5), the situation is inappropriately pessimistic. For it would lead in short order to a resistant population of size $g^{-1}(a/b)$. Happily this does not happen. What distinguishes the pesticide case from that of chemotherapy in terms of the kinetic model?

Some reflection reveals a plausible explanation--namely, the occurence of environmental shocks which reduce the size of both the resistant and susceptible populations. These might include late frost, drought, bird migrations, sudden intense rainfall, etc. Unfortunately, we have no similar phenomenon with chemotherapy (although such might be artificially introduced via, say, radiotherapy as a secondary treatment mode).

Let us then consider the model in (5) with the additional structure that at time intervals of length Δ a shock occurs with probability p. If a shock does occur, then the survival fraction of the susceptible population is f_s, that of the resistant is f_r. We may be rather confident that $f_s > f_r$ and/or $\lambda < 1$, since typically at time zero, $s \gg r$. In the untreated mode, unless the susceptible population has been completely destroyed, it will again dominate--the resistant being reduced back to a low fraction (but not to zero, since these populations typically have subareas of the ecosystem in which they are dominant).

Below, in Table 2, we consider a system which has been so overtreated that $r(0) = .99$ and $s(0) = .01$. We suppose that at this time, treatment is suspended. We tabulate

$\dfrac{r(t)}{r(t)+s(t)}$ for three simulations. In all cases, $\Delta = .01$, $p = .05$, $a = b = 1$, $g(s+r) = s+r$.

Space does not permit here the development of optimal control strategies for the pesticide management case. But, suffice it to say that environmental shocks make possible attractive interactive strategies.

TABLE 2

t	$f_s = f_r = .5$ $\lambda = \frac{1}{2}$ $\left[\frac{r(t)}{r(t)+s(t)}\right]$	$f_s = .5;\ f_r = .25$ $\lambda = 1$ $\left[\frac{r(t)}{r(t)+s(t)}\right]$	$f_s = .5;\ f_r = .25$ $\lambda = \frac{1}{2}$ $\left[\frac{r(t)}{r(t)+s(t)}\right]$
0	.990	.99	.99
5	.990	.961	.388
10	.955	.756	.010
15	.811	.162	.007
20	.327	.024	2.3×10^{-5}
25	.199	.001	8.5×10^{-7}

REFERENCES

1. Brown B., and Thompson, J (1971), "Combat Models as Applied to Radiotherapy," The Proceedings of the Seventeenth Conference of the Design of Experiments in Army Research Development and Testing, pp. 33-60

2. Brown, B., Thompson, J., Withers, R., Suit, H., Barkeley, T., (1974). "Theoretical Consideration of Dose Rate Factors Influencing Radiation Strategy," Radiology, V. 110, No. 1, pp. 197-202.

3. Brown, B. and Thompson, J. (1974). "A Rationale for Synchrony Strategies in Chemotherapy," SIAM Proceedings of the Symposium on Epidemiology, Cooke and Ludwig, editors, pp. 34-44.

4. Hokkanson, J., Brown, B., Drewinko, B., Alexanian, R., Thompson, J. (1977). "Tumor Growth Patterns in Multiple Myeloma," Cancer, V. 39, No. 3, pp. 1077-1084.

5. Laird, A., "Dynamics of Tumor Growth: Comparison of Growth Rates and Extrapolation of Growth Curve to one Cell," (1965). British Journal of Cancer, V. 19, pp. 278-291.

6. Tukey, J., Exploratory Data Analysis, (1977). Addison-Wesley, Reading, Massachusetts.

THE ROLE OF PHARMACOKINETICAL MODELS SIMULATION IN THE DRUGS DESIGN

R.Karba, F.Bremšak, D.Matko, * F.Kozjek, *A.Mrhar

Faculty of Electrical Engineering, *Faculty of Technology and Natural Sciences, Department of Pharmacy, University E. Kardelj, Ljubljana, Yugoslavia

The studies of pharmacokinetical properties of drugs are very important part of their pharmacological activities evaluation. So they can be used also as the criterium for the new and more appropriate drugs design. The present approach begins with the first pharmacokinetical model choice and continues with its evaluation and possible extension through the curve fitting procedure of the model response to the data obtained "in vivo" on the analog-hybrid computer. It gives as the results the drug concentration time responses in particular body compartments for the single and multiple dosing. By the aid of such results and known minimal effective and toxical concentration, the most appropriate dosage form as well as the optimal dosage regimen can be obtained. The successfulness and usefulness of the approach is also discussed and is illustrated by the example of the concrete drug design.

1. INTRODUCTION

The incorrect use of many drugs requires more and more profound studies of their pharmacological effects. Among them the pharmacokinetical studies are very important. Pharmacokinetics deals with the fate of the drug on its way through the human body and study the processes like absorption, distribution, metabolism and elimination. Only by the use of the drug which has besides the other requirements also the prescribed pharmacokinetical properties we can expect the results of therapy to be satisfactory. So we want to design dosage forms and dosage regimen necessary for the optimal application of drugs.

D. G. Lainiotis and N. S. Tzannes (eds.), Applications of Information and Control Systems, 294-301.

2. PHARMACOKINETICAL MODELLING IN DRUGS DESIGN

Studying the pharmocokinetical properties of the concrete drug the suitable pharmacokinetical model, which is the simple representation of the processes "in vivo", must be first chosen. This model is generated on the base of the previous knowledge about the real processes in the human body, taking into account some approximations. In the first phase of investigations the parameters of the model can be obtained from the "in vitro" tests, which are of course only the best possible approximations to the real processes which are very complicated. Chosen pharmacokinetical model can be described by the system of first order differential equations which represent the corresponding mathematical model. Such model can be successfully solved only by the aid of the computer which gives as the result the drug concentration time responses in particular body compartments. On the base of such results, known minimal effective concentration and multiple dosing simulation the onset, efficacy and duration of the pharmacological effect of drug can be evaluated as well as the suitability of the existent dosage regimen. We must take into account of course that the validity of the chosen model can be measured only on the base of the "in vivo" data. Such data are hardly attainable because of the problems of biological liquids analysis as well as because of the etical problems in the work with the volunteers. The suitability of the chosen model can be evaluated through the curve fitting procedure of the model response to the "in vivo" data curve, which must be generated from the enough frequent samples of blood from which the concentrations of drug must be determined by the aid of the enough accurate analytical methods. So the best model with corresponding parameters can be obtained which is generally the extension of the first chosen model. This model is the base of the appropriate dosage form and dosage regimen design considering besides the prescribed therapeutical limits also the available technology of the designed drug manufacturing. At last the successfulness of the approach must be evaluated from the "in vivo" data obtained from the clinical investigations of the designed drug effects. Because of his efficacy in solving the systems of differential equations and above all because of the simple extension of the model possibilities during the curve fitting procedure of the model response to the "in vivo" curve, the analog-hybrid computer is the most suitable for the simulation and identification of the pharmacokinetical models in the process of drugs design.

3. PHARMACOKINETICAL STUDY OF THE CONCRETE DRUG

In the case of the concrete bronhial dillatator pharmacokinetical study which we realised on the analog-hybrid computer EAI 580 we first chose the simple three compartment model including the compartments of gastrointestinal tract (U_{GI}), plasma (U_p) and urine (U_E) connected through the absorption (k_a) and elimination (k_e)

rate constants which is shown in Figure 1. This model was generated on the base of the previous knowledge about the real processes in the human body obtained from the several references (2,3,4). In the first phase of investigations the parameters of the model are generally obtained from the "in vitro" tests, while the total elimination rate constant has to be calculated from the attainable "in vivo" data from the references (1,2,3).

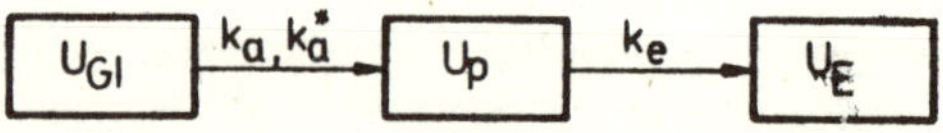

Figure 1. The simple three compartment pharmacokinetical model of the discussed drug.

The corresponding mathematical model is represented by the following differential equations:

$$\frac{dU_{GI}}{dt} = -k_a\, U_{GI}, \tag{3.1}$$

$$\frac{dU_P}{dt} = k_a\, U_{GI} - k_e\, U_P, \tag{3.2}$$

$$\frac{dU_E}{dt} = k_e\, U_P. \tag{3.3}$$

Analog simulation of the mathematical model was executed on an analog-hybrid computer by the aid of analog diagram shown in Figure 2.

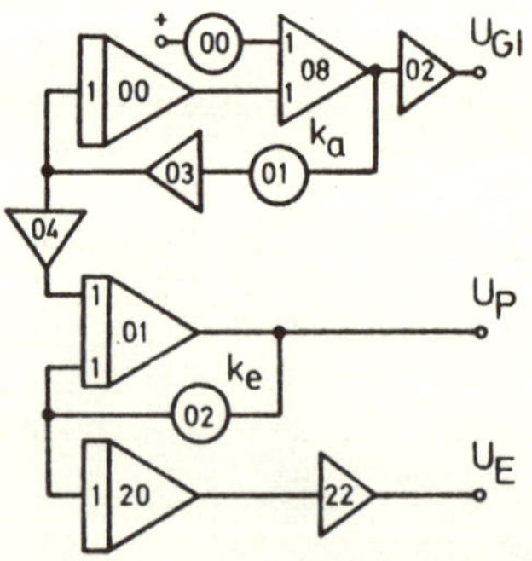

Figure 2. Analog diagram of the simple pharmacokinetical model.

With the simulation of the mentioned model using various doses the results shown in Figure 3 were obtained. From Figure 3 it can be seen that the existent dose (D_1) is insufficient to bring the

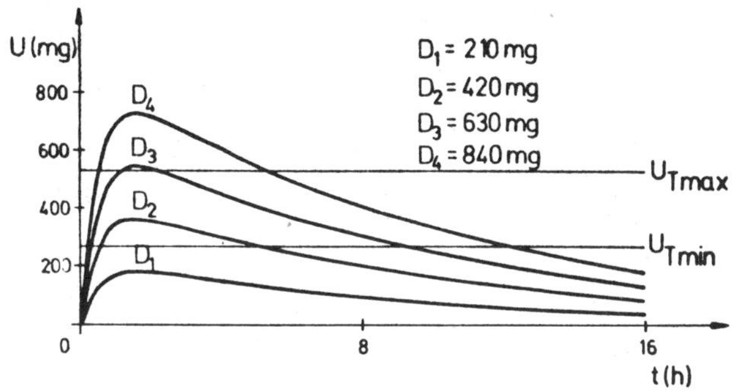

Figure 3. Drug quantity time responses in plasma for various doses.

drug quantity time response into its therapeutic range and that only the increasing of the dose is not advisable because of the relatively unfavourable pharmacokinetics of the drug. The simulation of the multiple dosing was realised by the aid of analog diagram in Figure 2 and the generator of the stairs function. Its analog diagram is shown in Figure 4.

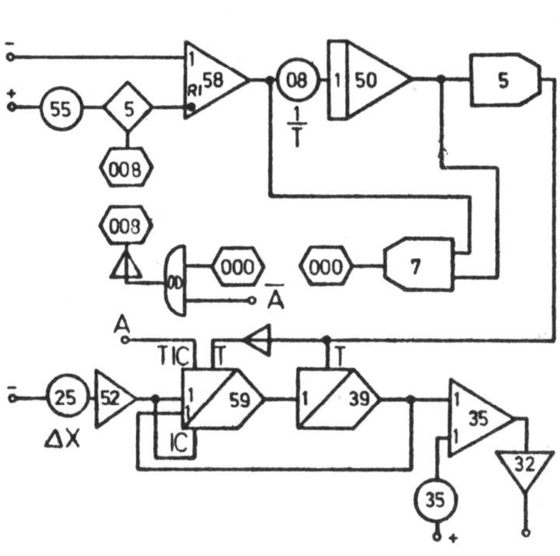

Figure 4. Analog diagram for the stairs function generator simulating the quantity and frequency of dosing.

So the results of the multiple dosing simulation are shown in Figure 5. From Figure 5 it is evident, that in spite of the phenomenon of cumulation the multiple dosing of 210 mg of drug each eight hours is insufficient taking into account the prescribed therapeutical limits. So the new dose of 350 mg was chosen which was not higher because of some technological limits. It is obvious,

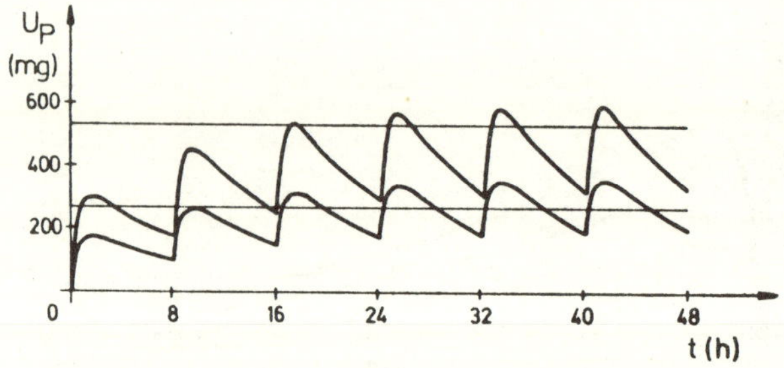

Figure 5. Drug quantity time responses in plasma after the application of 210 mg and 350 mg dose six times each eight hours.

owever, that the oscillations of the plasma levels are significant and that the introduction of the sustained release dosage form is justified (5). The simple one component sustained release dosage form was designed with $k_a=5k_e$ due to the technological possibilities in manufacturing the designed dosage form. The results of the multiple dosing simulation for this case can be seen in Figure 6.

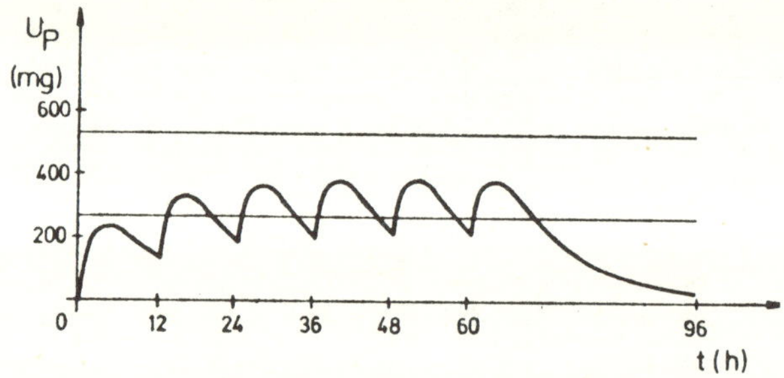

Figure 6. Drug quantity time response in plasma for the simple one component sustained release dosage form after the application of the dose of 350 mg six times each twelve hours.

From the Figure 6 we can conclude, that the oscillations of the plasma levels are lower in spite of the longer dosing interval and that the levels are within the therapeutical limits nearly whole time of observation except the first twentyfour hours.

3.1. Pharmacokinetical model suitability evaluation

The validity and suitability of the primarily chosen pharmacokinetical model can be evaluated only by the aid of the "in vivo" data with so caled "curve fitting" procedure (6). Because in our case the "in vivo" curve was lower than the simulated one we concluded that the pharmacokinetical model had to be extended.

Several possibilities of model extensions were studied and the possibility of the incomplete absorption was also checked. The best was the five compartment model with reversible side compartment, which represents the distribution of the drug in the tissue (U_D). The compartment of the dosage form was also added (U_T). Chosen pharmacokinetical model is shown in Figure 7, and its analog diagram is included in Figure 8.

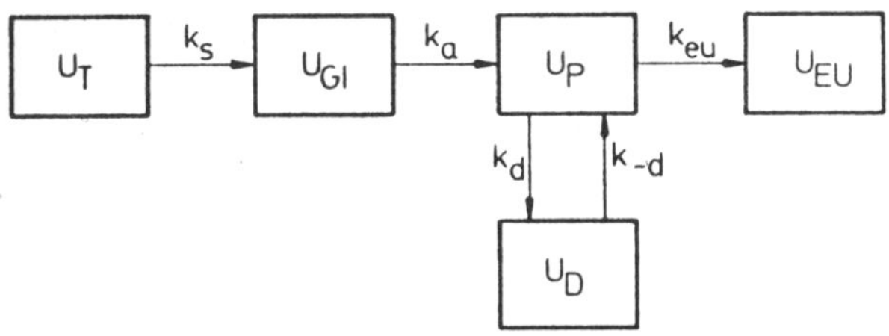

Figure 7. Extended pharmacokinetical model.

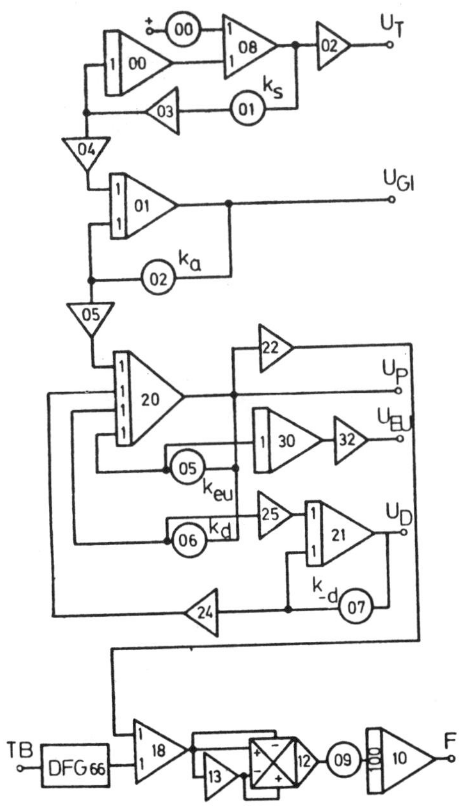

Figure 8. Analog diagram for the curve fitting procedure.

The results of the curve fitting procedure are shown in Figure 9.

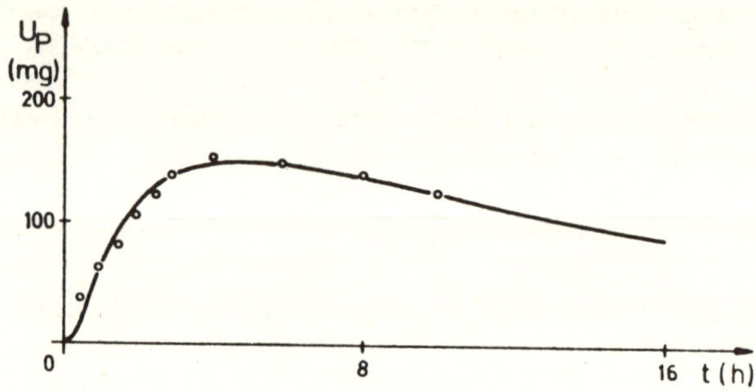

Figure 9. The "in vivo" data and the model response (plasma levels).

Figure 9 shows the very good fitting between the curves, so it can be concluded that the model was suitably chosen. The curve fitting procedure also showed, that the constants k_s, k_a and k_{eu} had to remain unchanged, while the distribution rate constants were identified.

3.2. The new dosage regimen design

By the aid of the multiple dosing simulation the optimal dosage regimen can be designed, what is shown in Figure 10.

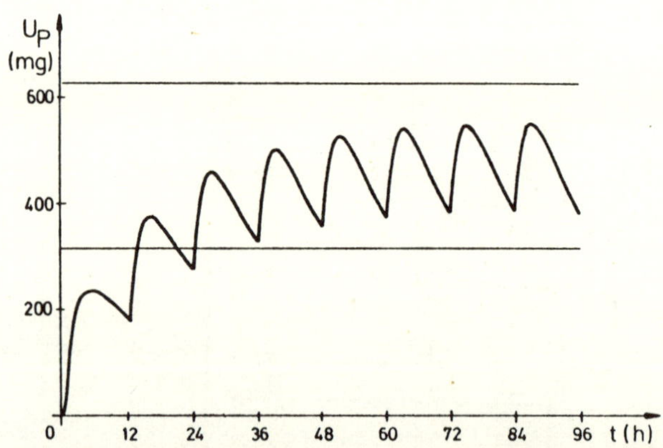

Figure 10. Drug quantity time response in plasma after the application of 600 mg dose eight times with single initial dose each twelve hours with the corresponding therapeutic range.

From Figure 10 it can be seen that the plasma levels maintain with relatively low oscillations within the therapeutical limits the whole time of observation and that the necessary dose of drug in the designed dosage form is 600 mg applicated each twelve hours.

4. CONCLUSION

The discussed drug design procedure on the base of pharmacokinetical studies executed by the aid of an analog-hybrid simulation and identification and "in vivo" tests seems to be successful and useful what is illustrated by the results of the mentioned example. With the introduction of some nonlinearities in the models they would be improved what would be another step to the optimal use of drugs. The implementation of the mentioned approach on the digital computer would enable the pharmacokinetical studies for the known model structure and the range of parameters for the individual patient also in the clinical praxis.

REFERENCES

(1) Wagner, J.G.: 1975, Clinical Pharmacokinetics, Drug Intelligence Publication, Hamilton.
(2) Jenne, J.W., Wyze, E., Rood, F.S., Mac Donnald, F.M.: 1972, Clin. Pharm. Ther., 13, 3, pp. 349-360.
(3) Mitenko, P.A., Ogilvie, R.I.: 1973, Clin. Pharm. Ther., 14, 4/1, pp. 509-513.
(4) Ogilvie, R.I.: 1978, Clin. Pharm., 3, pp. 267-293.
(5) Karba, R., Matko, D., Šega, P., Bremšak, F., Kozjek, F., Mrhar, A.: 1977, First Mediterranean Conference on Medical and Biological Engineering, Digest of Papers 2, Sorrento, Italy.
(6) Karba, R., Bremšak, F., Matko, D., Mrhar, A., Kozjek, F.: 1978, Preprints of Seventh Triennial IFAC Congress,1, Helsinki, Finland, pp. 557-564.

SECTION V

OPERATIONS RESEARCH, MANAGEMENT AND ECONOMICS

MULTI-LEVEL CONTROL SYSTEMS FOR INVENTORY MANAGEMENT

A. Bensoussan* and M. Crouhy**

*Université of Paris IX-Dauphine and INRIA, France
**CESA (HEC - ISA - CFC), France

ABSTRACT

The paper addresses the problem of determinating decentralized purchasing policies for a multi-installation inventory system, in the framework of modern control theory, in discrete time. It is first shown that an optimal solution exists for the general centralized problem. We then derive an optimal decentralized policy in the case with no ordering costs inside the inventory system ; the decoupling is obtained through a bottom-up determined penalty scheme. In the case of two installations with ordering costs, we can only derive an approximate solution for which an upper and a lower bounds are defined for the penalty imposed to installation 2.

INTRODUCTION

We consider a multi-level inventory management system, where there is a hierarchy of installations 1,...,N, with installation 1 being replenished from 2, with 2 being replenished from 3, etc... This problem is quite difficult when the demand is random. More precisely, it is in general impossible to obtain a solution of the problem which is fully decentralized. A seminal contribution is due to CLARK and SCARF [1], [2]. These articles were published twenty years ago, and since then little has been done to propose decentralized policies which are admissible from the point of view of information, and are optimal or quasi-optimal with respect to the global problem.

The objective of this article is to revist the results of CLARK and SCARF in the framework of modern control theory, in discrete

D. G. Lainiotis and N. S. Tzannes (eds.), Applications of Information and Control Systems, 305-315.

time. This leads us to derive a rigorous proof of the desired results. It is also a good intermediary step to the same problem in continuous time, which will be addressed in a forthcoming paper.

1. THE GLOBAL PROBLEM AND ITS CENTRALIZED SOLUTION

1.1. Setting of the problem

We consider a hierarchy of N installations arranged in series : $N \to N-1 \to \ldots \to 2 \to 1$ where the supplies of installation i is installation (i+1), and installation N orders from an outside source. Installation 1 is the only one in the system to face an external demand for a unique product at times 1,2,...,n, respectively $\xi_1, \xi_2, \ldots, \xi_n, \ldots$ where :

(1.1) ξ_n are random variables which are independent and identically distributed with a probability law $\mu(\xi)$, such that $\int |\xi| \, \mu(d\xi) < \infty$

These random variables are constructed on a probability space $(\Omega, \mathcal{A}, P)$. We set :

$$\begin{cases} \mathfrak{F}^n = \sigma \text{ algebra generated by } \xi_1, \ldots, \xi_n \,, \quad n \geq 1 \\ \mathfrak{F}^o = (\Omega, \Phi) \end{cases} \tag{1.2}$$

The decision variables v_n^i for i=1,...,N and n=0,... are the replenished orders placed at time n by installation i. They are random variables satisfying the constraints

$$v_n^i \geq 0 \,, \quad E\, v_n^i < +\infty \quad \text{and} \quad v_n^i \text{ is } \mathfrak{F}^n \text{ measurable.} \tag{1.3}$$

We denote by y_n^i the sum of the inventories on hand at installations i plus orders in transit to or inventories on hand at lower installations up to 1. It is convenient to call echelon i the group of installations from 1 to i. Thus, y_n^i is the inventory level of echelon i at time n. The orders v_n^i are placed at time n and delivery occurs at time n+1, which yields to the evolution equations :

$$\begin{cases} y_{n+1}^i = y_n^i + v_n^i - \xi_{n+1} \,. \\ y_o^i = x^i \end{cases} \tag{1.4}$$

We point out that y_n^i is the inventory level carried from time n to time n+1. Also, note that $y_n^{i+1} - y_n^i$ represents the physical inventory level at installation i+1, i=1,...,N-1, and y_n^1 is the inventory level at installation 1. The coupling between successive installations is defined as follows :

$$(1.5) \qquad v_n^i \leq y_n^{i+1} - y_n^i \quad , \quad i=1,\ldots,N-1$$

which means that installation i cannot place at time n, an order exceding the quantity available at installation i+1, at the same time. We impose no restriction to v_n^N, the orders of installation N being placed outside.

The quantities $x^1,\ldots,x^N$ are the initial echelon inventory levels. They are given numbers satisfying :

$$(1.6) \qquad x^1 \leq \ldots \leq x^N .$$

We denote by $\vec{v}^i$ the infinite vectors

$$\vec{v}^i = (v_o^i,\ldots,v_n^i,\ldots)$$
$$\vec{v} = (\vec{v}^1,\ldots,\vec{v}^N).$$

The objective function to be minimized at the central level is defined by :

$$(1.7) \qquad J_x(\vec{v}) = E \sum_{n=o}^{\infty} \sum_{i=1}^{N} (f^i(y_n^i) + c^i(v_n^i))\alpha^n ,$$

where $c^i(v)$ is the cost of ordering the quantity v at installation i. The interpretation of $f^i(x)$ is more involved except for i=1, for which :

$$f^1(x) = \left| \begin{array}{ll} \text{carrying cost} & \text{when } x \geq 0 \\ \text{shortage cost} & \text{when } x < 0. \end{array} \right.$$

For i=2,...,N, the argument of f^i is the inventory of echelon i. Therefore, the form (1.7) of the payoff function requires additional assumptions on the cost structure.

1.2. Centralized solution

Let $X = \{x \in R^N \mid x^1 \leq \ldots \leq x^N\}$. We define B_N as the set of functions $f : X \to R$, which are measurable and with linear growth, provided with the norm

$$||f||_N = \sup \frac{|f(x^1,\ldots,x^N)|}{1 + |x^1| + \ldots + |x^N|}$$

and C_N the subspace of B_N of continuous functions. We define an operator $\Phi \in \mathcal{L}(B_N;B_N)$ by setting :

(1.8) $$\Phi(f)(x^1,\dots,x^N) = \int f(x^1-\xi,x^2-\xi,\dots,x^N-\xi)\mu(d\xi).$$

It is easy to check that :

(1.9) $$\left| \begin{array}{l} \|\Phi\| \leq 1 + N\,E|\xi| \\ \|\Phi^n\| \leq 1 + nN\,E|\xi| \quad , \qquad \Phi \in \mathcal{L}(C_N;C_N). \end{array} \right.$$

We make the following assumptions :

(1.10) $$f^i(x^i) \in C_1 \;,\quad \sum_{i=1}^{N} f^i(x^i) \geq 0 \;,\ \text{for } x \in X$$

(1.11) $$\left| \begin{array}{l} c^i(v^i) = K^i\chi_{v^i>0} + c_o^i(v^i), \\ \text{where } c_o^i \geq 0 \text{ continuous } c_o^i(o) = 0 \\ c_o^N(v^N) \geq c^N v^N \text{ for } v^N \geq 1 \;,\ K^i \geq 0, \quad i=1,\dots,N \\ \alpha \in [0,1[. \end{array} \right.$$

Let $x \in X$. We write :

(1.12) $$\left| \begin{array}{l} \mathcal{V}_x = v^1,\dots,v^N \left| \begin{array}{l} 0 \leq v^i \leq x^{i+1}-x^i \;, \quad i=1,\dots,N-1 \\ 0 \leq v^N \end{array} \right. \\ \\ c(v) = \sum_i c^i(v^i) \;,\ f(x) = \sum_{i=1}^{N} f^i(x^i) \end{array} \right.$$

We introduce the Dynamic Programming equation :

(1.13) $$\left| \begin{array}{l} u(x) = f(x) + \inf_{v\in\mathcal{V}_x} \{c(v) + \alpha\Phi(u)(x+v)\} \\ \\ u \in B_N \;,\quad u \geq 0 \end{array} \right.$$

We also consider the linear equation :

(1.14) $$w(x) = f(x) + \alpha\Phi(w)(x) \quad , \qquad w \in C_N$$

which has one and only one solution.

Theorem 1.1

We assume (1.1), (1.10), (1.11). The set of functions u satisfying (1.13) is non empty and has a minimum solution u which is l.s.c., and a maximum solution $\bar{u}$, and $\bar{u} \leq w$,

$$(1.15) \qquad \underline{u}(x) = \min_{\vec{v}} J_x(\vec{v})$$

Proof : Consider the sequence :

$$(1.16) \qquad \begin{vmatrix} u_o = 0 \\ u_{n+1} = f + \inf\limits_{v \in \mathcal{V}_x} [c(v) + \alpha\Phi(u_n)(x+v)] \end{vmatrix}$$

Then $u_n \in C_N$, u_n is increasing and $u_n \leq w$; $u_n \uparrow u$. There exists a measurable mapping $\hat{v}_n(x) : X \to R^N$ such that $\hat{v}_n(x) \in \mathcal{V}_x \ \forall x$, and :

$$(1.17) \qquad u_{n+1}(x) = f(x) + c(\hat{v}_n(x)) + \alpha\Phi(u_n)(x + \hat{v}_n(x)) \quad \forall x.$$

Let $m \leq n$, we have from (1.17) and the monotonicity property :

$$u_{n+1}(x) \geq f(x) + c(\hat{v}_n(x)) + \alpha\Phi(u_m)(x + \hat{v}_n(x))$$

hence

$$(1.18) \qquad u(x) \geq f(x) + c(\hat{v}_n(x)) + \alpha\Phi(u_m)(x + \hat{v}_n(x))$$

But from the definition of $\mathcal{V}_x$ and the last assumption (1.11), we have :

$$|\hat{v}_n(x)| \leq c(1 + |x|)$$

For x fixed we can extract a subsequence $\hat{v}_{n_k} \to v^*$, and $v^* \in \mathcal{V}_x$.

Taking $n=n_k$ in (1.18) and letting $k \to \infty$, we obtain from the lower semi-continuity of the right hand side of (1.18) :

$$u(x) \geq f(x) + c(v^*(x)) + \alpha\Phi(u_m)(x + v^*(x))$$

and letting $m \to \infty$

$$(1.19) \qquad u(x) \geq f(x) + c(v^*) + \alpha\Phi(u)(x + v^*)$$

$$\geq f(x) + \inf_{v \geq 0} [c(v) + \alpha\Phi(u)(x+v)] \ .$$

On the other hand from (1.16) since $u_n \leq u$, we have :

$$u_{n+1} \leq f + \inf_{v\geq 0} [c(v) + \alpha\Phi(u)(x + v)]$$

and passing to the limit we obtain the reverse inequality (1.19), hence u is solution of (1.13). Il is the minimum solution, since if u is a solution, then, by induction $u_n \leq u$, hence $u \leq u$. It is l.s.c. since it is an increasing limit of a sequence of continuous functions. There exists a measurable mapping $\hat{v}(x) : X \to R^N$, such that $\hat{v}(x) \in \mathcal{V}_x\ \forall x$, realizing for any x the infinimum on the right side of (1.13). One can check that $\hat{v}$ defines an optimal feedback. The maximum solution $\bar{u}$ is obtained by the iterative scheme :

(1.20) $$u^o = w$$

$$u^{n+1} = f + \inf_{v\in\mathcal{V}_x} [c(v) + \alpha\Phi(u^n)(x + v)] .$$

Then, $u^n \downarrow \bar{u}$ which is a solution and the maximum one. Note that all solutions of (1.13) are less or equal to w.

□

2. DECENTRALIZED SOLUTION IN THE CASE WITH NO FIXED COSTS

2.1. Assumptions. Statement of the result

We assume here that

(2.1) $$f^i(x^i) \text{ convex } ; K^i=0 \ , \ c_o^i(v^i) = c_o^i v^i$$

for $i=1,\ldots,N-1$.

(2.2) $$\alpha\Phi(f^1)(x^1) + c_o^1 x^1 \to +\infty \text{ lorsque } |x^1| \to +\infty \ ; \ f^1 \geq 0.$$

We define $u^1(x^1)$ to be the solution of :

(2.3) $$u^1(x^1) = f^1(x^1) - c_o^1 x^1 + \inf_{y^1\geq x^1} \{c_o^1 y^1 + \alpha\Phi(u^1)(y^1)\}$$

which is the problem of echelon (or installation) 1, without constraints. There exists one and only one solution of (2.3) in the space C_1 , and u^1 is convex. More generally, knowing $u^i(x^i)$, $1 \leq i \leq N$ which is convex and belongs to C_1, define :

(2.4) $$z^i(x^i) = c_o^i x^i + \alpha\Phi(u^i)(x^i)$$

which is also convex. Assume that

(2.5) $$z^i(x^i) \to +\infty \text{ as } |x^i| \to +\infty$$

Then, z^i reaches its minimum and S^i is the smallest minimum. Define next :

(2.6) $$\Lambda^{i+1}(x^{i+1}) = (z^i(x^{i+1}) - z^i(S^i))\,\chi_{x^{i+1} \leq S^i}$$

and define u^{i+1} by equation :

$$(2.7)\quad u^{i+1}(x^{i+1}) = f^{i+1}(x^{i+1}) + \Lambda^{i+1}(x^{i+1}) - c_o^{i+1}x^{i+1} + \\ + \inf_{y^{i+1}\geq x^{i+1}} \{c_o^{i+1}y^{i+1} + \alpha\Phi(u^{i+1})(y^{i+1})\}$$

as long as $i+1 \leq N-1$,
and :

$$(2.8)\quad u^N(x^N) = f^N(x^N) + \Lambda^N(x^N) + \inf_{v^N\geq 0} \{C^N(V^N) + \alpha\Phi(u^N)(x^N+v^N)\}.$$

We note that because of (2.2), condition (2.5) is satisfied for $i=1$. We also note that $\Lambda^{i+1}(x^{i+1})$, $i=1,\ldots,N-1$ is convex.

We assume that :

$$(2.9)\quad \begin{cases} f^{i+1}(x^{i+1}) + \Lambda^{i+1}(x^{i+1}) \geq 0 & i=1,\ldots,N-1 \\ c_o^{i+1}\, x^{i+1} + \alpha\Phi(f^{i+1} + \Lambda^{i+1})(x^{i+1}) \to +\infty \text{ as } |x^{i+1}| \to +\infty & \\ & i=1,\ldots,N-2 \end{cases}$$

then (2.7) has one and only one solution in C^1 and (2.5) is satisfied at step $i+1$. For (2.8) we can guarantee the existence of a l.s.c. solution, which is the minimum among measurable ones and also a maximum solution. Define next for $i=1,\ldots,N-1$:

$$(2.10)\quad \hat{v}^i(x^i) = \begin{cases} S^i - x^i & \text{for } x^i \leq S^i \\ 0 & \text{for } x^i \geq S^i \end{cases}$$

and $\hat{v}^N(x^N)$ to be a measurable mapping which realizes the infinimum for any x^N, on the right hand side of (2.8). It exists if u^N is the minimum solution. Finally define :

$$(2.11)\quad \begin{cases} \hat{\hat{v}}^i(x) = \hat{v}^i(x^i) \wedge (x^{i+1} - x^i) & i=1,\ldots,N-1 \\ \hat{\hat{v}}^N(x) = \hat{v}^N(x^N) & \end{cases}$$

Then we can assert that :

Theorem 2.1. We make the assumptions of Theorem 1.1. and (2.1), (2.2), (2.9). Then :

$$\tilde{u}(x) = \sum_{i=1}^{N} u^i(x^i) \text{ is solution of (1.13) and}$$

$\hat{\hat{v}}$ defined by (2.11) is an optimal feedback.

2.2. Proof of theorem 2.2.

Set for $i=1,\ldots,N-1$

$$(2.12)\quad \tilde{u}^i(x^{i+1}) = f^i(x^i) - c_o^i x^i + \Lambda^i(x^i) + \inf_{x^i\leq y^i\leq x^{i+1}} z^i(y^i)$$

Then we have :

$$(2.13) \quad \tilde{u}^i(x^i,x^{i+1}) = u^i(x^i) + \Lambda^{i+1}(x^i) \qquad i=1,\dots,N-1.$$

Property (2.13) follows by direct checking. When $x^i \geq S^i$,

$$\tilde{u}^i(x^i,x^{i+1}) = f^i(x^i) = c_o^i x^i + z^i(x^i) = u^i(x^i)$$

and $\Lambda^{i+1}(x^i) = 0$. When $x^i < S^i$, then :

$$u^i(x^i) = f^i(x^i) + \Lambda^i(x^i) - c_o^i x^i + z^i(S^i).$$

If $x^{i+1} < S^i$, then :

$$\tilde{u}^i(x^i,x^{i+1}) = f^i(x^i) + \Lambda^i(x^i) - c_o^i x^i + z^i(x^{i+1})$$

and :

$$\Lambda^{i+1}(x^{i+1}) = z^i(x^{i+1}) - z^i(S^i).$$

If now $x^{i+1} \geq S^i > x^i$, then :

$$\tilde{u}^i(x^i,x^{i+1}) = f^i(x^i) + \Lambda^i(x^i) - c_o^i x^i + z^i(S^i)$$

and $\Lambda^{i+1} = 0$. Therefore (2.13) is proved.

To check that $\tilde{u}$ is a solution of (1.13), one evaluates the right hand side of (1.13), when u is replaced by u. Denote this right hand side by X, then :

$$X = \sum_{i=1}^{N-1} (\tilde{u}^i(x^i,x^{i+1}) - \Lambda^i(x^i)) + f^N(x^N) + $$

$$+ \inf_{v^N \geq 0} \{C^N(v^N) + \alpha\Phi(u^N)(x^N + v^N)\}$$

and by (2.13) :

$$= \sum_{i=1}^{N-1} u^i(x^i) + \Lambda^N(x^N) + f^N(x^N) + \inf_{v^N \geq 0} \{ \}$$

$$= \sum_{i=1}^{N} u^i(x^i).$$

We note that :

$$z^i(x^i+\hat{v}^i(x)) = \inf_{x^i \leq y^i \leq x^{i=1}} z^i(y^i)$$

from which it easily follows that $\hat{\hat{v}}$ is an optimal feedback, i.e. more precisely :

$$\tilde{u}(x) = \sum_{i=1}^{N} f^i(x^i) + \sum_{i=1}^{N} c^i(\hat{\hat{v}}^i(x) + \alpha\Phi(\tilde{u})(x+\hat{\hat{v}}(x))$$

□

3. DECENTRALIZED SOLUTION IN THE CASE WITH FIXED COSTS

3.1. Assumptions. The problem

In this section, we will consider only the case where N=2.
We assume (1.10), (1.11) and :

$$(3.1)\qquad \begin{cases} f^1(x^1) \text{ convex positive} \; ; \; c_o^1(v^1) = c_o^1 v^1, \\ \alpha\Phi(f^1)(x^1) + c_o^1 x^1 \to +\infty \quad \text{as } |x^1| \to +\infty \; ; \; K^1 > 0. \end{cases}$$

The problem of echelon 1 without constraints is governed by the equation :

$$(3.2)\qquad u^1(x^1) = f^1(x^1) - c_o^1 x^1 + \inf_{y^1 \geq x^1} \{c_o^1 y^1 + \alpha\Phi(u^1)(y^1) + K^1 \chi_{y^1 > x^1}\}$$

We set :

$$(3.3)\qquad z^1(x^1) = c_o^1 x^1 + \alpha\Phi(u^1)(x^1)$$

and z^1 is K^1 convexe, continuous and tends towards $+\infty$ as $|x^1| \to +\infty$.
Therefore, there exists numbers s,S $s \leq S$ such that :

$$(3.4)\qquad \zeta(x^1) = \inf_{y \geq x^1} \{z^1(y) + K^1 \chi_{y > x^1}\} = \begin{cases} z^1(s) & \text{when } x^1 \leq s \, , \\ z^1(x^1) & \text{when } x^1 \geq s, \end{cases}$$

S in the smallest minimum of z^1 and $z^1(s) = K^1 + z^1(S)$.
We define :

$$(3.5)\qquad \bar{\Lambda}(x^2) = (z^1(x^2) - z^1(S))\chi_{x^2 \leq S}$$

$$(3.6)\qquad \underline{\Lambda}(x^2) = (z^1(x^2) - z^1(s))\chi_{x^2 \leq s}$$

We next consider two problems corresponding to echelon 2, namely:

$$(3.7)\qquad \underline{u}^2(x^2) = f^2(x^2) + \bar{\Lambda}(x^2) + \inf_{v \geq 0} \{c^2(v) + \alpha\Phi(u^2)(x^2+v)\}$$

$$(3.8)\qquad \bar{u}^2(x^2) = f^2(x^2) + \underline{\Lambda}(x^2) + \inf_{v \geq 0} \{c^2(v) + \alpha\Phi(\bar{u}^2)(x^2+v)\}$$

We note that $\underline{\Lambda} \leq \bar{\Lambda}$. We assume that :

$$(3.9)\qquad f^2(x^2) + \underline{\Lambda}(x^2) \geq 0.$$

We then denote by $\underline{u}^2$ the minimum solution of (3.7), which is l.s.c. and by $\bar{u}^2$ the maximum solution of (3.9).

3.2. Estimate for the maximum solution

We can no more give an exact decomposition theorem as in section 2 .
We state an approximate decomposition result :

Theorem 3.1. We assume (1.10), (1.11) and (3.1), (3.2). We have

$$(3.10)\quad u^1(x^1)+\bar{u}^2(x^2) \le \bar{u}(x)$$

Proof : Set

$$\tilde{\zeta}(x^1,x^2) = \inf_{x^2\ge y\ge x^1} (z^1(y) + K^1\chi_{y>x^1})$$

then, we have :

$$(3.11)\quad \zeta(x^1) + \underline{\Lambda}(x^2) \le \tilde{\zeta}(x^1,x^2) \le \zeta(x^1) + \bar{\Lambda}(x^2).$$

This is done by direct checking as in theorem 2.1. Now computing the right hand side of (1.13) with u replaced by $u^1 + \bar{u}^2$ using the left side inequality (3.11), one checks that $u^1 + \bar{u}^2$ satisfies relation (1.13) with the sign = replaced by ≤ . Since $u^1+\bar{u}^2 \in B_2$, is positive, it is inferior or equal to the maximum solution $\bar{u}$ of (1.13).

□

3.3. Estimate for the minimum solution

We first state a technical Lemma, whose proof is omited for lack of space.

Lemma 3.1. : There exists a Borel measurable mapping $\sum(x^2)$, such that :

$$(3.12)\quad z^1(\textstyle\sum(x^2)) = \inf_{y\le x^2} z^1(y)$$

Let $\sigma(x^2)$ to be the smallest value such that :

$$(3.13)\quad z^1(\sigma(x^2)) = K^1 + z^1(\textstyle\sum(x^2))$$

then $x^2 \to \sigma(x^2)$ is Borel measurable.

Theorem 3.2. We make the assumptions of Theorem 3.1..Then we have:

$$(3.14)\quad \underline{u}(x) \le u^1(x^1) + \underline{u}^2(x^2) ,$$

and defining

$$(3.15)\quad v^1(x) = \begin{cases} \sum(x^2) - x^1 & \text{for } x^1 \le \sigma(x^2) \\ 0 & \text{for } x^1 > \sigma(x^2) \end{cases}$$

$v^2(x^2)$ = measurable map realizing the infinimum in (3.7) then $\hat{v} = (\hat{v}^1,\hat{v}^2)$ is admissible and :

$$(3.16)\quad f(x) + c(\hat{v}(x)) + \alpha\Phi(u^1+\underline{u}^2)(x+\hat{v}(x)) \le u^1+\underline{u}^2$$

Proof : Consider the increasing process u_n. Assume at some step that $u_n \le u^1 + \underline{u}^2$, then :

$$(3.17)\qquad u_{n+1} \le f + \inf_{v\in\mathcal{V}_x} \{c(v) + \alpha\Phi(u^1+\underline{u}^2)(x+v)\} + f^1(x^1) + f^2(x^2) - c_o^1x^1 + \tilde{\zeta}(x^1,x^2) + \inf_{v\ge 0}\{c^2(v) + \alpha\Phi(\underline{u}^2)(x+v)\}$$

and by the right hand inequality (3.11),

$$u_{n+1} \le f^1(x^1) + f^2(x^2) + \bar{\Lambda}(x^2) - c_o^1x^1 + \inf_{y\ge x^1}\{c_o^1y + \alpha\Phi(u^1)(y) + K^1\chi_{y>x^1}\} + \inf_{v\ge 0}\{c^2(v) + \alpha\Phi(\underline{u}^2)(x+v)\} = u^1 + \underline{u}^2$$

hence, (3.14). Moreover, the right hand side of (3.14) is equal to the left hand side of (3.16), hence, the desired result.

□

REFERENCES

[1] A.J. CLARK ; H. SCARF : Optimal Policies for a Multi-Echelon Inventory Problem, Management Science, 1960, 6(4), 475-490.

[2] A.J. CLARK ; H. SCARF : Approximate Solutions to a simple Multi-Echelon Inventory Problem, in Studies in Applied Probability and Management Science, K.J. Arrow, S. Karlin, H. Scarf eds., Stanford University Press, 1962, Chapter 5 , pp. 88-100.

MEASUREMENTS OPTIMIZATION OF MACHINES DIAGNOSIS MAINTENANCE

P. Drakatos, G. Massouros, A. Massinas

Professor, Lecture, Mech.Electrical Engineer
School of Engineering, University of Patras, Greece.

INTRODUCTION

The frequency spectrum of a time-variable signal, can be obtained by using the Fourier transform. Such a time-variable signal can be the vibration or sound signal of some part of an engine. The corresponding spectrum can be used as a tool to analyse the mechanical state of the examined part.[1]

Time wear of each part, will affect one or more characteristics of the time signal i.e. the displacement, the velocity or the acceleration of the vibration. As a consequence changes can be observed in the corresponding spectrums at the characteristic frequencies relative to each part of the engine.

Therefore by a simple comparison of spectrums received at different times, valuable information can be obtained about the wear of the rest of the remaining life of each part.

Many researchers have been engaged on this topic, and they have found that the spectrums are changing, as the parts are getting worn.

To obtain reliable results, a basic requirement is to place the transducer at the correct point and with the right direction, because the measured signal must be representitive only of the part under examination.[2]

This is a serious task, because practically the following problems are introduced:[3]

D. G. Lainiotis and N. S. Tzannes (eds.), Applications of Information and Control Systems, 316-325.

i) The engine of a vehicle is composed of many parts, hence a big number of measurements is required.
ii) Something it is difficult or even impossible (as in the case of the crank-shaft) to obtain the right measurement point.
iii) All the parts of the engine are fitted on the main body of the engine, which is the main carrier of the time signals. Therefore each recorder signal is composed of signals generated by the different parts of the engine. The contribution of each one of these signals to the final measured signal is different.
iv) By comparing the spectrums being recorded at different places of the engine, can be seen that all of them are generaly of a similar from. In other words the characteristic frequencies of the different parts of the engine are present in all spectrums. However the ammplitude of vibration is depedent on the measurement point. For example in fig. 1 the combustion event which corresponds to 30Hz, has an amplitude 93 db and 76 db at the centre of the cover of the pistion heads and at the Generator respectively.

Generally the amplitude of the vibration, is increasing, as the measurement point is closed to the source. For instance at the point (fig. 2) of the spectrum (17. 5 Hz) corresponding to the water pump, the amplitude is 66.5 db, 68 db, 74.5 db and 77 db, for measurement being taken at the 4th, 3th, 2nd and 1st cylinder respectively.

As it is seen, the real amplitudes of the vibration signal of each source, are dependent on the measurement point.[4]

If the above problems wouldn't exist, then very few measurements at each part would be sufficient, to record and identify the signals of each part.

These problems are introduced by the transfer function which exists between the source of the vibrations and the measurements point.

The present work is concerned with the experimental determination of the transfer function between the main parts of an engine and a specific measurement point which is regarded as the reference point.

ANALYSIS

The time signal f (t) generated by a source can be measured at two points x,y, of the engine. Consider that the recorded signals are $f_x(t)$ and $f_y(t)$ respectively. By calculating the corresponding Fourier tranforms the spectrums F_x (f) and F_y(f) are obtained then the ratio:

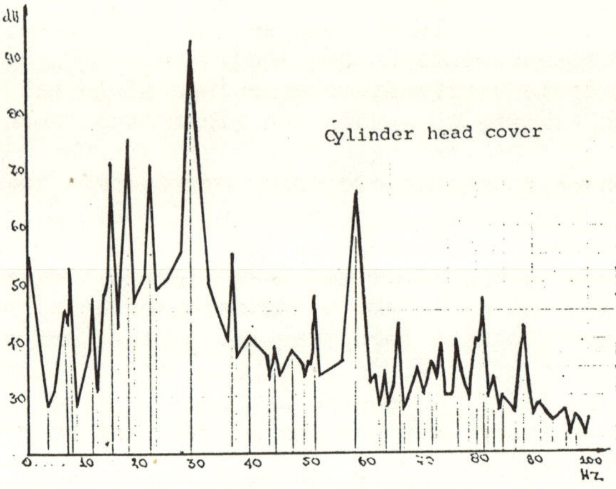

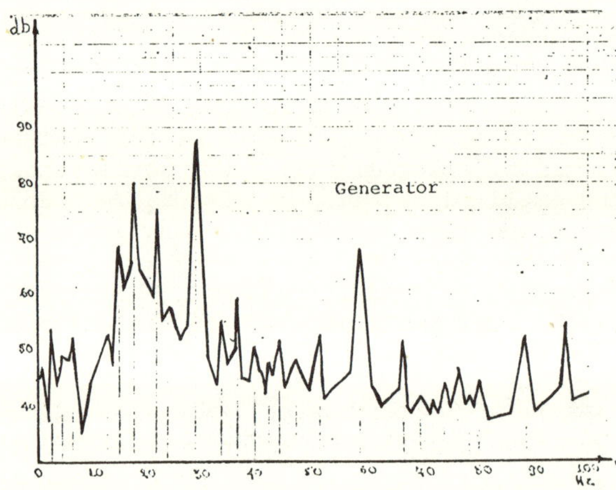

Fig. 1.

fig.1 Sprctrum of engine shown the combustion enent which corresponds to the 30 HZ

$$H_{xy}(f) = \frac{F_y(f)}{F_x(f)} \tag{1}$$

is defined as the transfer function from the point x to the point y or as the complex impedance seen by the source signal between these two points.

Equation (1), can be rewritten into the form:

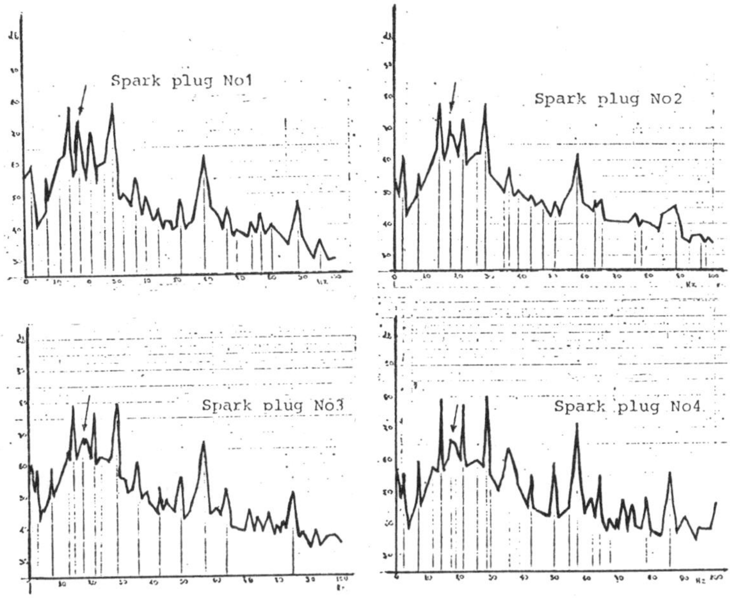

fig 2. Spectrum of the engine which shown the water pums at the point of the 17,5 HZ for measurements being taken at the 4th,3th,2nd,and 1st cylider respectinely.

$$F_y(f) = F_x(f).H_{xy}(f) \tag{2}$$

if the transfer function H_{xy} (f) of a physical system is equal to unit at the whole frequency domain[5], them

$$F_y(f) \approx F_x(f) \tag{3}$$

i.e. the measured signal is indepedent of the measurement point. If at the frequency region of the sources, the transfer function $H_{xy}(f)$ is equal to unity, them the source signal is indepent of the measurement point.

In a physical system many source are present namely $x_1, x_2 \ldots\ldots x_n$, each one having frequency spectrums F_{x_1} (f) F_{x_2} (f)...F_{x_n}(f)

respectively. Hence at the measurement point y, information is received from all the sources (fig. 3) i.e.

$$F_{y_i}(f) = F_{x_i}(f).\ H_{x_i y}(f) \tag{4}$$

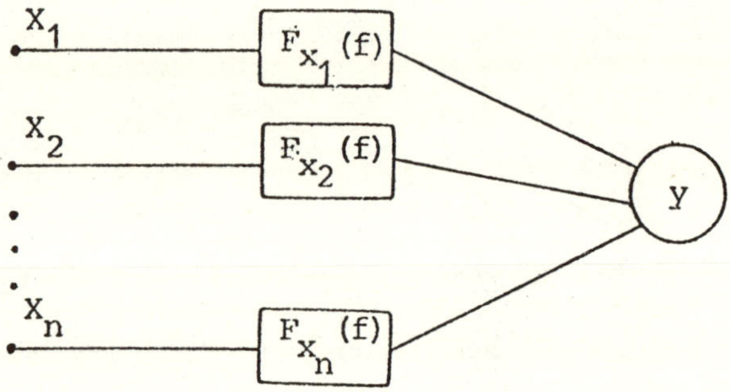

fig 3. Measured signals and signals of the sources in a physical system.

where

$F_{y_i}(f)$ the frequency spectrum of the signal generated from the sourse x_i being measured at y.

$F_{x_i}(f)$ the frequency spectrum of the source x_i.

To obtain the signals $F_{x_i}(f)$ of all the sources from measurements being taken at y, the transfer functions $H_{x_iy}(f)$ from the point x_i to y must be known beforehand.
Equation (4), can be written in the form:

$$\log F_{y_i}(f) = \log F_{x_i}(f) + \log H_{x_iy}(f) \qquad (5)$$

or

$$\log H_{x_iy}(f) = \log F_{y_i}(f) - \log F_{x_i}(f) \qquad (6)$$

Therefore, by measuring the functions $F_{y_i}(f)$ and $F_{x_i}(f)$, the transfer function can be calculated.

3. Experimental Procedure

At the present work, the transfer functions of different parts of the engine with respect to a common (reference) point for a specific type of vehicles are defined and examined.

The employed engines were of the type Willys-Overland Model MD. This engine, is a four-cylinder, four-stroke motor.

Analysis of the vibrations has been accomplished by using the Real-time analyser 2031 of Brüel and Kjaer.

By using this type of analysis, the measurement, analysis and transformation of the signals of different parts of the engine (piston, spark plugs, carburretor, etc) have been done. Care has been taken for the right selection of the transducer application

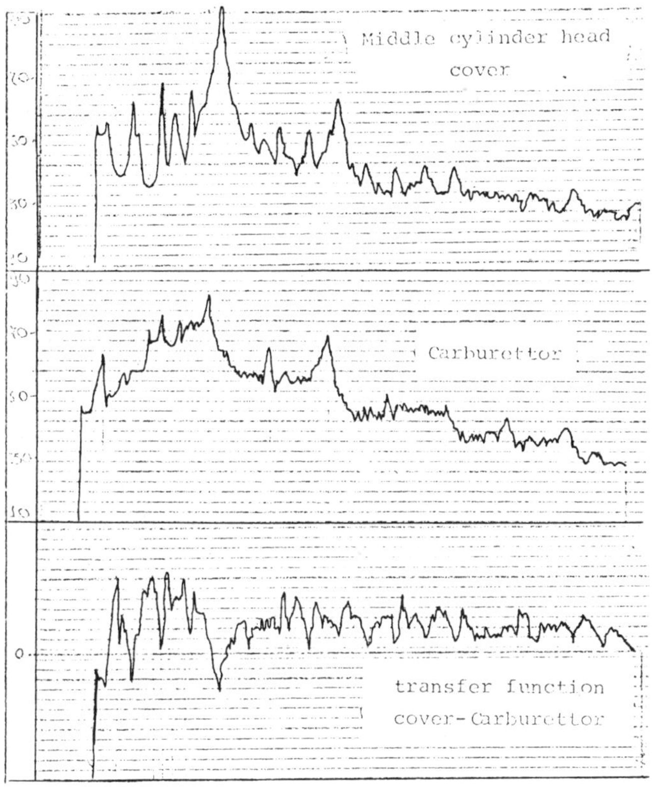

fig.4a The transfer function between cover and carburettor.

point.

Measurements have been taken at the main points of the engine. Each of the spectrums was received by the average of 32 spectrums. Some spectrums can be seen in fig. 4.

Finally, the transfer functions have been found for each one of the signal sources with respect to the reference point. A point at the middle of the top cover of the piston heads has been selected as the reference point. The measured transfer functions are that of the spark plugs, of the carburettor, of the middle of the main engine body, of the generator etc.

Examining those transfer functions, the following conclusions can be reached:[6]

i) The transfer functions of the spark plugs, of the carburettor, of the middle of the main engine body (where the crank-shaft is located) etc have the characteristic that their spectrums can be divided into two parts. For frequencies above 30Hz the transfer

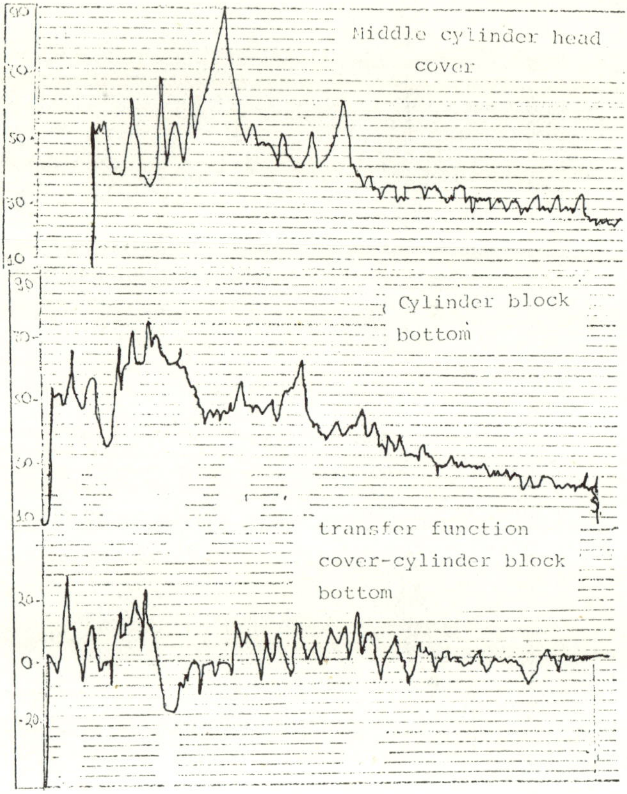

fig 4b. Transfer function between cover and cylinder block bottom.

function is almost constant, i.e. it is indepedent of the frequency. That means that the source signal is not distorted as it moves from the source to the reference point.

On the contrary for frequencies less that 30Hz. the amplitude of the transfer function is depedent on the frequency. The form of the amplitude variation is the same, for all the transfer functions.

As the transfer functions exhibit both positive and negative variations, then it can be considered that the reference point is somewhere close to the center of the sources of the engine.
ii) The generator transfer function, for frequencies above 30Hz, is rising as the frequency is increased, that happening at least up to 100Hz. There fore the mean value of the transfer function, at this part is not equal to unity as for other transfer functions.
iii) For constant speed, the transfer function is indepedent of the mechanical state of the engine while the measured signal is

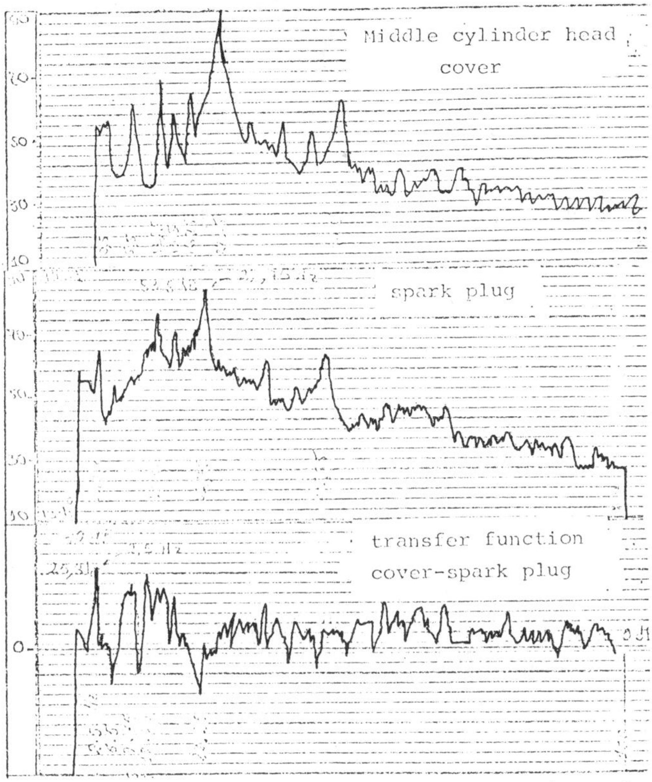

fig 4c. Transfer function between cover and spark plug.

a function of the engine state. It other words, knowing the transfer functions at the interested frequency region and the measurements received at the reference point, it is possible to findout the corresponding values of the different sources.

From the transfer functions corresponding to the different parts, to reference point, we can specify a coefficient of tranformation of the amplitute of vibration from the reference point to the various source of the engine.

This coefficient is a function of frequency and is shown in fig. 5.

CONCLUSIONS

From the experiments carried out, the following conclusions have been reached:

i) Information for frequencies above 30Hz, is indepedent of the

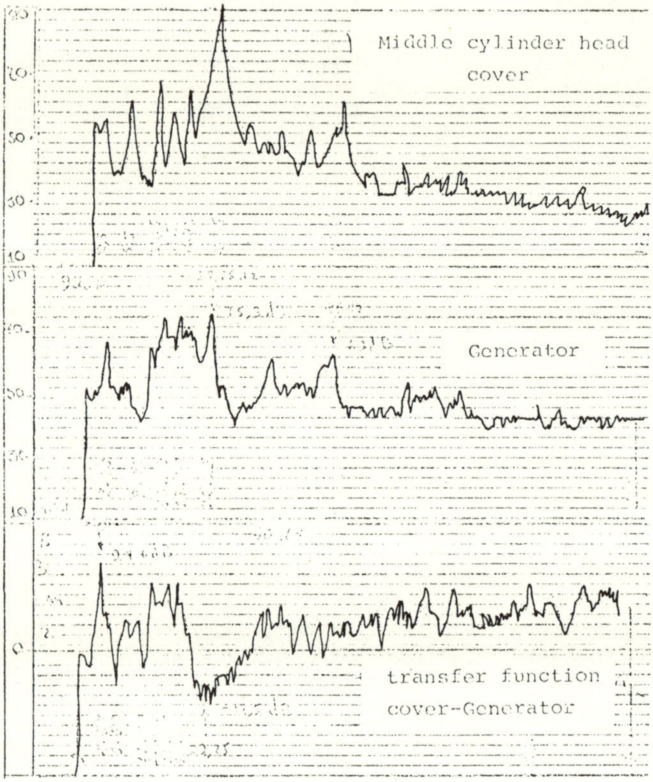

fig 4d. Trancfer function coner and Generator.

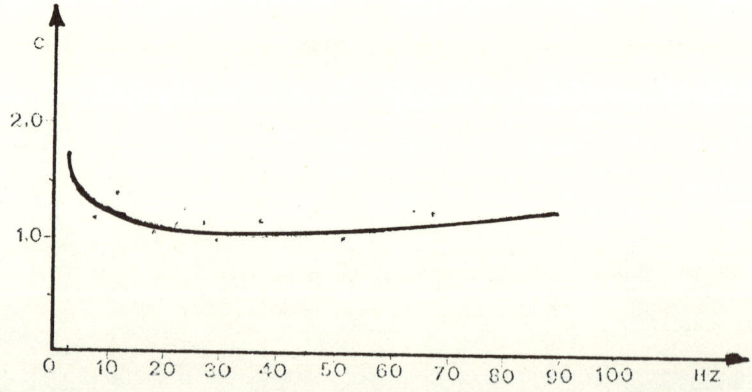

fig 5. Coefficient of transformation .

reference point. On the other hand for frequencies less than 30Hz all the measurements have to be referred back, to obtain the real values of the sources.

ii) The middle of the top cover of the piston heads of an engine, is a suitable point in terms of selecting information for all the sources inside an engine.

iii) Information about the dynamo cannot be received from the refference point. In this case, measurements have to be taken close to the dynamo.

REFERENCE

1. D. Harrison " Computer - based maintenance system for a medium - side company " UKMHMG Summaries N° 3, 1978

2. J.N. Butters " Holographic and speckle pattern methods of visual vibration monitoring " UKMHMG Summaries N°1, 1978

3. B.J. Stephenson " Predictive maintenance using vibration and sound monitoring techniques UKMHMG Summaries N° 1, 1978

4. D.J. Sherwin " Analysis of failure data censored by inspection optimization of inspection and replacement intervals " UKMHMG Summaries N° 1, 1978

5. J.J. Walker " A systems approach to condition monitoring " UKMHMG Summaries N° 1, 1978

6. A. E. Davies " Plant Performance criteria as a prerequisite to a health monitoring programme, UKMHMG Summaries, N° 1, 1978

A GRAPH COLOURATION TECHNIQUE

E. Burattini A. Massarotti A. Santaniello

Laboratorio di Cibernetica - Arco Felice - Italy

ABSTRACT

In this paper we propose, for the graph coloring problem, a heuristic algorithm according to which the choice priority relative to the node coloring is prescribed by its valence, that is by its compatibility degree with the coloration classes so far realized. We have compared our algorithm with the classical sequential coloring algorithm and with the Wood algorithm. Such comparison, carried out on a large sample of random graphs with different number of nodes and different densities, shows that our algorithm has a much better performance.

INTRODUCTION

The graph coloring problem (GCP) takes particular attention both for its conceptual simplicity and its large applicability in the field of Computer Science and Operational Research. Some examples are the problems of timetabling, scheduling, partitioning etc. (7,8). The GCP consists of finding an assignment of colours for the nodes of a graph G such that no pair of adjacent nodes has the same colour and the number of colours is minimum. This number is called "chromatic number" of G($\chi(G)$). Unfortunately, no efficient algorithm is known for this problem, that is no procedure was found which enabled us to achieve the optimal colouration of a graph in a time bounded by a polynomial in the number of the nodes. Viceversa it was shown that the graph coloring problem belongs to the class of the NP-complete problems (1,2), whose members are considered intractable from a computational point of view. Therefore, because of the impossibility of using exact algorithms for the NP-complete problems and since, luckily, many applications of these problems do not require exact solutions, a great development has

D. G. Lainiotis and N. S. Tzannes (eds.), Applications of Information and Control Systems, 326-333.

received the attempt of defining new approximation methods bringing to quasi-optimal solutions. The most well-known among them is the method of "guaranteed" approximation which consists of analyzing the behaviour of an algorithm in the worst case (5). According to that approach the purpose is to design, for a given problem, an algorithm such that changing the input of the problem, the eventual error, that the algorithm could make, is bounded by an "acceptable" constant indipendent from the problem size. This method has given excellent results for some well-known NP-complete problems, as for instance some scheduling problems, but for many others, among which the graph coloring, it gives quite unsatisfactory results (4). On the other hand when the worst case results show that an algorithm performs well even in the worst case, their relevance is obviuos. Viceversa, when they show that the algorithms can behave quite poorly, their practical significance is very doubtful, since we do not have any information about the incidence of worst cases on the distribution of all inputs set. This is indirectely corroborate by the study of the probabilistic behaviour of the approximation algorithms (3).
Let p_n be the probability that a given algorithm provides a solution quite unsatisfactory for an input of size n chosen at random. We say the algorithm to work properly, "almost everywhere" if $\Sigma\, p_n$ is finite, where the sum is carried out over any problems size (n=1,2,...). In the graph colouring case Karp (3) shoved that the classical algorithm of sequential coloring, introduced by Welsh and Powell (7), is "almost everywhere" quasi-optimal for graphs with constant density. In other words this procedure not only provides good solutions infinitely often, but also fails only finitely often. Unfortunately the conditions put by the probabilistic approach are of asymptotic kind and they say nothing about the behaviour of the algorithm for given problem instances. Therefore, apart from their theoretical relevance, the approaches to the analysis of approximation algorithms, briefly described before, cannot be considered as conclusive landmark to the practical solution of the graph colouring problem. Presently, the only evidence, mainly empirical, is based on the comparison among the different heuristic rather than on the measure of the nearness of the approximate solutions to the optimal ones. In this paper we propose, for the graph colouring problem a heuristic algorithm according to which the choice priority for the colouring node is fixed by its "valency", that is by its compatibility degree with the set of the partially filled colouration classes.
Comparing this procedure with the classical sequential colouration algorithm, and with the Wood algorithm, for a large sample of random graphs we obtain very encouraging results.

COLOURATION ALGORITHMS

Let G(V,E) be a graph with node set V and edge set E. We assume G has no loops or multiple edges. Let the degree of node v_i, in the graph G, $d(v_i)$, the number of adjacent nodes of v_i in G:

$$d(v_i)=|\{v_i \in V: \langle v_i,v_j\rangle \in E\}|$$

The most well-known method of sequential colouration of a graph, formally defined by Welsh and Powell, consists at first of arranging the node set of G in order of decreasing degree and of assigning the first node v_i to the first colouration class C_1. Then examinating the nodes in that order, one has to fill the class C_1 adding one node at a time and assuring that no pair od adjacent nodes happens to be assigned to that class. Since in that way it is obtained a maximal independent set one has to repeat the same procedure for the new colouration class C_2 and so forth until all the nodes are coloured.
For istance, let us consider the graph of fig.1:

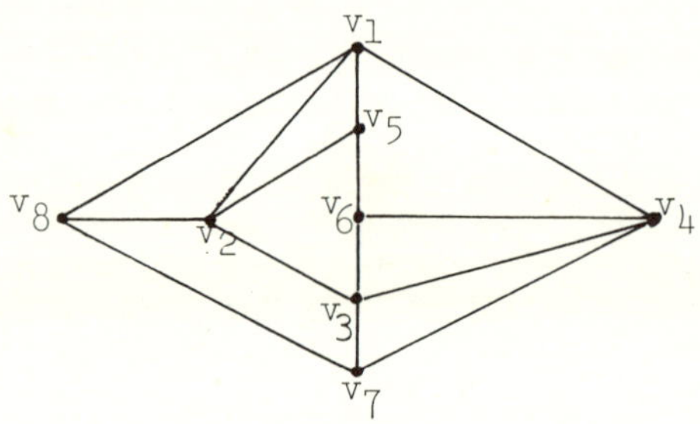

fig.1

A nodes arrangement required by the Welsh and Powell algorithm is: v_1, v_2, v_3, v_4, v_5, v_6, v_7, v_8. The resulting colouration is given by:

$$C_1:(v_1,v_3);\ C_2:(v_2,v_4);\ C_3:(v_5,v_7);\ C_4:(v_6,v_8).$$

Wood (8) introduced a different technique, often used in taxonomie, which consists of arranging all the pairs of nodes in decreasing similarity degree order, and of colouring every pairs according to that order. The similarity of a pair of nodes is defined by:

$$s(v_i,v_j) = \begin{cases} 0 & \text{if } \langle v_i,v_j\rangle \in E \\ |\{v_k \in V: \langle v_k,v_i\rangle \,\&\, \langle v_k,v_j\rangle \in E\}| & \text{if } \langle v_i,v_j\rangle \notin E \end{cases}$$

In other words the similarity defines the number of nodes v_k adjacent both to v_i and v_j, which are not adjacent among themselves. The procedure introduced by Wood acts so that every pair of nodes (v_i,v_j), chosen in order of decreasing similarity degree, be analyzed as follows:

1. Arrange the pairs of nodes in order of decreasing similarity degree. The pair of nodes of maximum similarity is put in class $\overline{C_1}$.
2. From the previously unconsidered pairs select one of maximum similarity (v_i,v_j).

3. If v_i and v_j are already coloured go to 2.
4. If the maximum similarity equal to 0, stop.
5. If v_i , belongs to C_h and v_j is uncoloured then
   if $<v_j,v_k> \notin E \ \forall \ v_k \in C_h$ then add v_j to C_h
   else go to 2.
6. If v_i and v_j are uncoloured then look for the first class C_h
   such that: $\forall \ v_k \in C_h$, $<v_i,v_k> \notin E$ & $<v_j,v_k> \notin E$
   if it exisits add the pair (v_i,v_j) to C_h.
7. Else (v_i,v_j) will be the first element of a new colouration class; go to 2.

All the nodes not coloured by this procedure will be coloured using the Welsh and Powell algorithm.
Applying the Wood's algorithm to the example of fig.1 we have the following arrangement for the pairs of nodes, all with the similarity degree equal to 2: (v_1,v_3), (v_1,v_6), (v_1,v_7), (v_2,v_4), (v_2,v_6), (v_2,v_7), (v_3,v_5), (v_3,v_8), (v_4,v_5), (v_4,v_8), (v_5,v_8), (v_6,v_7)
The resulting colouration is:

$C_1:(v_1,v_3)$; $C_2:(v_2,v_4)$; $C_3:(v_5,v_8)$; $C_4:(v_6,v_7)$

The colouration technique of graphs which we propose here is divided in two stages:

A - Look for the approximately maximum clique starting from node v_{i_1} of maximum degree. An approximately maximum clique can be obtained considering successively the subgraphs of G produced in the following manner.
    Let $\Gamma(v_{i_1})$ be the subgraph of G generated by the set of nodes of G adjacent to v_{i_1}. Let v_{i_2} be the node of $\Gamma(v_{i_j})$ of maximum degree and $\Gamma(v_{i_1},v_{i_2})$ the subgraph of $\Gamma(v_{i_1})$ generated by nodes of $\Gamma(v_{i_1})$ adjacent to v_{i_2} and so on.
    This process terminates at step p-th when the degree of v_{i_p} is equal to 0.
    The graph generated by the set $V_{MC}= \{v_{i_1},v_{i_2},\ldots\ldots,v_{i_p}\}$ is a maximal clique and therefore $\chi(G)\geq p$.
    Then any node v_{i_j} of the set V_{MC} is assigned to the colouration class C_j.

B - This stage is based on the notion of "valency" of a non-coloured node. The valency $\delta(v_k)$ of a node v_k indicates the compatibility degree of v_k with the set of colouration classes already realized, that is the number of classes to which v_k can be assigned. Let us remark that;
    a) $\delta(v_k)=0$ implies that v_k must be the first element of a new colouration class and for every node not adjacent to v_k the valency increases of 1.
    b) $\delta(v_k)>0$ and $v_k \in C_j$ implies that for every node adjacent to v_k and compatible with C_j the valency decreases of 1.

Our algorithm works as follows:

1. Look for the approximately maximum clique. Assign any node $\{v_{i_1},\ldots\ldots,v_{i_p}\}$ of the approximately maximum clique respectively to the colouration classes $C_1,\ldots,C_p$.

2. For the remaining uncoloured nodes look for the node v_k with minimum valency. If there are no more uncoloured node, stop.
3. If $\delta(v_k)=0$ put v_k in a new colouration class and increase of 1 the valency of nodes not adjacent to v_k; go to 2.
4. If $\delta(v_k)=1$ add v_k to the only class it is compatible with, let it be C_j, and decrease of 1 the valency of nodes adjacent to v_k and compatible with C_j ;go to 2.
5. If $\delta(v_k)>1$ add v_k to the class C_j containing the minimum number of nodes whose valency decreases; go to 2.

For the graph of fig.1 we have: $v_{i_1}=v_1$, $v_{i_2}=v_2$, $v_{i_3}=v_5$; $\chi(G)\geq 3$; the set of nodes of the approximately maximum clique is $V_{MC}=\{v_1, v_2, v_5\}$ and then $C_1:(v_1)$; $C_2:(v_2)$; $C_3:(v_5)$. Among the remaining nodes, v_8 is the node with minimum valency ($\delta(v_8)=1$) and is assigned to C_3 . The successive node with minimum valency is v_3 ($\delta(v_3)=2$) and is assigned again to C_3 since in this case only for one of its nodes the valency decreases. Eventually the colouration produced by our algorithm is the following: C_1: (v_1, v_6, v_7); C_2:(v_2, v_4); C_3:(v_5, v_8, v_3) which is the optimal one.

APPLICATION TO RANDOM GRAPHS AND CONCLUSIONS.

The comparison among the different colouration methods, previously discussed, based respectively on sequential coloring, on similarity and on valency, was carried out analyzing their behaviour on a large collection of random graphs generated by pseudo-random generators of random numbers. These generators starting from an assigned number of nodes, add edges to the graph until the prefixed density of graph is obtained. The analysis was carried out on groups of 100 random graphs for each of which we assigned the number of nodes (30, 40, 50, 60, 70,100) and the densities of graphs (25%, 50%, 75%)(in total 1800 experiments).
The results are condensed in fig.2. They show that the method of the valency works much better independently of graph density and more and more as greater is the number of nodes.
On the other hand this result is quantitatively verified by examining the confidence limits for percentage based on the binomial distribution of results and by using the relative statistical table (6). In fact, if we make the hypothesis that the results have a binomial distribution, for instance event A: solution obtained with valency method not worse than that with sequential colouring (similarity) method;event B: solution obtained with valency method worse than that with sequential colouring (similarity); we obtain that the observed percentages of incidence of the property for samples of size 100 have the values reported in Tab.1.
For instance, from these data, obtained using the table which gives the confidence limits for percentages (6) for samples of size 100 we derive, for 100 nodes with density of 75% the following properties:

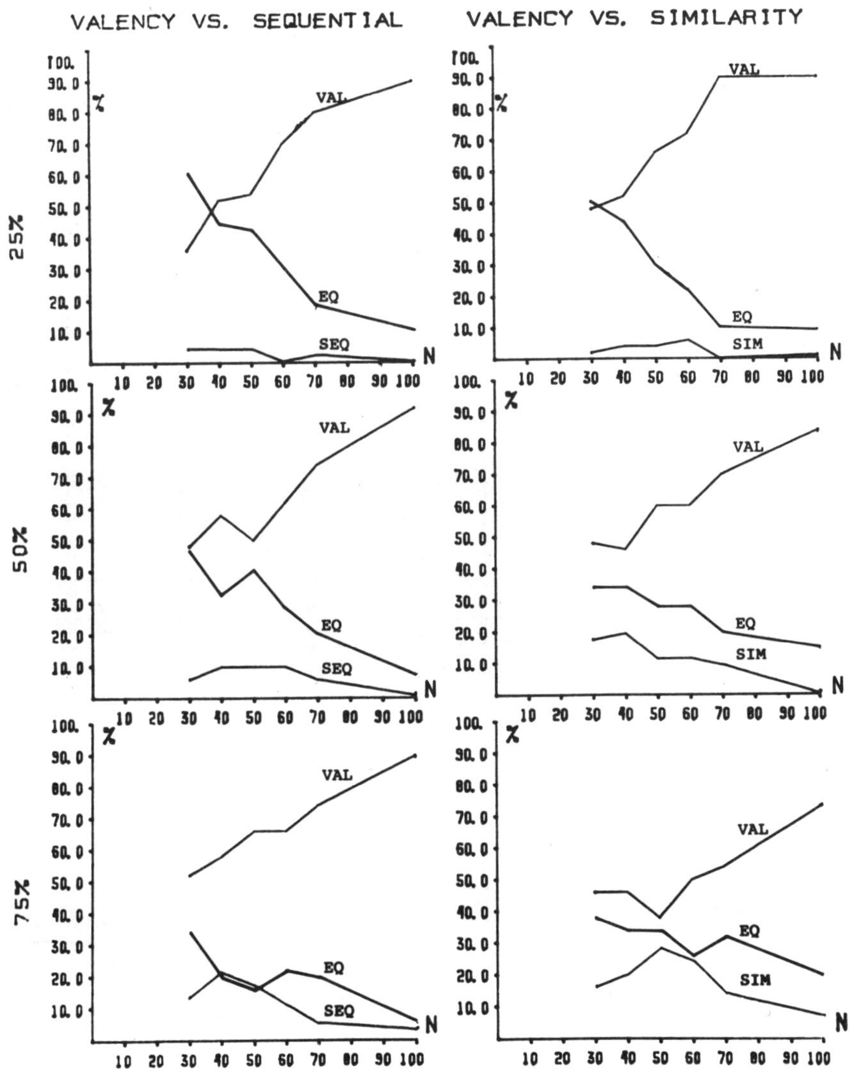
VALENCY VS. SEQUENTIAL
VALENCY VS. SIMILARITY
25%
50%
75%
%
N
VAL
EQ
SEQ
SIM
100.
90.0
80.0
70.0
60.0
50.0
40.0
30.0
20.0
10.0
10 20 30 40 50 60 70 80 90 100

fig. 2

TAB.1

NODE	VAL ≥	VAL <	RANGE %	SEQ ≥	SEQ <	RANGE %	VAL ≥	VAL <	RANGE %	SIM ≥	SIM <	RANGE %	
30	96	4	87.92-99.32	64	36	50.70-75.92	98	2	91.06-99.90	52	48	35.01-64.99	
40	96	4	87.92-99.32	48	52	35.01-61.19	96	4	87.92-99.32	48	52	35.01-61.19	
50	96	4	87.92-99.32	46	54	33.18-59.26	96	4	87.92-99.32	34	66	22.33-47.24	25%
60	100	0	94.84-100.0	30	70	18.90-43.06	94	6	85.07-98.44	28	72	17.25-40.91	
70	98	2	91.06-99.90	20	80	10.84-32.12	100	0	94.84-100.0	10	90	3.82-20.20	
100	100	0	94.84-100.0	10	90	3.82-20.20	99	1	92.79-100.0	10	90	3.82-20.20	
30	94	6	85.07-98.44	52	48	38.81-64.99	82	18	70.20-90.67	52	48	35.01-64.99	
40	90	10	79.80-96.18	42	58	29.45-55.33	80	20	67.88-89.66	54	46	40.74-66.87	
50	90	10	79.80-96.18	50	50	36.83-63.11	88	12	77.35-94.85	40	60	27.63-53.35	50%
60	90	10	79.80-96.18	38	62	25.85-51.32	88	12	77.35-94.85	40	60	27.63-53.35	
70	94	6	85.07-98.44	26	74	15.59-38.76	90	10	79.80-96.18	30	70	18.90-43.06	
100	99	1	92.79-100.0	8	92	3.51-15.16	99	1	92.79-100.0	16	84	7.89-27.49	
30	86	14	74.87-93.54	48	52	35.01-61.19	84	16	72.51-92.11	54	46	40.74-66.87	
40	78	22	65.65-87.59	42	58	29.45-55.33	80	20	67.88-89.16	54	46	40.74-66.87	
50	82	18	70.20-90.67	34	66	22.33-47.24	72	28	59.09-82.75	62	38	48.68-74.15	
60	88	12	77.35-94.85	34	66	22.33-47.24	76	24	63.43-86.02	50	50	36.89-63.11	75%
70	94	6	85.07-98.44	26	74	15.59-38.76	86	14	74.87-93.54	46	54	33.13-59.26	
100	96	4	87.92-99.32	10	90	3.82-20.20	93	7	83.72-97.92	27	73	16.42-39.84	

a) the valency method gives results better than or equal to those we obtain with the sequential colouring method in the range 87.92% 99.32% with probability of 0.99;
b) the valency method gives results better than or equal to that we obtain with similarity method in the range 83.72%-97.92% with probability 0.99. Similar results are obtained for samples of random graphs with densities 25% and 50%.
Moreover if we suppose a different binomial distribution of the results (i.e. event A': solution obtained with sequential coloring method (similarity) not worse than that obtained with the valency method; event B': solution obtained with sequential coloring method (similarity) worse than that obtained with valency method), we observe that the percentages of incidence of event A' for samples of size 100, for 100 nodes with densities of 75%, give the following results:
a') the sequential coloring method gives results better than or equal to those we obtain with the valency method in the range 3.82% 20.20% with probability 0.99;
b') the similarity method gives results better than or equal to those we obtaine with the valency method in the range 16.42%-39.84% with probability of 0.99.
In conclusion for the colouration of random graphs the method based on the valency show a behaviour significantly better than the classical methods and that is encouraging for further analysis of the performances of this algorithm applied to other classes of structured graphs which can rise from practical applications.

REFERENCES

1) Aho, A.V., Hopcroft, J.E., Ullman, J.D., 1974, The design and Analysis of Computer Algorithms, Addison Wesley, Reading, Mass.
2) Karp, R.M., 1972, Reducibility among combinatorial problems, In Complexity of Computer Computations, R.E.Miller & J.W.Thacther Eds., Plenum Press, New York, pp.85-104.
3) Karp, R.M., 1976, The probabilistic analysis of some combinatorial search algorithms, In Algorithms and Complexity, Academic Press, New York.
4) Johnson, D.S., 1974, Approximation algorithms for combinatorial problems, J. Comput. & Syst. Sciences, 9, pp.256-278.
5) Weide, B., 1977, A survey of analysis techniques for discrete algorithms, Comp. Surveys, vol.9, n.4, pp.291-313.
6) Rohlf, F.J., Sokal, R.R., 1969, Statistical Tables, W.H. Freman & Co., San Francisco.
7) Welsh, D.J.A., Powell, M.B., 1967, An upper bound for the chromatic number of a graph and its application to timetabling problems, The Computer Journal, 10, pp.85-86.
8) Wood, D.C., A technique for colouring a graph applicable to large scale timetabling, The Comp. Journ., 12, pp.317-319.

A NEW METHOD FOR THE ESTIMATION OF CONSTRUCTION COSTS USING A SPECIAL MATHEMATICAL MODEL OF MAN-MACHINE SYSTEM

A. Massinas, G. Massouros, P. Drakatos

Mechanical and Electrical Enginee,Lecturer,Professor.
school of Engineering, University of Patras, Greece.

ABSTRACT

In the present work we introduce a new method of determining the maximum efficiency of an earth moving machine (e.g. Buldozer) which can be used as basic information for the manager in his decision making.

INTRODUCTION

The introduction of constructions machines in various sectors in general had as a result to increase of their development rate.

As significant it is to decide on the mechanization of a given sector of production, it is also importance to select the method, of gathering information concering the appropriate machine (e.g. efficiency, production rate, maintenance conditions, e.t.c.) which can be used as a basis for the manager in his decision making for a specific work.

The efficiency and the production rate of earth-moving equipment can be determined either by statistical data on its operating results or analytically through the use of mathematical models.

Though the models for the analytical study have been compiled using data of the majority of machines, in no case can they approximate a specific machine and, therefore, data gathering during its operation can alone give its efficiency with absolute accuracy.

D. G. Lainiotis and N. S. Tzannes (eds.), Applications of Information and Control Systems, 334-343.

The rapid evolution of technology in the last thirty years has undoubtedly increased many times the ability of man to conceive and process information from his environment and to control it at the same time in the way he wants.

Among the factors that a manager of a business organization must take into consideration to make sound decisions are the factors related to production (e.g. machinery efficiency, efficiency indices, efficiency of human resources, productivity e.t.c.)

The manager can take the above information through an Management Information System within which various models will exist in each case.

One such model from which the manager will be able to draw information for the production rate of an earth moving machine as a function of its operating conditions is the one destribed below.

In the last three decades many experiments and studies have been made (5),(6) concerning the Man-Machine System. In those studies and experiments the interaction of Man-Machine was examined in each case in simple or complex system aiming either to man the systems or to reveal the behaviour of the human operator according to enviromental conditions of the system.

PARSON (3) (P.4-8) gives a list of experiments and studies which have been carried out over a period of time concering the man - machine system. RAO, B.K.N.JONESB, (4) have researched the psychophysical behaviour of individuals being in an environment of sinusoidal vibrations (e.g. vehicles, aircraft, e.t.c.) or random vibrations according to ISO criteria.

D.W. REPPERGER, K.A. SMILES (5) prove that there are important variations in the input-output characteristics of a close loop system when the operator is under different thermal conditions each time.

Finally, a large number of researches have been carried out only for human operators to make their work safer, more efficient and comfortable by improving the equipment, the working methods as well as the physical environment in which this work is carried out (2),(9). None of the existing studies concernes the efficincy of an integrated Man-Machine-Soil-Environment System that is to determine the efficiency of a machine in production units as a function of its conditions (e.g. age,power, speed, e.t.c.), the condition of the operator (e.g. skill, fatique, psychological conditions, e.t.c.) and working conditions (soil, environment).

Although several methods exist for determining the efficiency of a machine and the production rate, the authors, of this paper, however, believe that due to the inflationary trends in the world today, the fastest and most appropriate method is this which described in the present model.

ANALYSIS

The present work aims at determining a Mathematical Model of the integrated Man-Machine-Soil-Environment system which includes a large number of parameters and from which the optimum efficiency and the production rate of a machine (e.g. Buldozer) can be calculated. The whole study is based on data collected during the operation of five years and on data gathered from various experiments made adhoc.

For a full determination of the soil's influence on the efficiency and the production rate of the Man-Machine system many experiments were carried out on five different types of soil with different types of machines.

Thus the final formulation of the problem of this study is: for a given machine (type,life time, power, e.t.c.), operator (skill fatique, psychological conditions, e.t.c.) soil (Cohesion,shearing stress, Elasticity, e.t.c.) and environmental conditions (temperature, relative humisity, noise, vibrations,e.t.c.)what will be the maximum efficiency of the machine and the optimum production rate.

A method for solving this problem is using the Dimensional and Numerical Analysis through a specific computer program.

Out of the large number of independent and dependent variables which exist in the integrated Man-Machine-Soil-Environment system we select at the first stage those appearing in table I and using the Vaschy-Buckingham theorem [7], or theorem of the P_m terms.

We get the functional relationship that governs the Man-Machine-Soil-Environment System:

$$\Phi_1\left(\frac{r_i V_j}{N_j},\ R_s,\ \frac{N_s V_j^{1/2}}{\upsilon_s^{3/2}},\ F_e N_j \upsilon_s,\ \frac{V_L \upsilon_s^{3/2}}{V_j^{3/2}},\ t_p,\ \frac{f_g}{N_j},\ W_d,\right.$$

$$\frac{T_r \upsilon_s^{3/2}}{V_j^{1/2}}, \frac{C_i V_j}{N_j} \frac{E_i V_j}{N_j}, \vartheta_e, \frac{W_e \upsilon_s^2}{N_j V_j}, t_d, \frac{N_c}{\upsilon_s}, \frac{A}{\upsilon_s}, \frac{b_0}{W_\alpha},$$

$$\left. \frac{J_p \upsilon_s^{3/2}}{V_j^{1/2}} \right) = 0 \qquad (1)$$

Employing the general Curvilinear Regression equation and the method of least squares. We can determine equation (2).

$$V_{i,j,r,l,n} = \sum_{q=0}^{\nu} a_{q,i,m,r,l,n} \frac{N_i j}{r_i} \left[\prod_{m=2}^{m} P_{m,i,r,l,n} \right]^q \qquad (2)$$

Where:

$q = 0 \div v$	Power
$i = 1 \div 5$	Type of soil
$m = 1 \div 18$	Number of nondimensional terms P_m
$j = 1 \div 52$	Time in months
$r = 1 \div 5$	level of vibration (frequency)
$l = 1 \div 5$	level of Noise (frequency)
$n = 1 \div 5$	level of operator wage

The number of P_m terms in equation (2) and the Power of the Sum are determined by a special computer program using Least Squares equations, analysis of variance and recuring approximations by Comparative analysis
The Variables in the P_m terms are generally considered to be mathematical functions like:

Function of temperature

$$\vartheta_e(t) = \alpha_{10} + \sum_{n=1}^{52} \left[\alpha_{1n} \cos(0.523nt) + b_{1n} \sin(0.523nt) \right] \qquad (3)$$

Function of Relative Humidity

$$W_{es}(t) = \alpha_{20} + \sum_{n=1}^{52}\left[\alpha_{2n}\cos(0.523nt)+b_{2n}\sin(0.523nt)\right] \quad (4)$$

Operator skill factor function

$$t_d = f(t) = 0.25+1.22\,\frac{t}{100} + 11.5\,\left(\frac{t}{100}\right)^2 - 20\left(\frac{t}{100}\right)^3 \quad (5)$$

Function of Shearing Stress

$r_i = \alpha_i b_i^{C_i}$ (6) where α_i, b_i coef. e.t.c.

These functions are inserted in the P_m terms and subsequently in equation (2) to get the general form of equation (7).

$$Y_{i,j,r,l,n} = \sum_{q=0}^{\nu} a_{q,i,m,r,l,n}\left[\prod_{m=2}^{m}\prod_{k=1}^{V_1}\left(\sum_{s=1}^{5} b_{k,s,q,i,m,r,l,n}\, Z_{k,s,i,m,r,l,n}^{F_{k,s}} + C_{k,i,m,r,l,n}\right)\right]^q \quad (7)$$

where: $$Y_{i,j,r,l,n} = \frac{r_i V_{i,j,r,l,n}}{N_{i,j}} \quad (8)$$

$a_{q,i,m,r,l,n}$: Factors determined by least squares.

$b_{k,s,q,i,m,l,n}$: Factors arising from the product of the terms (7).

$S = 1 \div 5$: Number of terms of series.

$K = 1 \div V_1$: Number of multiplied sums.

$Z_{k,s,q,i,m,r,l,n} = f\,(U_s, t, C_i, E_i, K, \mathrm{Cos}\,(va), \sin\,(va)$

$F_{k,s}$: A function that gives the exponent of variable

Z depending on the P_m terms employed, at the variables function relationships.
$C_{k,q,i,m,r,l,n}$: Constants depending on the polynomial forms.

We call $$X_{i,m,r,l,n} = \prod_{k=1}^{V_1} \left(\sum_{s=1}^{5} b_{k,s,q,i,m,r,l,n} \cdot Z_{k,s,q,i,m,r,l,n}^{F_{k,s}} + C_{k,q,i,m,r,l,n} \right) \quad (9)$$

Solving equation (7) with respect $V_{i,j,r,l,n}$ we get the function of the efficiency.That is:

$$V_{i,j,r,l,n} = \frac{N_i^j}{r_i} Y_{i,j,r,l,n} = \sum_{q=0}^{\nu} a_{q,i,m,r,l,n} \cdot \frac{N_i^j}{r_i} \left[\prod_{m=2}^{m} X_{m,i,r,l,n} \right]^q \quad (10)$$

Limits within which the true value of the efficiency on the production rate probably lie,i.e. the confidence limits are calculated from the standard errors by means of the appropriate t-multipliers.

APPLICATION

Civen an Earth - moving machine type D7 with the following characteristics:
Machine age: first year
Machine power: 160 PS
Engine revolutions: 2000 RPM
Velocity of excavation: 60m/s
Blade capacity: 2,50 m^3
Machine Reliability: ist year 0.98, 2nd year 0.85 and 3rd year 0,74 working conditions of the machine.
Cohesion of soil: 0.12 kp/cm^2
Shearing stress of soil: 1.19 kp/cm^2
Elasticity Value: 24 kp/cm^2
Environment temperature: 18°C
Relative Humidity: 50%
Barometric pressure: 1.01 kp/cm^2
Level of wage factor of operator: $ 400/month

TABLE 1

Physical Parameters Coverning the Man-Machine System

1. INDEPENDENT VARIABLES	SYMBOL	DIMENSIONS
a. <u>Machine</u>		
type	t_p	$F^0L^0T^0D^0$
Age	J_p	$F^0L^0T\ D^0$
RPM	N_s	$F^0L^0T^{-1}D^0$
Velocity	U_s	$F^0L^0T^{-1}D^0$
Towing Power	F_e	$F\ L^0T^0D\ ^0$
Blade Capacity	V_L	$F^0L^3T^0D^0$
Reliability	R_E	$F^0L^0T^0D^0$
Operating Cost	b_o	$F^0L^0T^{-1}D$
b. <u>Man(operator)</u>		
Level of wage factor	W_a	$F^0L^0T^{-1}D$
Fatique	f_g	$F\ L\ T^{-1}D^0$
Work load (stress)	W_d	$F^0L^0T^0D^0$
Reaction Time	T_r	$F^0L^0T\ D^0$
Reliability	R_H	$F^0L^0T^0D^0$
c. <u>Soil</u>		
Cohesion	C_i	$F\ L^{-2}T^0D^0$
Shearing stress value	r_i	$F\ L^{-2}T^0D^0$
Electricity value	E_i	$F\ L^{-2}T^0D^0$
d. <u>Environment</u>		
Temperature	ϑ_e	$F^0L^0T^0D^0$
Relative Humidity	W_{es}	$F^0L^0T^0D^0$
Barometric Pressure	W_e	$F\ L^2T^0D^0$
2. DEPENDENT VARIABLES		
Machine Power	N_j	$F\ L\ T^{-1}D^0$
Efficiency of the machine	V_j	$F^0L^3T^{-1}D^0$
Operators Skill Factor	t_d	$F^0L^0T^0D^0$
Noise Displacement	N_c	$F^0L\ T^{-1}D^0$
Vibration Displacement of the machine	A_v	$F^0L\ T^{-1}D^0$

TABLE 2

Monthly Production Rate of an Earth-Moving Machine type D7

System Reliab R_s	J Months	V_{ij} (M^3/h)	Confidence Limits 90% Lower	90% Upper	95% Lower	95% Upper	97,5% Lower	97,5% Upper
	1	35.9	35.32	36.38	35.15	36.65	35.0	36.8
	2	31.9	30.32	32.38	31.15	32.65	31.0	32.8
	3	35.2	34.62	35.78	34.45	35.95	34.3	36.1
	4	37.6	37.02	38.18	36.85	38.35	36.7	38.5
	5	37.5	36.92	38.08	36.75	38.25	36.6	38.4
	6	34.7	34.12	35.28	33.95	35.45	33.8	35.6
	7	36.2	35.62	36.78	35.45	36.95	35.3	37.1
	8	35.1	34.52	35.68	34.35	35.85	34.2	36.0
R_s=0.90	9	39.3	38.72	39.88	38.55	40.05	38.4	40.2
	10	36.3	35.72	36.88	35.55	37.05	35.4	40.2
	11	37.3	36.72	37.88	36.55	38.05	36.4	40.2
	12	38.6	38.02	39.18	37.85	39.35	37.7	39.5
	13	40.4	39.82	40.98	39.65	41.15	39.5	41.3
	14	40.9	40.32	41.38	40.15	41.65	40.0	41.8
	15	41.2	40.62	41.78	40.45	41.95	40.3	42.1
	16	41.6	41.02	42.18	40.85	41.35	40.7	42.5
	17	41.8	41.22	42.38	41.05	42.55	40.9	42.7
	18	41.9	41.32	42.38	41.15	42.65	41.0	42.8
	19	42.1	41.52	42.68	41.35	42.85	41.2	43.0
	20	42.2	41.62	42.78	41.45	42.95	41.3	43.1
R_s=0.80	21	42.3	41.72	42.88	41.55	43.05	41.4	43.2
	22	42.1	41.52	42.68	41.35	42.85	41.2	43.0
	23	42.1	41.52	42.68	41.35	42.85	41.2	43.0
	24	42.1	41.52	42.68	41.35	42.85	41.2	43.0
	25	42.1	41.52	42.68	41.35	42.85	41.2	43.0
	26	41.7	41.12	42.28	40.95	42.45	40.8	42.6
	27	41.3	40.72	41.88	40.55	42.05	40.4	42.2
	28	40.8	40.22	41.38	40.55	41.55	39.9	41.7
	29	40.3	39.72	40.88	39.55	41.05	39.4	41.2
	30	39.7	39.12	40.28	38.95	40.45	38.8	40.6
	31	39.1	38.52	39.68	38.35	39.85	38.2	40.0
	32	38.5	37.92	39.08	37.75	39.25	37.6	39.4
R_s=0.70	33	37.8	37.22	38.38	37.05	38.55	36.9	38.7
	34	37.1	36.52	37.68	36.35	37.85	36.2	38.0
	35	36.4	35.82	36.98	36.65	37.15	35.5	37.3
	36	35.5	35.02	36.18	34.85	36.35	34.7	36.5

Fatique of the operator: 4.15 kcal/min
Reaction time of operator: 0.2 sec.
Reliability of the operator:ist year 0.92, 2nd 0.94,3nd 0.95
Noise Frequency: 45 Hz
Vibration Frequency: 6 Hz
Operating cost:

Fuel Price:	\$1 per galon
Lub-Oil price:	\$2 per Kgr
man-hour cost:	\$2 per hour

The monthly production rate or the afficiency of the machine during its first three years of operation is to be determined.

This can be achieved by solving equation (10) via a special computer program which results the 36 presicted values of V_{ij} (table 2) with different confidence limits.

CONCLUSION

From the solution of the model we conclude that the production rate of an earth-moving machine is in principle inversely proportional of the shearing stress of the soil, its configuration and depends on a multitude of variables (table 1), that are involved in a model each time production rate is to be determined. Finally, the results from the implementation of the model prove that it can be used in any integrated Man-Machine System.

REFERENCES

1. Drakatos P., Massinas. "Computerised of construction equipment cost estimating". AACE Transactions 1978.
2. Drakatos P., " The introduction of stohastic Models in the cost Estimating of construction industry" AACE Transactions,1977.
3. Parsons H.Me "Man-Machine System Experiments" John Hopkins Pross London, 1972.
4. Rao, B, Jones B " An Equal sensation study of seated subjects in three translational Modes" Dep.of Mechan, Engin. University of Birmigham, Ergonomics,1978,vol 21, No2, 123-134.
5. Repperger, K.A. smiles " A feature selection Approach in Evaluation of parameters changes on the Human operator under thermal stress. Ergonomics 1978, vol 21, No 21, No.1, 35-48.
6. Larry G. Richardos, Ira D. Jacobson "Ride quality Evaluation in ground Based Vehicles, Passenger Comfort Models for Buses and Trains". Ergonomics 1978, Vol. 21, No 6, 463-472.
7. Parsons Kenneth and Griffin M.J. "The Effect of Rotational Vibration in Roll and Ritch axes on the discomfort of seated subjects "Ergonomics, 1978, Vol 21, No 8, 615-625.

8. Murphy G. "Similitude in Engineering", N. York, Ronald Press Company, 1950.
9. A. S. Massinas, G. Massouros, and P. A. Drakatos "Cost Depreciating in Construction Equipment, 6th Interest Congress, 24-29 October, 1979, Garmisch-Parten Kirchen.

A COMPUTERISED METHOD FOR MACHINES MAINTENANCE MEASUREMENTS

P. Drakatos, G. Massouros, S. Diamantidis

Professor, Lecture, Research Engineer.
School of Engineering, University of Patras, Greece.

INTRODUCTION

In the recent years many researchers have been concerned with the problems of monitoring the state of an engine and predicting the wear of its moving parts, by analysing the vibration signals recorded at different points of the engine.[1]

The vibrations of the frequency spectrum of these signals can give valuable information about the general state of the engine.

General criteria correlating the state of the specific parts of the engine and their frequency spectrum, have already been published by ISO, DIN and VDI.

It is already known, that a strict relationship exists, between the mechanical state of the engine parts and the shape of the corresponding frequency spectrum.[2]

The subject of the current research is to obtain this relationship for all the mechanical parts. Specifically the most important subject is the interperation of each frequency spectrum,i.e. the accepatable vibration amplitude, the way is varied, and the diagnosis of the engine damages.

The vibration signal, is a vector and can be defined in terms of displacement, velocity and acceleration. Generaly, speaking, for low vibration frequencies, the signal used is that of the displacement, while for medium and high frequencies, the velocity and the acceleration are used respectively.

D. G. Lainiotis and N. S. Tzannes (eds.), Applications of Information and Control Systems, 344-351.

2. ANALYSIS

A problem in the interpretation of the frequency spectrum, is to define the frequencies which correspond at each specific part of the engine.

Each rotating or vibrating part has the frequency of rotation or vibration and its natural frequency defined from the equilibrium of the forces, torques momenta and energies[3].

Knowing in advance the vibration frequency and the natural frequency of each moving part, then the amplitude of the vibration in each frequency spectrum record can be found.

The recorded signals, under such changes can be compared with results been taken under normal operation. The changes between the corresponding spectrums will define the characteristic frequency of each part (fig. 1).

At the present work the spectrums of the characteristic quantities of a vibration i.e. displacement, velocity and acceleration have been examined.

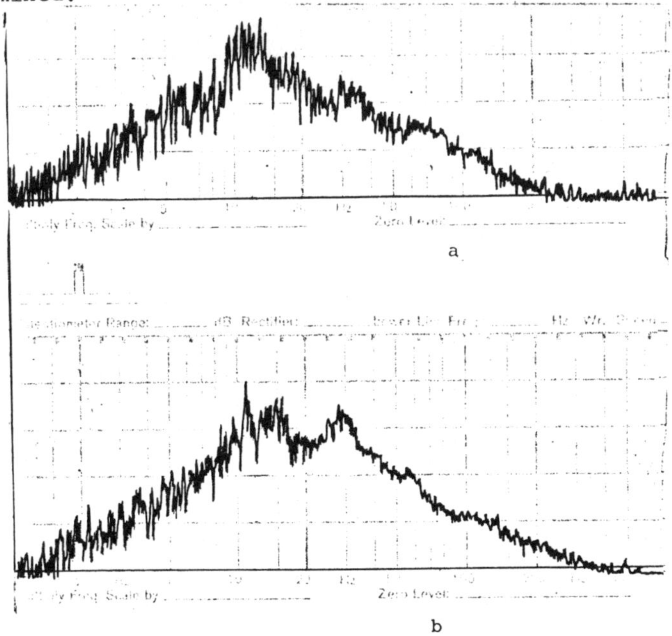

Fig. 1 Spectrums of vibrations changes for:
a. good operation conditions engine
b. bad operation conditions engine

It has been found that for the main frequencies at which the employed parts of engines are operating, the most useful spectrum is that of the velocity (in RPM).

The dominant frequency of the employed group of the FIAT 128 engine was 15Hz, which corresponds to 900 rpm. At the experriments carried out, engines of the same type have been used, but of different mechanical condition. Spectrums have been received with dominant frequencies at 7. 5Hz, 30Hz, 45Hz and 60Hz. These frequencies correspond to different parts of groups of parts of the engine.

From the basic frequency (15Hz), the vibration frequencies of the different parts can be found, i.e. of the pistons, valves, crank-shaft, cam-shaft, valve-springs, spindle-wheel, e.t.c.

The dispacement relationship of each of those parts can be related with that of the crank-shaft. Hence the frequency of their rotation of vibration is multiple or submultiple of the frequency of the crank-shaft (f_{cr}).

For example, the piston of a four-cycle engine reaches the upper dead point once, while it completes two trips in a full revolution of the crank-shaft. Them the measured frequency is equal to the frequency of rotation of the crank-shaft (f_{cr}).

- The cam-shaft is rotating with a speed equal to one half of that of the crank shaft, therefore $f_{cam} = \frac{1}{2} f_{cr} = 7.5$Hς.
- The combustion inside a four-cylinder, four-cycle engine occurs twice in a full turn of the crank-shaft, i.e. $f_{exp} = 2f_{cr} = 30$Hz.
- Each valve is reciprogating with a frequency, half of that of the crank-shaft, hence $f_v = \frac{1}{2} f_{cr} = 7,5$Hz.
- The mixture of fuel-air is passing through the carburettor twice in one turn of the crank-shaft, which means that $f_c = 2f_{cr} =$ $= 30$Hz.
- The Generator corresponds to $f_D = 27$Hz.
- The water pump and the ventilator corresponds to $f_f = 17.5$Hz.

The vibration frequencies of all the above parts can be tabulated into five groups as it is indicated in table 1.

TABLE 1

The vibration frequencies of the parts of a four-cylinder four-cycle engine. (Basic frequency of operation is 15Hz).

Frequency	Parts
f_1 = 7,5 Hz	Cam-shaft, Inlet valves,Outlet valves
f_2 = 15 Hz	Crank-shaft, Spindle-wheel, Piston
f_3 = 27 Hz	Generator
f_4 = 30 Hz	Combustion, Spark plugs, Carburettor
f_5 = 17,5 Hz	Water pump, Ventilator

Table 2 presents the basic and the superimposed harmonics amplitude (A) for each characteristic frequency.

The following nomenclature has been used:

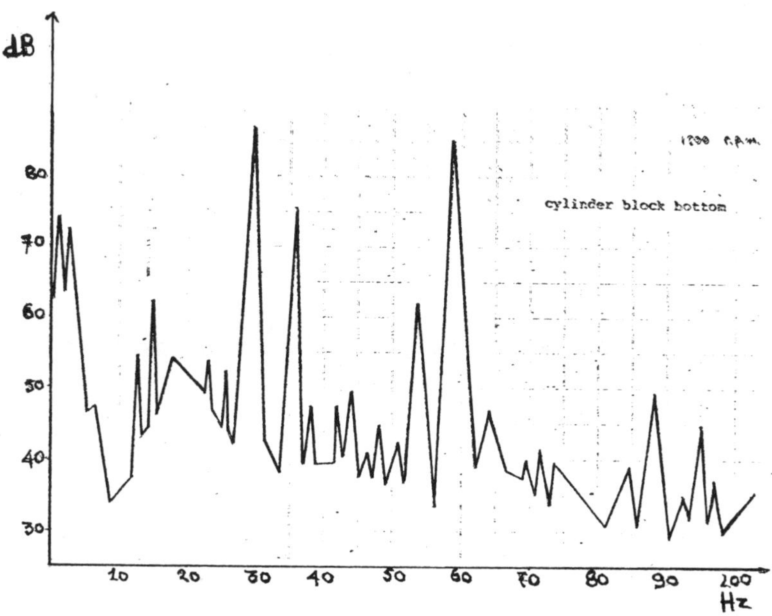

Fig. 2 Vibration spectrum taken by a real time analyser for a four cylinder, four stroke engine showing the characteristic frequencies.

f_{cam}	Cam-shaft	Frequency
f_{exp}	Combustion	"
f_{car}	Carburettor	"
f_{cr}	Crank-shaft	"
f_p	Piston	"
f_v	Valves	"
f_o	Dynamo	"

The second subscripts 0,1,2,3,.... refer to the basic, 1st,2nd, 3th,..... harmonic respectively. Using this table, the amplitudes of vibration at each characteristic frequency, in the case of a new engine can be found, by superimposing basic and harmonics amplitudes.

TABLE 2.

Superimposed harmonics amplitude (A) of an engine.

Frequency	The superimposed harmonics
f_1= 7,5 Hz	$A_{f_1} = A_{f_{cam_0}} + A_{f_{v_o}}$
f_2= 15 Hz	$A_{f_2} = A_{f_{cr_0}} + A_{f_{p0}} + A_{f_{cam_1}} + A_{f_{v_1}}$
f_3= 17,5Hz	$A_{f_3} = A_{f_{p_0}}$
f_4= 30 Hz	$A_{f_4} = A_{f_{exp_0}} + A_{f_{car_0}} + A_{f_{cr_1}} + A_{f_{p_1}} + A_{f_{cam3}} + A_{f_{v_3}}$
f_5= 22,5Hz	$A_{f_5} = A_{f_{cam_2}} + A_{f_{v_2}}$
f_6= 37,5Hz	$A_{f_6} = A_{f_{cam_4}} + A_{f_{v_4}}$
f_7= 45 Hz	$A_{f_7} = A_{f_{cr_2}} + A_{f_{p_2}} + A_{f_{cam_5}} + A_{f_{v_5}}$
f_8= 52,5Hz	$A_{f_8} = A_{f_{cam_6}} + A_{f_{v_6}}$
f_9= 60 Hz	$A_{f_9} = A_{f_{exp}} + A_{f_{car_1}} + A_{f_{cr_3}} + A_{f_{p_3}} + A_{f_{cam_7}} + A_{f_{v7}}$

The above characretistic frequencies can be easily detected in the dpectrums of fig. 2. It has to be taken into account that

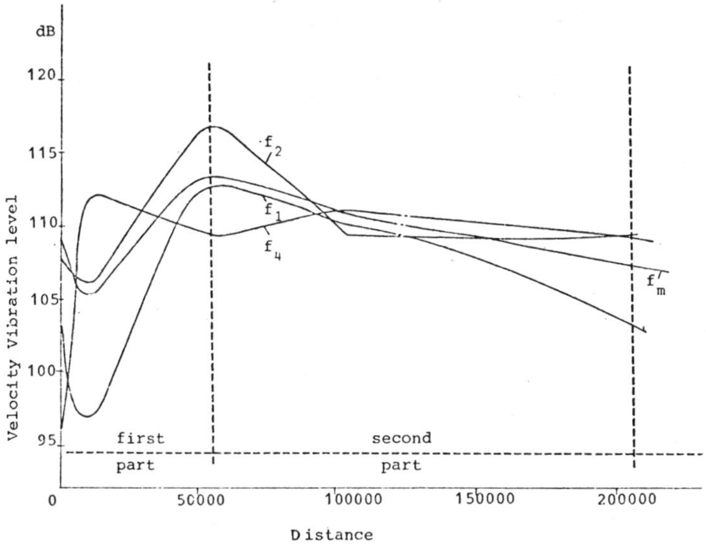

Fig. 3 Vehicle engine velocity vibration level change as a function of distance.

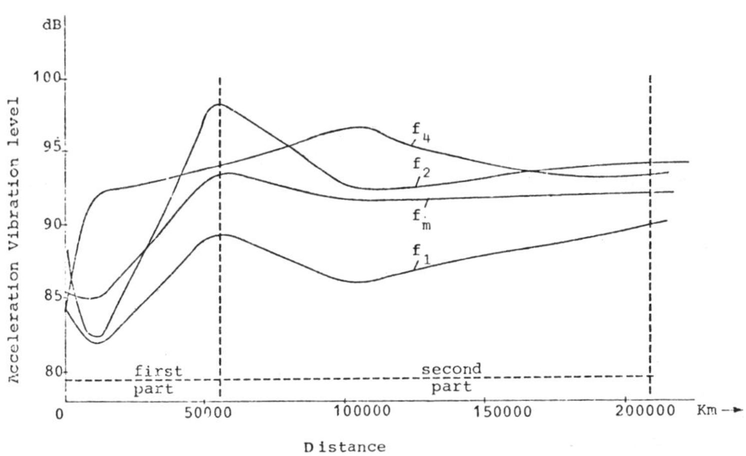

Fig. 4 Vehicle engine acceleration vibration level change as a function of distance.

frequencies shown in table 1, ate the basic harmonics. Besides those, their harmonics will be present in the spectrum. Taking into account that some basic frequencies (f_1,f_2,f_4) are already multiplies of other basic frequencies. Then at the basic frequencies f_2 and f_4, harmonics of lower basic frequencies will be superimposed to them.

The peaks (fig.1) occur at the points of the basic frequencies i.e. at f_2 = 15 Hz which is the characteristic frequency for the crank-shaft, spindle-wheel and the piston.

Another possible peak, is at the point where many harmonics are superimposed as for example at 45 Hz and 52,5 Hz.

Such an arrangement of the amplitude levels, can be seen in the spectrum.

3. EXPERIMENTAL PROCEDURE

Experiments have been carried out to a series of FIAT 128 engines, each one being at a different wear stage.[4]

Using the kilometer's indication, them relationships can be founf between the recorded variations of the velocity and a accelevation vibration and the distances covered by each vehicle.

The time signals have been measured and analyzed with the 2511 Brüel and Kjaer Real Time Analyser.

Also, it is used a computer of Hewlett-Packard No 97 and by means of numerical analysis, the following curves, (Fig.4.,5.) result:

The spectrum has been recorded directly to recorder. At the same time the arithetic results of the amplitude spectrum have been used for the analytic procedure.

It can be seen that the velocity and acceleration vibration is increasing relative to the covered distance i.e. the wear of the engine.

The diagrams shown in fig. 3.,4., shown the level of the velocity and acceleration vibration as a function of the covered distance, for each group of the engine parts. These curves can be divided into two parts.

The first part of the curve presents an increased level of vibration, which is decreasing gradually with the distance. This part corresponds to the running-in period of the engine.

In the second part of the curves, the vibrations are constant with respect to the distance.

Such a case corresponds to the health conditions of the engine by means of a computer program.(fig 5).

4. CONCLUSIONS

From the above discussion, the following conclusions can be reached:

1. Valuable information about the internal state of the engine, can be taken based on the vibration signals recorded at the engine surface. This information is a reliable way to monitor the operational stage of the engine.
2. A strict relationship exists between the wear of the engine, and the level of the vibration.
3. The reliability of the engine and the checking of its mechanical condition can be accomplished by comparing the results received during every periodic maintenance instant with those corresponding to good conditions engine.
4. An experimental model has been developed for an engine category.

REFERENCES

1. Paul Tokle and K. Rothang, " Monitoring and Diagnosing Process Deviation on Marine Diesel Enginess " UKMHMG Summaries N° 3 , 1978
2. D.N. Loynes " Condition Monitoring and manutenance " UKMHMG Summaries Nº 2 , 1978
3. P.M. Brown " Condition Monitoring of rolling element bearing by the shock pulse method " UKMHMG Summeries N° 2 , 1978
4. E. Downham " Vibration criteria for rotating machinery " UKMHMG Summeries N° 2, 1978
5. J. Bozich and J.F Greenfield, "Computer-managed monitoring plant machinery" UKMHMG Summeries N° 2,1978

WAYS OF A CYBERNETIC APPROACH OF AN ECONOMIC BRANCH-THE CHEMICAL INDUSTRY

Eugeniu Niculescu-Mizil

Computer Centre of Chemical Industry,
Bucharest, Romania

The paper deals with the ways of a cybernetic approach of the Romanian chemical industry.
The chemical industry is considered as a system consisting of interdependent subsystems. The subsystems are chemical enterprises and groups of enterprises. They are connected by information streams.
Several examples of mathematical formalization of the above mentioned system are presented.

SYSTEM PRESENTATION, ITS PLACE IN THE CYBERNETIC SYSTEM OF NATIONAL ECONOMY.

An economic system is considered from the cybernetic point of view, to belong to a higger system,i.e.,to the corresponding social macrosystem. The type of the economic system is related to the type of the social macrosystem to whom it belongs.
The existing economic activities and direct and backward connections indicate that any economic body (enterprise, branch or the entire economy) may be considered as a cybernetic system.
Due to the interrelated connections and to the permanent dynamic behaviour of the parameter variations, the economic system on the whole, may be considered the most complex social cybernetic subsystem.
In order to analyse the technological processes and the economic systems, by using modern scientific methods, cybernetic principles and categories must be

D. G. Lainiotis and N. S. Tzannes (eds.), Applications of Information and Control Systems, 352-363.

used, the most important being the automatic control theory.

Intensive trends in the use of cybernetic methods of analysis and forecast and modern concepts and techniques able to help the man in the conscious management of technological and economic processes with optimal results, are observed.

The dynamic behaviour, interconditioning and complexity of the economic systems represent characteristics of the cybernetic systems.

An economic system is in continuous movement, transformation and evolution and its parameters permanently change in relation with the development or the limitation of the activities. It is also an open system interconnected to some other external systems by mutual influences, receiving factors of influence depending or not on systems evolution.

The orientation to prescribed objectives, the own characteristic evolutions as well as their acceleration and deceleration due to its connections with other systems give the economic system the dynamic behaviour, the interaction and the complexity proper to the cybernetic systems.

The degree of complexity of an economic system is related to its position in the hierarchy of the economy, rising from the bottom to the top.

The national economy considered as a cybernetic system is comprising of subsystems (branches) one of them being the chemical industry. This subsystem is the object of our paper.

The concept of "cybernetic system" will be used to describe the obvious structural base of the management and control mechanisms and the complexity of the dynamic system which is the chemical industry. Together with the constituent elements, their connections and the relations with the environment, it represents a complete picture of a cybernetic system.

The technical and scientific revolution, the increasing development of raw materials chemical processing, the continuous search for new energy resources and for the efficient utilization of residual energies of the chemical plants, the realization of integrated plants, led, due to the intensive functional interdependencies of the chemical industry branch of the other economic branches of the national economy, to the development of extremely complex industrial chemical subsystems.

The cybernetic system, which is the chemical industry, is thus constituted of a multitude of interdependent subsystems.

Each element and subsystem belonging to the cybernetic

system of the chemical industry has a particular functional aim, i.e., a particular behaviour interdependent of the other elements, determining a certain character of rationality of the system topology.
The evaluation of the relationship between the system structure and its functions is of a great importance in the practice of management of the chemical industry.
The relationship between the cybernetic system of the chemical industry and the environment are intensly influenced by the organization of its elements and connections between them and have a determinist or probabilistic character. The system of chemical industry is dependant on the state organizations and on the central planning and synthesis bodies, on the one hand, and on the raw materials, energy utilities, equipment and Know-how suppliers and internal and external consumers of chemical products, on the other hand.

SYSTEM APPROACH OF THE CHEMICAL INDUSTRY BRANCH

The system approach of the chemical industry branch may be considered by taking into account several criteria, as for example :

A. The functions of the economic branch which is the chemical industry may be considered subsystems of the respective system.

As any economic-industrial cybernetic system, the chemical industry branch provides five classical functions :

- Research and development function consisting of planning, research and design of new processes and plants, development of production and permanent improvement of management organization in respect with the actual requirements of each stage of development.
- Production function, consisting of planning, organization, transport, maintenance repair and operation of plants and enterprises, activities with the purpose of obtaining the prescribed quantities of products within the prescribed quality parameters and economic efficiency.
- Commercial function, consisting of planning, organization, transport outside the plants, supply, delivery, import and export activities with the purpose of supplying the raw materials, additional materials, energy and other utilities necessary to

realize the products and to deliver them to consumers. This function is a finalization of the tendency which appear in the previous subsystem corresponding to the production function.
- Financial and accounting function, consisting of planning organization, accounting and financial activities regarding the provision, control and evidence of financial and material resources and of the economic results.
- Personnel function, consisting of planning, organization, training, recruitment and selection of specialists, wages computation and evidence and provision of working and living conditions of the employees.

A representation of the cybernetic system of the chemical industry based on the above mentioned functions is suggested in Figure 1.

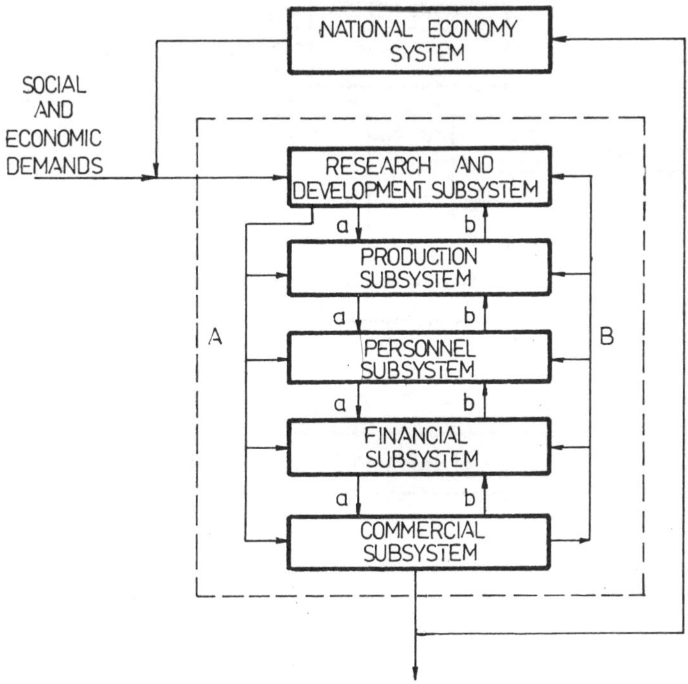

Figure 1. Chemical industry system based on functions of the economic branch.

The arrows indicate the direct (a) and backward (b) connections between two close subsystems and the

direct (A) and backward (B) connections between all subsystems.
Mutual influences of subsystems as well as an overall purpose of responding to the requirements of the society (inputs to the system) by social and economic achievements (outputs of the system) are observed. The outputs of this system represent the inputs to a larger system which is the national economy system.

B. The hierarchical levels of organization may be considered subsystems of the chemical industry system.

By this approach three main hierarchical levels of organization are put into evidence : the ministry of chemical industry, the industrial centrals and the enterprises (Figure 2).

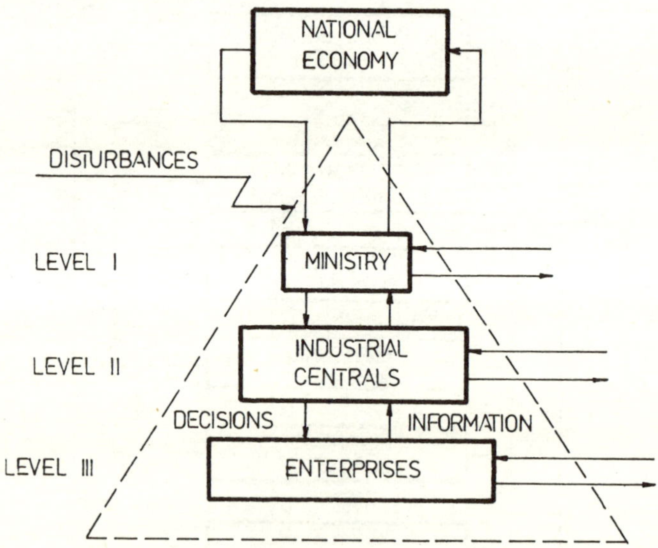

Figure 2. Chemical industry system based on hierarchical structure of organization.

Each level of organization corresponding to a subsystem, is based on the respective internal organization (the ministry with departments, sections, offices and research and designing institutes the industrial centrals with sections, offices and enterprises, and the enterprises with sections, offices, plants and workshops).

C. The hierarchical levels of management may be considered subsystems of the chemical industry system.

The main hierarchical levels of management are : the managed subsystem (operational and executive) and the control and management subsystem (decisional). (Figure 3).

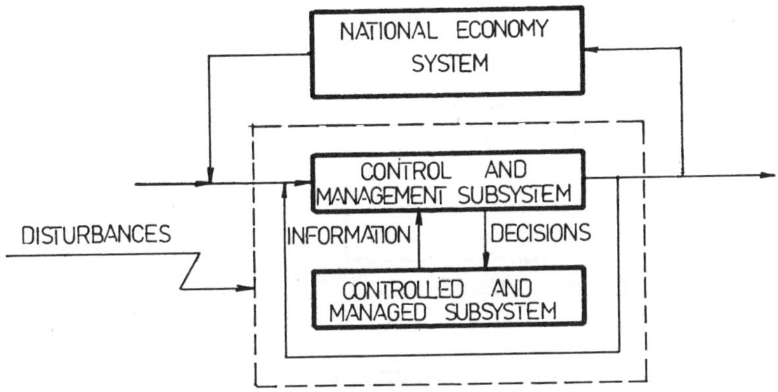

Figure 3. Chemical industry system based on hierarchical structure of management.

The controlled subsystem consists of chemical enterprises.
The control and management subsystem is decisional and consists of management boards of enterprises, industrial centrals and ministry.
The functional connection between the two above mentioned subsystems is achieved by transmission, processing and storage means.
Several characteristics of the above mentioned system are of particular importance.
The behaviour of the control and management subsystem is time varying because the state of the chemical industry system changes by a complex transformation process in order to attain the desired state, expressed by the objective function of the respective model.
Due to the aleatory character of the connections, various comportments of the managed subsystem occur at the same management actions.
In order to be stable and to improve its own functional parameters and behaviour, the system connections change in time in relation with the environmental

conditions. This determines the dynamic character of the system.
The existence of the backward connections between the subsystems determines a behaviour characterized by a selfcontrol mechanism (Figure 4). The adaptation to the environment, in this case, is made by maintaining unchanged the internal structure of the system.

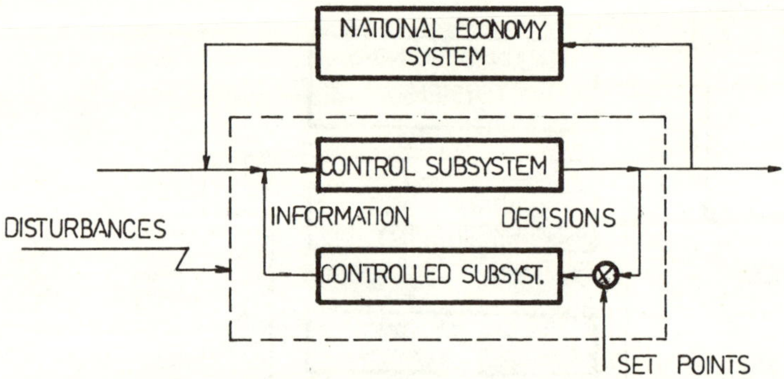

Figure 4. System selfcontrol.

A careful analysis of the system points out the possibility to modify, between certain limits, its economic structure as a necessary condition mainly to provide a suitable cooperation with all economic branches of the national economy and between the elements of the system itself.
The system is in this way selforganizing in order to satisfy the required modifications of the various financial funds, research or designing activity volumes, etc.

D. Both the hierarchical levels of organization and of management may be considered subsystems of the system.

The system consists of the above mentioned hierarchical levels of organization as well as of management (Figure 5).
The subsystems must be compatible from the point of view of both "management" and "organization" in order to avoid local optima which are different from the overall efficiency of the chemical industry system.
The ministry, the industrial centrals and the enterprises represent the first, the second and the third decisional levels, respectively. The plants (not indicated on the figure) represent the managed subsystems.

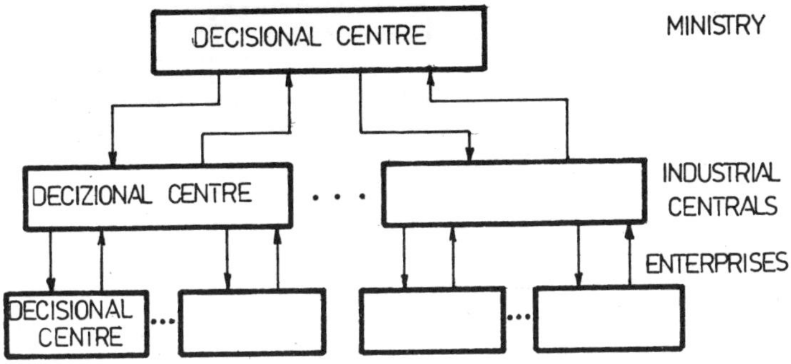

Figure 5. Chemical industry system based on hierarchical structure of organization and management.

An example of formalization.

The problem of formalizing of a two-level hierarchical management system may be stated in a generalized form.

Given :

$P: X \times \Omega \longrightarrow Y$	the output function
$G: X \times \Omega \times Y \longrightarrow V$	the performance function
$\mathcal{G}: \Omega \longrightarrow V$	the tolerance function
$X_a \subset X$	the input feasible solutions set,

to determine the inputs $x \in X_a$ such as for all $\omega \in \Omega$ the tolerance criterion should be satisfyied:

$$g(x, \omega, y) \leqslant \mathcal{G}(\omega)$$

where : Y is the process output set, Ω the disturbance set, ω the disturbance, x the input, X the input set, X_a the acceptable input set, P the output function of the controlled process, G the performance function, $\mathcal{G}$ the tolerance function, V the set of performance function values (G) . This hierarchical system including the chemical processes management systems is shown in Figure 6.

The modern techniques of operational research provide the development of mathematical models on all hierarchical levels of management. These models may be successfully used to optimize the chemical industry system considered as a cybernetic system.

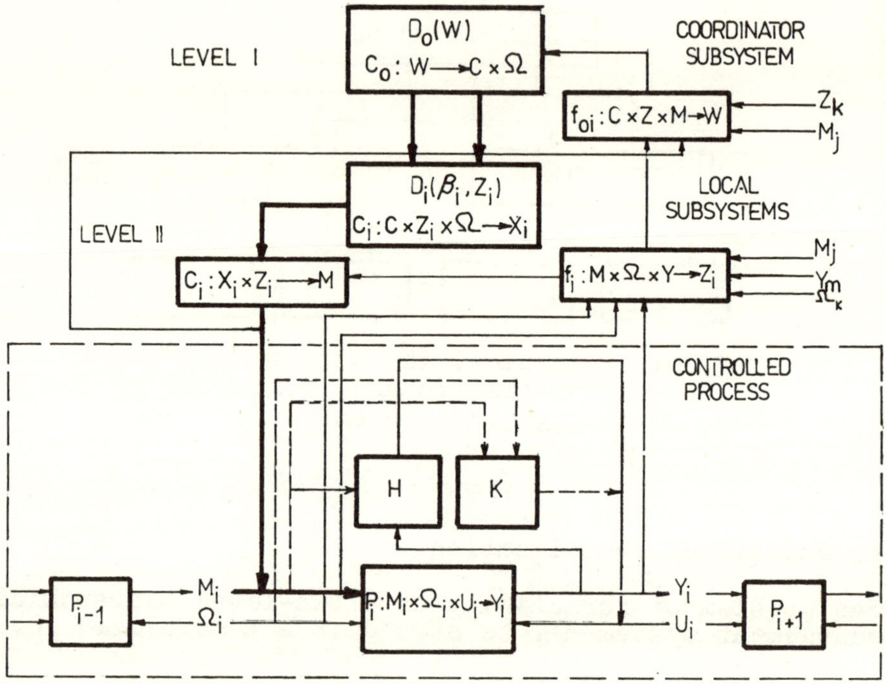

Figure 6. Two-level hierarchical management system.

The symbols presented in the Figure 6, are :

$P: M \times \Omega \longrightarrow Y$	the controlled chemical processes
$M = M_1 \times \ldots M_n ; M_i = \{m_i\}$	the controlled process inputs
$\Omega = \{\omega_i\}$	the disturbances set
$Y = \{y_i\}$	the process output set
$C_i: C \times Z_i \times \Omega_i \longrightarrow M_i$	the local control subsystems
$C = \{\beta_i\}$	the set of coordination inputs
$Z_i = \{z_i\}$	the set of feed-back input from the process
$C_o: W \longrightarrow C \times \Omega$	the coordination subsystem
$W = \{w_i\}$	the set of feed-back inputs from the control problems
$f_1: M \times \Omega \times Y \longrightarrow Z_i$	the information function regarding the process behaviour

$f_{oi}: C \times Z \times M \longrightarrow W$	the information function regarding the local control subprocesses
$P_i : M_i \times \Omega_i \times U_i \longrightarrow Y_i$	the decoupled subprocesses
$U_i = \{u_i\}$	the set of interactions of P_i subprocess
$Y_i = \{y_i\}$	the P_i subprocess output set
$H_i : M \times Y \longrightarrow U_i$	the coupling functions of P_i subprocesses
$K_i : M \times \Omega \longrightarrow U_i$	the interaction functions of P_i subprocesses
$d_o : W \longrightarrow X_o$	the decisional problem of the coordinator C_o.
$C_o : X \times X_o \longrightarrow C$	the implementation block of the coordinator
$D_o(w),\ w \in W$	the coordinator decision problems set
$\beta = C_o[w, d_o(w)] = C_o(w\ X_o)$	the coordination output from the coordinator
$d_i : C \times Z_i \longrightarrow X_i$	the decisional block of the local subsystems
$c_i : X_i \times Z_i \longrightarrow M_i$	the implementation block of local subsystems
X_i	the set of posible alternatives
$D_i(\beta_i, Z_i),\ \beta \in C$	the decisional problems family of local subsystems
$m_i = C_i[Z_i, d_i(\beta_i, Z_i)]$	the control outputs from the local subsystems to the process

The system approach and the modelling of chemical processes and plants.

The cybernetics has a particular importance for the improving of description of any technological process by system approach. The technological process consisting of all physical and chemical processes and the means for their realization, is considered as a system having input variables X_n (quantities and compositions of raw materials, yields of transformation and separation, operating conditions, equipment performances etc.), output variables Y_n (quantities and compositions of finite products, properties of finite products, etc.) and control variables C_n necessary to compensate the outside disturbances P_n (Figure 7).

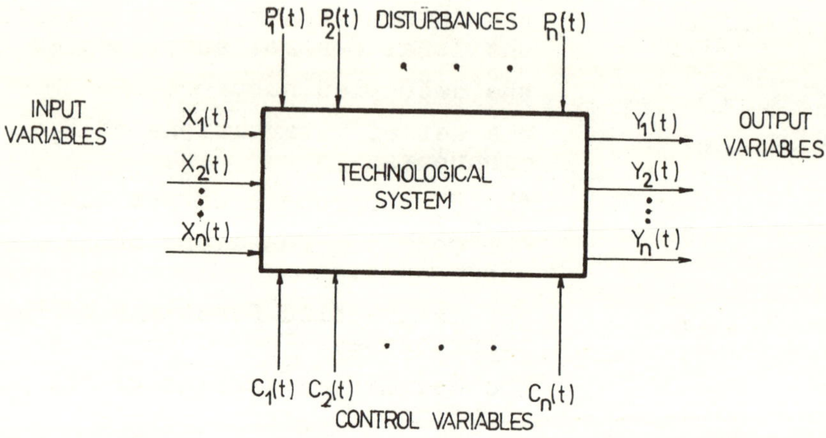

Figure 7. General model of chemical technological system.

The variables of the system are considered to be time-varying.
The qualitative and quantitative analysis of each subsystem corresponds to the following relation :

$$Y_n = f_n\ (X_n,\ P_n,\ C_n)$$

The development of mathematical models implies the following stages :
- analysis of the physical and chemical aspects of the process;
- mathematical formalization of the process by a set of constraints;
- testing the model;
- development of the objective function.

CONCLUSIONS

The analysis of the chemical industry by a system approach is of particular importance to the improvement of its functioning and development, for their optimization.
It requires the analysis of the following :
- priority of a subsystem over others;
- functional dependence of the subsystems;
- existing discrepancies between the objectives of the various decisional levels regarding the efficient functioning of the system;
- degree of freedom of the subsystems;
- duration of the managed subsystem behaviour;

- complexity of the decisional problems, mainly from the probabilistic point of view, of the supperior levels of decision;
- use of various methods of modelling depending on the subsystem character (determinist, probabilistic, statistic, linear, nonlinear, dynamic etc.).

In order to develop and solve efficiently the respective models, modern methods of cybernetics, systems and automatic control sciences, as well as powerfull analog and digital computers, must be used.

BIBLIOGRAPHY :

- Ciochină, M., Hierarchical multi-level system approach to the optimal control of a polyethylene plant, in:Modern trends in cybernetics and systems III-rd International Congress in Cybernetics, Springer Verlag, Stutgart, 1977.

- Florescu, M., Niculescu-Mizil, E., et.al., Cibernetică, informatică şi automatică în industria chimică, Editura tehnică, Bucureşti, 1979.

- Kafarov, V., Méthodes cybernétiques et Technologie chimique, Mir, Moscou, 1974.

- Niculescu-Mizil, E., Cibernetica şi aplicaţiile ei în tehnica şi economia industriei, Editura Academiei, Bucureşti, 1973.

CONTRIBUTION TO THE STUDY OF THE RELATIVE EVOLUTION OF THE FIRMS

J. PEYRON and J. FAVREL
I.N.S.A. de LYON - Département Informatique
G.R.E.C.O. Rhône-alpin "Analyse de Systèmes"
Bât. 502 69621 - VILLEURBANNE CEDEX FRANCE

ABSTRACT

In this paper, we emphasize a first difficulty when considering the evolution of the firm due to the bad quality of financial and economic informations about it and its environment. We denote also a second difficulty staying in the diversity of spaces to model the firm.

So, a tentative approach is developped by analyzing it as a complex system in order to better our knowledge about it. With this objective, we adopt a sort of relative approach to understand the evolution of the firm by means of data analysis methods.

They are used to do some factor rankings and partitions of a sample which allows us to track the trajectory of the firm relatively to its competitors'ones. So a dynamic model is defined not only do describe its previous evolution, but also to do some forecasts about the structure and the functioning of the firm.

INTRODUCTION

An understanding of the firm, necessary for managers as well as economits, presents two mains difficulties. On the one hand it is related to the definition of the studied system's boundaries and to the diversity of possible representation spaces when trying to explain the state of the firm at a certain time or its evolution during a given period, and on the other hand to the difference in methods and units of measure.

After briefly surveying various existing approaches to the understanding of the firm, we will show how data analysis methods can help to show the relative situation of a firm in comparison

D. G. Lainiotis and N. S. Tzannes (eds.), Applications of Information and Control Systems, 364-371.

with other firms and to show its evolution in this context. This will be demonstrated by means of various evolution models.

These models are explanatory ones but they could also be used as tools for assisting management and for economic forecasting.

1. MODELS OF THE FIRM

1.1. Existing models : a tentative survey

The accounting model is of course, the most widely used in spite of its imperfections owing to the method used for evaluating the firms assets and to the difficulties in homogenizing monetary measure units. However, it allows, in its current acceptance the so-called "double part" accounting method, the observation of the firm with two complementary and inseparable features for the study of a dynamic system :

- structural features, or situation of the firm, characterized by the balance sheet variables (level of assets and equities),
- functional features, or management of the firm, characterized by income statement variables (revenue and expense flows).

The financial theory of the firm is an extension of the accounting model adapted to its various financial functions. The derived models are more restricted but more reliable than the accounting one and especially allow us to take into account currency variations by using actualisation rates. In this category, we can find :

- the treasury models in a non-probabilistic future (BAUMOL's model) or stochastic model (MILLER and ORR's model) ; see (1) for a simplified statement,
- the investment and growth models (E. PENROSE (2) or J. LESOURNE (3)) and the planning and control models (Y.IJIRY (4)).

The corporate models of the firm (J. WANTY and J. FEDERWISH (5)) allows to build the graph of the explicit or implicit relationships between the variables characteristic of a firm (or a subsystem of it). In this category we can also include models such as NAYLOR's one (6).

The behavorial models include non-quantitative elements such as the influence of the organization, of the information flow, or of the hierarchical pressures between persons (see CYERT and MARCH (7) and application of BONINI's model (8)).

Finally, the relative evolution models, developped in the next sections use data analysis methods in order to elaborate firstly, factor classifications (J.P. BENZECRI's "analyse des correspondances" (9)) and then, partitions derived from these classifications, from a significant number of firms. With a small sample (less than 100 firms) partitions can be directly obtained by means of hierarchical classification methods (see algorithms of JOHNSON (10) or ROUX (11) or automatic classification methods (E. DIDAY's "Nuées dynamiques" algorithm (12)). An interesting application on the classification of the firms from the balance sheet structure is given by E. HURET (13). We can mention ALTMAN (14) who has also used the data analysis method for the study of behavior in the firm in order to predict possible bankruptcy.

1.2. Informational features

1.2.1. Systematic bias

Accounting information systems are submitted to legal and fiscal constraints and often give a false image of the reality of the firm. Particularly, the laws of accounting evaluation do not (or not entirely) take into account variations of monetary units over a period of time. Moreover, the evaluation of assets, very often does not correspond to the intrinsic value of the assets. For instance, stocks and shares are valued at their face value which is often very different from the stock exchange value. In the same way, some equipment can have a net book value equal to zero and still be used. Inversly, land and buildings have a market value very far from the net book value. We could make the same statement for inventories or outstanding debts.

1.2.2 Various spaces of representation of the firm

The firm is then a complex dynamic system which can be represented in a state space at a given time by means of the balance sheet which gives an incomplete picture of the firm. This is a representation in a structural space.

But during the period ranging between two successive state observations, the evolution of the firm can be observed by means of its management variables listed in the income statement. This is then the observation of the firm in the functional space. Finally, to complete the functional representation, we must also consider the financial flow in the statement of sources and applications of funds.

1.2.3 Necessity of a relative approach to understanding of the firm

The previous observations on reliability of information on the firm lead us on the one hand to study the firm relatively and not absolutely, i. e. to place it in comparison with other firms, and on the other hand to study it in a representation space limited to dimensions which allows us to provide a maximum of information on its relative situation. This space could be structural or functional or both by means of the ratios usually used in financial analysis.

2. UTILIZATION OF DATA ANALYSIS METHODS AS TOOLS FOR AN UNDERSTANDING OF THE FIRM

2.1. Factor analysis methods

These methods are used because they require less calculation time and costs than the direct partition methods such as clustering or hierarchical methods, particularily for the big population of E (up to several hundred elements).

There are several methods of this type, particularily correspondence analysis (9) which allows a reduction of the initial space i. e. passing from E(m, n) to E(m, n') with $n' < n$ while keeping a maximum of information on the initial space. The new space obtained is called the factor space.

It is then possible to partition set E from the classifications of the elements obtained on the factors in the E(m, n') space or in a subspace E(m, n") with $n'' < n'$ while retaining only the main factors. Particularily, we could carry out this partition in the first factor plan or in the first factor by defining equiprobable or equally ranging classes.

The economic signification of the factors is obtained through projection of the n initial variables on the n' factors of the new space.

The table below gives main factors obtained in the various representational spaces from a sample of 1500 french firms which are representative in dimensions (sales amount) and in economic sectors. This sample is extracted from a file from the "Centrale des Bilans" of the "Banque de France" from 1969 to 1975. Some application examples obtained from the classification of french economic activity sectors represented by their averages in structural and functional spaces are given in (15).

VARIABLES	AXIS 1	AXIS 2	AXIS 3	AXIS 4
Structural stu- study (balance sheet)	Capitalis- tic nature of the str.	Structure of the fi- xed assets	Structure of financ. source *	Structure of the cur- rent assets
Percent. of variance	40 %	25 %	15 %	12 %
Functional stu- dy (income statement)	Added value opposite to intermedia- te consump- tion	Salaries and wages	Cash-flow	Financial expenses
Percent. of variance	60 %	20 %	15 %	3 %
Synthetical study (using ratios)	Financial securities opposite to profitabi- lity	Total ad- ded value	Profit (from sa- les)	Value added through manpower op. v.a. through capital
Percent. of variance	40 %	22 %	16 %	12 %

Factor analysis interpretation

* Stockholders equities + long-term liabilities

2.2. Evolution models

2.2.1. Definition

Direct or indirect partitions are made, as described before, for r consecutive periods, from the sample of firms mentionned above (in 2.1.).

The partition obtained in the tth period of time defines a so-called belonging vector P(t), with dimension q = number of partition classes, where $P_i(t)$, i = 1... q, represents the numbers of firms that are in the ith class at time t : there are q possible classes. The evolution model is obtained by estimating the transition probability : $A_{ij}(t, t+1)$, from the ith class to the jth between time t and t+1, which is measured by the transition frequency $\frac{P_{ij}(t, t+1)}{P_i(t)}$, where $P_{ij}(t, t+1)$ represents the number of firms ranging from the ith class to the jth class between t and t+1.

The model has then the following form :

$$P(t+1) = P(t) \cdot A(t, t+1)$$

and the population of the jth class is then

$$P_j(t+1) = \sum_i p_i(t) \cdot A_{ij}(t, t+1)$$

where A is the transition matrix : $A = \left[A_{ij}\right]$.

2.2.2. Properties

The transition matrix A is stochastic since $0 < A_{ij} < 1$ by construction and $\sum_j A_{ij} = 1 \ \forall i$.

Furthermore, this matrix presents a particular structure in which the terms of the main diagnonal range from approximately 0.7 to 0.8 and characterize stationary firms. The under diagonal terms, ranging from approximately 0.1 to 0.2 represent the transitions of one class. Lastly, the transitions of two classes or more are often negligible (see example below).

We can also mention (15), that the model described thus is stationary during the studied period (1969-1975) i. e. A does not depend on the time. But this model has an order greater than 1 or, in other words, the transition between t and t+1 depends not only on the state of the firm at the time t but also on previous states. The model is then defined as an unhomogeneous markov chain.

As an example, we give the transition matrices of the first order estimated model provided in the structural space of the firms during the period 1971 to 1975.

$$A_{71-72}\% = \begin{bmatrix} 71.4 & 27.6 & 1 & 0 & 0 \\ 7.5 & 71.3 & 20.2 & 1.1 & 0 \\ 0.2 & 13.1 & 74.4 & 11.9 & 0.4 \\ 0 & 1.1 & 18.3 & 74.7 & 5.9 \\ 0 & 0 & 3.1 & 21.4 & 75.5 \end{bmatrix} \qquad A_{72-73}\% = \begin{bmatrix} 67.3 & 28.6 & 3.1 & 1 & 0 \\ 8.8 & 68.7 & 21.7 & 0.8 & 0 \\ 0 & 14.1 & 72.7 & 12.7 & 0 \\ 0 & 0.6 & 20.3 & 71.3 & 7.7 \\ 0 & 0 & 1 & 26.8 & 72.2 \end{bmatrix}$$

$$A_{73-74}\% = \begin{bmatrix} 68.3 & 27.7 & 4 & 0 & 0 \\ 8.1 & 70.3 & 20.6 & 0.8 & 0.3 \\ 0.5 & 13.3 & 72.1 & 13.2 & 0.9 \\ 0 & 0.3 & 22.8 & 71 & 5.8 \\ 0 & 0 & 1 & 28.3 & 70.7 \end{bmatrix} \qquad A_{74-75}\% = \begin{bmatrix} 73.3 & 24.8 & 1 & 1 & 0 \\ 6.4 & 70.7 & 20.9 & 1.4 & 0.6 \\ 0.5 & 13.6 & 71.7 & 13.8 & 0.3 \\ 0 & 1.7 & 23 & 69.3 & 6.1 \\ 0 & 0 & 1 & 25.8 & 73.2 \end{bmatrix}$$

We can define in a similar way a 2nd order estimated model taking into account the transitions during two consecutive periods : from t-1 to t and from t+1 to t. For term, the transition matrix then has A_{ijk} which represents the probability for a firm to be in the state of k at t+1 knowing that it was in the state of j at t and in the state of i at t+1 :

$$P_k(t+1) = \sum_i P_i(t-1) \sum_j A_{ijk}$$

2.3. Model defined by the evolution trajectory in a factor space

In this second sort or model, we can apply the aforementioned factor methods to the set E(n x p, m) of n firms which are studied during p consecutive years in the m-dimensioned space. This type of model was already used for the study of danish banking firms (17).

The projection of the p points representing firm Ei at p sucessive periods in the factor space defines a trajectory that characterizes the evolution by means of the factors defined by this analysis.

We can see the trajectory described by this method, for several french economic sectors featured by their mean ratios of the balance sheet structure, from 1971 to 1975 in (16).

CONCLUDING REMARKS

We have not solved the difficulties related to usual methods used for an understanding of the firm's situation and management. We have pointed out however, that the statistical approach, by means of data analysis methods allows a relative understanding of the firm's position in comparison with others, in a given representation space. These methods also permit an elaboration of descriptive evolution models which can be used as tools for management control and economic forecast. It seems that this approach, complementary of the usual accounting and financial methods, can provide additional interesting information to the decision maker in terms of management as well as economy.

REFERENCES

(1) S. Lacrampe, G. Causse, Méthodes de Gestion de la Trésorerie. Ed. Hommes et Techniques, Paris 1975.

(2) E. Penrose, The theory of the Growth of the firm. J. Wiley, New-York 1959.

(3) J. Lesourne, Modèles de Croissance des Entreprises Dunod, Paris 1973.

(4) Y. Ijiry, Objectifs et Contrôle de Gestion (translated of Management Goal and Accounting for Control) Dunod, Paris 1971.

(5) J Wanty, J. Federwish, Modèles globaux d'Economie d'Entreprise, Dunod, Paris 1970.

(6) T. H. Naylor, Computer Simulation Experiments with Models of Economic Systems. J. Wiley, New-York 1971.

(7) R. M. Cyert, J.G. March, A behavioral Theory of the firm Englewood Cliffs, New-York 1963.

(8) C.P. Bonini, Simulation of Information and Decision Systems in the Firm. Prentice Hall, New Jersey 1963.

(9) J.P. Benzecri, L'analyse des Données. Dunod Paris 1973.

(10) S.C. Johnson, Hierarchical Clustering Schemes. Psychometrica n° 32 1967.

(11) M. Roux, Un algorithme pour construire une Hiérarchie particulière. Thèse, Paris 1968.

(12) E. Diday, Optimisation et Classification automatique en reconnaissance de Formes. Bulletin I.R.I.A. n°12, Paris 1972.

(13) E. Huret, Structure des Bilans et Type de Croissance des Entreprises. Economie et Statistiques n° 50 1973.

(14) E.I. Altman, Financial Ratios, discriminant Analysis and the Prediction of corporate Bankruptcy. The J. of Fin. sept. 69

(15) J. Peyron, Approche statistique de la classification et de l'évolution des Entreprises. Thèse I.D. Lyon 1976.

(16) J. Peyron, J. Favrel, Etude du comportement conjoncturel des Entreprises. Col. Struct. Econ. et Econ. Lyon, mai 76.

(17) L. F. Pau, Feature Extraction in the Time Domain : Applic. to the Analalysis of financial Data and Strategie. Nato A.S.I. Pattern Recognition, Bandol 1975.

THE QUANDARIES OF SOCIO-ECONOMIC POLICY MODELING

John Henize

Gesellschaft für Mathematik
und Datenverarbeitung
5205 St. Augustin
W. Germany

ABSTRACT

This paper attempts to answer the question: Why have most socio-economic modeling efforts failed to produce useful results at the policy level, and how can we begin to to better? Previous failures are diagnosed as being due to a false emphasis on description and prediction rather than on explanation. The problem is then reduced to answering the fundamental question: How does one validly simplify? The various simplification dimensions and the tradeoffs between them are presented and explained. A major change in relative emphasis is recommended.

1. INTRODUCTION

Are socio-economic simulation models powerful tools than can aid immensely in helping to improve government policy decisions, or are they merely playtoys for bored academics who have nothing better to do? All of us who have been engaged in the business of constructing such models would like to think that the first of these is the case. Many businessmen and government officials however who have supported modeling efforts in the hope of obtaining some help for making better policy decisions think that the second is closer to the truth. Who is right?

The construction of models for decision analysis developed as an area of significant activity first during the Second World War. The results of this initial research were in some cases spectacular. The early successes of decision models in military

D. G. Lainiotis and N. S. Tzannes (eds.), Applications of Information and Control Systems, 372-384.

applications then led many to attempt to develop similar models elsewhere--for use in other areas of government planning as well as in private industry. At the lower levels of decision making, many of these models have proved themselves to be useful and beneficial--in industry, for instance, for inventory control and production control problems; in government, for such things as calculating future population levels and for predicting the probable financial costs of certain types of programs. Not all such applications have been successful, but the results on the average have been worthwhile.

At the higher policy-making levels of business and government, however, the results so far have been disappointing. Arthur Okun, former chairman of the U.S. Council of Economic Advisors, expressed the general disillusionment shared by most policy makers when he recently asserted:

> From where I sit, econometric models just don't reflect the real world.[1]

Arthur Burns, former chairman of the U.S. Federal Reserve Board, is if anything even more sceptical.

> I talk to people, not computers. A model's equations are not objective.[1]

The problems experienced by large socio-economic modeling efforts are by no means limited to the public sector. Large scale planning models in industry have encountered similar difficulties. For instance, a recent survey of U.S. industry has shown that

> examination of several corporations' experiences with strategic planning models reveals some disturbing facts. In a large number of firms, model development efforts have been reduced or stopped entirely. Many of the planning models which have been developed have not been implemented or are used only on infrequent occasions. Finally, most of the planning models which are being used are not significantly influencing the actual strategy formulation process within the firm.[2]

It would seem that the opposite should have been the case. The more complex and difficult the situation, the more useful it would seem that a model ought to be.[3] Answers to high-level policy questions require an understanding of the effects of many complex interrelationships. The difficulties of understanding and of correctly anticipating the results of various economic and social policies arise in large part because of the large number of multiple interrelationships among the many different parts of these systems. Everything depends on everything else. The supply of labor, the demand for labor, unemployment, wages, prices, inflation, personal income, consumer spending, savings, investment, corporate profits, tax receipts, government spending, imports, exports, labor union pressures, worker attitudes, labor force participation rates, migration of foreign workers, productivity, production levels, environmental pollution, energy consumption--all are intimately interrelated. It is impossible to accurately determine the future values of any one of these quantities without simultaneously knowing the values of all the others. And the future values of each of these depends on their own past and present values, as well as on the way they all interact to affect one another.

Accurate policy analysis requires, therefore: first, that one be able to correctly determine which variables are most important, and what the most important individual influences and relationships are by which each of the variables relates to each of the others; and secondly that we be able to correctly determine the end effects of the simultaneous interplay of all of these various influences interacting together. Carrying out this second task, purely mentally and intuitively, is simply impossible. There are so many mutually interacting factors that it is impossible for us to simultaneously keep all of them continually in mind. If however each of the individual influences were to be defined and represented by an equation in a computer model, then, for the computer, keeping track of all of these simultaneously interacting influences is quite easy. This is why computer models would seem to offer such incredible potential for helping us to improve the quality of our social and economic analyses and decisions.

But what a computer of course cannot do is to make the initial choice as to which variables and relationships should be represented in the model and which not. This choice must be properly made by humans. And it is due to inadequacies in the carrying out of this first task that the immense potential offered by computer models has as yet, at the higher policy-making levels, to be adequately realized.

What then is the problem? Why is it that most models still fail to contain the proper relationships and structures? The problem here arises from the necessity for simplification. We all know that the human mind is capable of dealing with only a very limited degree of complexity, and that providing a person with too much information can be equally bad or even worse than providing him with too little. It is sometimes naively believed, however, that large computer models can eliminate this constraining necessity for simplification. One simply programs all of the possible influences one can think of into an enormous computer model, lets the model run, and obtains therefrom the true prognosis of an ineluctable future.

This utopian notion of computer models however is not realistic. The amount of complexity with which one can competently deal, even with the help of computer models, is still quite limited. Computer models should indeed, if properly constructed, be able to help us deal with greater complexity than would otherwise be possible. But only with a somewhat greater, not an unlimited complexity. The possibility of computer modeling reduces but does not eliminate the necessity for simplification.

The limitations of time, resources, and human capacity always dictate that not every possible factor of any conceivable influence can be encompassed within one given analysis. This means that there are always simplifying assumptions that must be made--some factors which in the analysis must be neglected. How are these simplification decisions to be made?

2. THE DIMENSIONS OF SIMPLIFICATION

There are several dimensions to be considered:

- completeness
- simplicity (understandability)
- scope
- disaggregation
- theoretical soundness
- empirical grounding

Ideally it would seem desirable for an analysis to have all of these characteristics. It should contain a complete representation of all potentially significant factors, while yet being easy to use and simple to understand. It should have encompassing scope as to the number and type of different factors considered (i.e. there should be very few, if any, ceteris paribus assumptions), be highly disaggregated, have a complete intuitively and theoretically satisfying justification for all important structures, and have extensive empirical grounding establishing the value of each and every parameter.

Unfortunately the simultaneous attainment of all of these very desirable characteristics is not possible. It is not possible for an analysis both to completely encompass all potentially significant factors and also simultaneously be capable of being easily understood. But if an analysis is too complex for a policy maker to understand, he more likely than not will never put it to use, or if in lieu of other acceptable options he does attempt to put it to use, he is certain to apply the results incorrectly or in inappropriate situations. If on the other hand an analyis is simple enough to be easily understood, it stands the danger of being simply invalid, or if not invalid it at least will rarely be accorded adequate credibility by people who see factors missing from consideration which they personally consider to be important. Should we attempt to formulate policy analysis models that are so complete and comprehensive that no one can complain that something of particular interest to him has been left out, but which then are so complex that no one can possibly understand or appropriately use them? Or should we attempt to formulate models that are simple enough that they can be readily understood, but which then are open to criticism from those who are disturbed when they see factors of personal interest which have been left out?

It also is impossible for an analysis to have both a very extensive scope and a high degree of disaggregation. The number and the complexity of relationships to be considered quickly outruns all bounds of budget, time, and of human comprehension. Thus a choice must always be made. Is it better to have more extensive scope and less disaggregation? Or is it better to have less extensive scope and more disaggregation?

Another crucial problem is that we never have all the data we would like. There are often many factors that intuitively or theoretically would seem to be very important, but for which we have no, or little, or only very poor data. Should we include in our analysis only those factors for which we have good data, even at the expense of ignoring factors which intuitively or theoretically would seem to be very important? Or should we include all factors that intuitively or theoretically would seem to be the most important, even those for which we have little data?

These are the quandaries of modeling for policy analysis that we must somehow face up to and begin to deal with. Should we prefer completeness to simplicity, or simplicity to completeness? Should we prefer more disaggregation to larger scope, or larger scope to more disaggregation? Should we prefer complete empirical grounding to theoretical soundness, or theoretical soundness to complete empirical grounding?

3. EXPLANATORY POWER VERSUS PREDICTIVE ACCURACY

If our principal purpose is that of obtaining extremely accurate results, then we must have a very detailed and complex analysis, because in order to obtain this accuracy we must account not only for all the major influences but also for the minor ones as well. If on the other hand our principal purpose is that of achieving better insight and a better understanding of the problem, then a simpler analysis is preferable because the inclusion of minor influences serves only to confuse and detract our attention from those factors that are truly essential. There is thus a major tradeoff involved. A simpler model, if not so simple as to be simply false, has greater explanatory power but less predictive accuracy. A more complicated model, if the complexities are not merely superfluous, may have greater predictive accuracy but must always have significantly less explanatory power. Which is more important: explanatory power or predictive accuracy?

The answer depends on the problem that one is attempting to solve. For policy-makers, there seems to be no question but that they much prefer explanatory power (ease of comprehension) over miniscule precision. Arthur Okun has explained this with particular clarity:

> In public policy formulation, best estimate forecasting isn't the sole purpose of the quantified formalized approach. For one thing, the ability to make better policy is more heavily dependent on better hurricane warnings than on more accurate down-the-middle forecasts. The policymaker has a fairly flat utility function on whether GNP for this year is 1,140 or 1,145 billion dollars. What he really needs is to know that it is not going to be 1,135 or 1,155 billion dollars. He needs to appraise the possibility of major surprises.[4]

It also however is important to realize that it is by no means guaranteed that a larger model will necessarily produce more accurate results. Somewhat surprisingly, it is often more likely that a very large model will neglect to include relationships of crucial significance than is the case for a model of more moderate size. This is due to the fact that, with an extremely large model, one can easily lose the overview and hence the sense of proportion as to which relationships are of major significance and which are not. One can become so immeshed in modeling relationships of minor significance that he will overlook certain others that are crucial.

Thus it would appear that at least in the early stages of an analysis a model should be kept as small as possible and the emphasis placed more on obtaining greater explanatory power rather than on striving for extreme numerical accuracy. This however is more easily said than done, because keeping a model small requires that one make many simplifying assumptions. But if the simplifying assumptions made are the wrong ones, then the model's results, rather than being subject merely to small quantitative inaccuracies, may be degraded to the point of being qualitatively false and misleading. One must thus have some way of deciding whether a given simplifying assumption will lead to only minor inaccuracies or whether it will cause major error.

4. JUDGING THE VALIDITY OF SIMPLIFYING ASSUMPTIONS

There are two levels at which the validity of simplifying assumptions is to be judged: (1) that relating to the accuracy of individual equations, and (2) that relating to the validity of total system structure. The validity problems relating to the accuracy of individual equations are better understood and have been much more thoroughly researched than have the problems relating to the validity of total system structure. Also inaccuracies in the specification of individual equations are considerably less likely to lead to major qualitative errors in the model's results than are inadequacies in the representation of overall system structure. We thus will here concentrate our attention on the more difficult problem of validity standards at the level of overall system structure, and save discussion of the simpler and more manageable problems relating to the representational accuracy of individual equations for another place.

Relative to overall system structure there again are two basic dimensions to be considered: one being that of scope and the other being that of disaggregation.

4.1 The Scope Dimension of Structural Validity

Scope relates to the extent (or encompassment) of system boundaries--to the distinguishing of which variables and structures are to be considered as internal to the system from those which may be considered as belonging to the system environment, and again from those which need not be considered as being relevant to the analysis at all. The more factors encompassed by an analysis, the greater its scope. For instance, an inflation analysis that considers both the labor-cost determinants of inflation and the money supply determinants of inflation has greater scope than an analysis that considers only the labor cost determinants or one that considers only the money supply determinants. Similarly, an employment analysis that includes both labor supply factors and labor demand factors has greater scope than an employment analysis that includes only supply factors or one that includes only demand factors. The number of variables included however is not the only or even the most important determinant of the extent of system scope. The most important determinant of scope is that of the number and complexity of relationships between and among the various system variables. Thus an analysis that included the feedback (reverse causality) effects of current labor market conditions back on future levels of labor supply and labor demand would have greater scope than an analysis that included only the one-way relationships.

Few people realize how crucial to the success of an analysis it is just how one differentiates between what is system and what is environment. But this is the single most important decision that an analyst has to make. Ths choice is not, as many seem to think, arbitrary. If one inappropriately considers a variable to be part of the environment (i.e. exogenous) when it should have been considered to be part of the system (i.e. endogenous), then he may end up with completely erroneous results. On the other hand, if he considers a variable to part of the system, which should best have been considered to be part of the environment, then he makes the analysis needlessly complicated--thereby both reducing its understandability and usefulness, and increasing its cost.

But how is this decision (as to what to include and what not to include) to be made? Rules for the proper determination of system boundaries are presented in Henize [4] and thus need not be repeated here. A few comments however should be made relative to the general problems involved. First of all, questions relating to system boundary can be adequately answered only if we are clear as to why we want to carry out the analysis in the first place. If our only interest is that of academic curiosity, then there simply are no validity criteria--other than the purely subjsctive ones of what seems currently to best stimulate our own personal amusement. Meaningful validity criteria can be found only if our interest goes beyond mere academic curiosity to that of wanting to help someone to deal better with some real problem. Now the first task in dealing better with any problem is that of correct diagnosis. If the causes of the problem are incorrectly diagnosed, then any remedies proposed are unlikely to be effective, and may even be harmful. A good diagnosis is nothing more or less than a good explanation of the causes of the problem, and a good explanation always involves a selection of the factors that in this case are viewed as being relevant.[5] By thus selecting, it also simplifies by implicitly telling us what factors we may ignore as being irrelevant. Simplification is a necessary element of all understanding and therefore of all diagnosis. Once a good understanding (explanation, diagnosis) of the problem has been achieved, then and only then will it be possible to develop proper remedies for dealing with that problem.

A good explanation requires that the causes of system behavior lie within and not outside the analytical boundaries of the system. Otherwise our analysis says nothing about what has caused this behavior and therefore nothing about how such behavior may be modified or controlled. Now, the only type of system that is capable of self-created behavior is one containing mutual interdependencies or feedback. No structure in which all influence relationships are merely one-way, in which there is no mutual interdependency, can possibly exhibit any behavior not already to be found in the inputs to that system. Without the explicit representation of feedback, there can be no proper diagnosis and no valid evaluation of policy.

This would seem to be so obvious that the reader may wonder why it is even worth saying. The reason it is worth saying is that at least 95 percent of all socio-economic models currently in existence or under construction do not satisfy this most fundamental and most obvious of all possible validity criteria. The behavior of these models is determined principally by exogenous input variables and not by structures that are endogenous to the models themselves.

But why is it that the builders of such models have neglected to include the feedback relationships required to satisfy this all-important endogenization requirement? There are two principal reasons why this has been so. First, many analysts have so occupied themselves with the modeling of non-essential details that they have either overlooked or have simply not found time to include the feedback relationships that really are crucial--or putting the matter in more colloquial terms, they have not been able to see the forest for the trees. But secondly, even when they have been aware of the important feedback structures that ought to have been included, they frequently have deliberately neglected to explicitly include them--because of the immense difficulties encountered in getting these beastly things to behave in any reasonable fashion.

Constructing models that contain feedback is no easy task. Precisely because they do generate their own behavior, feedback structures have the unfortunate habit of often developing a mind of their own and of not behaving at all in the way the model builder would like them to. Without a thorough grounding and long and arduous training in the analysis of feedback structures, and without considerable practical experience with the construction and use of such, one simply cannot construct feedback models that will generate even plausible, much less accurate, behavior. It is for this reason for instance that most large econometric models, if simulated over more than say 8 to 10 periods, begin to exhibit severe stability problems. When one observes that there is no econometric curriculum in the world today that provides adequate training in feedback analysis, it is not difficult to understand why. It therefore should come as no surprise to discover that most such models are inadequate.

4.2 The Disaggregation Dimension of Structural Validity

In addition to the problem of scope there is the problem of determining the proper level of disaggregation to be applied. Disaggregation relates to the separate consideration of different groups of a particular type of entity, whereby each group differs from each of the others in terms of one or more characteristics of special interest to the particular analysis in question. The entities to be disaggregated may be industries, geopolitical regions, energy sources, types of energy consumption, occupations, households, or in general populations of any sort (physical, biological, or social).

In general (from the point of view of economy and understandability--i.e. of maximal simplicity and of minimal cost), the rule is that one should not disaggregate more than is absolutely necessary. But how much is absolutely necessary? The answer depends on how important changes in composition are in terms of their effects on model results. Keynes for instance has argued in his

Treatise on Money that in order to properly explain changes in the velocity of circulation it is necessary to individually account for the separate sectoral components of transactions demand rather than having aggregate transactions demand depend simply on aggregate GNP. This is because, Keynes argues,

> the observed facts do not necessarily require for their explanation a substantial fluctuation in true velocities--for they can also be explained by fluctuations in the proportions of deposits employed for different uses. Thus the observed increase during slumps may be regarded as due to an increase of transactions relating to business deposits B, and part as a corroboration of our previous expectation that the value of transactions relating to the industrial circulation will increase relatively to income deposits during profit periods and decrease relatively during loss periods.[6]

In this case, we see that failing to separately account for each of these two types of demand deposits would lead to significant error. But there are many other cases where the effects of composition changes may not be nearly so significant as one might initially expect.

How important it is whether or not changes in composition are explicitly accounted for depends on the nature of the policies that a model is supposed to be able to evaluate. If the policies to be evaluated are ones that cause major changes in the composition of the population in question, and the resulting changes in composition are ones that will have a major impact on the outputs of principal concern, then these changes in composition must indeed be explicitly accounted for. If, on the other hand, any of the above conditions do not hold, then one may validly forego explicit consideration of composition effects without fear of causing major errors in model results.

For instance, the fact that the economy is composed of many different industries that differ from one another in the nature of their production technology and amount and type of factor usage does not in and of itself require that all of these differences be explicitly accounted for, unless the composition of the economy (the percentage contribution of each different sector to total output) undergoes major changes over the period over which the effects of various policies are to be evaluated. If the composition of the economy does not undergo major changes, then aggregate analysis based on the average value across all industries will provide quite adequate results. In addition, even if the composition of the economy does undergo significant changes, aggregate analysis may still be adequate, as long as the change in composition is relatively constant--that is, if every policy simulation of the model assumes the same change in industrial composition to have occurred. If this is the case, then the average value across all industries can simply be adjusted to include the average composition effect, and an aggregate analysis will still provide quite adequate results. It is thus only for the case where different policy simulations of the model lead to significantly different types of industry composition where a more aggregate analysis would end up producing inadequate results.

Why is this an important consideration? Why not simply disaggregate to be sure that changes in composition are adequately accounted for, independently of whether their influence is great or only minor? The answer is that to explicitly account for changes in composition is an extremely time consuming and expensive task, and if one does this, he rarely has time or money enough remaining to account for other factors that may be much more important. Thus the basic rule: Never disaggregate more than is absolutely necessary for the purposes of the specific policy evaluation in question.

5. CONCLUDING REMARKS

There is not space here to argue the other remaining issue of theoretical soundness versus empirical grounding. But if one accepts explanatory power as the principal goal, then it is not difficult to see the absolute necessity of adequate scope and of theoretical soundness. Disaggregation can be desirable and useful and may sometimes be necessary, but if carried out at the expense of inadequate scope, then it is surely harmful. Empirical grounding is always important, but we must never use the problems of inadequate data as an excuse for unrealistic or inappropriate structures, i.e. for bad theory.

Most previous socio-economic modeling efforts have tended to emphasize prediction over explanation, complexity over simplicity, disaggregation over adequacy of scope, and empirical grounding over theoretical soundness. I think that these priorities have been false. We must first emphasize explanation, if we are to have any hope of eventually obtaining better prediction. We must obtain adequacy of scope before disaggregation can become truly meaningful. Empirical grounding is useful only when applied to structures that are theoretically sound. We thus must begin I think to change our priorities. We must begin to emphasize explanation over prediction, simplicity over complexity, adequacy of scope over disaggregation, and theoretical soundness over empirical gounding.

The problems still to be solved obviously are great. Yet I think the problems that have retarded the successful construction and application of socio-economic policy models can be solved, if we can admit our problems and begin to face up to them. If we as model builders want to create the trust and confidence necessary for a successful cooperation with policy makers, then we must begin to deserve it with better critiques fo our own and of other people's models. Criticism is the wellspring not only of democracy but of science as well. Without criticism there can be no improvement. If we can begin to be more critical of ourselves, then I think that we will be able to solve the important problems remaining and to finally begin to fulfill the great promise that models, if they are good ones, should be able to offer policy makers in helping them to better understand and to deal with the most important present and future problems of our society.

NOTES

1. As cited in Cameron [1] pp. 104-105.
2. Hall [2] p. 33.
3. See also Henize [3].
4. Okun [6] p. 360.
5. There are two basic dimensions (in addition to the logical ones of consistancy, etc.) along which model validity is to be evaluated. One is that relating to whether or not the variables and relationships contained within a model correspond in some meaningful and accurate way with what actually exists in reality. The significance of this dimension is obvious. But there is also a second dimension which many tend to forget, or at least not to adequately appreciate the significance of. This second dimension is that of whether the variables and relationships the model contains, even if accurate representations of certain given aspects of reality, are in fact the ones and the only ones that should have been included--i.e. whether or not the model fails to contain variables and relationships that should have been included, and whether or not it contains unnecessary variables and relationships that should best have been left out. It is principally with regard to the second of these two validity dimensions that most modeling efforts have come to grief. In fact there seem to be surprisingly few modelers who even recognize it as being an important problem. Indeed, as long as one stays in the realm of mere description it is not a problem. The question does not even arise. It is only after one advances from the realm of mere description to the realm of explanation that this second validity dimension becomes meaningful and indeed crucial.
6. Keynes [5] pp. 37-38.

REFERENCES

1. Cameron, Juan: The Economic Modelers Vie for Washington's Ear, Fortune (November 20, 1978) pp. 102-105.

2. Hall, William K.: Strategic planning models: are top managers really finding them useful?, Journal of Business Policy, Vol. 3, No. 2 (1973) pp. 33-42.

3. Henize, John: Toward a Better Understanding of Social Systems, Proceedings of the IEEE, Special Issue on Social Systems Engineering, Vol. 63, No. 3 (March 1975) pp. 380-390.

4. ______: Modeling Large-Scale Socio-Economic Systems: Can we begin to do better?, Large Scale Systems 1 (1980).

5. Keynes, John Maynard: A Treatise on Money, in Collected Writings Vol. VI, Macmillan, London, 1971.

6. Okun, Arthur: Uses of models for policy formation, in G. Fromm and L. Klein, Eds., The Brookings Model: Perspectives and Recent Developments, North-Holland/Elsevier, Amsterdam, 1975, pp. 361-373.

SECTION VI

MULTIDISCIPLINARY APPLICATIONS

OF

INFORMATION SYSTEMS

STABILITY ANALYSIS FOR SOME ALGORITHMS

Marie COTTRELL, Jean-Claude FORT, and Gérard MALGOUYRES

Université PARIS XI, ERA 532, ORSAY - FRANCE

From large deviations to the law of large numbers we get a fast simulation method of the importance sampling's type and in the same way an approximation diffusion formula. We apply these two results for studying the stability of ALOHA system in two cases.

I - INTRODUCTION

In this note, we are specifically interested in fiability problems in multi-access channel shared by a large number of users. First, we introduce discrete time independent increments processes modelizing ALOHA systems. Asymptotically they can be viewed as small perturbations of an ordinary differential equation, with non zero variance at the equilibrium point. We study the average exit time from a "safe area" around this stable equilibrium point in the unstable case, the queues of the law of the time before a successful transmission in the stable one. Precisely, we present two results

1) An efficient method of simulation (of the Importance Sampling's type),
2) An actual calculation of the mean exit time by a diffusion approximation of a transformed process.

II - ALOHA

ALOHA systems are multi-access channel sharing algorithms for satellite-type transmission, which is to say that the length of information backand forth trip from the user to the terminal is large compared to the transmitted information packages length (See [1]).

a. We are interested in discrete time algorithms messages with

D. G. Lainiotis and N. S. Tzannes (eds.), Applications of Information and Control Systems, 387-394.

duration equal to the discretization. In this case (several simultaneous emitted messages which are lost) terminals involved toss heads or tails at each moment, with a fixed bias p, to know they are going to retransmit or not. We can define the process $\mathcal{N}(t)$= the number of blocked terminals at time t. Assuming the source is a Poisson process with parameter β, $\mathcal{N}(t)$ is a transient Markov chain over N whose transition probabilities are:
$\text{Prob}(\mathcal{N}(t+1)=j/\mathcal{N}(t)=i)=P_{ij}=0$ if $j\le i-2$, $ip\,(1-p)^{i-1}e^{-\beta}$ if $j=i-1$, $(1-p)^i\beta e^{-\beta}+(1-ip(1-p)^{i-1})e^{-\beta}$ if $j=i$, $\beta e^{-\beta}(1-(1-p)^i)$ if $j=i+1$, $\beta^{j-i}\,e^{-\beta}/(j-i)!$ if $j\ge i+2$. Let μ_i the distribution of the jump at i, and $f(i)=E(\mu_i)$. The deterministic mean process $\mathcal{N}(t)$ has an asymptotically stable equilibrium point n_o, and unstable equilibrium point $n_c>n_o$. A trajectory ω of $\mathcal{N}(t)$ takes a very long time before reaching n_c (time of saturation). We are interested in the mean time before the process will pass through a point n_b from which the probability of going further to n_c is sufficiently large. We will exam two cases : $n_b<n_c$ far away from n_o and n_c, and $n_b=n_c$.

b. Instead of a fixed p retransmission probability we can choose p depending on the number n of blocked terminals. Thus the system becomes stable for a good choice of p(n) (See [2]). For instance, we can take $p(t) = (1-\beta)/(n(t)-\beta)$. Then $\mathcal{N}(t)$ is a positive recurrent Markov chain if and only if $\beta<1/e$. If we consider a message getting in the system when the number of blocked is N_o, we evalue the probability $P(N_o,T)$ that this message will have to wait a time $>T$, before a successful transmission. This probability is a good evaluation test of the stability of the system and is directly related to a probability of buffer memory overflows. For this "marked" message, we define a new process $\mathcal{N}(t)$ which is the number of blocked terminals at instant t, if at this instant the marked message is still in the system. There exists now a probability of death = $e^{-\beta}(p(i))(1-p(i))^{i-1}\neq 0$ (if $\mathcal{N}(t)=i$). We will call ν_i the jump distribution at i of the process $\mathcal{N}(t)$ ($||\nu_i||<1$).

III - LAW OF LARGE NUMBERS - LARGE DEVIATIONS

For studying the process $\mathcal{N}(t)$, the important remark is that the size of the process's jumps (the standard deviation of μ_i) is small compared to the characteristic dimensions of the evolution of μ_i as a function of i (for example n_c-n_o). It is therefore legitimate not to consider this chain alone, but as being part of a fixed - drift family with homothetically smaller and smaller jumps which can be studied asymptotically.
To realize this family, time and space must be made continuous. Let us introduce a parameter $0<\varepsilon\le 1$, the process $X^\varepsilon(t)$ jumps at moment εN, with the transition probabilities

$$P^\varepsilon_{x,x+\varepsilon j} = P_{x,x+j} \text{ (interpolate between the } P_{i,i+j}$$

We will now give results valid for $\varepsilon \to 0$; according to the initial remark, they give good indications on the case $\varepsilon = 1$.

<u>Law of large numbers</u> : Let be T fixed, $X^{\varepsilon}(0)=x_o$ fixed; the trajectories $X^{\varepsilon}(t)$, $0 \le t \le T$, converge uniformly, with a probability converging on 1, on the trajectory starting from x_o of the differential equation (deterministic) $dZ/dt=f(Z)$ (where $f(x)=E\mu_x=\Sigma jP_{x,x+j}$). Specifically, for every neighbourhood V of n_o, there exists an S so that for every T>s, the trajectories $X^{\varepsilon}(t)$, $S \le t \le T$, remain inside V, with a probability converging on 1 when ε converges on 0.
<u>Central limit theorem</u> (See [3]) : Let be T fixed, $X^{\varepsilon}(0) = n_o$ for simplification, the process $(1/\sqrt{\varepsilon})(X^{\varepsilon}(t)-n_o)$, $0 \le t \le T$, converges in distribution, when $\varepsilon \to 0$, on the Ornstein-Uhlenbeck process $d\xi = -b\xi dt + \sigma dw$ where $-b=f'(0)$, σ = standard deviation of the law μ_{n_o}.

<u>Cramer's theorem (Chernoff). Action Integrals</u>. Let δ be fixed small, T fixed and $\phi \in C^1([0,T], \mathbb{R})$ fixed. Let R^{ε} the probability of the set of trajectories starting from $\phi(0)$ so that $\forall t, 0 \le t \le T, |X^{\varepsilon}(t)-\phi(t)| \le \delta$
We have $-\varepsilon \log R^{\varepsilon} \xrightarrow[\varepsilon \to 0]{} \int_0^T h_{\phi(t)}(\phi'(t))dt = I(\phi)$ where h_x is the Cramer transformed of μ_x, defined by $h_x(v)=\sup_{\alpha}(\alpha v - \log\hat{\mu}_x(\alpha))$, ($\hat{\mu}_x$ is the Laplace transformed of α_x). <u>$I(\phi)$ is said to be the action integral along ϕ.</u>

<u>Corollary</u> : Let $n_o < y < x_o < z \le n_c$ and let R^{ε} be the probability of the set of trajectories starting from x_o, and exiting from [y,z] through z. We have when $\varepsilon \to 0$: $-\varepsilon \operatorname{Log} R^{\varepsilon} \to \inf\{I(\phi)/\phi(0)=x_o, \phi(T)=z\}$
It can be shown that the infimum is reached when ϕ is a solution ϕ_{opt} of the differential equation.

$$d\phi/dt=\hat{\mu}'_x(\lambda(x))=\chi(x) \text{ and that } I(\phi_{opt})=\int_{x_o}^{z} \lambda(x)dx$$

where $\lambda(x)$ is solution of $\hat{\mu}_x(\lambda(x))=1$
And therefore we have as $\varepsilon \to 0$, $-\varepsilon \operatorname{Log} R^{\varepsilon} \to \int_{x_o}^{z} \lambda(x)dx$

<u>REMARQUE</u> : The over-bounding of R^{ε} is well know (Chernoff's inequality), the minoration can be obtained by changing the P^{ε} - Probabilities into $\tilde{P}^{\varepsilon}$ obtained from the transition probabilities (recalling $\hat{\mu}_x(\lambda(x))=1$) by $\tilde{P}^{\varepsilon}_{x,y}=\exp(\lambda(x)(y-x))P^{\varepsilon}_{x,y}$ ($y \in x+\varepsilon N$).
If so the law of large numbers states that R^{ε} is almost equal to 1, and one can easily bound, over the set of trajectories which interests us, the Radon-Nikodym derivate $dP^{\varepsilon}/d\tilde{P}^{\varepsilon}$. If a $\tilde{P}^{\varepsilon}$ typical trajectory ω is such that for Θ small, but large compared to ε, on each time interval $[j\Theta,(j+1)\Theta]$ the empirical jumps distribution of ω is close to $\tilde{P}^{\varepsilon}_{\omega(j\Theta),\cdot}$, going back to the corollary, we can deduce from this, that the "typical" trajectories of the $\tilde{P}^{\varepsilon}$ process constitute a subset of the set of trajectories exiting through z,

and whose P^{ε} probability is almost equal to R^{ε}, as $\varepsilon\to 0$, $\Theta\to 0$ and $\Theta/\varepsilon \to \infty$.

IV - APPLICATION TO A FAST SIMULATION METHOD

a. Unstable ALOHA system

Assume we have got $\hat{p}$, estimator of the P-probability p for a trajectory starting from n_o to pass through n_b before passing again through n_o, and $\hat{t}$ an estimator of the mean time to return to n_o. Then (neglecting the duration of the last part of the trajectory), $\hat{\tau}=\hat{t}/\hat{p}$ is an estimator of τ, expectation of the first time of passage through n_b. We can obtain $\hat{t}$ by a direct simulation with the Pij transition probabilities, but to estimate $\hat{p}$ in such a way need a very large computation time. The last remark of § III leads us to the following conclusion : exchanging the P^{ε}-probability into $\tilde{P}^{\varepsilon}$-probability ensure that all the trajectories issued from n_o will reach n_b (or n_c) very quickly. Taking $\varepsilon=1$, we actuate a simulation using the $\tilde{P}_{ij}$ probabilities. Starting from n_o, every trajectory is divided into parts going back to n_o, and into a last part joining n_o to n_b without going back through n_o. Let $\omega_k(k=1,\ldots,N)$ be the simulated parts stopped at their natural terms either n_o or n_b. One can easily calculate :

$L_k = \text{Log}\,\frac{dP}{d\tilde{P}}(\omega_k) = \sum_t \lambda(\omega_k(t))(\omega_k(t+1) - \omega_k(t))$ and estimate p by :

$$\hat{p}=\frac{\Sigma \exp L_k \text{ (for the } \omega_k \text{ stopping at } n_b)}{\sum_{k=1,N} \exp L_k}$$

Thus $\hat{p}$ is obtained as the mean of an independent N sample image law of $\tilde{P}$ by $dP/d\tilde{P}$. An asymptotically evaluation of $\log\int_A [dP/d\tilde{P}(\omega)]^2 dP(\omega)$ (where $A = \{\text{traj. stopping at } n_b\}$) shows that this change of probabilities is the best among all the exponential ones given by $P^{\varepsilon,\ell}_{x,y} = e^{\ell(x)(y-x)}P^{\varepsilon}_{x,y}/\hat{\mu}_x(\ell(x))$, in that sense that it minimizes the variance of $\hat{p}$. We have, when $\varepsilon\to 0$ $\text{Log}\int_A (\frac{dP^{\varepsilon}}{dP^{\varepsilon,\ell}}(\omega))^2\, dP^{\varepsilon,\ell}(\omega)\sim$

$$\max_{\Phi} \int_{n_o}^{n_b} d\Phi \left[\frac{-h_{\Phi(t)}(\Phi'(H)+\log \hat{\mu}_{\Phi(t)}(\ell(\Phi(t))- \ell(\Phi(t))}{\Phi'(t)}\right]$$

The maximizing Φ is solution of $d\Phi/dt=\chi(\Phi)=\hat{\mu}'_{\Phi(t)}(\ell(\Phi(t))$. Finally, we must minimize in $\ell(\Phi(t))$, and we obtain that for $\ell(x)$ so that $\text{Log}\,\hat{\mu}_x(\ell(x))=0$, that is to say $\ell(x)=\lambda(x)$.

b. Stabilized ALOHA system

We define $\tau(\Phi)$ as death time on the path $\Phi(\Phi: \mathbb{R}^+\to\mathbb{R}^+$ of class C^1). $A(T,N_o)=\{\Phi/\Phi(0)=N_o, \tau(\Phi)\geq T\}$, and $P(N_o,T)$ = Proba $(A(T,N_o))$. We can make space and time continuous as in § III, and let $\eta^{\varepsilon}(t)$ be the corresponding process and $P^{\varepsilon}(N_o,T)$ the researched probability. An easy generalization of the quoted theorems shows that when $\varepsilon\to 0$, $-\varepsilon\,\text{Log}\,P^{\varepsilon}(N_o,T) \to \min\{I(N_o,T,\Phi) / \Phi\in A(T,N_o)\}$ with

$I(N_o,T,\phi)=\int_0^T h_{\Phi(u)}(\Phi'(u))du$. Let $\Psi(x,s)=\inf\{\int_{T-s}^T h_{\Phi(u)}(\Phi'(u))du/\Phi(T)=x\}$ It verifies the equation of dynamic programming. Let $\lambda=\frac{\partial\Psi}{\partial x}(x,s)$ and $L(x,\lambda)=-Log(\hat{\vartheta}_x(-\lambda))$, and we have $\frac{\partial\Psi(x,s)}{\partial s}=L(x,\lambda)$. The λ over the minimizing trajectory is interesting, for it represents the rate of exponential modification one must include in the jump distribution to make absolutely typical those trajectories only conditionnaly typical since living a time > T. The optimal $\phi(\phi_{opt})$ verifies : $\phi_{opt}(0)=N_o$ and $\frac{\partial\Psi}{\partial x}(\phi_{opt}(T),T)=0$. The optimal ϕ are the level curves (in the (x,λ) plane) of L, that is to say are solutions of $L(x,\lambda)=C$, $C\in\mathbb{R}$. Therefore $I(N_o,T,\phi_{opt})=S+CT$, with S = area of the domain included between $L(x,\lambda)=C$, the x-axis, the line $x=N_o$. The constante C depends on the time T. When $\lambda=0$, at $x=N_f$, it is the process's death. We can use these results for calculating $p(N_o,T)$ by a fast simulation, and we note that the "optimal" trajectories are those which have a constante death probability.

c. Two diffusion approximation formulas

Let $\mathcal{N}(t)$ be the Markov chain defined before. Let y,x_o as shown: $n_o<y<x_o<<n_b<n_c$. We can divide our computation into 3 parts as we did in the simulation method. The same principles are used in both calculus and simulation. First, we evaluate by diffusion approximation the $\tilde{P}$-probability $\tilde{R}$ of exiting from $[y,n_b]$ through n_b, starting from x_o. Second, we give a formula for $dP/d\tilde{P}(\omega)$ along a $\tilde{P}$-typical trajectory. Third, we use a Ornstein-Uhlenbeck approximation near n_o.

At this step we get a formula for $\tau(n_o,n_b)$, average time before passing through n_b starting from n_o, with n_b far from n_o and n_c. If this result is still good for n_b not too far from n_c, we use a "repulsion" diffusion approximation near n_c and we obtain a formula for $\tau(n_o,n_c)$. As in [4], we obtain (setting $\hat{\mu}_x(\lambda)=\hat{\mu}(\lambda,x)$)

$$\boxed{\tau(n_o,n_b)=\sqrt{\frac{\pi}{b_o}}\ \sigma_o\frac{\exp I(n_b)}{\frac{\partial\hat{\mu}}{\partial\lambda}(\lambda(n_b),n_b)}\exp H(n_b)}$$

where $b_o=-\frac{\partial\hat{\mu}}{\partial\lambda}(0,n_o)$, $\sigma_o=\frac{\partial^2\hat{\mu}}{\partial\lambda^2}(0,n_o)$, $I(n_b)=\int_{n_o}^{n_b}\lambda(x)\,dx$ and

$$H(n_b)=\int_{n_o}^{n_b}dx\left[\frac{\hat{\mu}_{\lambda,x}(\lambda,x)}{\hat{\mu}_\lambda(\lambda,x)}-\frac{1}{2}\hat{\mu}_x(\lambda,x)\left(1+\frac{\hat{\mu}_{\lambda\lambda}(\lambda,x)}{\hat{\mu}_\lambda^2(\lambda,x)}\right)\right]$$

(using the subscripts for derivation) where $\lambda=\lambda(x)$.

This formula is valid for n_b sufficiently close to n_c, so that $E(\mu_x)=f(x)\simeq f'(n_c)(x-n_c)$ for $n_b\le x\le n_c$. Let $b_c=f'(n_c)=\frac{\partial\hat{\mu}}{\partial\lambda}(0,n_c)$. We can use on (n_b,n_c) a diffusion approximation with drift $f(x)=b_c(x-n_c)$ and variance $\sigma_c^2=\frac{\partial^2\hat{\mu}}{\partial\lambda^2}(0,n_c)$. As above we get the frontier conditions for the average time $\tau(n_o,n_c)$ used to pass through n_c starting from n_o . If we get $\tau(x,n_c)$ = average time for reaching n_c starting from x, $\frac{\partial\tau(n_b,n_c)}{\partial x} = \frac{2 \exp H(n_b) b_c(n_b-n_c)}{\sigma_c^2}$ and $\frac{1}{2}\sigma_\varepsilon^2 \frac{\partial^2\tau}{\partial x^2}(x,n_c) + b_c(x-n_c)\frac{\partial\tau}{\partial x}(x,n_c) = -1$. Resolving these equations and using approximations we get, neglecting the first passage through n_b, $\tau(n_o,n_c) \simeq (\pi\sigma_o/\sigma_c\sqrt{b_o b_c}) \exp I(n_c) \exp H(n_c)$.

V - NUMERICAL RESULTS

a. We tested two methods on a very simple example where every thing can be computed with a pencil. S_t is a Markov chain with transition probability : $p_{i,i-1}=r(i)$ if $-10<i\leq n_c$, $p_{-10,-9}=1$, $p_{n_c,n_c} = 1$. We choose $r(i) = 0.5+\varepsilon i(n_c-i)$. For various values of ε, n_o, n_c we calculate $\tau(n_o,n_c)$.

a) analytically using the transition matrix; b) by diffusion formulas, c) by fast simulation.

b. For the unstable ALOHA system, we have used diffusion formulas and fast simulation.

c. For stable ALOHA system (with control) we get $P(N_o,T)$ for differents values of N_o and T, using fast simulation.

a. - Example : We fixed $n_c = 50$

n_b \ ε	10^{-4}		2.10^{-4}		3.10^{-4}		4.10^{-4}	
20	$2.8\ 10^3$	$2.8\ 10^3$	$2.\ 10^4$	$2.\ 10^4$	$2.3\ 10^5$	$2.4\ 10^5$	$3.8\ 10^6$	$4.3\ 10^6$
25	$9.1\ 10^3$	$9.3\ 10^3$	$2.3\ 10^5$	$2.4\ 10^5$	$1.\ 10^7$	$1.1\ 10^7$	$8.\ 10^9$	$9.1\ 10^9$
30	$3.1\ 10^4$	$3.3\ 10^4$	$2.9\ 10^6$	$3.\ 10^6$	$5.\ 10^9$	$5.7\ 10^9$	$1.8\ 10^{11}$	$2.1\ 10^{11}$
$n_b=n_c$	$1.4\ 10^6$	$1.3\ 10^6$	$3.7\ 10^8$	$3.5.10^8$	$1.9\ 10^{13}$	$1.8\ 10^{13}$	$2.2\ 10^{17}$	$2.1\ 10^{17}$
	Anal.	Diff.	Anal.	Diff.	Anal.	Diff.	Anal.	Diff.

b. – We taked β=0,30 and β=0,27

K is a parameter such that the probability p = 2/(K+1)

	K	n_o	n_c	$\hat{t}$	β	τ	Diff.
β=0.30	4	4	30	8.75	$1.8\,10^{-4}$	$4.8\,10^{4}$	$4.7\ 10^{4}$
	5	5	37	9.0	$6.0\,10^{-5}$	$1.5\,10^{5}$	$1.6\ 10^{5}$
	6	6	45	10.5	$1.9\,10^{-5}$	$5.5\,10^{5}$	$5.6\ 10^{5}$
	7	7	52	11.0	$6.0\,10^{-6}$	$1.8\,10^{6}$	$1.8\ 10^{6}$
	8	8	60	11.3	$1.9\,10^{-6}$	$5.9\,10^{6}$	$6.2\ 10^{6}$
	9	9	67	11.8	$6.7\,10^{-7}$	$1.8\,10^{7}$	$1.9\ 10^{7}$
	10	10	75	13.1	$2.2\,10^{-7}$	$5.9\,10^{7}$	$6.4\ 10^{7}$
β=0.27	3	2	26	5.6	$3.3\,10^{-5}$	$1.7\,10^{5}$	$1.6\ 10^{5}$
	4	3	35	7.1	$4.3\,10^{-6}$	$1.6\,10^{6}$	$1.8\ 10^{6}$
	5	4	44	8.6	$5.5\,10^{-7}$	$1.6\,10^{7}$	$1.5\ 10^{7}$
	6	4	52	8.2	$6.8\,10^{-8}$	$1.2\,10^{8}$	$1.2\ 10^{8}$
	7	5	61	9.4	$9.5\,10^{-9}$	$1.0\,10^{9}$	$1.0\ 10^{9}$
	8	6	70	10.7	$1.3\,10^{-9}$	$8.2\,10^{9}$	$8.4\ 10^{9}$

c. – Stable ALOHA

N_o : initial number in the system	Stationnary probability	$P(T\geq 40 \mid N_o)$	$P(T\geq 60 \mid N_o)$
0	0.6251	4.4 10^{-3}	1.5 10^{-3}
1	0.0311	11.1 "	4.4 "
2	0.0868	23.8 "	10.2 "
3	0.0608	44.0 "	16.2 "
4	0.0460	59.3 "	25.2 "
5	0.0348	85.8 "	39.4 "
6	0.0265	11.5 10^{-2}	54.7 "
7	0.0202	16.8 "	70.8 "
8	0.0155	19.1 "	93.2 "
9	0.0119	23.1 "	11.5 10^{-2}
10	0.0091	27.5 "	15.6 "
11	0.0070	28.4 "	17.2 "
12	0.0054	34.5 "	19.9 "
13	0.0012	37.8 "	23.8 "
14	0.0032	40.8 "	25.7 "
15	0.0025	43.7 "	30.6 "
16	0.0019	45.8 "	32.1 "
17	0.0015	48.0 "	35.3 "
18	0.0011	51.4 "	38.1 "
19	0.0009	52.5 "	40.2 "

$P(T\geq 40)=40.10^{-3} \pm 5\ \%$

$P(T\geq 60)=20.10^{-3} \pm 5\ \%$

It appears that when the process suffer a lot of small jumps, the large deviation diffusion procedure, which is markedly quicker than the direct matrix-inversion formulas, give very good numerical results. The fast simulation method should win when jumps are not so small, but nevertheless numerous (due for instance to the dimensionality of the states space).

REFERENCES

[1] N. ABRAMSON, The throughput of Packets Broadcasting Channels. TEEE Transactions on communications Vol. Com 25, n° 1, january 77.

[2] G. FAYOLLE, E. GELENBE, J. LABETOULLE, Stability and Optimal Control of the Packet Switching Broadcast Channels JACM

[3] R.Z. KHASMINSKII, On stochastic processes defined by differential equations with a small parameter. Theory of probability and its applications 1966. Vol XI n° 62, pp 211-277.

[4] G. RUGET, Some tools for studying channel shaving Algorithms Coun CISM Udine 1979 (springer).

[5] A.D. VENTSEL, Rough limit theorems on large deviations for Markov stochastic processes II. Theory of probability and its applications 21 (1976), pp 499-512.

[6] R. AZENCOTT, G. RUGET, Mélanges d'équations différentielles, grands écarts à la loi des grands nombres. Z.Wahrscheinlichkei theorie verm. Gebiete 38 (1977), pp 1-54.

ALGORITHMS FOR SYMBOLIC COMPUTATION IN CHEMISTRY: COMPUTING THE ACYCLIC POLYNOMIAL OF MOLECULAR GRAPH

B. Džonova-Jerman-Blažič, B. Mohar, N. Trinajstić*

J.Stefan Institute, E.Kardelj University of Ljubljana, Ljubljana, and *Rudjer Boskovic Institute, Zagreb, Yugoslavia

ABSTRACT. The algorithm, named Decomposition graph algorithm and its implementation in a computer program is presented. The program computes the coefficients of the acyclic polynomial, the graph and the acyclic spectra of the molecular graph. The generated results were used in the studies on the relations between molecular topology and molecular behavior of a special class of chemical species. On the basis of these studies an amount of theoretical rules were derived and collected under the name Topological Theory of Aromacity.

1. INTRODUCTION

Technique from the mathematical fields of combinatorial analysis, group theory and graph theory have made substantial contributions in solving different problems in chemistry. Since the early works of Polya [1] and Henze and Blair [2] on the enumeration of structural isomers, there has been growing community of researchers interested in developing applications of combinatorics and graph theory to problems of describing molecular structure [3] and molecular behavior [4]. Most of these application represent sophisticated adaptation of known mathematical structure to chemical systems.

The most successful and most often applicated mathematical model was taken from graph theory. The structural formulae of chemical compounds as form of representation is very close to the notion of graph. This analogy suggested a simple translation between the graph theory and constitutional chemistry. This translation produced the molecules to be represented as planar graphs, the chemical bonds as edges and the atoms as vertices and furthermore, all mathematical apparatus and proof techniques from graph theory to be used in the studies of molecular structural features. The representation of the molecules as graphs was used also, as background in setting

D. G. Lainiotis and N. S. Tzannes (eds.), Applications of Information and Control Systems, 395-403.

up of some sophisticated computer programs which deal with the representation and manipulation of chemical structures. Among these programs we should mention Dendral [5], Synchem [6], Chemics [7], etc. The essential part of these programs are some basic algorithms which performe various symbolic manipulations. The purpose of this paper is to present an algorithm called, decomposition graph algorithm, which belongs to the same class of algorithms. The algorithm, under the name procedure ACYCPOLYN represents the essential part of the computer program GRAPH. The program GRAPH was designed in order to support the studies on the relations between molecular topology and molecular behavior of a special class of chemical species. On the basis of these studies, some very interesting relations were discovered and an amount of theoretical consideration produced [25]. The results and the derived theoretical rules were collected under the name Topological Theory of Aromacity (TTA)[8]. The TTA lean on the Topological Resonance Energy (TRE) Model. The decomposition graph algorithm represents the essential part in one of the methods for evaluating the values of TRE. In the following section we give brief outline of the meaning of TRE in order to elucidate the application of the algorithm.

2. TOPOLOGICAL RESONANCE ENERGY MODEL

The representation of the molecules in the form of molecular graph and the underlying theory was used as theoretical background in developing the Topological Resonance Energy Model, which belongs to the chemistry of conjugated systems [9]. Conjugated systems are known class of chemical compounds which include conjugated hydrocarbons, heterocycles and Möbious structures [10]. The essential parameters in the TRE Model are the values of the TRE, which are used to predict the nature of various chemical systems as are: molecules, ions and radicals [11]. TRE values were applicated in the estimation of the following chemical properties [12,25,26]: reactivity, stability, aromacity. The comparative investigations [13] with the computed TRE values and the evaluated experimental results, showed a great agreement between the theoretically calculated values and experimental findings [27].

Resonance Energy (RE) is a traditional index in the Molecular Orbital Theory of Conjugated systems [14]. Resonance Energy parameter was used in the classification of the conjugated systems according to their aromacity [15]. The RE index had also a great impact on the works of several generation of organic chemists engaged in organic synthesis. RE is defined as the difference between the total pi electron energy, E_{pi}, of a given molecule and of a corresponding hypothetical reference structure:

$$RE = E_{pi} \text{ (conjugated molecule)} - E_{pi} \text{ (reference structure)} \qquad (1)$$

The reference structure is nonexisting entity and its choice is arbitrary. In the Topological Resonance Energy Model the RE value is derived from the topological characteristics of the molecular graph. The reference structure is connected only with the configuration of the molecular graph, and thus the name.

The molecular graph in TRE model is defined as follows:
A graph G (V,X) is a molecular graph, the elements of the set V are atoms and elements of the set X are chemical bonds. The molecular graph is planar graph and corresponds to pi – electron network [16] of the conjugated system, Fig. 1.

Fig. 1: structural formula of molecular graph of 3,4–dimethylenecyclobutene

The molecular graph is labelled and is represented by adjacency matrix defined as follows:

$$a_{i,j} = \begin{cases} k, & \text{if vertices i and j are incident, and the edge connecting i and j has weight k,} \\ h, & \text{if } i = j \text{ and the vertex i has a loop with weight h, 0 otherwise.} \end{cases}$$

In the TRE model conjugated systems are represented as weighted graphs. Every atom different from carbon is represented by a loop, and every edge joining a vertex with loop is weighted with $k \neq 1$.

The E_{pi} of the reference structure in the TRE Model contains by definition only the acyclic contributions of the E_{pi} of the conjugated system or expressed with the formula:

$$E_{pi}(\text{conjugated system}) = \sum_{j=1}^{N} b_j x_j\,;\quad E_{pi}(\text{reference structure}) = \sum_{j=1}^{N} b_j x_j^{ac} \qquad (2)$$

$$TRE = \sum_{j=1}^{N} b_j (x_j - x_j^{ac}) \qquad (3)$$

where x_j $(j = 1,2, \ldots N)$ are the roots of the characteristic polynomial of the molecular graph [17], b_j is the occupancy number of the j-th molecular ortital, x_j^{ac} $(j = 1,2,\ldots N)$ are the roots of the acyclic polynomial (the characteristic polynomial of the reference structure). The acyclic polynomial Q(G,x) is a polynomial obtained by considering only the linear, non-cyclic elements of the molecular graph. The computing of the roots of the characteristic polynomial of a graph can be performed by using the adjacency matrix and some standard computer program developed specially for this purpose. The computing of the roots of acyclic polynomial is conditioned with the construction of the polynomial, or in other words with the computing of the coefficients of the polynomial, as the structure of the reference molecular is not known.

3. COMPUTING THE ACYCLIC POLYNOMIAL OF MOLECULAR GRAPH

One of the methods for computing the coefficients of the acyclic polynomial is by using the Sachs theorem [18], adapted for chemical application [19].

The characteristic polynomial of an arbitrary graph by use of Sachs theorem can be expressed in the following form:

$$P(G,x) = \sum_{i=0}^{N} \sum_{L \in L_i} (-1)^{c(L)} 2^{r(L)} x^{N-i} \tag{4}$$

where c(L) denotes the total number of components in the Sachs graph L; Sachs graph of a graph G is every subgraph of the graph G which contains only disjoint cycles. Cycles with length 1 are loops, with lenth 2 are edges, cycles with lenth more than 2 have the usual notion, [21], r(L) denotes the total number of cycles with length 3 or more in the graph L and L_i is the set of all possible Sachs graphs of the graph G with i number of vertices. The acyclic polynomial by definition, contains only acyclic contribution of the molecular graph, by omitting the cyclic contributions in the expression (4) the following form is obtained:

$$Q(G,x) = \sum_{i=0}^{N} \sum_{\substack{L \in L_i \\ r(L)=0}} (-1)^{c(L)} x^{N-i} \tag{5}$$

The expression (5) enables the computation of the acyclic polynomial by counting all possible Sachs graphs of the graph G. The construction of the acyclic polynomial by using (the Sachs theorem) the expression (5) is time consuming and very unpractical, even by aid of computer. The number of Sachs graphs increases very quickly with the size of the system. Therefore, instead of counting the Sachs graphs, more efficient algorithm and suitable for computer implementation algorithm was developed. The algorithm was constructed by use of two recursive relations, the derivation of which was based on the results obtained by Heilbronner [21], Hosoya [22] and Lovasz and Pelikan [23]. The relations are:

$$Q(G,x) = Q(G - e,x) - w_{i,j}Q(G - i - j,x) \tag{6}$$

were e is an edge in the graph G which joins the vertices i and j, G − e is subraph of G, obtained by omitting the edge e,G − i − j is subgraph of G, obtained by deleting the vertices i and j and all incident edges, $w_{i,j}$ is defined as follows:

$$w_{i,j} = \begin{cases} k^2, & \text{if } i \neq j \\ h, & \text{if } i = j \end{cases} \tag{7}$$

The other recursive relation follows as consequence of the expression (6):

$$Q(G,x) = xQ(G - i) - w_{i,j}Q(G - i - j,x) \tag{8}$$

The notion of the terms in (8) are the same as in (6).

Let us, as an illustration to derive the Q(G,x) by applying the recursive relations (6) and (8) on the following example:

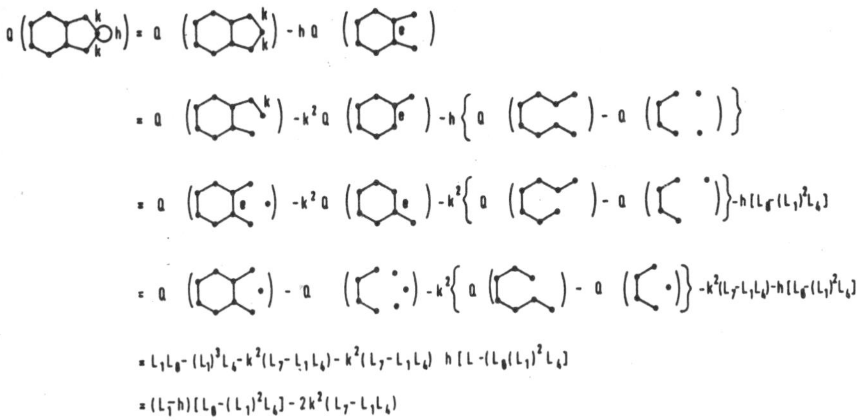

Fig. 2: Derivation of Q(G,x) for isobenzofuran–like systems

L_N denotes the characteristic polynomial of a chain with N vertices.

4. DECOMPOSITION GRAPH ALGORITHM

The decomposition graph algorithm was implemented in the computer program GRAPH, as procedure ACYCPOLYNOMIAL. The program GRAPH computes the coefficients of the acyclic polynomial and the acyclic and characteristic spectrum of an arbitrary graph. The program is written in the programming languages PASCAL and FORTRAN and is running on the CDC CYBER 72 computer. The program is consisted of the following six modules: INIT, READDATA, EIGENVAL, ACYCPOLYN, FINDROOTS, WR RESULTS.
The input data have the following form:

{ identification of the molecule}
{ n} N
{ edge$_1$ }/{ edge$_2$ }/ .... /{ edge$_k$ } { E

Identification of the molecule is an arbitrary sequence of characters, usually the name of the molecule

{ n } denotes the number of vertices,

edges are represent in the following form:

{ i j , weight } ,

i and j are vertices connected with observed edge and weight is the weight $w_{1,2}$. Representation of the loops is the same. Vertices of the graph are labelled as usual from 1 to n.

The program generates output file with following information:
– identification of the molecule
– number of vertices,
– list of all vertices, edges and weights,
– coefficients of acyclic polynomial,
– acyclic and characteristic spectrum of the graph listed in increasing order.
In the following we give brief description of the functions, which particular modul performe:
INIT: as the name shows, the modul INIT initializes the starting conditions
READDATA: reads data from the input file and stores them in adequate form (for examples the weights of the edges are translated and stored in the form of square matrix).

```
Procedure ACYCPOLYN var G: graph, var Q: polynomial

      G1, G2: graph
      Q1, Q2: polynomial
      i, j vertices

begin
      if connected (G) then
         if chain (G) then
            CHAINPOLYNOMIAL (G,Q)
         else
               chooseedge (G,i,j)
               G1 ← G
               deleteedge (G1,i,j)
               ACYCPOLYN (G1,Q1)
               G2 ← G
               deletevertices (G2,i,j)
               ACYCPOLYN (G2,Q2)
               Q ← Q1 − w_i,j Q2
         end if
      else
         G1 ← G
         Q1 ← 1
         repeat
               G2 ← component (G1)
               ACYCPOLYN (G2, Q2)
               G1 ← G1 − G2
               Q1 ← Q1 Q2
         until G1 = empty graph
         Q ← Q1
      end if
end
```

Fig. 3: Graph decomposition algorithm

EIGENVAL: computes the eigenvalues of the adjacency matrix of the molecular graph. The computation is performed by use of the Hausholder transformation and QR method for symmetric matrix.
ACYCPOLYN: is the computer implementation of the decomposition graph algorithm and is presented on the Fig. 3. This modul generates the coefficients of the acyclic polynomial.
FINDROOTS: computes the roots of the acyclic polynomial using standard computer programs (NAG, Library, [28]).
WR RESULTS: generates on the output file information about the results of computation.

During the execution of the procedure ACYCPOLYN, two new subgraphs (G_1, G_2) are generated, which are stored and decomposed in the second execution. The procedure run until the graphs with structure of chains and cycles are obtained. In that point the procedure CHAINPOLYNOMIAL is called and the coefficient of the acyclic polynomial computed.

We briefly describe the procedure CHAINPOLYNOMIAL represented on Fig.4. After the application of the expression (6) on graphs with structure of chain follow the relations:

$$Q(L_N,x) = xQ(L_{N-1},x) - h_n Q(L_{N-1},x) - k_{N-1}^2 Q(L_{N-2},x) \quad (9)$$

$$Q(L_N,x) = (x - h_N) Q(L_{N-1},x) - k_{N-1}^2 Q(L_{N-2},x)\,; \quad n = 2,3, \ldots, N \quad (10)$$

The boundary conditions are:

$$Q(L_1,x) = x - h_1\,;\; Q(L_o,x) = 1 \quad (11)$$

The recursive relations (9) and (10) represent the main part of the procedure CHAINPOLYNOMIAL.

```
Procedure CHAINPOLYNOMIAL
(p, q, r : polynomials)
    p ← 1.0
    q ← x − h
for: 2 to n do
    begin
        r ← q
        q ← (x − h_i) q − k_{i−1}^2 p
        p ← r
    end
acycpolynomial ← q
```

Fig. 4: Algorithm for computing acyclic polynomials of cycles and chains

Cycles with N vertices, have the same structure as chain with one exception, vertices 1 and N are connected. After ommitting the edge joining the vertices 1 and N; and after application of the expression (6) follows the relation:

$$Q(C_N,x) = Q(L_N,x) - k_N^2 \cdot L_{N-2},x) \; ; \qquad (12)$$

where $Q(L_N,x)$ is a polynomial of chain with N vertices and $Q(L_{N-2},x)$ is a polynomial of chain without the vertices 1 and N. Both terms can be computed by the procedure Chainpolynomial.

5. CONCLUSION

The acyclic and characteristic polynomials of the majority of different chemical species as are represented in the works of Ilic and Trinajstic [24], [13], were computed easy and considerabily fast. After four or six successive heuristically gouverned decomposition steps, in spite of the exponential complexity of the algorithm (at least $2^{N/2}$), chains and symple cycles were produced and the time needed for computation was not critical. The heuristic was used in the form of following rule: "delete the edge joining the vertex with the highest degree". The computational complexity was decreased by dividing the generation of the polynomial into two parts, one of which was with linear dependency:
– decomposition of the graph into chains and cycles,
– computing the polynomial of chains (linear complexity).

REFERENCES

1. Polya, G.: 1936, Helvetica Chim. Acta 19, 22.
2. Henze, H.R., Blair, C.M.: 1931, JACS 53, 3042.
3. Nourse, J.G.: 1975, ibid, 97, 4594.
4. Wipke, W.T., Dyott, T.M.: 1974, ibid, 96, 4834.
5. Lederbert , J. et al.: 1969, ibid, 91, 2973.
6. Masinter, L.M., Sridharan, N.S., Lederberg J., Smith, D.H.: 1974, ibid, 96, 7702.
7. Kudo, Y., Sasaki, S.I.: 1976, J.Chem.Inf.Comp.Sc., 16, 43.
8. Gutman, I., Milun, M., Trinajstić, N.: 1977, JACS, 99, 1692.
9. Garratt, P.J.: 1971, "Aromacity", McGraw-Hill, New York, N.Y., 118.
10. Lewis, D., Peters, D.: 1975, "Facts and Theories of Aromacity", Mac Millan, London.
11. Trinajstić, N.: 1977, (Int.J.Quantum Chem.Symp. S11) 469.
12. Milun, M., Trinajstić, N.: 1977, CCA, 49, 101.
13. Ilić, P., Jerman-Blažič-Džonova, B., Mohar, B., Trinajstić, N.: 1979, CCA, 52, 35.
14. Dewar, M.J.S.: 1969, "The molecular Orbital Theory of Organic Chemistry", McGraw Hill, New York, 73.
15. Hess, B.A., Schaad, L.J.: 1971, JACS, 93, 305.
16. Trinajstić,N.,1977,CCA, 49, 593.
17. Gutman, I., Milun, M., Trinajstić, N.: 1976, CCA, 48, 87.
18. Sachs, A.: 1964, Publ. Math. (Debrecen), 11, 119.

19. Graovac, A., Gutman, I., Trinajstić, N., Živković, T.: 1972, Theoret.Chim. Acta, 26, 67.
20. Balaban, A.T.: 1976, Chemical Applications of Graph Theory, Academic Press, London, 230.
21. Heilbronner, E.: 1954, Helv. Chim. Acta, 36, 170.
22. Hosoya, H.: 1971, Bull.Chem.Soc. Japan, 44, 2332.
23. Lovasz, L., Pelikan, J.: 1973, Period Math. Hung., 3, 175.
24. Ilić, P., Trinajstić, N.: work in preparation.
25. Gutman, I., Milun, M., Trinajstić, N.: 1975, Mathematical Chemistry, 1, 171.
26. Aihara, J.-i: 1978, Bull,Chem. Japan, 51, 1788.
27. Hess, B.A., Jr., Schaad, L.J., and Agranat, I.: 1978, J.Amer.Chem.Soc., 100, 5268.
28. NAG Fortran Library Manual, Mark 6 (NAG Ltd, 1977).

ECONOMICAL WAVE ACTIVE FILTERS

G.Alexiou, I.Haritantis, Prof.T.Deliyannis

Electronics Laboratory, University of Patras - Greece

Abstract: New wave active networks simulating LC ladder filters are derived, which are more economical than most of the previously proposed wave active filters, (up to 50% more economical than the initially proposed wave active filters). This method of simulation is applicable to any type of doubly resistively terminated LC ladders. Experimental results show, that the new filters can be used for operation at frequencies higher than nearly all the other LC ladder simulating active networks. It is also shown, that they have low sensitivity in the passband.

1. INTRODUCTION

Wave Active Filters (WAF) (1),(2) are active RC circuits, which have been proposed as alternatives to other types of simulation of doubly resistively terminated LC ladder filters. The procedure in this simulation approach is (2) to split the passive ladder into one-element L or C two-ports and substitute each of them by a proper Wave Active Equivalent (WAE). The determination of the WAE is achieved through the use of the scattering parameters of the passive one-element two-port. Successive WAEs are connected in such a way that the reflected wave at one port of a WAE becomes the incident wave in the port of the neighbouring WAE that faces it and vice-versa (cross-cascade connection). WAFs are now considered a special case of the Linear Transformation Filters (3). They have low sensitivity and can work satisfactory at higher frequencies than almost all passive ladder simulating active filters.

However, they possess an important disadvantage, namely, they employ a large number of operational amplifiers. This is really a

D. G. Lainiotis and N. S. Tzannes (eds.), Applications of Information and Control Systems, 404-408.

drawback both from the economy and the power consumption points of view.

In this paper we show, that this large number can be reduced by as much as fifty percent, and that the WAFs thus obtained retain the low sensitivity and the suitability for high frequency operation of the original WAFs.

2. REDUCTION OF THE NUMBER OF OPERATIONAL AMPLIFIERS IN WAFs

In the original WAF design the passive ladder is split into one-element two-ports. The WAE of each of these elementary inductor or capacitor two-ports, employes two operational amplifiers apart from passive elements. Thus, it follows, that for a passive ladder with N storage elements, the corresponding WAF will employ about 2N operational amplifiers. This number is too large by any standards. However, it can be reduced in most cases, as follows: In WAF theory the scattering parameters of each WAE is formulated in terms of the parameters of the chain matrix of the elementary two-port. There is no problem in formulating the scattering parameters of a two-port consistingof more than one elements. But it is very difficult, if not impossible, to determine from those the WAE circuit, if the passive elements in the two-port are of both types i.e. L's and C's. However, if all elements in the passive two-port are of the same kind, the determination of its WAE is trivial if the storage elements are capacitors, and quite easy if they are inductors.

This theory can be extended in the case of an n-port subnetwork consisting of elements of one kind. Its scattering parameters can be easily formulated in terms of the y-parameters of the passive subnetwork. It is then quite simple to determine the WAE subnetwork following the rules set for the two-port subnetworks.

Thus in a manner similar to that of Gorski-Popiel's (4), we split the passive ladder into n-port subnetworks containing as many elements of one kind as possible. For each subnetwork we determine its WAE, and finally connect all WA subnetworks properly to form the overall WAF.

The method will be further explained by means of an example. Consider the 8th order symmetrical bandpass LC ladder filter shown in fig 1a. This is split in L and C two-port subnetworks in the manner indicated by the broken lines. The source resistor R_S and the load resistor R_L can be included in the elementary two-ports consisted of capacitors C_1 and C_4 respectively. Next the scattering parameters of each two-port subnetwork in terms of its y-parameters are determined in the manner described in (5). From these the WAE of each subnetwork is obtained. This is straight forward in the case of a C-subnetwork, whose WAE will contain the C-subnetwork intact, the two-port normalization resistances and two-operational amplifiers each arranged to give a gain of two. The latter are used to perform the transformation of variables at the ports. In the case of the L-subnetwork the numerator and denominator of each scattering parameter is divided by a proper complex scaling factor to obtain active RC realization of the corresponding WAE. With this complex impedance scaling all inductances are turned to resistances, while the two port-normalization resistances to capacitances. The WAE is then completed by connecting two operational amplifiers to the RC subnetwork in the usual WAF manner. Finally the WAE subnetworks are connected in cross-cascade and the overall WAF is as shown in fig 1b. This employs 8 operational amplifiers instead of 16, which will be required if the circuit is designed according to the original WAF procedure. The filter in fig 1b was denormalized to centre frequency at 1kHz and built using components whose values were to within 1% of their nominal values. It was found to work quite satisfactory without any component adjustment.

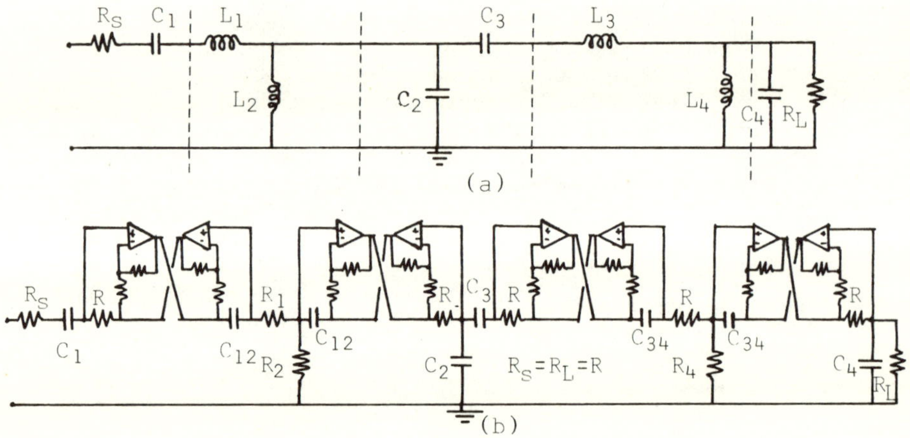

Fig 1. a) 8th order symmetrical bandpass LC filter.
b) The new wave equivalent active circuit.

The method described above has been applied to the simulation of several other LC ladder filters, such as a 5th order elliptic lowpass, an eight order minimum inductance ladder where a four-port L-subnetwork has to be simulated e.t.c. In all cases the circuits obtained employed about half the number of operational amplifiers required in the corresponding circuits of the initial WAFs.

3. SENSITIVITY AND HIGH FREQUENCY OPERATION

The sensitivity of WAFs, which are obtained by the method

explained above, was studied in terms of the Schoeffler (6) sensitivity measure. This was calculated for a 4th order Butterworth bandpass filter, which was realized as a LC ladder, a WAF according to the original design and as a WAF with reduced number of operational amplifiers (4 instead of 8 amplifiers in the original WAF). Only passive components were considered in this calculation. Results are shown in fig 2. It can be seen that the sensitivity of the new WAF is quite close to that of the original WAF. As it was expected, both WAFs have higher sensitivity than the passive filter, but in the passband they are much less sensitive than the corresponding cascade filter.

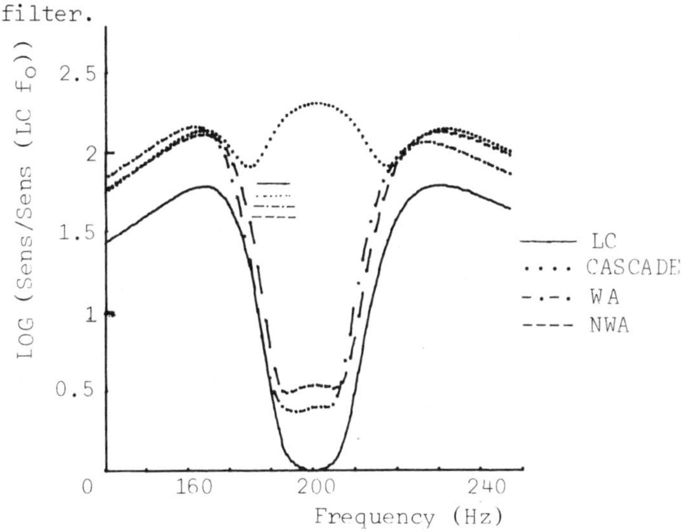

Fig 2. Sensitivity curves for various circuits realizing a 4th-order Butterworth bandpass function.

The suitability of the new WAFs for operation at high frequencies was examined next. A 5th order elliptic lowpass filter was built for a cut-off frequency at 100kHz. The new WAF employed 6 operational amplifiers plus 1 for zero output impedance (against 14 required in the original WAF). The amplifiers had a nominal f_T of 10MHz. The measured response was very satisfactory in the passband but in the stopband the minimum attenuation was about 7dBs lower than the expected one.

CONCLUSIONS

WAFs with reduced numbers of operational amplifiers can be obtained, thus saving up to fifty percent in active elements. These new WAFs have low sensitivity to passive element values variation and can be used for operation at higher than usually frequencies, say up to 100kHz. In our current work we examine the performance of WA resonators under practical conditions in an attempt to study the effect of the non-ideal operational amplifier on the response of highly selective WAFs.

REFERENCES

1. Wupper, H., Meerkötter, K.: "New active filter synthesis based on scattering parameters" 1975, IEEE Trans. on Circuits and Systems, CAS-22, pp.594-602.

2. Haritantis, I., Constantinides, A., Deliyannis, T.: "Wave active filters" 1976, Proc. of IEE, Vol.123, No.7, pp.676-682.

3. Constantinides, A., Dimopoulos, H.: "Active RC filters derivable from LC-Ladder filters via linear transformations" 1976, Elect. Circuits and Systems, Vol.1, No.1, pp.17-21.

4. Gorski-Popiel, J.: "RC-Active synthesis using PICs" 1967, Elect. Letters, Vol.3, No.8, pp.381-382.

5. Alexiou, G., Haritantis, I., Deliyannis, T.: "New economical wave active filters" 1979, Proc. of 4th I.S.N.T., Ljubljana, pp.197-202.

6. Schoeffler, J.D.: "The synthesis of minimum sensitivity networks" 1964, IEEE Tranc. on Circuit Theory, Vol.CT-11, No.2, pp.271-276.

ECONOMICAL MULTIPLE FEEDBACK ACTIVE RC NETWORKS

S.Fotopoulos and Prof. T.Deliyannis

Electronics Laboratory,University of Patras - Greece

Abstract: A method is presented for reducing the number of operational amplifiers in certain multiple feedback active RC networks. This is accomplished by substituting first-order active lowpass sections in certain positions in the prototype network by RC sections. It is shown, that the networks thus obtained retain the low sensitivity properties of the multiple feedback active RC networks. Thus these networks are more economical in active elements than LC ladder simulating active networks, with which they have comparable sensitivities.

1. INTRODUCTION

Low sensitivity RC active filters are usually designed nowdays in one of the following ways: a) Cascading second order sections. b) Simulating a doubly resistively terminated LC ladder filter. c) Realizing the filter function in a multiple loop feedback structure.

In this communication we present some preliminary results from a study of the third type of RC active filter realization. In particular we are interested in the design of lowpass multiple-loop feedback circuits with reduced numbers of operational amplifiers.This will lead to a reduction in the power consumption of the circuit also.

2. MULTIPLE-LOOP FEEDBACK STRUCTURES

All multiple-loop feedback active RC filters consist of a cascade connection of first and/or second order stages and of a

D. G. Lainiotis and N. S. Tzannes (eds.), Applications of Information and Control Systems, 409-416.

feedback scheme applied to the cascade arrangement. If this feedback scheme corresponds to all possible loops being closed a filte results, which is called the minimum sensitivity multiple-loop feedback filter.

For reasons of design complexity, which, after all, does not offer substantial advantages, this complex feedback scheme is not used in practical design. Instead, simplified feedback schemes can be applied, which make the design more easily formulatable and the study of the filter much simpler.

Such simplified structures are, in the order of their first appearance in literature, as follows:

The Leap-frog structure (1).
The Coupled Biquad (CB) (2).
The Primary Resonator Block (PRB) (3).
The Follow the Leader Feddback (4).
The Shifted Companion Form (5).
The Inverse Follow the Leader Feedback (6).
The Minimum Sensitivity Feedback (7).

The design of a frequency symmetric bandpass filter as a multiple-loop feedback circuit usually proceeds by first designing the lowpass prototype and then applying the lowpass to bandpass transformation. Then the functional blocks in the cascade are equally tuned resonators. To take into account the losses of the resonators i.e. their finite Q factor, it is necessary to start from a lowpass prototype with lossy integrators.

Multiparameter sensitivity studies have shown (4) that, at least for sixth-order Butterworth bandpass filters, the sensitivity in the passband of multiple-loop feedback filters, when properly designed, is much lower than the sensitivity of the corresponding cascade filter, i.e. the filter resulting from the cascade connection of second order stages. Also this sensitivity, although higher than that of the corresponding doubly resistively terminated LC ladder filter, is comparable to ladder simulating RC active filters and therefore acceptable in practice. However, in the stopband and transition band the cascade filter performs equally well with, if not better than, most of the other filters.

3. LOWPASS MULTIPLE-LOOP FEEDBACK CIRCUITS

The greater demand for sharp bandpass filters has somehow overshadowed the study of lowpass multiple-loop feedback circuits. When such filters required, multiple-loop feedback circuits can be improved from the economy view-point. In some cases they may show improvement in sensitivity also. This is shown in the rest of this paper.

Lowpass multiple-loop feedback circuits can be obtained from the previously referred to multiple-loop schemes, if the functional blocks in the cascade are integrators. This requires at least N+1 operational amplifiers for a Nth order all pole lowpass filter. A method is given here for reducing this number of operational amplifiers.

All that the method amounts to, is to substitute as many active stages in the cascade as possible by RC lowpass stages, as it is shown in figure 1. This is possible under the condition that the integrators are lossy. In figure 2a the block diagram is shown of a multiple-loop feedback circuit realizing a third order filter function. The actual filter, which results after substituting stage T_1 by a RC lowpass circuit is shown in fig 2b. Component values will be obtained from the prototype lowpass circuit. In fact, if

$$T_1 = T_2 = T_3 = \frac{1}{s+\alpha} \tag{1}$$

and if F_{12}, F_{23}, F_{13} and α are known quantities, component values will be obtained from the following equations:

$$\left(\frac{1}{R_1} + \frac{1}{R_4} + \frac{1}{R_6} + \frac{1}{R_7}\right)\frac{1}{C_1} = \frac{1}{R_2 C_2} = \frac{1}{R_3 C_3} = \alpha \tag{2}$$

$$\frac{1}{R_6 C_1 R_4 C_2} = F_{12} \tag{3}$$

$$\frac{1+\frac{R_9}{R_8}}{R_5 C_2 R_3 C_3} = F_{23} \tag{4}$$

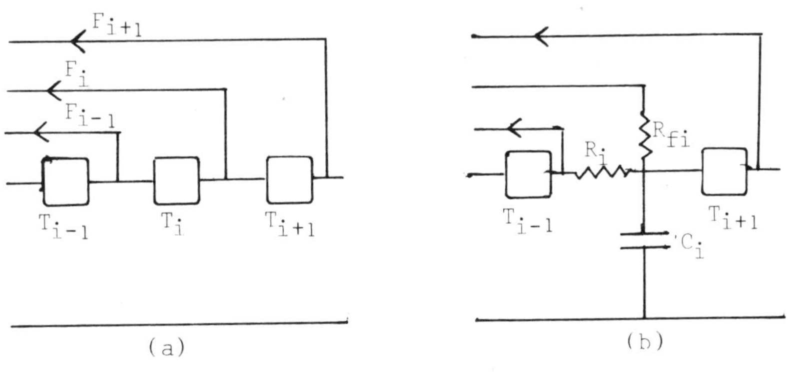

Fig 1.a)Part of a multiple-loop feedback network. b) An active lossy integrator stage T_i has been substituted by a passive RC stage.

$$\frac{1+\frac{R_9}{R_8}}{R_7C_1R_4C_2R_3C_3} = F_{13} \tag{5}$$

Another two conditions regarding the overall gain G and the maximisation of the dynamic range can give two additional equations. Thus we have

$$G = \frac{1+\frac{R_9}{R_8}}{R_1C_1R_4C_2R_3C_3} \tag{6}$$

$$V_{2max} = V_{3max} \tag{7}$$

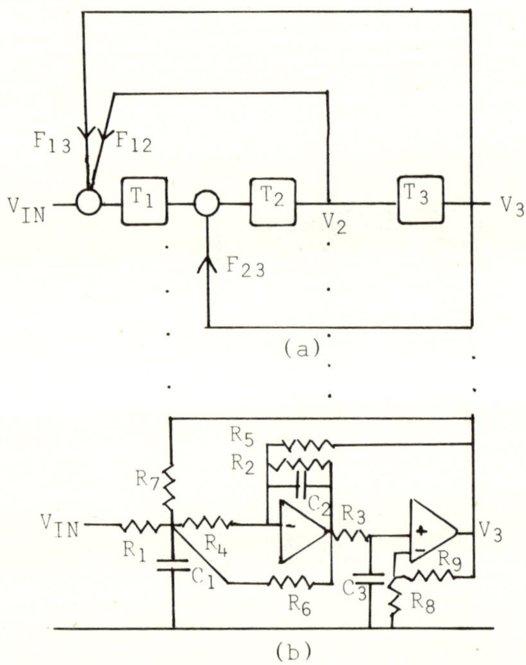

Fig 2. a) Block Diagram of a multiple-loop feedback circuit of 3rd order. b) A 3rd order all pole filter with reduced number of operational amplifier.

There exist some more degrees of freedom, which can give additional equations. However we have found experimentally that their consideration does not affect the sensitivity substantially, since this is primarily determined by the component values of the prototype multiple-loop feedback lowpass circuit.

Theoretical and experimental response of this circuit, built with components whose values were selected to within 1% of their nominal values, is shown in fig 3 for a cut-off frequency at 1kHz.

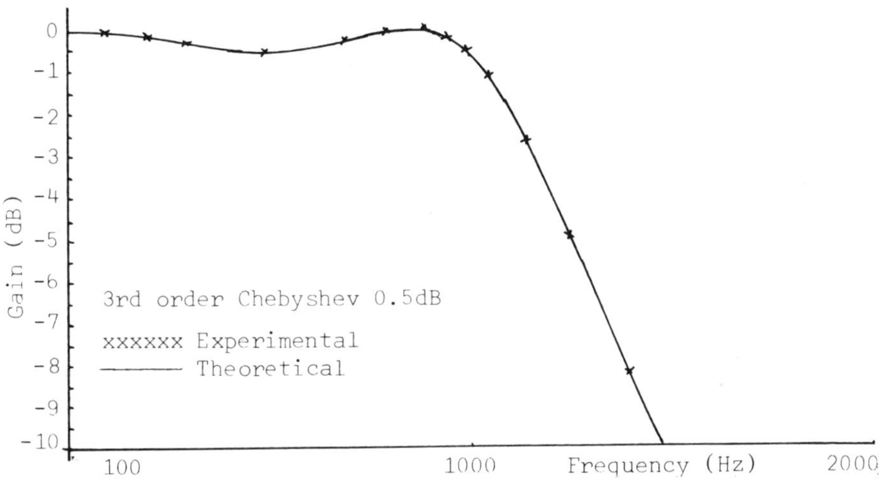

Fig 3. Theoretical and experimental response of the network in fig 2b realizing the 0.5dB ripple 3rd order Chebyshev lowpass filter function.

This design method is valid for filters of any order and for any multiple-loop feedback topology. However, of particular interest are the Leap-frog(or Modified Leap-frog) circuits. In each Leap-frog circuit one can substitute the first and last stages in the cascade, these being lossy integrators, by RC circuits. Thus for the 3rd order Leap-frog shown in block diagram in fig 4a the circuit in fig 4b results, which employs only one operational amplifier. It is interesting to note that this circuit has been independently obtained (8) through the simulation of the corresponding doubly resistively terminated LC ladder filter.

In Modified Leap-frog circuits (MLF), in which intermediate functional blocks are also lossy integrators, the substitution described above can again be applied. Thus the 5th order modified Leap-frog circuit (9) in fig 5a can be modified as in fig 5b, obtained by substituting stages 1,3 and 5 in the MLF prototype by RC stages. It can be seen that only two operational amplifiers are employed in this circuit compared to seven in the original MLF circuit.

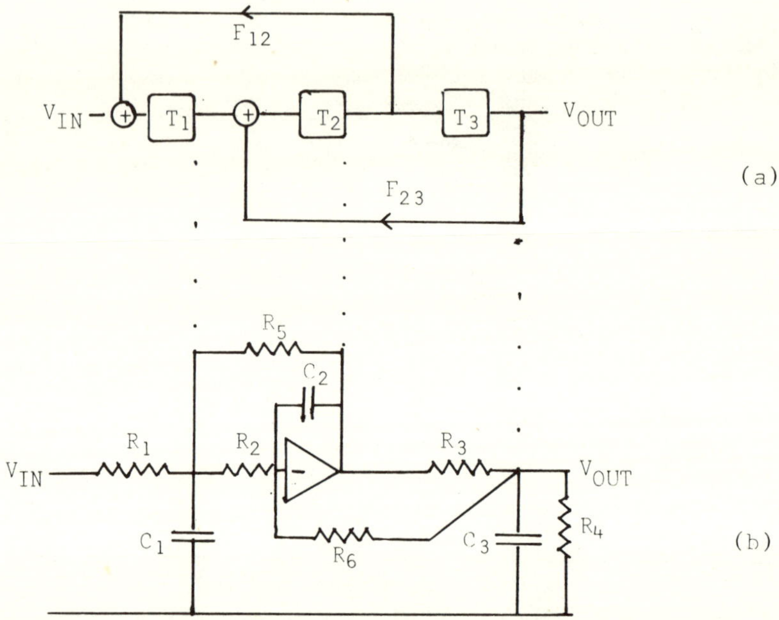

Fig 4. a) Leap-frog topology for a 3rd order function.
b) An economical equivalent realization.

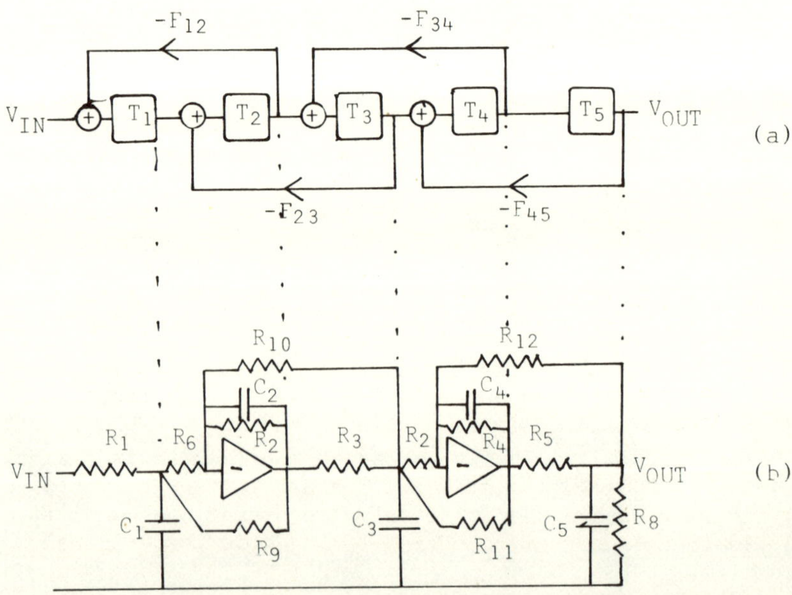

Fig 5. a) Leap-frog structure for a 5th order filter function.
b) An economical equivalent realization.

4. SENSITIVITY

In order to investigate whether the low sensitivity property of the multiple-loop feedback prototype has been retained in the new networks, the Schoeffler (10) sensitivity measure was calculated for the two new networks, and was compared to the sensitivity of the corresponding Leap-frog circuits. For this reason the network in fig 5b was designed to realize the 5th order Butterworth lowpass function. Results are shown in fig 6 for the network in fig 2a. It can be seen that the sensitivity of the new circuit is slightly lower than the sensitivity of the corresponding Leap-frog circuit. Similar to these are the sensitivity curves for the network in fig 5b and the corresponding Leap-frog circuit. In both cases the picture would have been more in favour of the new networks, if the effect of the non-idealness of the active elements had also been considered.

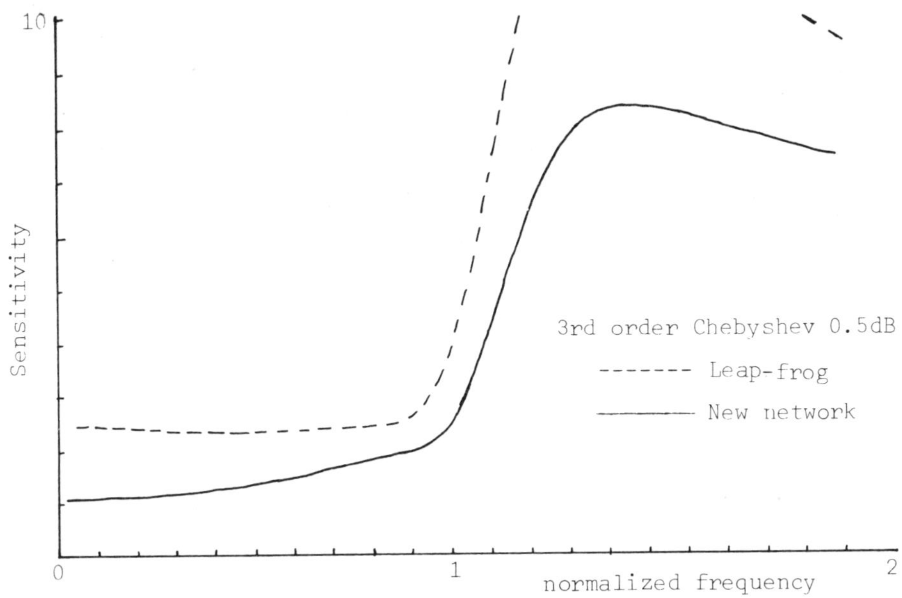

Fig 6. Schoeffler sensitivity measure for the Leap-frog and the circuit in fig 2b both realizing the 0.5dB ripple 3rd order Chebyshev filter function.

5. CONCLUSIONS

It has been demonstrated here that multiple-loop feedback lowpass circuits can be designed with fewer active elements than their corresponding prototype circuits. This is achieved by substituting lossy integrator stages by RC stages. The networks thus obtained retain the low sensitivity property of their parental multiple-loop feedback circuits and, in some cases, may be even better than those from the sensitivity point of view. It is not clear to the authors why this is so. It is believed that it is due to the same reason that multiple-loop feedback filters have low sensitivity. However the latter may sometimes have high sensitivities, which means, that it is not their topology only that grants them the low sensitivity property, but also their proper design.

REFERENCES

1. Girling,F.E.J. and Good,E.F.: "The Leap-frog or active ladder synthesis" 1970, Wireless World, pp.341-345.
2. Tow,J. and Kuo,Y.L.: "Coupled biquad active filters" April 1972 Proc.IEEE Int.Symp.on Circuit Theory, pp.164-167.
3. Hurting,G.: "Filter network having negative feedback loops" March 1973, U.S. Patent 3720881.
4. Laker,K.R. and Ghausi,M.S.: "Synthesis of a low sensitivity multiloop feedback active RC filter" March 1977, IEEE Trans. Circuits and Systems, Vol.CAS-21, pp.252-259.
5. Tow,J.: "Design and evaluation of shifted companion form active filters" March 1975,Bell Syst.Tech.J.,Vol.54,No.3,pp.545-568.
6. Perry,D.J.: "New multiple feedback active RC network" August 1975, Electron. Letters, Vol.11, No.16, pp.364-365.
7. Laker,K., Schauman,R. and Ghausi,M.: "Multiple-loop feedback topologies for the design of low sensitivity active filters" Jan.1979, IEEE Trans. on Circuits and Systems, Vol.CAS-26,No.1 pp.1-21.
8. Haritantis,I.: "Signal flow filters" 1978 European Conference on circuits theory and design, Lausanne, pp.68-72.
9. Wise,B.: "Active filters using a predistortion technique" January 1968, Electronics Letters, Vol.4, No.2, pp.25-26.
10. Schoeffler,J.: "The synthesis of minimum sensitivity networks" January 1964, IEEE Trans.on Circuit Theory, Vol.CT-11, No.2 pp.271-276.

A DESCRIPTION LANGUAGE AND ROUTING ALGORITHM FOR C.A.D.

DORN Roman, DIVJAK Saša, ZRIMEC Miran

FAKULTETA ZA ELEKTROTEHNIKO, LJUBLJANA

The usefulness of a design system is conditioned by the method of entering the input data. A problem describing language described in this paper is used to specify to characteristics of the problem such as board data, elements of the circuit and their positions on the board, connections between pins of the elements, etc. An interpreter for this language was implemented. The interpreter converts the input data into data structures of the design system. Described are the data structures and an algorithm for routing and placement. The results of this algorithm can be grafphically displayed. All the programs of the design system run under RT-11 on the PDP 11/34 with a graphic terminal.

INTRODUCTION

In order to produce a usefull tool for CAD, we have divided our problem into two parts. The first is the processor for the board description language. In the second part, the routing algorithm and output are programmed.

The input language is used for the inscription of characteristics of the problem such as physical data about the printed board (size, length, number of layers), type and symbol identifier of the composing elements, its interconnections and the definition on the entering data working modę (interactive batch). In the second part, the Lee's algorithm is programmed. This algorithm finds out the shortest patch between two connection points (pins) on the board. This is then done for the

D. G. Lainiotis and N. S. Tzannes (eds.), Applications of Information and Control Systems, 417-422.

whole set of connection points of the handled circuit.

DESCRIPTION LANGUAGE

In general, it can be said that the usefulness of a design system is conditioned by the method of entering data. Our goal has been to define a language as close to the designer engineer as possible. Let us see the conditions that we have set for our language:

- an unformed simple language which uniformly describes the problem;
- a system library with standard elements (dual-in--line with 14, 16, .. pins, TO05, resistance of 1/4 W, etc.);
- a simple way of adding new elements to the system library;
- description of placement and orientation of each element;
- a definition of special connections (battery);
- a simple definition of interconnections;
- simple implementation on a small computer.

The inscription of input data is done in many sections. Every section begins with a keyword, followed by the data of a defined type written in statements. The delimiter between same statements is (;). The succession of sections and their number are not determined. The BNT form of the input language can be seen at the end of the paper.

The sailient features of keywords are:

COMMENT - an optional comment can be written,
DIMENSION - inscription of the board dimensions and grid scale,
ELEMENTS - symbol name assignment and type definition of the circuit components,
STRUCTURE - connection between two pins or set of pins,
INTERACTIV- interactive mode of operation,
BATCH - file name where the input data are stored,
LOCATION - placement and orientation of the components,
ADD - adding of new elements into system library,
END - end of input data.

The interpreter for previously described language converts data into two files. In the first file are stored the elements, their position and orientation on the board, in the second all connection point with appropriate number of interconnections to which they belong.

ROUTING

These two files are the input data of routing routine. In this the Lee's algorithm was programmed. In any routing problem, the first constraints to be considered are those imposed on the dimension of the conductors which are to be laid out. We set two variables: conductor width and spacing between two neighbor conductors on the board. These two together define the increment spacing in rectangular grid by which the board is presented. The request that all conductors follow the lines of this grid automatically ensures that the width and spacing requirements are met. To meet the requirement of a minimum of a total connection length, the algorithm begins by simple interconnections of the pair of points which are the closest together.

The output of this routing routine are rectangular matrices of connections for each layer of the board. In this matrices, all placement data about connections are stored. Each connection has one natural number and all elements of matrices of equal value represent the placement of this connection on the layer of the board.

The graphical output of these matrices has not yet been realised.

CONCLUSION

The design system described in this paper is written in FORTRAN IV and implemented on PDP 11/34 computer. In view of the requirement to design boards with 25.000 elements of grid, the whole system is divided into five programs and all communication between them goes over files.

The engineer oriented input language is easy to learn. The use of this design system is useful in complex circuits and in multilayer rechnique. The amount of time seved by the use of these programs is more than 75 %. In appendix some input data and results of the programs are shown.

APPENDIX

The BNF form of the description language.

```
<description of problem> ::= { <sections> } END
<sections> ::= <comm.sec.> ! <elem.sec.> ! <loc.sec.> !
               <struc.sec.> ! <dim.sec.> ! <int.sec.> !
               <batch.sec.>
<comm.sec.> ::= COMMENT <string of words without key>
<elem.sec.> ::= ELEMENTS { <elem.stat.> }
<elem.stat.> ::= <log.ident.> : <typ of element>;
<log.ident.> ::= <one elem.> { , <one elem.> }
<one elem.> ::= <element> ! <more elem.>
<element> ::= <ident> <number>
<ident> ::= R ! C ! IC ! T ! X ! Y
<number> ::= [ <one num.> ]
<one num.> ::= 1 ! 2 ! 3 ! 4 ! 5 ! 6 ! 7 ! 8 ! 9 ! 0
<more elem.> ::= <ident> <beg.num.> . <end.num.>
<beg.num.> ::= <number>
<end.num.> ::= <number>
<typ of element> ::= DL14 ! DL16 ! R1/4 ! <or any new >
<struc.sec.> ::= STRUCTURE { <struc.stat.> }
<struc.stat.> ::= <connec.> { + <connec.> } ;
<connec.> ::= <pin num.> ( <element> { , <element> } )
<pin num.> ::= <number>
<loc.sec.> ::= LOCATION { <loc.stat.> }
<loc.stat.> ::= <element> ( <xloc> : <yloc> : <rot> ) ;
<xloc> ::= <blank> ! <number>
<yloc> ::= <blank> ! <number>
<rot> ::= <blank> ! 1 ! 2 ! 3 ! 4
<blank> ::= < any string of blanks >
<dim.sec.> ::= DIMENSION { <dim.stat.> }
<dim.stat.> ::= <lenght> : <bright> : <num.lay.>
<lenght> ::= <blank> ! <number>
<bright> ::= <blank> ! <number>
<num.lay.> ::= <blank> ! <number>
<int.sec.> ::= INTERACTIV
<batch.sec> ::= BATCH <file name of input data>
```

```
ELEMENTS
        R1,4 : R1/4; IC1 : DL14;
        C1, C2 : C1/4; X : CONECT2;
STRUCTURE
        1(R1,2)+14(IC1)+8(IC1)+1(X);
        2(R2,C2)+13(IC1) +11(IC1);
        1(R3)+2(IC1)+6(IC1); 2(R1)+3(IC1);
        5(IC1)+1(IC1); 2(C1,R4); 1(R4)+7(IC1);
        4(IC1)+10(IC1); 9(IC1)+12(IC1)+2(X);
LOCATION
        IC1(3:7:1); X(12:14:1); R1(:8:); R3(3::);
DIMENSION
        32:28: 2;
END
```

some output:

FIRST LEVEL

XXX	1	2	3	4	5	6	7	8	9	10	11	12	14	15
1	0	0	0	0	0	0	0	0	0	0	0	0	0	0
2	0	10	10	10	10	10	10	10	10	10	10	10	10	10
3	0	0	10	8	8	0	0	0	0	0	0	0	0	10
4	0	0	10	8	9	8	5	9	0	0	0	0	0	10
5	0	8	0	8	9	5	5	9	0	0	0	0	0	10
6	0	8	8	8	9	5	0	9	9	9	9	9	0	10
7	0	8	2	7	6	5	4	7	1	0	0	9	0	10
8	0	8	2	7	6	0	4	7	1	6	6	9	10	10
9	0	8	2	7	0	0	4	7	0	0	0	9	0	10
10	0	8	0	7	0	0	4	0	0	0	0	9	0	10
11	0	8	0	7	0	0	4	0	0	0	0	9	0	10
12	0	8	0	7	0	0	0	0	0	8	8	9	10	10
13	0	0	0	7	0	0	3	3	3	0	0	9	0	10
14	0	0	7	7	0	0	0	0	0	0	0	9	0	10

SECOND LEVEL

XXX	1	2	3	4	5	6	7	8	9	10	11	12	14	15
1	0	0	0	0	0	0	0	0	0	0	0	0	0	0
2	0	0	0	0	0	0	0	0	0	0	0	0	0	0
3	0	0	0	0	0	0	0	0	0	0	0	0	0	0
4	0	0	0	0	0	0	0	0	0	0	0	0	0	0
5	0	0	0	0	0	0	0	0	0	0	0	0	0	0
6	0	0	0	0	9	9	9	9	0	0	0	0	0	0
7	0	0	0	0	0	0	0	0	0	0	0	0	0	0
8	0	0	0	0	6	6	6	6	6	6	0	0	0	0
9	0	0	0	7	7	7	7	7	0	0	0	0	0	0
10	0	0	0	0	0	0	0	0	0	0	0	0	0	0
11	0	0	0	0	0	0	0	0	0	0	0	0	0	0
12	0	8	8	8	8	8	8	8	8	8	0	0	0	0
13	0	0	0	0	0	0	0	0	0	0	0	0	0	0
14	0	0	0	0	0	0	0	0	0	0	0	0	0	0

REFERENCES

/1/ AKERS S.B.: A modification of Lee's path connection algorithm. IEEE-Computers, Vol. EC-16, february 67.

/2/ BREUER A.M.: Digital system design automation: languages, simulation & data base. Computer science press, Inc. Woodland Hills, Californija 75.

/3/ BREUER A.M.: Recent developments in the automated design and analysis of digital systems. IEEE-Proc., Vol. 60, january 72.

/4/ DORN R.: Definicija vhodnega jezika za simulacijo digitalnih vezij in programski modul za njegovo procesiranje. Zbornik del INFORMATICA, Bled 76.

/5/ KERNIGHAN B.W., LIN S.: An efficient heuristic procedure for partitioning graphs. The Bell system tech. j., Vol. 49, 1970.

/6/ LEE C.Y.: An algorithm for path connections and its applications. IEEE-Computers, Vol. EC-10, september 61.

OPTIMAL CONTROL OF FERMENTATION PROCESS USING MICROPROCESSORS

A. S. Fawzy and O.R. Hinton

Electronics Lab., The University, Canterbury, Kent, UK.

ABSTRACT
This paper shows the advantages of using microprocessors in the control of fermentation processes. A linear model of the system (single cell protein production process) is considered. An optimal feedback controller is determined which maintains the productivity constant when the system is subjected to disturbances in the influent substrate concentration and organisms concentration. Simulation results are presented for the accessible and inaccessible states.

1. INTRODUCTION

Fermentation is a process that results in the formation of alcohols or organic acid; it occurs characteristically in plant material as a consequence of the breakdown of carbohydrates, the predominant organic compounds in plant tissue. Fermentation systems are characterised by the complexity of their response to a wide variety of environmental variables. Most of these factors are independently variable, giving rise to an almost infinite number of possible growth conditions. The application of computers to fermenters is not new. The reason why such systems have not found more widespread applications in industrial research and production is no doubt because they have been very expensive to produce and require highly specialised knowledge to use. Microprocessors open up the possibility of economically monitoring and controlling several parameters simultaneously. The advantages arise from four characteristics of microprocessors. These characteristics are: (i) low cost; (ii) small physical size, which means they can be located where space is a severe limitation; (iii) the low cost and small size make it possible to use

D. G. Lainiotis and N. S. Tzannes (eds.), Applications of Information and Control Systems, 423-432.

several of them with one fermenter - this can lead to more efficient and complex control while simplifying operation of the system compared with the use of large computers; and (iv) a microprocessor system can be "tailor made" for specific process problems without greatly increasing cost.

A linear model of single cell protein system [1] is considered. An optimal feedback controller is determined which maintains the productivity constant when the system is subjected to disturbances in the influent substrate and organisms concentration. Two control schemes are presented. The first scheme assumes all the states are accessible for feedback control and the feedback gains are obtained from the solutions of the well known Riccati equation [2]. In the second one, feedback gains are determined from the output equation instead of constructing an estimator to estimate the unmeasured states [3,4]. Actually, this is more reasonable since all states are not practically accessible for measurement. An algorithm given in [5] for computing the optimal output feedback gains is based on averaging a performance function obtained for a linearly independent set of initial states. A modification of this algorithm [6,7] is then developed which avoids the solution of nonlinear algebraic equations at each iteration. Simulation results and conclusions are presented.

2. DESCRIPTION OF THE PROCESS

The fermentation process studied in this paper is that used for making single cell protein. Specifically, the constants below refer to the organism pseudomonas; this process is being studied both theoretically, and practically in a laboratory experimental fermenter controlled by a Motorola M6800 microprocessor system. Such micro-organisms require a carbon-energy source (methanol), oxygen, nitrogen source (usually ammonia), and some mineral nutrients. The control effort is exerted on the liquid flow rate (u) to maintain the substrate concentration (x_1), organisms concentration (x_2) and productivity (y) at desired values when the system is subjected to step disturbances in the influent substrate and organisms concentration. It will be assumed that the growth chamber is provided with separate control systems for temperature, pressure, PH, foam and dissolved oxygen.

A dynamic mathematical model of a fermentation system was developed by employing a Monod model [1]. For the system we are interested in studying here, it is known experimentally that optimal conditions (resulting in maximum yield) are as follows [8,9,10]:

$\mu_{max} = 0.55,\ hr^{-1}$ $\qquad V = 2,\ l$

$D = 0.2,\ hr^{-1}$ $\qquad Y = 0.4,\ g/g$

$S_s = 0.05,\ g/l$ $\qquad X_s = 0.78,\ g/l$

where μ_{max} = maximum specific growth rate

D = dilution rate

Y = yield factor

S_s, X_s = the desired steady state conditions of substrate and organisms concentration, respectively.

The following values are assigned to constants and parameters used in this work

$S^o = 2\ g/l$ $\qquad X^o = 0.01\ g/l$

$K_s = 0.7\ mg/l$ $\qquad K_D = 0.002\ hr^{-1}$

where S^o, X^o = Concentration of substrate and organisms in the influent stream

K_s = saturation constant

K_D = specific endogeneous microbial attrition rate.

Then, a linearised form of the Monod model may be written [1]

$$\begin{aligned} \dot{x} &= Ax + Bu + DW \\ y &= Cx + E \end{aligned} \tag{1}$$

With initial conditions

$$x_i(0) = 0, \qquad i = 1,2$$

where

$x = [x_1\ x_2]'$, the deviations of the state variables from the desired states

u = the deviation of the control variable from the desired value necessary to maintain the desired states

y = the productivity

W = Unit step disturbances in the influent substrate and organisms concentration.

$$A = \begin{bmatrix} -2.46 & -2.712 \\ 1.46 & 1.702 \end{bmatrix}, \quad B = \begin{bmatrix} 2.437 \\ -2.406 \end{bmatrix}, \quad D = \begin{bmatrix} -0.975 \\ 0.9625 \end{bmatrix}$$

$$C = \begin{bmatrix} 0 & 0.2 \end{bmatrix}, \quad E = \begin{bmatrix} 0.001 \end{bmatrix} \tag{2}$$

The dot denotes the time derivative and the prime denotes the transposition.

3. DESIGN OF AN OPTIMAL CONTROLLER OF FERMENTATION PROCESS

Two control schemes are presented. The first scheme assumes all the states are accessible for feedback control. In the second one, output feedback gains are determined from the output equation rather than state equation.

3.1 Optimal controller when all states are accessible

Before designing an optimal controller to any system, the controllability and observability conditions [11] must be checked. For this system, it is found that these conditions are satisfied.

Next it is desired to find the control, u, which minimises the following cost function:

$$J = \frac{1}{2}\int_{0}^{T}\left[x'(\theta)Qx(\theta) + Ru^{2}(\theta)\right]d\theta \tag{3}$$

where

$Q = \begin{bmatrix} q_{11} & 0 \\ 0 & q_{22} \end{bmatrix}$, positive semi-definite weighting state matrix

$R = [r]$, positive definite control weighting matrix

$\theta =$, dimensionless time

$T =$, the fixed terminal time.

Hence, for the linear quadractic problem, that is, for the system equation and objective function as defined by eqs. (1) and (2), respectively, the optimal proportional control derived by the maximum principle is [2,12]

$$u(\theta) = -R^{-1}B'\left[K(\theta)\ x(\theta) - Z(\theta)\right]$$

or

$$u(\theta) = -\frac{1}{r}B'\left[K(\theta)\ x(\theta) - Z(\theta)\right] \tag{4}$$

where $K(\theta)$ and $Z(\theta)$ are solutions of

$$\dot{K} = -K(\theta)A - A'K(\theta) + K(\theta)BR^{-1}B'K(\theta) - Q, \quad K(T) = 0 \tag{5}$$

$$\dot{Z} = \left[A' - K(\theta)BR^{-1}B'\right]Z(\theta) - K(\theta)D \quad , \quad Z(T) = 0 \tag{6}$$

Eq. (5) is known as the matrix Riccati equation and can be solved for $K(\theta)$ backward in time with a boundary condition $K(T) = 0$, where T is the final time.

For the case when $T \to \infty$ (1), the Riccati equation (5) tends to a steady state value, that is, the right hand side of eq. (5) goes to zero. The case of $T \to \infty$ is justified because T is generally left unspecified in process control, thus implying a semi-infinite process duration. Hence, for the case $T \to \infty$, eq. (5) yields an algebraic equation

$$KA + A'K - KBR^{-1}B'K + Q = 0 \tag{7}$$

The solution is obtained by an iterative algorithm [2].

The solution vector Z is found to be [13]:

$$Z = -[KBR^{-1}B' - A']^{-1} KD \tag{8}$$

Eq. (1) can be written as

$$\begin{aligned} \dot{x} &= (A - BR^{-1}B'K)x + BR^{-1}B'Z + D \\ y &= Cx + E \end{aligned} \tag{9}$$

and the optimal proportional control is obtained from

$$u = -\frac{1}{r} B'[Kx + (KBR^{-1}B' - A')^{-1} KD] \tag{10}$$

It should be noted that even though the above formulation is in terms of unconstrained x and u, it should be recognised that both x and u can be maintained within physical constraints by simply varying the weighting matrices Q and R.

3.2 Optimal controller for inaccessible states

In this case the designer does not have a complete set of state variables directly available for linear feedback control. Moreover, he may wish to generate the control variables directly by taking linear combinations of the available output variables, instead of first reconstructing the states via a Kalman filter [3]. Then the optimal control problem may be stated as follows. Find the output feedback matrix F such that

$$u = -Fy \tag{11}$$

minimises the quadratic performance index (3).

In [5], the optimal output feedback controller is determined for a linear system free from disturbance input. The single cell protein production process as described in (1) and (2) is subjected to step disturbances in the influent substrate and organisms concentration. Thus, we may consider the disturbance input as a new state and a new state vector is defined as:

$$X = [x,\ W]'$$

Then (1) and (2) may be written as

$$\dot{X} = A_1X + B_1u$$

and

$$y = C_1X \tag{12}$$

where

$$A_1 = \left[\begin{array}{c|c} A & D \\ \hline \multicolumn{2}{c}{0} \end{array}\right], \quad B_1 = \left[\begin{array}{c} B \\ \hline 0 \end{array}\right], \quad C_1 = \left[\begin{array}{c|c} C & E \end{array}\right] \tag{13}$$

The performance index (3) becomes

$$J = \frac{1}{2}\int_0^\infty [X'Q_1X + Ru^2]d\theta \tag{14}$$

where

$$Q_1 = \left[\begin{array}{c|c} Q & 0 \\ \hline \multicolumn{2}{c}{0} \end{array}\right]$$

Substituting (11) in (12), we get

$$\dot{X} = A_cX(\theta) \tag{15}$$

where

$$A_c = A_1 - B_1FC_1 \tag{16}$$

is the closed loop system matrix which is assumed to be stable.

If we consider the initial states of (15) to be a random variable with zero-mean and covariance matrix X_o, then according to [5], we find

$$J = \frac{1}{2}\,\mathrm{tr}[KX_o] \tag{17}$$

where K is the solution of

$$KA_c + A_c'K + Q_1 + C_1'F'RFC_1 = 0 \tag{18}$$

and

$$\partial J/\partial F = RFC_1LC_1' - B_1'KLC_1' \tag{19}$$

where L is the solution of the linear algebraic equation

$$LA_c' + A_cL + X_o = 0 \tag{20}$$

The optimal feedback gain F is determined [6], by direct minimisation of J, e.g. using the gradient method [14], while satisfying (18) and (20).

Thus the algorithm may proceed as follows:-

1. Choose a starting value of F such that A_c is stable.
2. Determine K and L from (18) and (20).
3. Evaluate (19), if $\partial J/\partial F < \epsilon$, where ϵ is a prescribed accuracy measure, then the solution is obtained. If not, a gradient method may be employed to obtain F by direct minimisation of J.

4. SIMULATION RESULTS AND CONCLUSION

Figure 1 shows the effect, using control scheme A (when the states are assumed accessible) of taking the value of matrix R constant and varying the weighting matrix Q. The graphs show the optimal response of the system to known constant step disturbances in influent substrate and organisms concentration. The results show that, when the organism concentration is weighted more heavily than the substrate concentration (as in a single cell protein production process), it can be well maintained close to the desired value. In the example in the figures q_{11} is maintained arbitrarily constant at 1, while q_{22} is varied within the physical bounds on the flow rate, namely by the critical washout flow rate and/or the pumping capacity.

For control scheme B (when states are assumed inaccessible), it is assumed that matrix Q is the identity, matrix R is 1, and X_0 is the identity matrix. Figure 2 shows the optimal trajectories for the control variable, states and productivity, along with the case when all states are assumed measurable (for the case of $q_{11} = q_{22} = 1$). It is quite obvious from this figure, that the response of the states and producitivity for the accessible state system are better than the inaccessible state. The value of the steady state errors are smaller and the speed of responses are faster. From this figure also, the deviation of the control variable (flow rate) in state accessible system is less than the inaccessible state, and in the meantime it results in a better productivity. The performance index for state accessible case is found to be J = 1.278, but in the inaccessible state case is J = 3.338.

In conclusion, if we can measure the states of the system, not only is the design of the optimal controller simpler, but also the performance is superior to a system which has unmeasurable states.

Acknowledgement

The authors would like to thank Dr. P. Wymer, Biological Laboratories, University of Kent, Canterbury, England, for all his assistance in helping us to achieve this work.

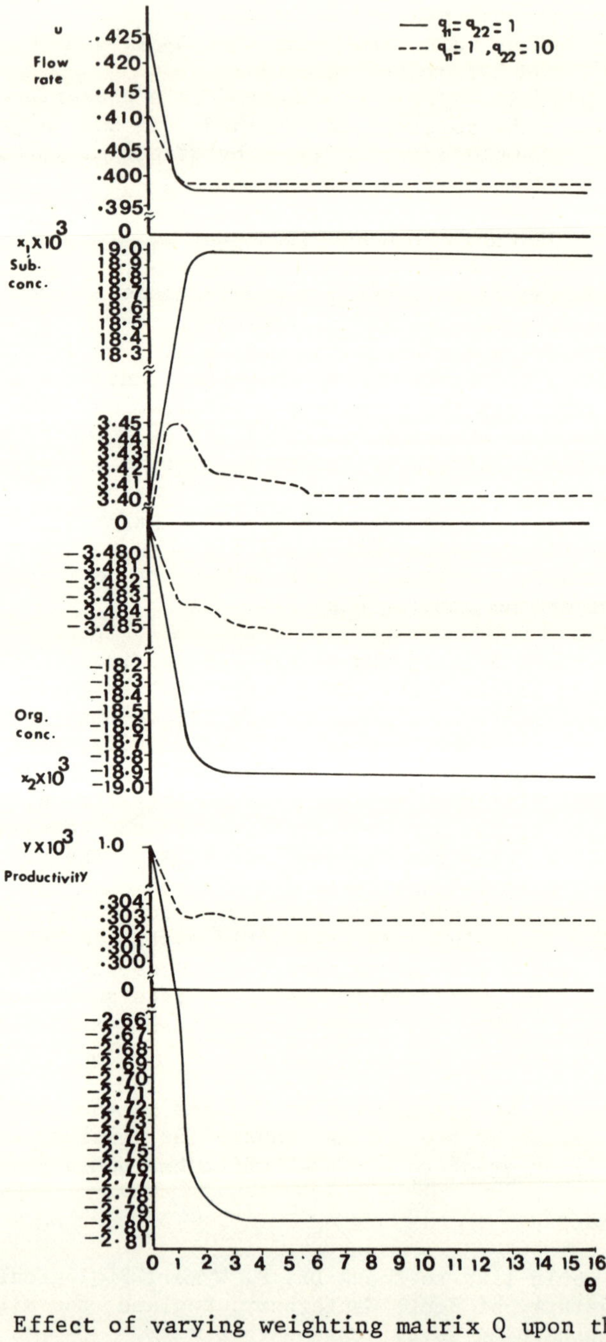

Figure 1 Effect of varying weighting matrix Q upon the response of system to unit step disturbances in influent substrate and organisms concentration using control scheme **A.**

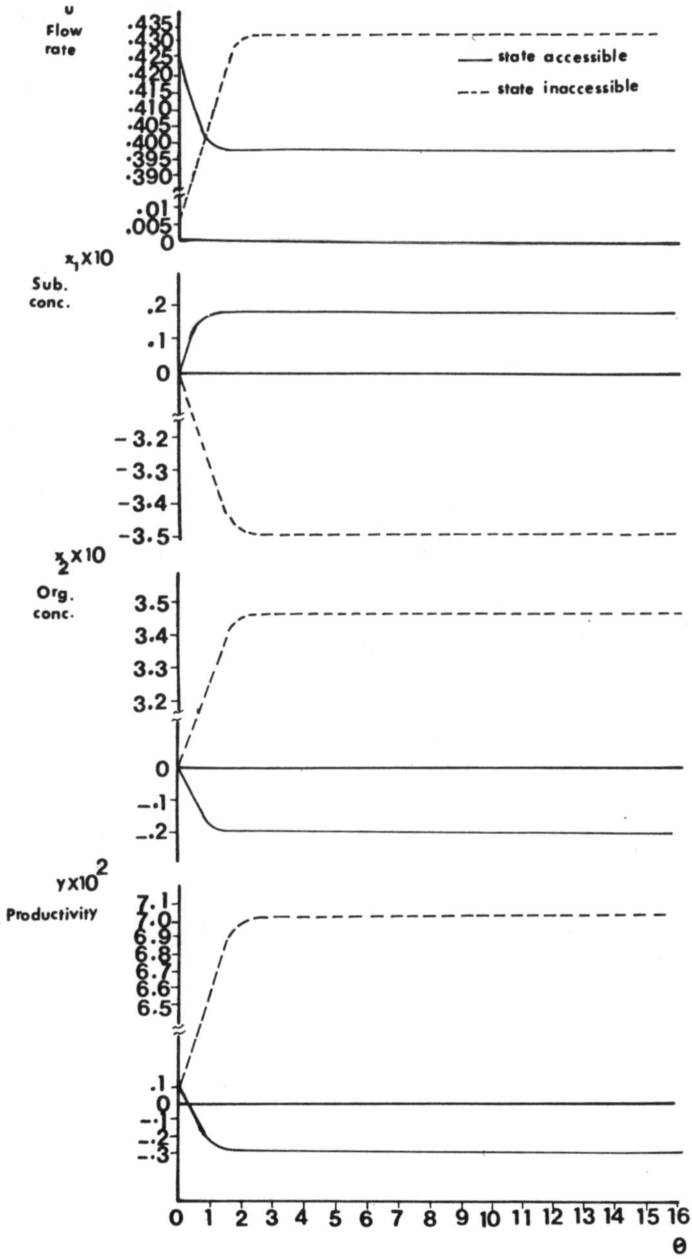

Figure 2 Optimal system response to step disturbances in influent substrate and organisms concentration using control scheme A (when $q_{11} = q_{22} = 1$) and control scheme B.

References

(1) Fan, L.T., Shah, P.S., and Erickson, L.E.: 1972, Ferment. Technol. Today, Proc. IV.IFS, Koyoto, pp. 167-177.

(2) Puri, N., and Gruver, W.A.: 1967, Proceedings of the Joint Automatic Control Conference, pp. 335-344.

(3) Meditch, J.S.: 1969, *Stochastic optimal linear estimation and control,* McGraw-Hill.

(4) Fairman, F.W.: 1977, IEEE Trans. Automat. Contr., 22, pp. 673-675.

(5) Levine, W.S., and Athans, M.: 1970, IEEE Trans. Automat. Contr., 15, pp. 44-48.

(6) Choi, S., and Sirisena, H.: 1974, IEEE Trans. Automat. Contr., 19, pp. 257-258.

(7) Choi, S., and Sirisena, H.: 1977, IEEE Trans. Automat. Contr., 22, pp. 134-136.

(8) Kim, J.H., and Ryu, D.Y.: 1976, J. Ferment. Technol., 54, pp. 427-436.

(9) Harrison, D.E.: 1976, Chem. Tech., 6, pp. 570-574.

(10) Pirt, S.J.: 1975, *Principles of microbe and cell cultivation,* Oxford blackwell scientific.

(11) Ogata, K.: 1970, *Modern control engineering,* Prentice Hall.

(12) Sage, A.P.: 1977, *Optimum systems control,* Prentice Hall.

(13) Fawzy, A.S.: 1977, *Optimisation of industrial processes,* M.Sc. thesis, Ainshams University, Cairo, Egypt.

(14) Hestenes, M.R.: 1969, J. Optimiz. Theory Appl., 4, pp. 303-320.

OPTIMIZATION OF THE STRUCTURE OF A MULTIPURPOSE CHEMICAL SYSTEM

Leonard Badea

Designing-Institute for the Chemical Industry
- I.I.T.P.I.C. - Bucharest, Romania

A multipurpose chemical system is a system performing sequentially the purposes of several singlepurpose systems.It consists of multipurpose and singlepurpose subsystems.
The paper deals with the description of a model for the optimization of the structure of a multipurpose system,when the distribution in time of the purposes is known.
The model is based on the interdependence of the variables corresponding to the singlepurpose subsystems and of those corresponding to the multipurpose system. The set of constraints consists of linear equations and inequations representing the material balance of the multipurpose system and of nonlinear equations representing the identity of composition of the material streams.The objective function is nonlinear and represents the maximum benefit obtainable from the multipurpose system or the minimum operating costs.

Optimization of the structure of a chemical system is often very complicated when the constants of the respective model are time varying.
Changes in the initial basic design data during the later stages of the process development,due to the improvement of technologies and fluctuations of raw material and finite product demands,involve the design of plants able to perform sequentially several processes or alternatives of a process.The respective systems are multipurpose systems,the purpose repre-

D. G. Lainiotis and N. S. Tzannes (eds.), Applications of Information and Control Systems, 433-441.

senting mainly a finite product,a raw material,a process or a design alternate of a process.
A multipurpose system is a system performing sequentially the purposes of several singlepurpose systems. In the general case it consists of multipurpose subsystems.
A multipurpose subsystem is a subsystem performing sequentially the purposes of several singlepurpose subsystems which are compatible from the manufacturing and operating point of view.It may be considered an assembly of the respective singlepurpose subsystems. In a limit case a multipurpose subsystem with a single purpose is a singlepurpose subsystem.
A singlepurpose subsystem represents a stage of raw material and intermediate product transformation or intermediate and finite products separation corresponding to the respective purpose.
Optimization of the structure of a multipurpose system is in most cases very complicated due to the interdependencies of the variables corresponding to the singl purpose subsystems and of those corresponding to the multipurpose system.

The object of this paper is the description of a model to determine the optimal structure of a multipurpose system,when the time distribution of the purposes is known.It represents a development of a model previousl published (1).
As shown in Figure 1 a multipurpose subsystem may be fed,for each purpose,with material streams originating from outside the system (fresh raw materials),recycled streams and streams resulting from other subsystems (non-transformed raw materials and intermediate products).From the subsystem,material streams distributed outside the system (intermediate products and finite products),recycled streams and streams feeding other subsystems may result.
Each stream feeding or resulting from the subsystem consists of a multitude of components (species).
The optimization criterion is the maximum benefit or the minimum operating costs of the multipurpose system.

The objective function corresponding to the multipurpose system consisting of N singlepurpose subsystems assembled in N' multipurpose subsystems is:

$$f_b(\mathrm{opt})(1,2,..,p)=$$

$$
\begin{aligned}
\max \Bigg[& \sum_{i} \sum_{j_u=1}^{N} \sum_{p_u=1}^{p} v_{gi(j_u 0(p_u))} g_{i(j_u 0(p_u))}^{+} \\
& + \sum_{i} \sum_{j_v=1}^{N} \sum_{\substack{j'_v=1 \\ j'_v \neq j_v}}^{N} \sum_{p_v=1}^{p} v_{gi(j_v j'_v 0(p_v))} g_{i(j_v j'_v 0(p_v))}^{-} \\
& - \sum_{J=1}^{N'} v'_{Sg^{o}i(J)} \sum_{i} g^{o}_{i(J)} \Big(\sum_{i} g_{i(J)} \Big/ \sum_{i} g^{o}_{i(J)} \Big)^{a_J} - \\
& - \sum_{j_v=1}^{N} \sum_{j'_v=1}^{j_v-1} \sum_{k} \sum_{p_v=1}^{p} v'_{Sgi(j_v j'_v(p_v))k} \sum_{i} g_{i(j_v j'_v(p_v))k}^{-} \\
& - \sum_{i} \sum_{j_r=1}^{N} \sum_{p_r=1}^{p} v_{gi(0 j_r(p_r))} g_{i(0 j_r(p_r))} \Bigg] \qquad (1.1)
\end{aligned}
$$

$i=A,B,..,W \quad J=1,2,..,N' \quad j_r,j_u,j_v,j'_v=1,2,..,N$

$k=k(j_v j'_v(p_v))=1,2,... \quad p_r,p_u,p_v=1,2,..,p=A,B,..,P$

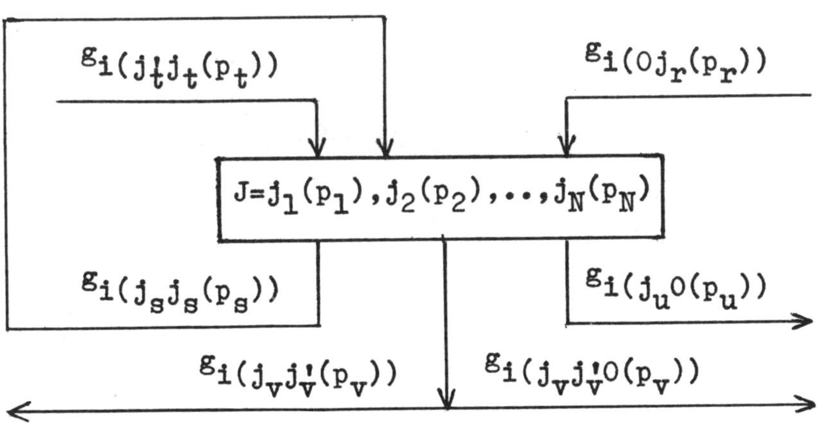

Figure 1. Multipurpose subsystem.

or

$$f_c'(opt)(1,2,..,p)=\min \left[\sum_{J=1}^{N'} v_{Sg^o i(J)}' \sum_i g_{i(J)}^o \left(\sum_i g_{i(J)}/\sum_i g_{i(J)}^o\right)^{a_J} + \sum_{j_v=1}^{N} \sum_{j_v'=1}^{j_v-1} \sum_k \sum_{p_v=1}^{p} v_{Sgi(j_v j_v'(p_v))k}' \sum_i g_{i(j_v j_v'(p_v))k} + \sum_i \sum_{j_r=1}^{N} \sum_{p_r=1}^{p} v_{gi(0j_r(p_r))} g_{i(0j_r(p_r))}\right] \quad (1.2)$$

$i=A,B,...,W \quad J=1,2,..,N' \quad j_r,j_u,j_v,j_v'=1,2,..,N$
$k=k(j_v j_v'(p_v))=1,2,... \quad p_r,p_u,p_v=1,2,..,p=A,B,..,P$

The constraints are:

$$g_{i(j_z(p_z))}=g_{i(0j_z(p_z))}+\sum_{j_z'=1}^{N} g_{i(j_z'j_z(p_z))} \quad (2.1)$$

$$g_{i(j_z j_z(p_z))}=c_{i(j_z j_z(p_z))}\sum_i g_{i(j_z(p_z))}= \quad (2.2.1)$$

$$=c_{i(j_z j_z(p_z))}^{(i')} g_{i'(j_z(p_z))}= \quad (2.2.2)$$

$$=c_{i(j_z j_z(p_z))}^{(i)} g_{i(j_z(p_z))} \quad (2.2.3)$$

$$g_{i(j_z j_z'(p_z))}+g_{i(j_z j_z'0(p_z))}=(c_{i(j_z j_z'(p_z))}+$$

$$+c_{i(j_z j_z'0(p_z))})\sum_i g_{i(j_z(p_z))}= \quad (2.3.1)$$

$$=(c_{i(j_z j_z'(p_z))}^{(i')}+$$

$$+c_{i(j_z j_z'0(p_z))}^{(i')})g_{i'(j_z(p_z))}= \quad (2.3.2)$$

$$=(c^{(i)}_{i(j_z j'_z(p_z))}+$$

$$+c^{(i)}_{i(j_z j'_z 0(p_z))})g_{i(j_z(p_z))} \quad (2.3.3)$$

$$g_{i(j_z 0(p_z))}=c_{i(j_z 0(p_z))}\sum_i g_{i(j_z(p_z))}= \quad (2.4.1)$$

$$=c^{(i')}_{i(j_z 0(p_z))}g_{i'(j_z(p_z))}= \quad (2.4.2)$$

$$=c^{(i)}_{i(j_z 0(p_z))}g_{i(j_z(p_z))} \quad (2.4.3)$$

$$g_{i(j_z(p_z))}=h_{i(j_z(p_z))}\sum_i g_{i(j_z(p_z))}= \quad (2.5.1)$$

$$=g_{i(0j_z(p_z))}/h^{(i)}_{i(j_z(p_z))} \quad (2.5.2)$$

$$g_{i(0j_z(p_z))} \leqslant \text{const.} \quad (2.6)$$

$$g_{i(j_z 0(p_z))} \geqslant \text{const.} \quad (2.7)$$

$$g_{i(j_z j'_z(p_z))k}/\sum_i g_{i(j_z j'_z(p_z))k}=$$

$$=g_{i(j_z j'_z 0(p_z))k}/\sum_i g_{i(j_z j'_z 0(p_z))k} \quad (2.8)$$

$$\sum_i g_{i(J)}/t(J)d(J) \geqslant \sum_i g_{i(j_w(p_w))}/t(p_w)d(j_w(p_w)) \quad (2.9)$$

$i=A,B,..,W$ $J=1,2,..,N'$ $j_z,j_w,j'_z=1,2,..,N$
$p_z,p_w=1,2,..,p=A,B,..,P$ $i'=A$ or B or ... or W
$k=k(j_z j'_z(p_z))=1,2,...$

where: $f_b(opt)(1,2,..,p)$ is the maximum benefit obtainable from the multipurpose system performing sequentially the purposes 1,2,..,p;
$g_{i(j_u 0(p_u))}$ -the quantity of component i resulting from the subsystem j_u, per-

forming the purpose p_u, and being distributed outside the system as finite product;

$g_{i(j_v j'_v(p_v))}$, $g_{i(j_v j'_v 0(p_v))}$ - the quantity of component i resulting from the subsystem j_v, performing the purpose p_v, and feeding the subsystem j'_v or being distributed outside the system as intermediate product, respectively;

$g_{i(0 j_r(p_r))}$ - the quantity of component i feeding the subsystem j_r, performing the purpose p_r, and originating from outside the system;

$\sum_i g_{i(J)}$ - the total quantity of components i feeding the multipurpose subsystem J, performing the purposes $p_1, p_2, .., p_N$ of the single-purpose subsystems $j_1, j_2, .., j_N$ (capacity of the multipurpose subsystem J);

$\sum_i g^o_{i(J)}$ - the standard capacity of the multipurpose subsystem J;

$v_{gi(0 j_r(p_r))}$, $v_{gi(j_u 0(p_u))}$, $v_{gi(j_v j'_v 0(p_v))}$ - the selling price of the quantity unit $g_{i(0 j_r(p_r))}$, $g_{i(j_u 0(p_u))}$ and $g_{i(j_v j'_v 0(p_v))}$, respectively;

$v'_{Sg^o i(J)}$ - the operating costs in the multipurpose subsystem J of the quantity unit $\sum_i g^o_{i(J)}$;

$v'_{Sgi(j_v j'_v(p_v))k}$ - the additional operating costs for the recycling of the quantity unit $\sum_i g_{i(j_v j'_v(p_v))k}$;

a_J - the scale-up exponent of the capacity of the multipurpose subsystem J;

$f'_c(opt)(1,2,..,p)$ - the minimum operating costs of the multipurpose system performing sequentially the purposes 1,2,..,p;

$g_{i(j_z(p_z))}$ - the quantity of component i feeding the subsystem j_z, performing the purpose p_z;

$\sum_i g_{i(j_z(p_z))}$ - the total quantity of components i feeding the subsystem j_z (capacity of the subsystem j_z);

$g_{i(j_z j_z(p_z))}$ - the quantity of component i resulting from the subsystem j_z and being recycled;

$\sum_i g_{i(j_z j_z'(p_z))k}, \sum_i g_{i(j_z j_z' 0(p_z))k}$ - the quantity of material stream k consisting of components i resulting from the subsystem j_z and feeding the subsystem j_z' or being distributed outside the system as intermediate products, respectively;

$g_{i(j_z j_z'(p_z))k}, g_{i(j_z j_z' 0(p_z))k}$ - the quantity of component i of the material stream k resulting from subsystem j_z and feeding the subsystem j_z' or being distributed outside the system as intermediate product, respectively;

$c_{i(j_z j_z(p_z))}, c_{i(j_z j_z'(p_z))}, c_{i(j_z j_z' 0(p_z))}$, $c_{i(j_z 0(p_z))}$ - the transformation and separation fraction in the subsystem j_z, at one pass, of the total quantity of components i feeding the subsystem, into the quantity of component i being recycled, feeding the subsystem j_z', being distributed outside the system as intermediate product, and being distributed outside the system as finite product, respectively;

$c^{(i')}_{i(j_z j_z(p_z))}, c^{(i')}_{i(j_z j_z'(p_z))}, c^{(i')}_{i(j_z j_z' 0(p_z))}$, $c^{(i')}_{i(j_z 0(p_z))}$ - the transformation and separation fraction in the subsystem j_z, at one pass, of the quantity of component i' feeding the subsystem, into the quantity of component i being recycled, feeding the subsystem j_z', being distributed outside the system as intermediate product, and being distributed outside the system as finite product, respectively;

$c^{(i)}_{i(j_z j_z(p_z))}, c^{(i)}_{i(j_z j'_z(p_z))}, c^{(i)}_{i(j_z j'_z O(p_z))},$

$c^{(i)}_{i(j_z O(p_z))}$ - the separation fraction in the subsystem j_z, at one pass, of the quantity of component i feeding the subsystem, into the quantity of component i being recycled, feeding the subsystem j'_z, being distributed outside the system as intermediate product, and being distributed outside the system as finite product, respectively;

$h_{i(j_z(p_z))}$ - the ratio of the quantity of component i to the total quantity of components i feeding the subsystem j_z;

$h^{(i)}_{i(j_z(p_z))}$ - the ratio of the quantity of component i originating from outside the system to the quantity of component i feeding the subsystem j_z;

t(J)- the operating period of the multipurpose subsystem J;

$t(p_w)$- the operating period of the subsystem j_w, performing the purpose p_w;

d(J)- the density of the components feeding the multipurpose subsystem J;

$d(j_w(p_w))$- the density of the components feeding the subsystem j_w;

p_r, p_u, p_v, p_z, p_w are reffering to the purpose of the subsystem j_r, j_u, j_v, j_z and j_w, respectively.

The linear constraints (2.1)-(2.7) are expressing the material balance of the singlepurpose subsystems (2). The nonlinear constraints (2.8) are expressing the identity of composition of the material streams (3). The linear constraints (2.9) are expressing the dependence of the multipurpose subsystems on the singlepurpose subsystems. The dependence is expressed by the respective volumetric capacities. This condition is essential for the multipurpose system operation.

The model may be solved by nonlinear programming using an algorithm based on the decomposition of the nonlinear problem in several linear problems which are sequentially solved by an algorithm Simplex (4).

Applications of the model reffer mainly to the design,

in the initial stages of the process development,of flexible plants able to perform adequately several processes or alternatives of a process.

References

1. Badea, L.; 1978, Rev. Chim.(Bucureşti),29, pp. 427-432.
2. Badea, L.: 1978, Bucureşti, Ed. tehnică, "Modelarea bilanţurilor de materiale ale schemelor complexe".
3. Badea, L.: 1978, Rev. Chim.(Bucureşti),29, pp. 841-843.
4. Belcea, I.: 1979, Rev. Chim.(Bucureşti),30, pp. 57-59.

MINI-MICROCOMPUTER SOFTWARE FOR PROCESS CONTROL

G. Perdikaris, H. Schulz, and S. Weingarth

University of Wisconsin-Parkside, Kenosha, WI 53141

Abstract - The organization and design characteristics of a general purpose program suitable for real-time process control applications are presented and analyzed. An algorithm for solving the process control problem is described and flowcharted. Computer implementation of the algorithm using assembly language coding and integer arithmetic for the PDP-11 minicomputer is discussed and evaluated.

1. INTRODUCTION

The objective of this paper is the development of an appropriate software scheme for implementing computer control in the laboratory. The justification for the programming concepts presented stems from the belief that recent advances and continued improvement in mini-microcomputer technology, coupled with the advantages offered by digital computer control, will result in considerable increase of applications in the area.(1)

Generally speaking, a computer control system consists of the process to be controlled, the digital control law, or algorithm, stored in the computer, and the computer itself with its associated software and interface hardware. Such a control system is shown in Figure 1, where $G_c(z)$ and $G_p(s)$ represent the digital controller and process transfer functions, respectively. In addition, r represents the reference input signal, e the actuating error signal, m the manipulation variable, and c the controlled variable, or output. The computer interface may consist of analog-to-digital (A/D) and digital-to-analog (D/A) converters, or other types of suitable devices.

D. G. Lainiotis and N. S. Tzannes (eds.), Applications of Information and Control Systems, 442-446.

To implement computer control in the laboratory, the computer samples the process output every T seconds, and then converts the sampled value to an internal representation, such as a word, appropriate for the particular computer. Next, the error signal, which represents the difference between the reference input and controlled output, is determined and fed into the digital controller to calculate the manipulation. Finally, the manipulation is sent to the process via a D/A converter, or other appropriate interface. It is important that all these events are calculated in a period of time which is small in relation to the sampling time T in order for the control application to be successful and efficient.

2. THE APPLICATION PROGRAM, AN OVERVIEW

The proposed algorithm for solving the process control problem in the laboratory is described by the flowchart of Figure 2. The depicted solution is designed with flexibility, convenience, and efficiency in mind.

The assumed form of the digital controller is given by its 5th-order transfer function:

$$G_c(z) = \frac{K_o(A_o + A_1z^{-1} + \dots + A_5Z^{-5})}{(1 + B_1z^{-1} + \dots + B_5z^{-5})} \cdot \qquad (1)$$

Equivalently, the difference equation form of the control law is

$$m_n = K_o(A_oe_n + A_1e_{n-1} + \dots + A_5e_{n-5})$$

$$- (B_1m_{n-1} + \dots + B_5m_{n-5}). \qquad (2)$$

The parameters A_o - A_5 and B_1 - B_5 represent the controller coefficients, and K_o denotes the controller gain which can prove helpful when tuning the control algorithm on line. For instance, increasing the value of K_o normally tightens the response of the control loop, while decreasing K_o tends to reduce any ringing, or ripple, that may be present in the system response. The value of K_o is initially set to one, and any changing of K_o results in automatically multiplying the A coefficients by K_o and, thus, updating them.

The assembly language program coded from the flowchart of Figure 2 includes detailed instructions, which are optional. The operator may choose to command the system either from the teletype, or by using the A/D converter. The A/D converter feature makes it possible, for instance, to determine and evaluate the frequency response characteristics of the computer control system by simply using as reference the sampled output of a sine generator. Additional input includes the sign of the output to a positive reference, the maximum value, or limit, allowed for the manipulation, the sampling time, and the parameter values for $A_0 - A_5$ and $B_1 - B_5$. Finally, the type of control, namely open-loop or closed-loop, may also be selected.(2)

After the initial input of data, the program enters the command mode in which any of the parameter values may be examined and/or changed. Typing a G (for GO) starts the real-time control process, while typing an S (STOP) interrupts it.

The new manipulation, as defined by the control law, is calculated within the clock service routine. All other functions, such as changing or examining parameter values, are of lower priority and are performed in the background as the clock is interrupting.

It can be seen from Eq. (2) that the only term of the difference equation that depends on the present time is the product A_0e_n. All other terms depend on previous values of m and e. This suggests that the term SUM shown in the flowchart can be calculated ahead of time. When the program is first started, for instance, the term SUM is equal to zero and the new SUM can be calculated for the next interrupt as soon as the present manipulation is moved out. This scheme ensures that the time between sampling the process output and sending out the new manipulation, is kept minimum.

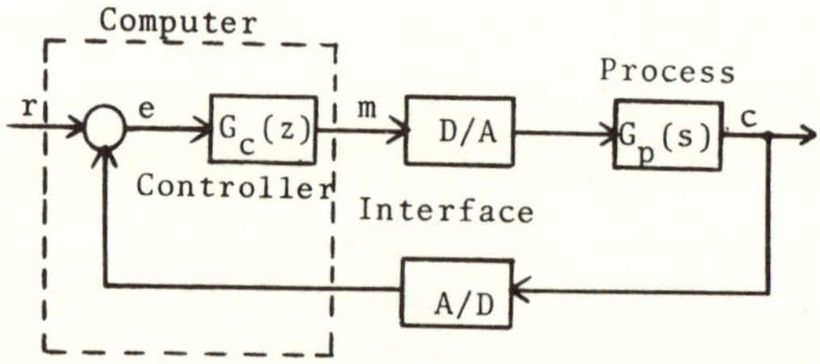

Figure 1. A Computer Control System.

3. A PDP-11 IMPLEMENTATION

The programming scheme illustrated in Figure 2, has been implemented on a PDP-11, 16-bit minicomputer using integer arithmetic. The computer interface shown in Figure 1, consists of a 10-bit-plus-sign A/D and an 11-bit-plus-sign D/A converter subsystems. For convenience the interrupt timing is kept by a programmable real-time clock operated in the repeat-interrupt mode. The selected clock rate is 10 KHz, crystal-controlled. For in-

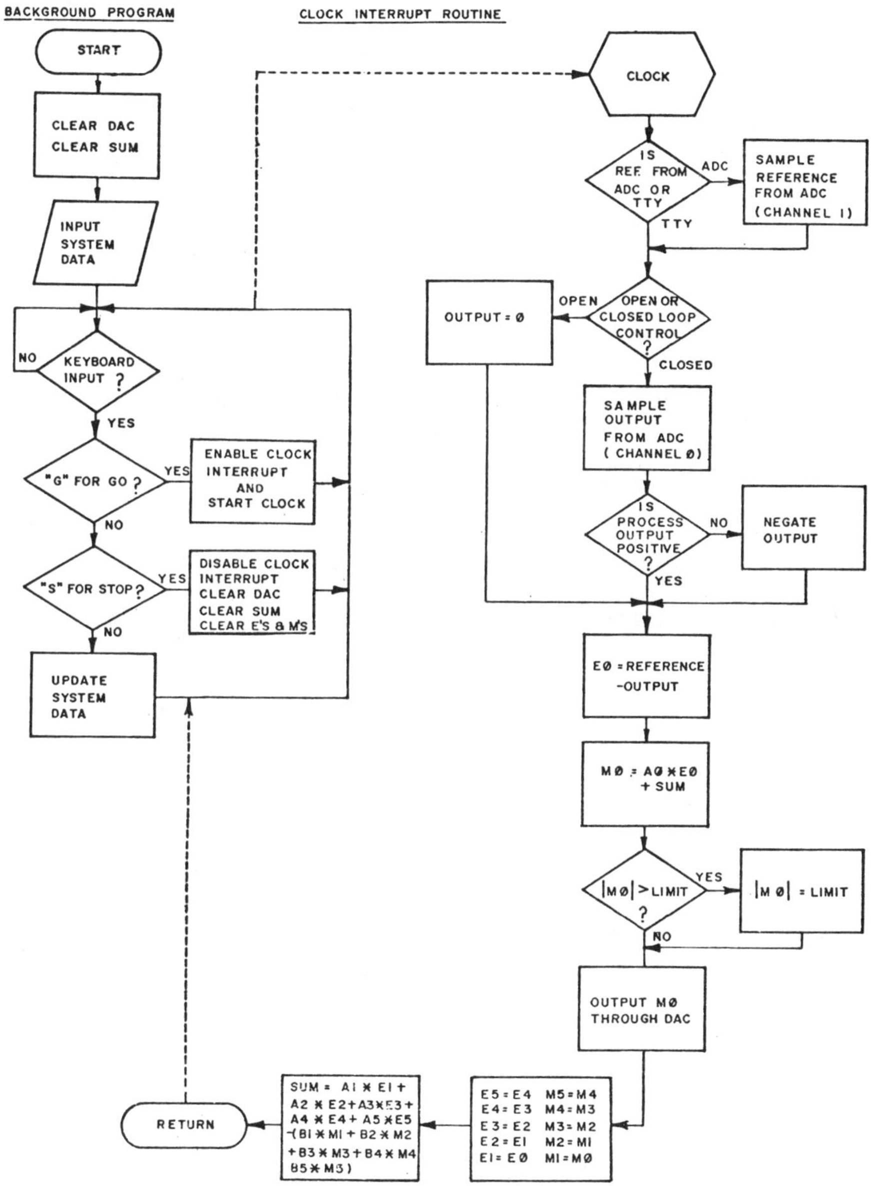

Figure 2. Flowchart for Implementing Process Control Algorithm.

stance, to specify a sampling time of 1 sec for the control loop, the integer value of 10,000 must be loaded into the clock's counter.

The A and B coefficients of a well-behaved control law range in values that fit in one 16-bit word with a fixed decimal point, that is $\pm$ 32.767. Two 16-bit words, or 32 bits can hold values ranging between $\pm$ 2.1475*10^9, while the 11-bit-plus-sign D/A converter can hold values between $\pm$ 2047. It can be shown that if all the terms of Eq. (2) are multiplied out to calculate a possible double-precision result, and then are added up in double precision, the sum will not underflow or overflow. For instance, considering the worst possible case, the maximum value of the m's is 2047, of the e's is 4094, and of the A's and B's the maximum value is 32,767. The sum then becomes

$$(32767)(2047)(5) + (32767)(4094)(6) = 1.1403*10^9,$$

which value will fit in a double-precision word.

After adding up the sum, the new manipulation is determined by dividing by 1000, because of the fixed decimal points of the A and B coefficients. The value of 1.1403*10^9/1000 obviously will not fit in a single 16-bit word, but this is not a problem. The worst case example is not realistic but serves to illustrate that the calculations can not overflow a double precision, or 32-bit word. After dividing by 1000, the result should not be much more than 2047, which is the limit of the $\pm$ 10 volt D/A converter. In conclusion, the intermediate calculations should not cause underflow/overflow problems. In addition, if after division by 1000 the result does not fit in one 16-bit word, something else is wrong and it should be corrected.

4. CONCLUSIONS

The digital controller form of Eq. (1) has proven to be versatile and practical for implementing controllers of 5th-order, or less. This controller may be used as is, or it may be easily modified to handle such special purpose controllers, for instance, as the proportional-integral-derivative (PID).

The PDP-11 assembly coding includes double precision multiply and single precision divide subroutines that can be replaced for processors with hardware multiply/divide capability.

REFERENCES

1. Harrison, T.J., Minicomputers in Industrial Control, ISA, 1978.
2. Perdikaris, G.A., "Computer Control for Deadbeat Response," Proceedings of the IEEE WESCON, Los Angeles, CA, 1978.

RADIAL AND ANGULAR SPECTRUM IN GEOPHYSICAL MAP ANALYSIS

P.S. Naidu, D.C. Mishra*

Indian Inst. of Science, Bangalore, India
(present address: Philips GmbH Forschungs-
laboratorium Hamburg, 2000 Hamburg 54, FRG)

From power spectrum of a geophysical map (gravity or magnetic field map) we define two new functions, namely, radial spectrum and angular spectrum by averaging the power spectrum estimate over annular and angular sectors. The radial spectrum contains useful information on depth to sources while the angular spectrum bears information on the orientation of the sources.

1. RADIAL AND ANGULAR SPECTRUM

A geophysical map of a physical quantity such as gravity or magnetic field may be modelled as a two dimensional (2-D) random field [1]. For stationary random field spectrum is an important attribute. The spectrum function is necessarily a decreasing function; and this decay rate turns out to be a useful piece of information to bring out the average decay rate. We define radial spectrum:

$$R_x(s) = \frac{1}{2\pi} \int_0^{2\pi} S_x(s \cos \theta, s \sin \theta) d\theta \qquad (1)$$

where $S_x(u,v)$ is 2-D spectrum of a random field, $X(x,y)$ and $s = (u^2+v^2)^{1/2}$ where u and v refer to two frequency variables of 2-D spectrum. $R_x(s)$ is now function of single variable s, radial frequency. The average decay rate is given by derivative of $R_x(s)$ with respect to s.

*Laboratory for Geofysik, Aarhus University,
8200 Aarhus N, Denmark.

D. G. Lainiotis and N. S. Tzannes (eds.), Applications of Information and Control Systems, 447-454.

The angular spectrum is defined as follows:

$$A_x(\theta) = \frac{1}{\Delta s} \int_{s_1}^{s_2} \frac{S_x(s \cos \theta, s \sin \theta) ds}{R_x(s)} \tag{2}$$

where $\Delta s = s_2 - s_1$ and $s_2 > s_1$. Note that from the definition of angular spectrum it follows that,

$$\frac{1}{2\pi} \int_{0}^{2\pi} A_x(\theta) = 1$$

and similarly from the definition of radial spectrum it follows that

$$\frac{1}{2\pi} \int_{0}^{\infty} s\, R_x(s)\, ds = \text{variance of } X(x,y).$$

Further, on account of symmetry of 2-D spectrum, namely,

$$S_x(u,v) = S_x(-u,-v)$$
$$S_x(-u,v) = S_x(u,-v)$$

the angular spectrum will have an additional property,

$$A_x(\theta) = A_x(\pi+\theta) \quad . \tag{3}$$

Since spectrum is positive real, the radial spectrum and angular spectrum are also positive real.

2. GEOPHYSICAL FIELDS

We shall now show how we can use the quantities defined in the previous section by considering simplified but useful models of density and susceptibility variation. The first model that we consider is a point mass placed at h units below the surface of observation.

The Fourier transform of vertical component of gravitational field is given [2] by

$$\bar{F}_z(u,v) = 2\pi Gm \exp(-sh)\exp(-j(ux_o+vy_o)) \tag{4}$$

where G is gravitational constant, m is mass and (x_o,y_o) refer to location of the point mass. From (4) the spectrum of the gravitational field is given by:

$$S_x(u,v) = |\bar{F}_z(u,v)|^2 = 4\pi^2G^2m^2 \exp(-2hs), \tag{5}$$

and hence the radial spectrum

$$R_x(s) = (2\pi)^2G^2m^2 \exp(-2hs) . \tag{6}$$

Since the spectrum is a function of radial frequency alone the angular spectrum is simply a constant

$$A_x(\theta) = 1 . \tag{7}$$

From (6) it is clear that the decay rate of logarithm of radial spectrum $\partial \ln R_x(s)/\partial s$ = constant = - 2h. This property of radial spectrum is, therefore, useful in estimating depth to the point mass.

The role of angular spectrum is best shown by considering a dipole. The Fourier transform of the vertical component of magnetic field due to a horizontal magnetic dipole is given by

$$\bar{F}_T(u,v) = 2\pi\ j(u\ m_x+vm_y)\exp(-sh) \exp(-j(ux_o+uy_o)) \tag{8}$$

where m_x and m_y are two components of dipole moment. The spectrum of the magnetic field will be given by:

$$S_x(u,v) = (2\pi)^2(u\ m_x+vm_y)^2\exp(-2hs) \tag{9}$$

The radial spectrum may be obtained from (1) using (9),

$$R_x(s) = (2\pi)^2(m_x^2+m_y^2)\exp(-2hs)s^2 . \tag{10}$$

Note that $\ln\{R_x(s)/_s2\}$ decays logarithmically at a rate equal to -2h. The angular spectrum may be shown to be

$$A_x(\theta) = (\cos\theta + \tan\alpha \sin\theta)^2/(1 + \tan^2\alpha) \qquad (11)$$

where $\tan\alpha = m_y/m_x$, that is α is direction of the dipole. It is easy to show that $A_x(\theta) = \max$ at $\theta = \alpha$ and $A_x(\theta) = \min$ at $\theta = \alpha + \pi/2$. Therefore, the angular spectrum points in the direction of dipole.

Next we consider a more complex situation: A thin horizontal sheet of surface mass density $\rho(x,y)$ at depth h units below the observation plane. The Fourier transform of the gravitational field is given by [1, 2]

$$F_z(u,v) = 2\pi G\, \bar{\rho}(u,v)\exp(-sh) \qquad (12)$$

where $\bar{\rho}(u,v)$ is Fourier transform of surface density $\rho(x,y)$. The radial spectrum and angular spectrum of the gravitational field can be derived as in case of simple point mass or dipole model above,

$$R_x(s) = 4\pi^2G^2\, R_\rho(s)\exp(-2sh)$$

and

$$A_x(\theta) = A_\rho(\theta) \qquad (13)$$

where $R_\rho(s)$ and $A(\theta)$ refer to radial spectrum and angular spectrum of $\rho(x,y)$ respectively. Let us assume that $R_\rho(s)$ can be approximated by an exponential function

$$R_\rho(s) = \exp(-2\alpha s) \qquad (14)$$

where $\alpha \geq o$ is a constant. Under above approximation the decay rate of $\ln R_x(s)$ shall be proportional to $-2(h+\alpha)$. The depth estimate from the slope of $\ln R_x(s)$ plot will be in error by an amount equal to α. The angular spectrum is independent of depth to the sheet. It can, however, be related to major directions in the density variation. Closed form expressions for angular spectrum due to even simple geometries are difficult to obtain, however, it is relatively simple to compute numerically.

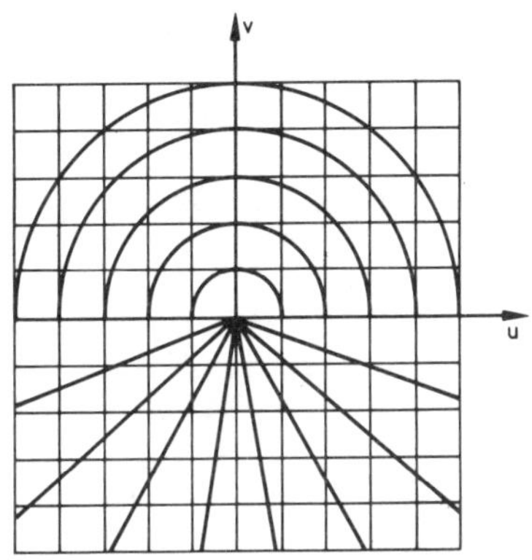

Figure 1. Concentric rings and angular sectors are drawn in spectral plane for computation of radial and angular spectrum

3. COMPUTATIONAL ASPECTS

The radial and angular spectrums are best computed directly from DFT coefficients. A square data matrix is derived from a geophysical map by uniform sampling. Next 2-D DFT coefficients are computed using FFT techniques. The DFT coefficients are squared and rearranged such that DC frequency is at the center and normal symmetry is preserved. A set of concentric rings with radii 1, 2, 3, ... are drawn centered at DC frequency (Fig. 1). All DFT coefficients lying inside a ring are averaged yielding an estimate of radial spectrum at frequency equal to mean of inner and outer radii.

For estimation of angular spectrum the frequency plane is divided into several angular sectors as shown in Fig. 1. The DFT coefficients are normalized with respect to radial spectrum. All normalized squared DFT coefficients lying inside a sector are averaged yielding an estimate of angular spectrum. Sometimes it is necessary to average over a portion of the sector only, for example, over all points lying within a radial frequency interval s_1 to s_2 ($s_2 > s_1$).

4. FIELD EXAMPLE

We have processed a bouger anomaly map of Central India (area lying between 80° - 83° E and 21,75° - 24,50° N [3]. The above area covers two major tectonic units of India geology - namely, Godwana Basins (South Rewa and Mahanadi Basins) in Southern part, and Vindhyan Basin in the Northern part separated by Narmada-Sone lineament. The Mahanadi Basin follows an NW-SE trend while the Rewa Basin follows E-W trend. The sampling interval was 2 miles (3.22 km). A data matrix of 96 x 96 was obtained. Both radial spectrum and angular spectrum were computed as described in section 3. In Fig. 2 we present the radial spectrum plotted on a semilog graph paper. It is remarkable

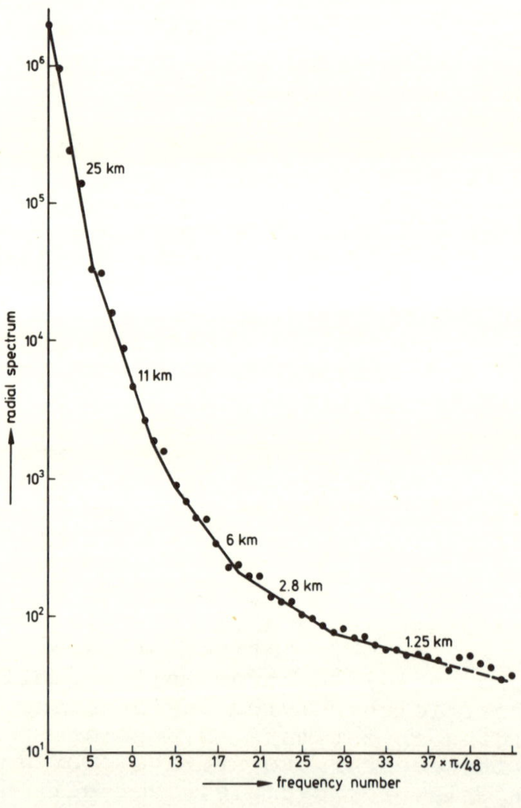

Figure 2. Radial spectrum of bouger anomaly map of Central India

to note that the spectrum decays over six decades before it gets submerged into white noise. The radial spectrum can be approximated by five linear segments. From the slopes we obtain the following depth estimates: 25., 11., 6., 2.8, and 1.25 km. We have not applied any correction to take into account the source radial spectrum (see eq. (14)). Further discussion on significance of the estimated depths may be found in [4].

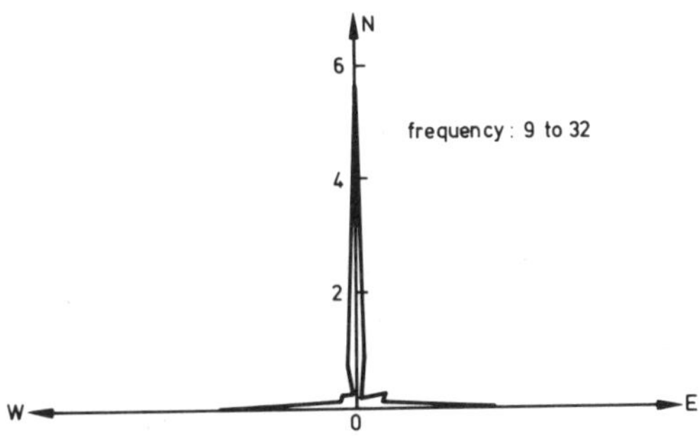

Figure 3. Angular spectrum averaged over frequency interval, 9 to 32

We have computed angular spectrum over four frequency intervals 1-8, 9-16, 17-24, 25-32. While the angular spectrums over 9-16, 17-24, and 25-32 were quantitatively similar, the angular spectrum over 1-8 was quite different. In Fig. 3 we present averaged angular spectrum over 9-32 frequencies and in Fig. 4 we present

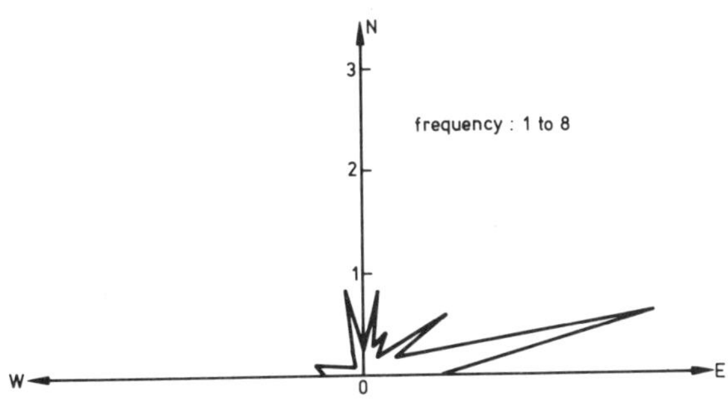

Figure 4. Angular spectrum averaged over a frequency interval, 1 to 8.

angular spectrum over 1-8 frequencies. From the orientation of main lobe in Fig. 3 we conclude that the major strike direction is E-W corresponding to that of Rewa Basin and from Fig. 4 the major strike direction is NNW-SSE, which is quite close to that of Mahanadi Basin.

Finally, it may be concluded that both radial and angular spectrums, derived from 2-D spectrum provide useful geophysical information.

Acknowledgement

Financial support from Alexander von Humboldt Stiftung to P.S. Naidu for participation in the conference is gratefully acknowledged. The work was done while the authors were with their parent institutions, Indian Institute of Science, Bangalore and National Geophysical Research Institute, Hyderabad, India, respectively.

References

[1] Naidu, P.S.: 1968 Spectrum of the potential field due to randomly distributed sources, Geophysics, vol. 33, p. 337-345.
[2] Naidu, P.S., and Mishra, D.C.: 1979 Digital processing of geophysical maps - unpublished manuscript.
[3] Kailasam, L.N.: 1976 Geophysical studies of major sedimentary basis of the India Craton. Textono-Physics, vol. 36, p. 225-245.
[4] Mishra, D.C. and Tiwari, R.K.: 1979 Spectral Study of bouger anomaly map of a rift valley and adjacent areas, presented at 41st Annual Conference of European Association of Exploration Geophysicists, Hamburg, June 1979.

GROWTH AND EXTINCTION IN RANDOM ENVIRONMENT

Amelia G. Nobile and Luigi M.Ricciardi

Istituto di Scienze dell'Informazione,Università di Salerno, Salerno, Italy

ABSTRACT

The evolution of a population whose size is described by a first order stochastic differential equation is investigated. Starting with a deterministic regulated growth model, a parameterization is made that implies random fluctuations of the population's carrying capacity. Contrary to other instances discussed in the current literature, for a Gompertz growth model it is proved that persistence of the population in the environment is independent of the intensity of the environmental fluctuations.

1. INTRODUCTION

In previous publications the effect of random environmental fluctuations has been investigated to shed some light on the ultimate fate of a population subject to a regulated growth process of the logistic type (1) ÷ (6). While referring to the quoted works for the general background, here we limit ourselves to mentioning that the apparently contradictory conclusions drawn by some authors on the chance of extinction for a logistically evolving population are entirely originated by the overlooked and yet fundamental difference existing between the two considered stochastic growth equations (5).

The purpose of this paper is twofold: On the one hand we shall investigate the role played by environmental fluctuations for a different type of regulated growth model; on the other hand

D. G. Lainiotis and N. S. Tzannes (eds.), Applications of Information and Control Systems, 455-465.

we shall pinpoint the relation existing between intrinsically discontinuous and intrinsically continuous growth processes of a given type whenever a diffusion approximation is invoked.

2. GOMPERTZ GROWTH PROCESS IN RANDOM ENVIRONMENT

Let x(t) denote the population size at time t, x_0 the initial number of individuals and α the population's intrinsic growth rate. The deterministic equation

$$\begin{aligned} \frac{dx}{dt} &= \alpha x - \beta x \ln x \\ x(0) &= x_0 , \end{aligned} \tag{2.1}$$

where $\beta > 0$ is a regulation parameter, defines the so called Gompertz growth law. This may be used to model a population with overlapping generations in continuous time. From (2.1) we obtain:

$$x(t) = \exp\left[\frac{\alpha}{\beta} - \left(\frac{\alpha}{\beta} - \ln x_0\right) \exp(-\beta t)\right], \tag{2.2}$$

showing that the value $K \equiv \exp(\alpha/\beta)$ is asymptotically attained by the population size. K is thus the environment's carrying capacity, the non trivial equilibrium point for the population size.

A discrete time "analogue" of (2.1) can be obtained in two fundamentally different ways by setting $t = n\tau$ $(n = 0, 1, 2, \ldots)$ with $\tau > 0$. A first way consists of substituting Eq.(2.1) with the difference equation

$$\begin{aligned} y_{(n+1)\tau} - y_{n\tau} &= (\alpha\tau - \beta\tau \ln y_{n\tau})\, y_{n\tau} \qquad (n = 0, 1, \ldots) \\ y_0 &= x_0 , \end{aligned} \tag{2.3}$$

where $y_{k\tau}$ denotes the population size at time $t = k\tau$. The second way, more natural if the population undergoes on intrinsically continuous growth process, leads us to the following system of difference equations:

$$x_{(n+1)\tau} - x_{n\tau} = x_{n\tau}\left\{\exp\left[\frac{\beta \ln x_{n\tau} - \alpha}{\beta}\left(e^{-\beta\tau} - 1\right)\right] - 1\right\}$$

$$(n = 0, 1, 2, \ldots), \qquad (2.4)$$

where $x_{k\tau} = x(k\tau)$ with the r.h.s. obtained by setting $t = k\tau$ in (2.2). Although (2.3) and (2.4) express two alternative discrete analogues of the continuous time Gompertz growth law, indistinguishable of one another in the limit as $\tau \to 0$, they are conceptually quite different from one another. Indeed, (2.3) is an approximation of the continuous growth equation, whereas (2.4) is obtained by approximating the solution of the growth equation.

Starting from (2.3) and (2.4) we now construct random processes by allowing for random fluctuations of the intrinsic fertility. Along the lines outlined in (5), we thus look at $\alpha\tau$ (the intrinsic relative change in population size during the generation time τ) as the sequence $\theta_0, \theta_\tau, \theta_{2\tau}, \ldots$ of independent identically distributed (i.i.d.) random variables defined as follows:

$$P_r\left\{\theta_{n\tau} = \sigma\sqrt{\tau}\right\} = \frac{1}{2} + \frac{\alpha\sqrt{\tau}}{2\sigma}, \qquad P_r\left\{\theta_{n\tau} = -\sigma\sqrt{\tau}\right\} = \frac{1}{2} - \frac{\alpha\sqrt{\tau}}{2\sigma}, \qquad (2.5)$$

$\sigma > 0$ being an arbitrary constant expressing the magnitude of the considered random fluctuations. From (2.5) we easily find for all n:

$$E[\theta_{n\tau}] = \alpha\tau$$

$$E[\theta_{n\tau}^2] = \sigma^2\tau \qquad (2.6)$$

$$E[\theta_{n\tau}^{2+p}] = \frac{\sigma^{2+p}\tau^{1+p/2}}{2}\left\{1 + (-1)^p + \frac{\alpha\sqrt{\tau}}{\sigma}\left[1 + (-1)^{p+1}\right]\right\}$$

$$= o(\tau) \qquad (p = 1, 2, \ldots).$$

With definition (2.5), from (2.3) we obtain the following analogue in random environment :

$$Y_{(n+1)\tau} - Y_{n\tau} = \left(\Theta_{n\tau} - \beta\tau \ln Y_{n\tau}\right) Y_{n\tau} \qquad (n = 0,1,2,\ldots)$$

$$\text{Prob}\{Y_0 = x_0\} = 1, \qquad (2.7)$$

whereas from (2.4) we get:

$$X_{(n+1)\tau} - X_{n\tau} = X_{n\tau}\left\{\exp\left[\frac{\beta\tau \ln X_{n\tau} - \Theta_{n\tau}}{\beta\tau}\left(e^{-\beta\tau} - 1\right)\right] - 1\right\}$$

$$(n = 0, 1, 2, \ldots)$$

$$\text{Prob}\{X_0 = x_0\} = 1. \qquad (2.8)$$

Making use of relations (2.6), from (2.7) we easily obtain:

$$\tau^{-1} E\{Y_{(n+1)\tau} - Y_{n\tau} \mid Y_{n\tau} = x\} = \alpha x - \beta x \ln x$$

$$\tau^{-1} E\{[Y_{(n+1)\tau} - Y_{n\tau}]^2 \mid Y_{n\tau} = x\} = \sigma^2 x^2 + \beta^2 \tau x^2 \ln^2 x$$

$$- 2\alpha\beta\tau x^2 \ln x$$

$$\tau^{-1} E\{[Y_{(n+1)\tau} - Y_{n\tau}]^{2+p} \mid Y_{n\tau} = x\} = x^{2+p}\, o(\tau)$$

$$(p = 1, 2, \ldots). \qquad (2.9)$$

Note that the l.h.s. of (2.9) are the moments of the increment of the process over one generation time, conditional on $Y_{n\tau} = x$. Hence, as $\tau \to 0$ $Y_{n\tau}$ converges to the diffusion process $Y(t)$ whose drift $A_1(x)$ and infinitesimal variance $A_2(x)$ are given by:

$$A_1(x) = \alpha x - \beta x \ln x \qquad (2.10)$$
$$A_2(x) = \sigma^2 x^2 .$$

Let us now return to (2.8). In this case the evaluation of the moments of the population size over one generation, conditional on $X_{m\tau} = x$, is rather cumbersome and special care has to be paid to the involved approximations. In the following, only the essential steps of the calculations are sketched, whereas the long intermidiate algebra is omitted. The first step consists of defining the random variables:

$$\xi_{m\tau} = \left[\exp\left\{\left[\beta\tau \ln X_{m\tau} - \theta_{m\tau}\right]\left[\exp(-\beta\tau) - 1\right] / \beta\tau\right\} \mid X_{m\tau} = x\right]. \qquad (2.11)$$

Setting

$$\varepsilon = \sigma\tau^{1/2}, \qquad (2.12)$$

due to (2.5) the following scheme defines $\xi_{m\tau}$:

$$\Pr\left\{\xi_{m\tau} = \exp\left[\frac{\beta\varepsilon \ln x - \sigma^2}{\beta\varepsilon}\left(e^{-\beta\varepsilon^2/\sigma^2} - 1\right)\right]\right\} = \frac{1}{2} + \frac{\alpha\varepsilon}{2\sigma^2}$$

$$\Pr\left\{\xi_{m\tau} = \exp\left[\frac{\beta\varepsilon \ln x + \sigma^2}{\beta\varepsilon}\left(e^{-\beta\varepsilon^2/\sigma^2} - 1\right)\right]\right\} = \frac{1}{2} - \frac{\alpha\varepsilon}{2\sigma^2} \qquad (2.13)$$

With these definitions we have:

$$E\left\{\left[X_{(m+1)\tau} - X_{m\tau}\right]^r \mid X_{m\tau} = x\right\} = x^r\, E\left\{(\xi_{m\tau} - 1)^r\right\} \qquad (r = 1, 2, \ldots), \qquad (2.14)$$

so that the moments of $\xi_{m\tau}-1$ have to be calculated to determine the conditional moments of the change in population size over one generation. Let us first calculate $E\{\xi_{m\tau}\}$. To this purpose, we make use of (2.13) and then expand the exponentials to the second order in ε to obtain:

$$E(\xi_{m\tau}) = x^{\exp(-\beta\varepsilon^2/\sigma^2)-1}\left[1+\frac{\varepsilon^2}{2}+\frac{\alpha\varepsilon^2}{\sigma^2}+o(\varepsilon^2)\right] \tag{2.15}$$

Hence:

$$E\{X_{(m+1)\tau}-X_{m\tau}\mid X_{m\tau}=x\} = x\left\{x^{\exp(-\beta\varepsilon^2/\sigma^2)-1}\left[1+\frac{\varepsilon^2}{2}+\frac{\alpha\varepsilon^2}{\sigma^2}+o(\varepsilon^2)\right]-1\right\}. \tag{2.16}$$

From (2.16), after some calculations we obtain:

$$\lim_{\tau\to 0}\tau^{-1}E\{X_{(m+1)\tau}-X_{m\tau}\mid X_{m\tau}=x\} = x\left(\alpha+\frac{\sigma^2}{2}-\beta\ln x\right) \tag{2.17}$$

By a similar procedure, one can then prove the following result:

$$\lim_{\tau\to 0}\tau^{-1}E\{[X_{(m+1)\tau}-X_{m\tau}]^2\mid X_{m\tau}=x\} = \sigma^2x^2 \tag{2.18}$$

The calculation of the higher order infinitesimal moments is more cumbersome. One starts from (2.14) and write:

$$E\{[X_{(m+1)\tau}-X_{m\tau}]^{2+p}\mid X_{m\tau}=x\} = x^{2+p}\sum_{k=0}^{2+p}\binom{2+p}{k}(-1)^{2+p-k}E(\xi^k_{m\tau})$$

$$(p=1,2,\ldots). \tag{2.19}$$

Making then use of (2.13) the $E(\xi_{m\tau}^{k})$ can be written. After disregarding infinitesimal quantities of order higher than $\varepsilon^2 = \sigma^2\tau$, one obtains:

$$E\left\{\left[X_{(m+1)\tau} - X_{m\tau}\right]^{2+p} \mid X_{m\tau} = x\right\} = x^{2+p}\left\{(-1)^{p} + (-1)^{p+1}\cdot\right.$$

$$(2+p)\,x^{\exp(-\beta\varepsilon^2/\sigma^2)-1}\left[1 + \frac{\varepsilon^2}{2} + \frac{\alpha\varepsilon^2}{\sigma^2}\right] + \sum_{k=0}^{p}(-1)^{p-k}\cdot$$

$$\binom{p+2}{k+2}\left[x^{\exp(-\beta\varepsilon^2/\sigma^2)-1}\right]^{k+2}\cdot\left[1 + \varepsilon\left(\frac{\varepsilon}{2} + \frac{\alpha\varepsilon}{\sigma^2}\right)\cdot\right.$$

$$(k+2) + \varepsilon^2\binom{k+2}{2} + o(\varepsilon^2)$$

$$(p = 1, 2, \ldots). \qquad (2.20)$$

Hence, on account of the identities

$$\sum_{k=0}^{p}(-1)^{p-k}\binom{p+2}{k+2}(k+2) = (-1)^{p}(p+2)$$

$$\sum_{k=0}^{p}(-1)^{p-k}\binom{p+2}{k+2}\binom{k+2}{2} = 0 \qquad (2.21)$$

from (2.20) the following result can be proved:

$$\lim_{\tau\to 0}\tau^{-1}E\left\{\left[X_{(m+1)\tau} - X_{m\tau}\right]^{2+p} \mid X_{m\tau} = x\right\} = 0 \qquad (p = 1, 2, \ldots). \qquad (2.22)$$

We have thus proved that as $\tau \to 0$ $X_{m\tau}$ converges to the diffusion process $X(t)$ having drift and infinitesimal variance given by:

$$B_1(x) = \left(\alpha + \frac{\sigma^2}{2} - \beta \ln x\right) x$$
$$B_2(x) = \sigma^2 x^2. \tag{2.23}$$

A glance at (2.23) and (2.10) indicates that processes $X(t)$ and $Y(t)$ differ due to the extraterm $\sigma^2 x/2$ present in the drift of $X(t)$. This difference is a consequence of the procedure followed to derive $Y(t)$ and $X(t)$. The former arises by substituting a discrete approximation of equation (2.1) by a difference equation and then changing it into a stochastic equation and taking the limit as the generation time goes to zero. The latter is the continuous limit of the solution to the difference equation (2.8) obtained by discretizing the solution of (2.1). It should be mentioned here that this is the manifestation of a property holding for models described by equations of more general form. However, this will be proved elsewhere.

Returning to the diffusion processes (2.10) and (2.23), let us determine their steady state distributions $W(x)$. Using the known formula

$$W(x) = \frac{c}{B(x)} \exp\left\{2 \int^x \frac{A(y)}{B(y)}\, dy\right\}, \tag{2.24}$$

holding for a diffusion process having drift $A(x)$ and infinitesimal variance $B(x)$ and specifying the arbitrary constant c by means of the normalization condition over the interval $(0,\infty)$ one is led to the following expression:

$$W(x) = \left(\frac{\beta}{\pi\sigma^2}\right)^{1/2} x^{-2} \exp\left\{-\frac{\sigma^2}{4\beta}\left(\frac{2m}{\sigma^2} - 1\right)^2 + \frac{2m}{\sigma^2} \ln x - \frac{\beta}{\sigma^2} \ln^2 x\right\} \tag{2.25}$$

where $m = \alpha$ for $Y(t)$ whereas $m = \alpha + \frac{\sigma^2}{2}$ for $X(t)$.

In both cases $W(x)$ is a lognormal distribution having mean $\exp\left\{\frac{m}{\beta}-\frac{\sigma^2}{4\beta}\right\}$ and variance $e^{\frac{2m}{\beta}}\left[1-e^{-\frac{\sigma^2}{2\beta}}\right]$. Making use of (2.25), the probability of ultimate extinction

$$P(x_0) \equiv \lim_{t\to\infty} \Pr\left\{ Z(t) \leq 0 \mid Z(0) = x_0 \right\} \qquad (2.26)$$

is immediately seen to be vanishing for both cases $Z(t) \equiv Y(t)$ and $Z(t) \equiv X(t)$. This implies that both models for the Gompertz growth law in random environment lead to persistence of the population.

Quite different is this conclusion from that of (2) in which a logistic growth process in random environment is studied. Starting from equation

$$\frac{dx}{dt} = \alpha x - \beta x^2 \qquad (2.27)$$
$$x(0) = x_0 ,$$

he sets:

$$\alpha = \alpha + \Lambda(t) \qquad (2.28)$$

where $\Lambda(t)$ is a stationary delta-correlated normal process with intensity σ^2. Hence, he models the population growth by a stochastic differential equation and obtains the diffusion process having drift and infinitesimal variance given by:

$$\begin{aligned} C_1(x) &= \alpha x - \beta x^2 \\ C_2(x) &= \sigma^2 x^2 \end{aligned} \qquad (2.29)$$

This process can be viewed as the continuous approximation of the stochastic difference equation

$$Z_{(n+1)\tau} - Z_{n\tau} = \theta_{n\tau} Z_{n\tau} - \beta\tau Z_{n\tau}^2 , \qquad (2.30)$$

corresponding to eq. (2.7) in the Gompertz case. However, for this case from (2.24) and (2.29) it follows that $W(x)$ only exists if $\alpha > \sigma^2/2$. This implies that the ultimate extinction probability (2.26) is null only in the carrying capacity (α/β) is sufficiently large compared with the magnitude of the environmental fluctuations. This remark had led May to the conjecture that, independently of the growth model, some such relation has to exist between carrying capacity and magnitude of environmental fluctuations in order for the population to persist in the given environment. Successively (3) it was pointed out that such conjecture does not hold for the logistic case if the stochastic differential equation originating from (2.27) and (2.28) is integrated by means of the Stratonovich calculus, which corresponds to writing a stochastic difference equation by a procedure similar to that leading to (2.8) and by considering the continuous limit case of its solution. It thus appeared that the validity of the conjecture was dependent upon the calculus used to construct the continuous diffusion processes or, equivalently (5), upon the alternative way of defining and solving the approximating stochastic difference equations. From the foregoing considerations it follows that also this second conjecture does not hold since the Gompertz growth process in random environment in no case ultimate extinction is a certain event. One can finally investigate the behavior of the sample paths by arbitrarily fixing two states $\varepsilon < x_0 < \eta$ and calculating the probabilities $P(\varepsilon \mid x_0)$ and $P(\eta \mid x_0)$ that the process reaches for the first time ε and η, respectively. While omitting the calculations at this time, we limit ourselves to mentioning that for the Gompertz models there results:

$$P(\varepsilon \mid x_0) = P(\eta \mid x_0) = 1$$

for both $X(t)$ and $Y(t)$. This is a qualitatively different result from the one holding for the logistic growth model where once more an explicit role is played by the magnitude of the carrying capacity as compared to the magnitude of the environmental fluctuations (7).

ACKNOWLEDGEMENTS. Work supported in part by the Group for Mathematical Information Sciences of the National Reseach Council.

REFERENCES

1. May,R.M.: 1971,Theoret.Population Biology 8, pp.49-63

2. May,R.M.: 1973, Princeton, University Press

3. Feldman, M.W., Roughgarden, j.: 1975, Theoret.Population Biology 7, pp.197-207

4. Keiding, N.: 1975, Theoret.Population Biology 8, pp.49-63

5. Ricciardi, L.M.: 1979, Biological Cybernetics 32, pp.95-99

6. Capocelli,R.M., Ricciardi, L.M.: 1979, Journal of Cybernetics 9, pp.297-312

7. Ricciardi,L.M.: 1977, Lecture Notes in Biomathematics 14

SITING OF NUCLEAR POWER PLANT FACILITIES: THE ENERGY - ENVIRONMENT INTERFACE

G. J. VACHTSEVANOS AND P. N. PARASKEVOPOULOS

School of Engineering
Democritus University of Thrace
Xanthi - Greece

ABSTRACT The selection and evaluation of sites for nuclear power plants, throughout the world, have become increasingly difficult in recent years as pressure from various societal segments has resulted in strict consideration of the institutional,environmenta safety, socio-economic and engineering factors affecting the siti construction and operation of such facilities.
Systems analysis approaches have been used in establishing the acceptability of candidate sites and facility designs.
This paper focuses upon techniques for determining the environment impact of proposed nuclear facilities.
The methodology is based upon the development of accurate and detaild mathematical models for predicting the impact of cooling water intake and thermal effluent discharge from nuclear plants.
Case studies from the application of these techniques to nuclear facilities along the Eastern United States coastline are described Model results and available data indicate close agreement.
Finally, a methodology for the integration of such analytical algorithms into an overall decision making tool is outlined.

1. INTRODUCTION Worldwide concern has been evident in recent year for the development of alternative energy sources because of the uncertain availability and the spiraling cost of conventional foss fuels. Nuclear energy, as an alternative source, has been finding extensive use in developed and developing countries accompanied with an increased concern over safety, environmental and social issues. Of paramount importance is the problem of siting nuclear facilities with its multitude of critical safety, environmental and socio-economic parameters. The immense complexity of the situation facing the decision maker leads to a realization of the need

D. G. Lainiotis and N. S. Tzannes (eds.), Applications of Information and Control Systems, 466-474.

for distinguishing and characterizing the hierarchy of problems and the approaches required for their effective solution.

Tools of systems analysis have found extensive use in nuclear facilities siting. Decision analysis and multi-attribute theory[1], the use of exclusionary criteria[2] and probabilistic methods[3] as well as the utility of critical pathway models and log-normal frequency distributions[4] have assisted in establishing systematic methodologies essential for the solution of the siting problem.

2. THE SITE SELECTION METHODOLOGY A comprehensive site selection process presents a formidable task to the decision makers. From the selection of potential sites through detailed site review and preliminary site recommendations, the path is most difficult, uncertain and often susceptible to severe criticism.The relativity of even a systematic procedure must be recognized at the outset and the enormous number of "fuzzy" and overlapping factors influencing the decision making must be delineated and accepted

The site selection methodology combines selective screening to narrow down the search area and a classification and rating scheme to rank siting possibilities in order of preference for detailed consideration. This cimbination for large - sized regional studies appears to be most cost effective in that the more detailed and costly field investigations are directed to a relatively small number of specific siting areas.

The selective serening procedure involves a series of steps. The first ones, characterized as country-wide screening below, use a gross screening mesh, primarily surface criteria and an overlay technique to narrow down the search area. The candidate regions are further ranked on the basis of information available on potential cooling water availability, seismotectonic characteristics, meteorology, land use, demography, accessibility, transmission line access and routings and topography. Candidate regions are selected on the basis of a selective screening and ranking technique. In the final step of the selective screening procedure, ecology is added as a decision parameter and the procedure is repeated in order to delineate candidate siting areas. This concludes the "regional" approach leading to an aggregate of possible sites to be evaluated in detail for their suitability to host a nuclear power plant facility. We summarize below the basic procedural steps used in the selective screening philosophy.

SELECTIVE SCREENING PROCEDURE

(a) Country - Wide Screening

- Land Availability
- Water Availability
- Seismotectonic Areas

(b) Candidate Regions Screening

- Hydrology
- Geology
- Land Use
- Meteorology
- Accessibility
- Transmission Lines
- Demography
- Topography

(c) Candidate Siting Areas Screening

- Hydrology
- Geology
- Land Use
- Meteorology
- Accessibility
- Transmission Lines
- Demography
- Topography
- Ecology

At this point in the site selection process, the candidate sites are screened in an attempt to determine a recommended ranking. The systematic methodology for the quantification and evaluation of alternative site options includes the following elements. A decision matrix, consisting of a vertical tabulation of the natural social and economic characteristics affecting the decision making mechanism and a horizontal listing of the potential sites; a multi-disciplinary group of professionals is convened after each had examined the sites and obtained first - hand in depth knowledge of such factors in their relevant areas of expertise which would influence the decision making process. Expert opinion modeling is here substituting for the exact laws of causality found in the physical sciences. Given also the fact that single experts may hold incorrect opinions, it is logical to seek increased accuracy through the opinions of a collection of experts. Thus a representative statistic of the experts opinions is chosen which, it is hoped, is more accurate than the opinion of the average expert; it is somehow corrected for individual bias and misinformation. The technique, known as Delphi[5] , has found numerous applications in technological forecasting. The group is charged with the task to determine collectively a numerical rating of all the natural, social and economic characteristics in accordance with their relative importance in meeting the project objectives.
With the rating, weights are assigned to each impact factor on a scale from 1 to 10, with the most significant factor carrying the maximum weight; a grading system is then developed to evaluate the alternate sites; each site is evaluated for each characteristic

on a scale from -10 to +10; -10 indicating severe impact and +10 indicating little or no impact. The grade assignment is carried out as follow:

For quantifiable factors, such as capital or operating costs, the site carrying the minimum cost estimate is assigned the maximum grade of 10. With this as a basis, the remaining alternatives are assigned grades using an inverse cost grade relationship. For factors which are not easily quantifiable, the grading is accomplished in a similar manner on the basis of selected criteria. For each factor, its weight is multiplied by the grades (for each alternative) and the results tabulated as elements of the decision matrix. These matrix elements are summed for each alternative, i.e along a column and the largest score designates the best alternative.

The procedure, as outlined above, incorporates a combined cost/benefit-risk analysis- the grades associated with risk factors carrying a negative sign. Any factors that do <u>not</u> affect the alternatives per se are not part of the vertical breakdown; they are included though in the general write-up and discussion.

This general model for the quantification of alternatives may be "tuned" further (particularly in cases of "equal" alternatives) via an incremental cost/benefit analysis technique.

The basic siting considerations accounted for in this methodology are listed below:

BASIC SITING CONSIDERATIONS

(a) Institutional

- (i) Required service date or on - line availability
- (ii) System reliability requirements
- (iii) Size and number of units/sites
- (iv) Search area boundaries

(b) Engineering

- (i) Safety
  - . Geology (Seismic)
  - . Hydrology (Flooding and Effluent Dispersion)
  - . Demography
  - . Meteorology (Dilution/Dispersion)
- (ii) Functional
  - . Cooling - Water Availability
  - . Geology (Foundation, Soil Characteristics)
  - . Accessibility (People, Materials and Components, Transmission Grid)

(c) Environmental
- (i) Ecological Sensitivity (Site, Transmission Corridors, Site Environs)
  - . Terrestrial
  - . Aquatic
- (ii) Land Uses (Site and Environs)
  - . Compatibility
  - . Dedicated Lands
  - . Areas of Historic and Archaelogical Significance
  - . Water Quantities and Qualities
  - . Climatology
  - . Demography
  - . Aesthetics

(d) Economic
- (i) Land Costs
- (ii) Cooling - System Alternatives
- (iii) Site Preparation Costs (Geology and Topography)
- (iv) Transmission Line Corridors
- (v) Site - Dictated Special Engineered Safequards

(e) Socio - Economic
- (i) Land - Owner Dislocations
- (ii) Competitive Use of Resources (Water and Land)
- (iii) Community Attitudes and Public Acceptance
- (iv) Economic Influence on Existing Life Styles

3. DETAILED SITING CONSIDERATIONS Among many detailed studies, field surveys, etc, required for the assessment of the potential impact of nuclear power plant operation upon the natural characteristics of the candidate sites are those related to environmental questions. In recent decades we have witnessed a proliferation of electric generating facilities along the banks of rivers, estuaries and other water bodies strategically located to utilize readily available cooling water for plant condenser needs. Major impacts of power plants using river water in once-through cooling relate to thermal effluent discharges and the entrainment and impingement of fish at the cooling water intake structure. Several mathematical modeling tools are available for the near-field impact assessment of the plant effluent discharges. Far - field, real - time hydrodynamic information is obtained from an unsteady, laterally - averaged two-dimensional hydrodynamic subsystem model[6]. Previous estuarine hydrodynamic models have been concerned with one - dimen-

sional (cross-sectionally averaged) behavior, either steady state, i.e. tidally averaged, or unsteady state. These efforts are unable to model the two dominant dispersal mechanisms of large coastal plain estuaries: gravitational circulation and shear dispersion. In addition, of course, such models are not designed to give information on vertical or lateral structure such as density stratification and velocity shear. The simplest approach which adequately takes account of some of these effects requires an unsteady-state two-dimensional model obtained by laterally integrating the governing three-dimensional equations.

The hydrodynamic subsystem model has a low degree of empiricism, thus claiming good predictive reliability for hydrodynamics and salinity transport. The model involves the solution by numerical means, of the laterally integrated time-dependent equations of motion for a region extending through the entire length of the Hudson Estuary. The numerical model computes the complete vertical and longitudinal distribution of unsteady velocity components. Intratidal computed flows and velocities, to be used as inputs to the population model, are everywhere in good agreement with field data.

The far-field temperature distribution profiles are obtained from a thermal energy balance equation similar to the one used for the salinity intrusion characteristics.

The biological system model[7] is formulated as a set of nonlinear partial differential equations in the age, temporal and special dimensions. The life cycle model has been applied to the striped bass population in a reach of the Hudson River extending from the George Washington Bridge (Mile Point 10, with Mile Point zero at the Battery) to Coxsackie(Mile Point 120).

This model reach is the location of major existing or planned power generating stations with a combined rated capacity of 4541 mw and a total plant flow (once-through-cooling) of 7272 cfs.

The basic life cycle for striped bass in the Hudson River consists of several stages: the first stage is the egg stage, which for any individual organism, lasts from 1.5 days to 3.0 days, depending on prevailing river temperatures. A large percentage of the vast number of eggs produced do not survice, being subjected to natural mortality via settlement into bottom muds, lack of fertilization, predation, e.t.c. They are also subject to entrainment and some presumed resultant mortality, when within the influence of plant intakes.

Those eggs which survive are transferred into yolksac larvae, the next stage of the life cycle. The yolk-sac larval stage lasts approximately 5-8 days. Similar behavior of natural and plant-induced mortality as well as passage of survivors into the next stage occurs from the yolk-sac to the post yolk-sac and juvenile stages. The post yolk-sac stage lasts for about 20-22 days. Migration mechanisms are also operative here. Plant induced mortality includ impingement on fish screens for the juvenile stages as well as entrainment into the circulating water system for larval and the first juvenile stages. The term "juveniles' as used here represents only that stage of the life cycle between the larval stage and completion of the first year of life. These "juvenils" are potenti ly subject to damage by the plant. After reaching the age of one year, the fish are no longer subject to entrainment nor for the most part, impingement.

Therefore, these immature fish have been lumped together with the mature adult population. Some five-year-old females are sexually mature and so, from that point on, adults are contributing to the egg complement, thus completing the cycle.
For an arbitrary element in space, the population balance is:

$$\frac{\partial c^k(x,a,t)}{\partial t} + \frac{\partial c^k(x,a,t)}{\partial a} + \frac{1}{A}\frac{\partial\left[c^k(x,a,t)\ Q^k(x,t)\right]}{\partial x}$$

$$= \frac{1}{A}\frac{\partial}{\partial x}\left[EA\frac{\partial c^k(x,a,t)}{\partial x}\right] + M^k(x,a,t)$$

$$-E_R^k(c,a)\ c^k(x,a,t) - KD(c,a)\ c^k(x,a,t) \qquad (1)$$

Where $c^k(x,a,t)$ represents the concentration (number of fish/volume) at a point in the population space defined by the longitudinal distance x, age a and time t in layer k along the vertical. The third and fourth terms on the left hand side of Equation (1) account for the net advective and despersive transport of organisms respectively. The population dynamics are described in terms of recruitment, loss and reproduction of individuals. Thus, the second left hand side term of Equation (1) represents individuals entering from a different life stage because of aging, M represents the migration fuction, E_R is the loss due to entrainment and/or impingement of organisms and KD represents the mortality rate.

Young-of-the year stripers are primarily influenced by water transport, whereas later stages are influenced by migratory preferences and fishing stresses. To predict the cropping of fish populations by power plants on the Hudson, the model requires data concerning the location of the plants, the plant flow rates and the rate at which various life stages of the fish will be drawn into the plant. Entrainment applies to the egg, larvae and early juvenile stages, in which the fish can pass through mesh screens and be subjected to mecahnical pressure and temperature stresses. Impingement applies to the later juvenile stages in which the fish cannot pass the travelling screens, and may be caught on them. The reduction of the population of any stage by entrainment or impingement is modelled empirically on the basis of extensive experimental evidence. In the ecosystem under consideration, the ability of the system to withstand the impact of power plant operations is determined by the density dependent mechanisms inherent in the natural regulation of the striped bass population. The presence of density dependent, i.e. compensatory phenomena, in the natural control of biological population has been demonstrated in numerous studies. In the model presented here, density dependent effects are incorporated through a functional dependence of the natural mortality rate of striped bass on their population density. The expression chosen to describe natural mortality in the striped bass biological system is of the form:

$$\text{Rate of mortality(fish/unit volume/day)} = KD(c).c \qquad (2)$$

Where c is the number of fish per unit volume of water and KD the decay rate, day^{-1}. This expression recognizes the tendency of the system (homeostasis or "biofeedback") to compensate itself when driven in either the direction of increased population or in the direction of decreased population. The functional form chosen for the variation of KD is:

$$KD = KE + (KE - KO)\left[\frac{c-C_s}{C_s}\right]^3 \qquad (3)$$

in which KE is the conventional first order or equilibrium rate, KD is the minimum unit mortality rate consistent with the species biology and C_s is the equilibrium population level. Thus, for the larvae and juvenile stages, KD is varied with the prevailing concentration so that the rate of mortality in these stages increases with increasing concentration (due to crowding, food availability, predation, etc.) and decreases with decreasing concentration for the converse reasons.

The numerical solution of the non-linear parabolic population equation is based upon a Crank-Nicolson implicit scheme. This procedure avoids stability problems and achieves second-order accuracy in the temporal direction. Verification of models of fish populations is, in general, a difficult task due to a paucity of reliable field data. However, in recent years, intensive biological sampling of the Hudson has been carried out, and there is a growing body of information concerning the striped bass population in the river. Beginning with an initial egg production the model moves through the entire series of life stages and provides a complete description of striped bass life history by accounting for the fate of all surviving organisms. The model generates egg, larvae and juvenile distributions in terms of the aging process as a function of river location and time, for a given layer of the river. Good agreement is observed between model predictions and available field data.

4. CONCLUSIONS A methodology, based upon the Delphi technique, has been presented for the analytical evaluation of the impact of natural and socio-economic parameters upon the siting of nuclear facilities. Such systems techniques have found extensive use in technological forecasting. In the detailed siting considerations, the dynamic coupling of the hydrodynamic and biological/thermal subsystem models has resulted in a highly accurate predictive tool for assessing power plant impacts. The model reliability is reduced whenever such interactive mechanisms are neglected.

The usefullness of such models is indisputable in testing the effectiveness of possible mitigating measures and deciding upon the most suitable of a number of alternative sites.

REFERENCES

1. KEENLEY, R.L., NAIR, K., "Decision analysis for the siting of nuclear power plants - the relevance of multi attribute theory" Proc. IEEE Special Issue on Social Systems Engineering, 1975.

2. NAIR, K., BROGAN, G.E., CLUFF, L.S., IDRISS, I.M., and MAO, K.T. "An approach to siting nuclear power plants: the relevance of earthquakes, faults and decision analysis"Proc. of the IAEA Symposium Vienna, 1975, pp 435-467.

3.JACCARINO, E, GIULIANI, P. and ZAFFIRO C. "Impact of external events on site evaluation" Proc.of the IAEA Symposium on Siting of Nuclear Facilities, Vienna, 1975, pp 427-431.

4. WAITE, D.A. and DENHAM, D.H., "Use of critical Pathway models and Log-Normal Frequency Distributions for siting nuclear faciliti Proc. of the IAEA Symposium on Siting of Nuclear Facilities, Vienna 1975 pp 407-415.

5. DALKEY, N.C., "Delphi", Rand Corporation, P-3704, 1969.

6. LEONARD, B.P., VACHTSEVANOS, G.J and ABOOD, K.A., "Unsteady-Stat Two-dimensional Salinity Intrusion Model for a Estuary" Proc. of International Conference on Applied Numerical Modeling, Southampton England, 1977.

7. VACHTSEVANOS, G.J., "Ecological Modeling:An Assessment Methodology for Power Plant Impacts". Proc. of International Conference on Cybernetics and Control, Japan, 1978, pp 634-637.

STREAM QUALITY MODELLING: A DISCUSSION

Mohamed F. Hassan* and Mohamed I. Younis**

* Asst. Prof. Cairo University, Faculty of Engineering, Giza-Egypt.

** Res. Assoc. Prof. Head, Systems and Information Sciences Lab., NRC, Giza-Egypt.

ABSTRACT

In this paper, different models for water qualtiy in streams are discussed. These models are used either to analyse water constituents or to solve the problem of treating effluent in order to keep standards of water quality. Specifically, two models are considered in detail. The first one is used to analyse water quality in stream while the second is used to control polluted reaches within reasonable bounds.
The mathematical relationships used in each are developed and the advantages as well as the limitations of each model are also discussed.

D. G. Lainiotis and N. S. Tzannes (eds.), Applications of Information and Control Systems, 475-496.

1. Introduction.

Research on the modelling and control of water quality in rivers and stream has been reported widely. The reason for this is that, with the extension of civilization, water in streams is used for industrial purposes and effluents are discharged again in rivers. This, of course disturbs the ecological balance in the system which may effect life (fish die for example). What is essentially required in a river which is also being used either as a source for portable supply or for recreation is to maintain pollution levels in polluted reaches within reasonable bounds consistent for both the community needs and for maintaining a satisfactory ecological balance as well as standards of water quality.

During the past few years, numerous mathematical models have been applied to simulate water quality conditions in rivers [1]. These models vary considerably in the degree of resolution (refinement)with which they represent the physical world. Low resolution models represent general trends for a few linked dependent variables over a limited set of boundary conditions. An example of the use of this type of model is the application of the Streeter-Phelps equation to represent average monthly dissolved oxygen (DO) concentrations in the San Antonio River [2]. High resolution models may be used to represent the responses of a large number of linked dependent variables over a much wider set of boundary conditions, for example, the application of the HEC model [3] simulating 20 constituents in the Boise River system. As the order of resolution of a model increases, so does the difficulty and the cost of its application. The selection of a model for a particular situation requires value judgements of acceptable trade-offs between the practicability and economy of model application and the amount and refinement of information to be provided by the model responses. The current trend is towards the application of the higher resolution models.

The purpose of this paper is to discuss the most recently developed models for water quality analysis and control. The rest of the paper is devided into four sections. In section (2), the dynamic models for constituents balance in a stream are discussed. Three different models which can be used to simulate water quality constituents in a stream are given and the general feature of each is clarified. In section (3) we consider in more details QUAL II model which can be used in stream quality analysis. A description of the model is given as well as its general relationships. Water constituents, reactions and interactions are given in details and the advantages as well as the model limitations from our point of view are discussed at the end of this section.

A simplified version of QUAL II model which can be used to simulate the dynamic behaviour of DO - BOD balance in polluted rivers is considered in section (4). The mathematical equations used in this model are derived and a suitable cost function which can be used for the purpose of controlling the levels of DO and BOD in polluted streams to maintain water quality standards is also proposed.

Finally in section (5) the problem of controlling DO in a river system is discussed. Different approaches are proposed to solve this optimization problem for a multi-reach river system in both deterministic and stochastic cases.

The advantages and limitations of this model are also discussed at the end of this section.

2. Dynamic Models for constituents balance in a stream:

Since most methods of systems analysis as well as of control system synthesis currently available to the control engineer require parametric models of the dynamic system under consideration, it is clear that these approaches necessitate the construction of a dynamic model for the in-stream interaction of the chemical and biochemical constituents. And if this model is to be useful, it should possess the following properties:

1- It should be a truly dynamic model, being capable of accepting time varying input functions and operating upon them to give time varying responses.

2- It should be as simple as possible consistent with the ability to adequately characterise the important dynamic and steady state aspects of the system behaviour.

3- If possible, it should provide a reasonable mathematical approximation of the physico-chemical changes occuring in the river system and should be verified against real data collected from the river over an extended period of time.

This section of the paper discusses briefly the various models that have been proposed in the published literature and in view of the available field data against which the model can be verified.

Research on the modelling of constituents interaction in a river system has been dominated by the classical model of Streeter and Phelps [4] which first appeared in a seminar paper published in 1925, and which has since been improved in various ways by other investigators such as Dabbins [5] and Camp [6]. Theoretically, the quantity of fluid mixing in stream provides a convenient parameter by which to distinguish between the presently available physico-chemical models;

i) Partial mixing: a second order partial differential equation model represented by :

$$\frac{\partial \underline{c}\,(z,t)}{\partial t} = \underline{f}\left[\frac{\partial^2 \underline{c}\,(z,t)}{\partial z^2}, \frac{\partial \underline{c}\,(z,t)}{\partial z}, \underline{c}\,(z,t)\right] \quad \text{----(2-1)}$$

$z_0 \leq z \leq z_1$; $t_0 \leq t \leq t_f$

ii) No mixing: a first order partial differential equation model represented by:

$$\frac{\partial \underline{c}\,(z,t)}{\partial t} = \underline{f}_2\left[\frac{\partial \underline{c}\,(z,t)}{\partial z}, \underline{c}\,(z,t)\right] \quad \text{-------(2-2)}$$

$z_0 \leq z \leq z_1$; $t \leq t \leq t_f$

iii) perfect mixing: a first order ordinary differential equation with associated transportation delay T_d to allow for distributed effects between spatial points Z_0 & Z_1

$$\frac{d\underline{c}\,(z_1,t)}{dt} = \underline{f}_3[\,\underline{c}\,(z_1,t)\,,\,\underline{c}\,(z_0,t-T_d)\,] \quad \text{-------(2-3)}$$

$t_0 \leq t \leq t_f$

In these equations t is the independent valiable of time, z is the independent valiable of distance, $\underline{c}$ is a vector of dependent concentration variables and f_1 , f_2 , f_3 are vector functions in the variabes, as shown. z_0 and z_1 represent the spatial boundaries of the system while t_0 and t_f the operational time interval.

From the theoretical point of view, model (i) would appear to represent the system dynamics most accurately, but for control and practical dynamic investigations of the lumped paremeter model (iii) seems the most attractive one.

As we shall show, in the rest of the paper, that QUAL II model is actually used to analyse the constituents changes within a river system and is based on type (i) model while the models developed by Young et al [7] Tamura [8], Hassan et al [9,10,11] and which are used to develop an automatic control system for maintaining pollution levels within reasonable bounds are actually based on model (iii).

3. Stream quality analysis by QUAL II model :

A model which has been widely applied in the past years is the moderable resolution QUAL II model. The original model QUAL I was developed by the Texes Water Development Board [12] to simulate conservative constituents, temperature, biochemical oxygen demand (BOD) and dissolved oxygen (DO) in a one dimensional steady flow river. The U.S. Environmental Protection Agency developed the model by adding additional constituents (ammonia, nitrate,coliform,phosphate,and algae) through a variety of consultants during planning studies on a number of river basins [13-15]. The resulting models are generally refered to as "QUAL II" although a number of different versions now exist. Most of these models have retained the original QUAL I model structure, the solution algorithm, the hydraulic subroutines and much of the input format.

What is actually used in QUAL II to represent the constituents changes within a stream is the type (i) model described in section (2). The fundamental one dimensional transport equation used in the model is given by :

$$(A_x\, d_x)\frac{\partial c}{\partial t} = \frac{\partial\left(A_x D_L \frac{\partial c}{\partial \chi}\right)}{\partial \chi}\,dx - \frac{\partial (A_x \bar{u} c)}{\partial \chi}\,dx + A_x\,dx\,\frac{dc}{dt} \pm s \quad (3\text{-}1)$$

Where c = concentration (M/L^3)
χ = distance (L)
T = time (T)
A_x = cross section area (L^2)
D_L = dispertion coefficient (L^2/T)
$\bar{u}$ = stream velocity (L/T)
s = source or sink (M/T)

Equation (3-1) is the traditional one dimensional mass balance equation representing the temporal change in concentrations of constituents within the river. The first term on the right hand side is the dispertion term and represents the transport of material by turbulence. The second term represents advection of constituents by the flowing water. The third term describes the individual constituent changes due to biochemical transformation while the fourth one represents the input or loss of constituent mass caused by external sources.

There are two terms that deserve special attention, These are the two derivatives that describe the local gradients and individual constituent changes. Under steady state conditions, the local derivative became equal to zero, in other words :

$$\frac{\partial c}{\partial t} = 0 \text{ ---------------------------} (3\text{-}2)$$

Changes that occur to individual constituents or particles independent of advection, dispertion and waste inputs are defined

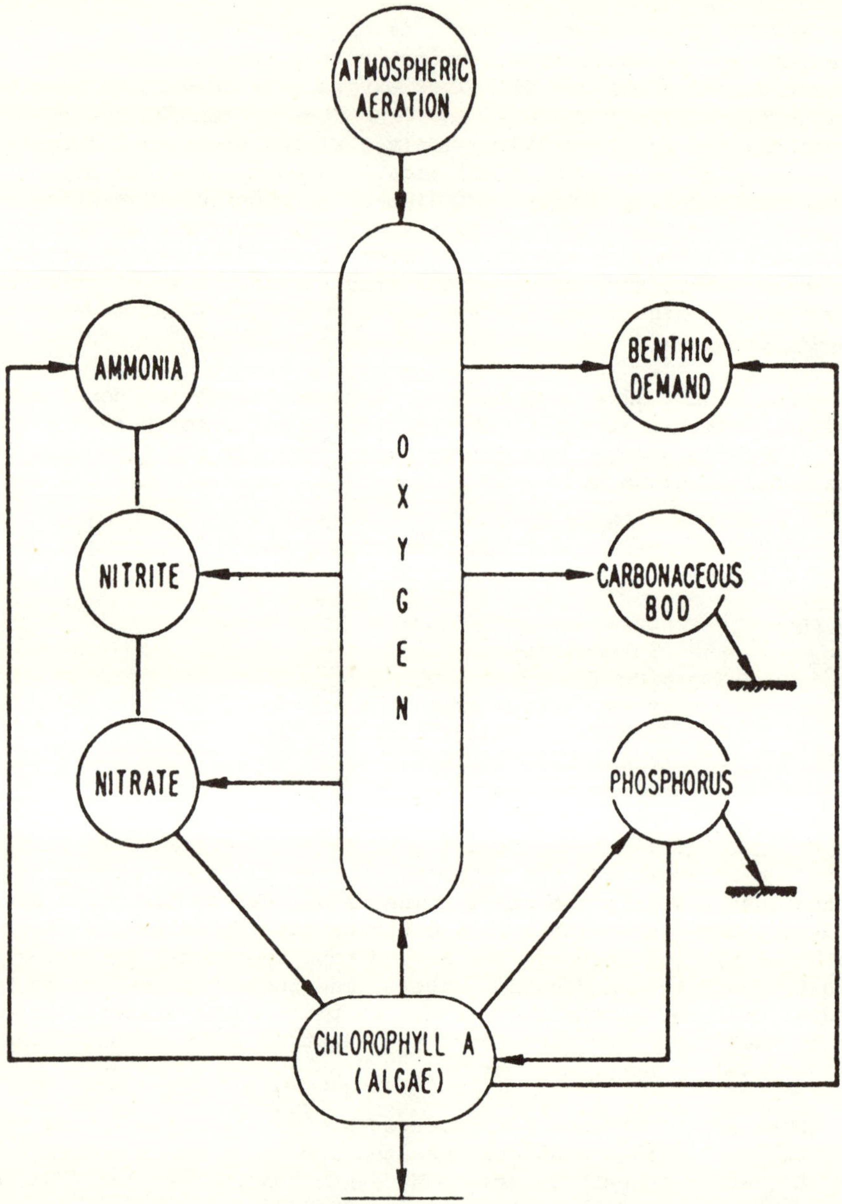
ATMOSPHERIC AERATION
AMMONIA
OXYGEN
BENTHIC DEMAND
NITRITE
CARBONACEOUS BOD
NITRATE
PHOSPHORUS
CHLOROPHYLL A (ALGAE)

Figure (1)

by the term :

$$\frac{dc}{dt} = \text{individual constituents changes} \text{-----}(3\text{-}3)$$

These changes include the physical, chemical and biological reactions and interactions that occur in the stream. Examples of these are reaction, algae respiration and photosynthesis and coliform die-off.

The basic differences between QUAL - I in its original form and QUAL II is that by means of QUAL II ; one can solve steady state problems directly plus it includes all the complex reactions and interactions of the nonconservative constituents (chlorophyll a, Nitrogen , Ammonia, Nitrate, Nitrite, phosphorus, Carbonaceous BOD , Benthic Oxygen Demand , Dissolved Oxygen, coliforms radioactive Material, Conservative Substances).

(3-1) General Model Relationships [16] :
QUAL - I in its original form had the capability to simulate conservative constituents. Thus the necessary modifications were directed towards the development of a model that could simulate the nonconservative constituents listed above. Of this list, QUAL I already had the capability to simulate carbonaceous BOD and dissolved oxygen as dependent constituents with first order Kinetics.

To accommodate the other constituents, QUAL I is modified to include the major interactions of the nutrient cycles, algae production, benthic oxygen demand, carboneous oxygen uptake and their effect on the behaviour of dissolved oxygen. Figure (1) illustrates the conceptualization of this model. It should be noted that the arrows on this figure indicate the direction of normal system progression in a moderately polluted environment; the direction may be reversed in some circumstances for some constituents. As an example of process reversal, consider that under normal conditions oxygen will be transferred from the atmosphere into solution and thus into the oxygen resources of the stream. Under conditions of oxygen supersaturation, however, which might occur as a result of algae photosynthesis, oxygen might actually be driven from solution, opposite to the indicated direction of the flow path.

(3-2) Constituent reactions and interactions :

The following paragraphs define the mathematical relationships that describe the individual reactions and interactions among the constituents treated.

(3-2-1) Chlorophyll a (Phytoplanktonic Algae) :

Chlorophyll *a* was considered to be directly proportional to the concentration of phytoplanktonic algae biomass. For the purpose of this model algae biomass was converted to chlorophyll *a* by the simple relationship.

$$c = \alpha_o A \qquad \text{(3-4)}$$

Where c = chlorophyll *a* concentration
A = algae biomass concentration
α_o = a conversion factor

The differential equation that governs the growth and production of algae (chlorophyll *a*) is formulated according to the following relationship :

$$\frac{dA}{dt} = \mu A - \rho A - \frac{\sigma_1}{D} A \qquad \text{(3-5)}$$

Where : A = algae biomass

T = time

μ = the local specific growth rate of algae as defined below, which is temperature dependent.

ρ = the local respiration rate of algae which is temperature dependent.

σ_1 = the local settling rate of algae

D = average stream depth.

Now, the local specific growth rate of algae is known to be coupled to availability of required nutrients and light. The standardformulation for the local specific growth rate in a stream takes the form :

$$\mu = \hat{\mu} \frac{N_3}{N_3 + K_n} \cdot \frac{P}{P + K_p} : \frac{1}{\lambda D} \ln \frac{K_L + L}{K_L + L e^{-\lambda D}} \qquad \text{(3-6)}$$

Where : $\hat{\mu}$ = the max specific growth rate
N_3 = the local concentration of nitrate nitrogen.
P = the local concentration of orthophosphate.
L = the local intensity of light.
λ = the light extinction coefficient in the river.
K_n, K_p, K_L = empirical half-saturation constants (temperature dependent).

(3-2-2) Nitrogen Cycle :

The nitrogen cycle contains three components. The differential equations governing transformation of nitrogen from one form to another are given below :

a) Ammonia Nitrogen :

$$\frac{dN_1}{dt} = \alpha_1 \rho A - \beta_1 N_1 + \frac{\alpha_3}{A_x} \qquad \text{----------(3-7)}$$

Where : N_1 = the concentration of ammonia nitrogen as nitrogen.

β_1 = rate constant for the biological oxidation of ammonia nitrogen, temperature dependent,

α_1 = the fraction of respired algae biomass which is resolubilized as ammonia nitrogen by bacterial action.
α_3 = the benthos source rate for ammonia nitrogen
A_x = average stream cross-section area
and other terms are as previously defined.

b) Nitrite Nitrogen :

$$\frac{dN_2}{dt} = \beta_1 N_1 - \beta_2 N_2 \qquad \text{---------------(3-8)}$$

Where : N_2 = the concentration of nitrite nitrogen as nitrogen.
β_2 = rate constant for the oxidation of nitrite nitrogen, temperature dependent.
and other terms are as previously defined.

c) Nitrate Nitrogen :

$$\frac{dN_3}{dt} = \beta_2 N_2 - \alpha_1 \mu A \qquad \text{-----------------(3-9)}$$

1 Note the coupling that exists between the conversion of nitrate and the production of algae to close the loop indicated in figure (1).

(3-2-3) Phosphorus cycle :

The formulation of the phosphorus cycle is less complex than the nitrogen cycle because themodel considers only the interaction of phosphorus and algae plus a sink term. Correspondingly, the differential equation describing the distribution can be written as :

$$\frac{dp}{dt} = \alpha_2 \beta A - \alpha_2 \mu A + \frac{\sigma_2}{A_x} \qquad \text{-------(3-10)}$$

Where : P = the concentration of orthophosphate as phosphorus
α_2= the fraction of algae biomass that is phpsphorus
σ_2= the benthos source rate for phosphorus
and all other terms are as previously defined.

(3-2-4) Carbonaceous BOD :

The rate of change of carbonaceous BOD in the stream is formulated as a first order reaction according to the formula :

$$\frac{dL_1}{dt} = - K_1 L_1 - K_3 L_1 \qquad \text{-----------(3-11)}$$

Where : L_1 = the concentration of carbonaceous BOD
K_1 = the rate of decay of carbonaceous BOD (temperature dependent).
K_3 = the rate of loss of carbonaceous BOD due to settling.

Note that while the change in BOD is expressed by equation (3-11) the oxygen demand exerted as a result of the change is only $K_1 L_1$.

(3-2-5) Benthic oxygen demand :

The benthic oxygen demand is considered to be fixed demand i.e.

$$\frac{dL_2}{dt} = \frac{K_4}{A_x} \qquad \text{---------------(3-12)}$$

Where : L_2 = benthic oxygen demand
K_4 = constant benthic uptake.

(3-2-6) Dissolved oxygen :

The differentional equation that describes the rate of oxygen in the model is written in the form :

$$\frac{do}{dt} = K_2 \; (0^*-0) + (\alpha_3\mu - \alpha_4\rho) \; A - K_1 L_1 - \frac{K_4}{A_x} - \alpha_5\beta_1 \; N_1 - \alpha_6\beta_2 N_2 \quad (3\text{-}13)$$

Where : 0 = the concentration of dissolved oxygen.
0* = the saturation concentration of dissolved oxygen at local temperature and pressure.
α_3 = the rate of oxygen production per unit of algae (photosynthesis).
α_4 = the rate of oxygen uptake per unit of algae respired.
α_5 = the rate of oxygen uptake per unit of ammonia oxidation.
α_6 = the rate of oxygen uptake per unit of nitrite nitrogen oxidation.
K_2 = the aeration rate in accordance with the Fickian diffusion analogy.

(3-2-7) Coliforms :

The differentional equation that describes the die-off of coliforms in the stream is :

$$\frac{dF}{dt} = -K_5 F \qquad \text{-------------}(3\text{-}14)$$

Where : F = the concentration of coliforms
K_5 = coliform die-off rate.

(3-2-8) Radionuclides :

The differentional equation which describes the changes in the radionuclides is given by :

$$\frac{dR}{dt} = -K_r R - K_a R \qquad \text{-------------}(3\text{-}15)$$

Where : R = concentration of radionuclides
K_r = radioactive decay rate
K_a = radioactive absorption rate.

Note :

All rate constants and other factors (except the saturation concentration of oxygen) that are known to be temperature dependent are formulated according to the relationship :

$$X_T = X_{T_s} \, \theta^{(T-T_s)}$$

Where : X_T = the value of the variable at the local temperature T.
X_{Ts} = the vaue of the variable at the standard temperature T_s.
θ = an empirical constant for each system variable.

The application of the QUAL II river model in some streams indicated several interest model limitations, the most significant being flow discontinuities at large diversions. Grenney et al [17] modified the model algorithm as well as the model program to eliminate the following limitations :

1- Mass balance was not maintained at a major tributary junction.

2- The solution algorithm became unstable when large diversions (flow discontinuities) were presented in the system.

3- Time varying boundary conditions were not available.

4- The manual assembly of a designation (flag) for each space element was tedious and time consuming.

(3-3) Discussion :

1- The developed QUAL II model is an extremely useful one for water quality analysis in streams.

It gives us the possibility of analysing both biochemical constituents as well as conservative minirals in either dynamic or steady state situations.

2- In polluted streams it is very difficult to use that model for controlling purposes. This is due to the fact that it is a distributed parameter model and control engineers will be faced with computational burden when trying to apply control theory on that model.

In the next section we shall discuss a simplified version of water quality models which fulfill control engineer requirements and on which one can apply the well developed control theory to get satisfactory responses of the concentration of water quality constituents in order to maintain ecological balance as well as water quality standards.

4- A simple dynamic model for DO-BOD balance in a Non-tidal river :

In industrial societies, rivers and streams are used as pumps for sewages. It was found that a good measure for the qualtiy of a stream can be obtained from two main factors, namely the instream biological oxygen demand (BOD) and the disolved oxygen (DO). If the Do falls below certain level or the BOD rises above certain level fish die for example.

One possible method for maintaining the DO in a stream at some prescribed satisfactory level is to control the level of BOD in the effluent discharges from sewage works adjacent to the river system. Actually sewage is treated at sewage stations prior to discharge into the stream and in general sewage works operates at a fixed level of treatment which would be determined by the level of BOD in the sewage which can be safly absorbed in the stream. However, the ecological balance of the river is often disturbed by unknown perturbations and it becomes necessary to vary the BCD content of the sewage in order to bring the river quality back to the desired values. It is the problem of control of DO and BOD levels in a river system.

(4-1) Model description [7]

In this model, a river or a stream is divided into a number of reaches where a reach is defined as a streach of a river of some convenient length which receives one major controlled effluent discharge from a sewage treatment facility. In chemical engineering terms a reach is considered as an analog to a continuous stirred tank reactor as shown in figures (2) & (3).

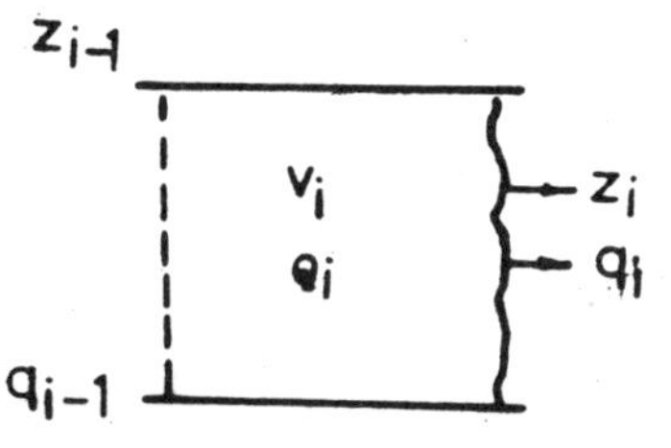

Fig. 2

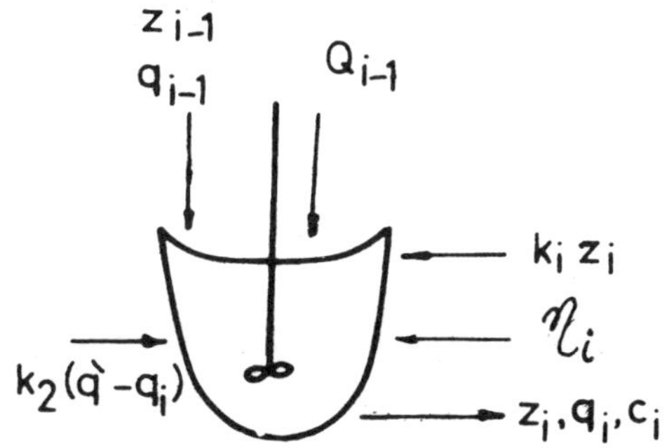

Fig. 3

Thus, for each reach a second order state variable model can be derived from a two component mass balance over the CSTR section and which describe the BOD-DO relationship at some average point in the reach, These are given by :

For reach i :

BOD :

$$\frac{dz_i}{dt} = -(K_{1i}+K_{2i}+\frac{Q_i + Q_E}{V_i})\, z_i + \frac{Q_{i-1}}{V_i}\, z_{i-1} + L_A + \frac{m_i Q_E}{V_i} \qquad (4\text{-}1)$$

Do :

$$\frac{dq_i}{dt} = (k_{3i} + \frac{Q_i+Q_E}{V_i})\, q_i - K_{1i} z_i + \frac{Q_{i-1}}{V_i}\, q_{i-1} + K_{2i} q_i s - \frac{m_i}{V_i} \qquad (4\text{-}2)$$

Where : z_i = output (downstream) BOD in mg/l

q_i = output DO in mg/l

z_{i-1} = input BOD in mg/l

q_{i-1} = input DO in mg/l

q_i^s = is the saturation concentration of DO in mg/l

Q_E = is the flow rate of effluent in mg/l

Q_i, Q_{i-1} = are the stream flow rates in reaches i & i-1 in million gallons/day.

V_i = the volum of water in reach i in million gallons.

K_{1i} = BOD decay rate $(day)^{-1}$ in reach i

K_{3i} = DO reaeration rate $(day)^{-1}$ in reach i

K_{2i} = is the rate of sedimentation rate constant $(day)^{-1}$.

$\frac{\eta_i}{V_i}$ = is the removel DO due to sludge reqirements in mg/l $(day)^{-1}$

L_A = is the mean rate of addition of BOD to the reach by local run-off in mg/l/day.

m_i = is the concentration of BOD in the effluent in mg/l.

In equations (4-1) and (4-2) q_{i-1} , z_{i-1} can be written in three different ways by taking into account the transportation delay between the adjacent reaches [8] :

a) no delay model :

$$z_{i-1} = z_{i-1}(t)$$

$$q_{i-1} = q_{i-1}(t)$$

b) pure delay model :

$$z_{i-1} = z_{i-1}(t-\tau)$$

$$q_{i-1} = q_{i-1}(t-\tau)$$

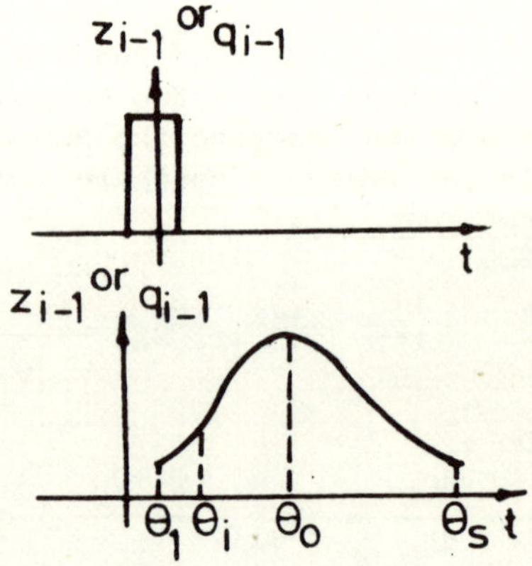

Fig.(4)

c) distributed delay model :

$$z_{i-1} = \sum_{j=1}^{s} aj \quad z_{i-1}(t - \Theta_j)$$

$$q_{i-1} = \sum_{j=1}^{s} aj \quad z_{i-1}(t - \Theta_j)$$

$$\sum_{j=1}^{s} aj = 1 \quad \Theta_1 < \Theta_2 < \ldots \Theta_s \text{ mean } \Theta_j = \Theta_0$$

(4-2) Steady state consideration :

Before considering the control problem it is necessary to define the desired values of the controlled variables which any controller should try. These desired values should be cosistance with the dynamic of the system . This is attained by setting $\frac{dz}{dt} = 0$; $\frac{dq}{dt} = 0$ which gives the desired values of z_i^*, z_{i-1}^*, q_i^*, q_{i-1}^* , m_i^*, m_{i-1}^* ,..etc.

(4-3) Cost functions [18]

What is actually required on a river is to maintain pollution levels within reasonable values to satisfy Community needs and ecological balance. For this reason we have chosen the following cost function which has to be minimized:

$$J = \frac{1}{2}\sum_{i=1}^{N} \int_{t_0}^{t_f} (z_i - z_i^*)^2 s_{i1} + (q_i - q_i^*)^2 s_{i2} + r_i (m_i - m_i^*)^2 \, dt$$

Where N is the number of reaches

Minimizing such a cost ensures that :

1- Minimizing the money to change the control which is a result of the term $r_i(m_i - m_i^*)^2$

2- z_i and q_i must not exceed an upper and lower limits $(z_i^* \pm z_i^{**})$ and $(q_i^* \pm q_i^{**})$ this can be ensured by the proper choise of the weights s_{i1} and s_{i2}.

One has to note that the exact method of implementing the effluent discharge control is open. In terms of the control analysis this can be done by either one of the following two methods :

Method 1 Using some form of variable treatment of the sewage effluent so that the level of BOD entering the river in the effluent could be modified in the manner specified by the control law.

Mehtod 2 If the control variable is considered as the effluent flow rate rather than BOD content, then it is possible that the effluent treated in some standard level, may be stored in tanks. This will be discharged in a controller manner as required by the control law. Such a control system would make the system to be an inherent time variable parameters.

5- Control of DO in a river system :

(5-1) Preliminary Control studies :

During the past few years there have been a number of publications concerned with the control of DO and BOD levels in river systems.

At the one extrem Dysart [19] uses discrete dynamic programming to solve the simpler static optimization problem for a multi-reach systemwhile at the other, Tarassov et al [20] consider the dynamic optimal control of the full distributed parameter system using in-stream aeration from a pattern of submerged pipes. Kendrik et al [21] using a lumped parameter model similar to that given by equations (4-1) and (4-2) suggest a rather complicated state variable solution to the dynamic optimization problem in the multi-reach case while Naito et al [22] apply discrete dynamic programming to a similar dynamic optimization problem with variable flow rates. More recently, Variaya [23] uses the DO-BOD system as an example to show the application of duality theorem.

Perhaps the most striking aspect of all the previous published work on the control of DO-BOD levels in a river is the almost total emphasis on optimal solutions. Now while it seems reasonable that optimal control theory should be applied to static problems such as those connected with the planning of future investment in sewage treatment facilities (see, e.g. Newsome et al [24])it is difficult to see why there has been such exclusive interest in optimal operational control, especially since it would appear that simpler andmore realistic control strategies have received no attention at all. In an attempt to correct this situation a little, therefore, we will consider the solution to the problem of maintaining a prescribed level of DO in a multi-reach river system.

(5-2) Control of DO in a multi-reach river system :

To get the optimal control of DO in a multi-reach river system, one has to chose a model for z_{i-1} and q_{i-1} in equations (4-1) and (4-2) . The most realistic one is that of distributed delay nature. Two possible procedures have been considered to solve this problem. The first one has been given by Tamura [8]. In that approach the discrete version of the model given by equations (4-1) and (4-2) and written for multi-reach system is considered. Tamura developed a two and three levels techniques based on the decomposition of a mathematical programming problem. The advantage of these methods is that the dimensionality of the problem is not increased due to the transportation time delay considered in the model since at the lower level one solves a single variable equation rather than a vector one.

However, the applicability of these techniques necessitates that the ponderation matrices for the states as well as the controls in the criterian functional must be diagonal as well as positive definit.

In addition, the solution of the problem results in an open loop control rather than a closed loop one. What is actually required is to get an on-line regulation of the sewage discharge of BOD from the multiple sewage works on a polluted river and this regulation must be independent of the initial disturbances in the system. This is the problem of getting a closed loop controller.

To solve this problem for large scale interconnected dynamical systems, Hassan et al [25] have developed a decomposition-coordination technique which results in a single or two level closed loop controllers. To solve the pollution control problem of a river system, Hassan, et al [10] considered six reaches of the river Cam near Cambridge. They approximated each time delay element by a second order differential equation and finally they got a 52 state variable model to be controlled. The technique developed in [25] is successfully applied on this problem and a two level optimal closed loop controller is derived which returns the BOD and DO to their desired levels in an optimal manner.

Remarks :

1. The continuously varying control levels, in time, which is derived from this model is too complex and too expensive to be considered for practical implementation. One can overcome this difficulty by using a discrete model in time which is realistic for a river system since it has a long time constant (about a day).

2. To overcome the transmission of information between the coordinator level and the subsystems (reaches) Hassan [26] has developed mainly three different techniques which yield to a decentralized robust controller. The applicability of these techniques has resulted in a stable and satisfactory response and the implementation of these controllers needs no transfer of information between the reaches. The advantage of this is that transmission of information between the subsystems and the

corrdinator may be more expensive than accepting certain levels of suboptimality.

3. In the derivation of these controllers it was assumed that:

a) All the state variables are measurable. Actually, the measurements of DO are easy to do while that of BOD is difficult and needs approximately four days.

b) It was assumed that the model is deterministic and that the system has no input disturbances as well as the measurements are not corrupted by measurement errors. However, these are not realistic assumptions since normally any system is usually corrupted by noises at both the inputs and outputs.

To overcome these drawbacks, Hassan, et al [11,27] have considered the stochastic version of that model in which it was assumed that the system is contaminated by white gaussion input disturbances and we measure,at the output, only the concentration of DO which is corrupted by white gaussian error of measurements. The optimal closed loop stochastic controller is calculated using two different ways :

i) Using the duality theorem between estimation and control [28], the optimal constant Kalman filter gain matrix for the 52 state variable model is calculated using the two level technique developed in [25] and by using the separation principle between estimation and control theoptimal closed loop stochastic controller is developed [11].

ii) Using the decentralized computational algorithm for the time varying global Kalman filter developed by Hassan et al [25] as well as the seperation principle, Hassan et al [27] have succeeded to get the optimal responses of the river pollution control problem.

(5-3) Discussion :

1- The developed model for the concentration of BOD and DO in polluted streams is extremely useful for pollution control purposes. This is because the model is linear, controllable, and observable . Thus, one can use the well established and developed techniquesof control theory to get the requirements that have to be satisfied (e.g., stability, optimal regulation, etc.)

2- The disadvantage of this model is that it considers the effects of constituents concentration in the adjacent upperstream reach on the reach under study and this is modeled by taking into consideration the transportation time delay. A more realistic model in which this difficulty does not appear can be derived for the steady state case by dividing the river into reaches and segments which have, for example, the same hydraulic properties and following a drop of water from the extremely upperstream point to the extremely downstream one. In this case, and due to the continuity property, the final results which one obtains at the end of one reach or segment is taken as the initial condition for the adjacent downstream one.

6- Conclusion :

In this paper, modelling of water quality in streams is taken into consideration in some details. Our discussion is actually oriented to two different models. The first one, which is called QUAL II model, is described and its mathematical relationships are developed. It is shown that this model is the most accurate one for water quality analysis purposes. However, it is very difficult to effectuate any control or management strategies from this model since it will cause computational burden. On the other hand, a simplified version of the QUAL II model which can be used to simulate the dynamic behaviour of DO-BOD balance in a river system is also developed. It is shown that this model is controllable as well as observable and it can be used successfully to get optimal control policies in order to maintain the levels of BOD and DO in polluted reaches within standard boundaries specified by the community. However, as it is shown, the model is very simple, ignores all other factors affecting DO and BOD levels and cannot give a good picture about all other water quality constituents in a stream. What is actually required is to develop a model which fulfills the two requirements. Specifically, we need a model which can be used for analyzing water quality in streams accurately and from which one can develop control or management policies for treating effluents discharged in rivers in industrial societies to maintain the ecological balance in polluted streams. This will be the purpose of our future study.

References :

1. Grenney , W.J.,et al. "Water Quality Relationships to Flow-stream and Estuaries" in "Methodologies for the Determination of Stream Resource Flow Requirements : An Assessment." Edited by C.B. Stalnaker, U.S. Fish and Wildlife Service, Office of Biololgical Services, Western Water Allocation, Washington, D.C.

2. Texes Water Development Board, "Simulation of Water Quality in Streams and Canals," Program Documentation and Users Manual, EPA-OWP-TEX-DOSAGI, NTIS (1970).

3. Chen, C.W.,and J.Wells. "Boise River Water Quality Ecologic Model for Urban Planning Study." Tetra Tech. No. TC-368 and TC -605. Tetra Tech., Inc., Lafayette, Calif. (1966)

4. Streeter, H.W.and Phelps, E.B. A Study of the Pollution and Natural Purification of the Ohio River. Bulletin No. 146, U.S. Public Health Service, 1925.

5. Dobbins, W.E. B.O.D. and Oxygen Relationships in Streams. A.S.C.E., J. Sanit Eng. Div., Vol. 90, No. SA3, June 1966.

6. Camp, T.R. Field Estimates of Oxygen Balance Parameters. A.S.C.E. J. Sanit. Eng. Div., Vol. 91, No. SA5, Oct. 1965.

7. Young, P., Beck, B. and Singh, M.G. The Modelling and Control of Pollution in a River System, Report CUED/B-Control/ TR32, University ofCambridge, 1972.

8. Tamura, H. A Discrete Dynamic Model with Distributed Transport Delays and its Hierarchical Optimization for Preserving Stream Quality. IEEE, SMC,Vol. SMC 6, No. 5, Sept., 1976

9. Singh, M.G. and Hassan, M.F. Closed Loop Hierarchical Control for River Pollution Automatica, Vol. 12, PP 261-266, 1976.

10.Hassan, M.F., Hurteau H., Singh, M.G. and Titli, A. A New Three Level Algorithm for River Pollution Control. Proc. IFAC Symposium on Systems Approaches to Development Cairo, Egypt. 1977.

11.Hasssan, M.F., Hurteau, R., Singh, M.G. and Titli, A. Stochastic Hierarchical Control of a Large Scale River System. Procs. of the VII IFAC World Congress. Helsinki, June 12-16, 1978.

12. Texes Water Development Board. "QUAL I Program Documentation and Users Manual," EPA-OWP-TEX-QUAL I, NTIS (1970).

13. Environmental Dynamics, Inc. "Utah Lake-Jordan River Basin Modelling Project Documentation and Sensitivity Analysis Report," Environmental Dynamics, Inc., Los Angeles, Calif., 1973.

14. Grenney, W.J.,t al. Model Used for Water Quality Management in a Waste Load Allocation Study of the Bear River. Utah Water Research Lab., Logan, Utah. Report PRWA 20-1 (1976).

15. Grenny, W.J.et al. Modifications to the QUAL I River Model for Application to the Schuylkill River. Final Report to Teledyne Isotopes, Inc.,Westwood, N.J. (1973).

16. Computer Program Documentation for the Stream Quality Model QUAL II, prepared for the Environmental Protection Agency, Systems Development Branch, Washington, D.C.

17.Grenny,W.J.,et al. Characteristics fo the Solution Algorithms for the QUAL II River Model. Journal Water Pollution Control Federation, Washington, D.C. 1976.

18. Singh,M.G. River Pollution Control. Int. Journal of Syst. Sci. 1975, Vol. 6, No. 1, 9-21.

19. Dysalt, D.C.,III. The Use of Dynamic programming in Regional Water Quality Planning. Presented at A.A.P.S.E. Systems Analysis Conference, June 1969.

20. Tarassov, V.J., Perlis, H. T. and Davidson, B. Optimization of a Class of River Aeration Problems by the Use of Multivariable Distributed Paremeter Control Theory. Water Resources Research, June 1969.

21. Kendrick, D. A., Rao, and Wells, C.H. Optimal Operation of a System of Waste Water Treatment Facilities. IEEE Symposium on Adaptive Processes. 1970.

22. Nairo, M., Takamsatsu, T. and Tamura, H. Optimum Planning of Sewage Treatment Systems for Preserving Stream Quality. Proc. 5th IFAC Congress, Paris, 1972.

23. Varaiya, P. P. Notes on Optimisation. New York. Van Nostrend 1972.

24. Newsome, D. H., Bowden, K. and Green, J. A. The Trent Mathematical Model: a) Construction Proc. I.W.E. Conference on Advanced Techniques in River Basin Management, Birmingham 1972.

25. Singh, M.G., Hassan, M.F. and Titli, A. Multilevel Feedback control for Interconnected Dynamical Systems Using the Prediction Principle. IEEE Trans. SMC., Vol. SMC 6, April 1976.

26. Hassan, M.F. Optimisation et Commande Hitrarchisee des Systems Dynamiques Interconnecte. These d'etat (Science) Toulouse 1978, France, No. d'ordre 816.

27. Hassan, M.F., Singh, M.G. and Titli, A. Computer Aided Design of the Stochastic Controlles for a Large Scale River System. IFAC Symp. "Computer Aided Design of Control Systems," Zurich, August 1979.

28. Sage, A.P. Optimum Systems Control. Prentice-Hall. Inc., Englewood Cliffs, N.J. 1968.

29. Hassan, M.F., Salut, G. Singh, M.g. and Titli; A. A Decentralize Computationed Algorithm for the Global Kalmar Filter. IEEE Trans. A.C. Special Issue on Decentralized Control and Large Scale Systems, A.C. 23, No. 2, April 1978.

ON A MATHEMATICAL MODEL IN THE POPULATION DYNAMICS

Claudia Lidia Simionescu

Department of Mathematics, Univ. of Braşov
2200 Braşov, R.S. Romania

ABSTRACT

The paper deals with a distributed delay equation for the population growth. Previously obtained results by the author are briefly reviewed, then numerical approximations for the solutions via finite differences and Galerkin methods are shown.

1. MATHEMATICAL MODEL

Any scientist dealing with models in ecology is conscious of how risky is to extrapolate results from laboratory to the real world, far riskier than in engineering or physics.
Nevertheless, mathematics can provide some service to ecology by delineating the logical consequences of simple models which, while oversimplified forms of the reality, suggest some aspects of the population behaviour.
Starting from the very early model by T.Maltus in 1798 many discrete and continuous equations for state description of a population have been **considered** (see the survay on population dynamics [1]).
Our discussion will focus on one of them, the delay balance equation of Volterra [6]. This is the nonlinear integro-differential equation:

D. G. Lainiotis and N. S. Tzannes (eds.), Applications of Information and Control Systems, 497-504.

$$N'(t) - \lambda N(t) + \lambda c^{-1} N(t) \int_0^\infty N(t-s)\, d\alpha(s) = 0\,, \quad t \in R \,. \qquad (1)$$

where N(t) is a measure of the population size at time t , $c > 0$ is the carrying capacity of the species, $\lambda > 0$ is the exponential birth rate for the unrestrained population and the integral term represents the hereditary effects of past population sizes on the growth rate.
The solution for (1) hase to obay to the asymptotic low

$$\lim_{t \to \infty} \frac{N(t)}{c} = 1 \qquad (2)$$

In earlier papers [2] , [3] we deald with a control problem for (1) and with the problem of getting a solution when large hereditary influences can be taken into account.
In fact, we considered a linear perturbed equation derived from (1) and we looked for a perturbation the way that the solution for the linear equation approaches as well as possible the solution for (1). This turned out to be an optimal control problem which we solved by standard methods. We obtained the solution under un integral representation.
However, the expression giving N(t) is quite difficult to compute and for the simulations we have been done, we had to make some simplifying assumptions.
The aim of our work here is to indicate two numerical methods for approximating N(t) , one by space discretization and the other one by means of finite differences.

2. THE LINEARIZED MODEL

Let us consider (1) under the form

$$N'(t) - \lambda N(t) + \lambda \int_0^\infty N(t-s)\, s^n ds = \lambda F(u) \qquad (3)$$

which means we allow hereditary influences to be as significant as necessary.
From linearizing (1) by means of a perturbation term, we get the:

Optimal Control Problem : If the state equation is

$$N'(t) - \lambda N(t) + \lambda \int_0^\infty N(t-s) s^n ds = \lambda F(u) \quad (3)$$

find a control u , such that the functional

$$J(u) = \left(1 - \frac{N(t)}{c}\right) \int_0^\infty N(t-s) s^n ds - F(u) \quad (4)$$

attaines its infimum.

Let V and H be two Hilbert spaces, $V \subset H$, V dense in H, the injection being continuous. Equation (3) can be written under the variational form $(n=0)$, $u \in U$

$$\left(N'(t) - \lambda N(t) + \lambda \int_0^\infty N(t-s) ds, v\right)_H + \left(1 - \frac{N(t)}{c}, v\right)_V = \left(h(t) + \lambda F(u), v\right)_H, v \in V \quad (5)$$

understood in the sense of scalar distributions. Denote by $D^k(p; E)$ the Sobolev space with weight where k is an integer, p a C^1 function having the property [5]

$$\exists\, p_0 > 0 ,\ p'(t) > p_0 ,\ \forall t \in R$$

and E a Hibert space. Assume $h(t) + \lambda F(u) \in D^k(p, H)$. Then from [3] , we know that there exist a positive function $G(t) \in C^1$, $t \in R$ and for each integer k , the positive function g_k, $g_k \in C^1$, such that if $p'(t) \geqslant g_k(t)$,

$\forall t \in R$ and $p'(t) + G'(t) \geqslant p_0$, $\forall t \in R$, we have a unique solution for (3),

$$N(t) \in D^k(p+G; V) \cap D^k(p; H) \quad (6)$$

and in norm

$$\|N(t)\|^2_{V;p+G+g_k,k} + \|N(t)\|^2_{H;p+g_k,k} \leq \left(h(t) + \lambda F(u), N(t)\right)_{H;p+g_k,k} \quad (7)$$

Since (5) has to hold for any $v \in V$, it can be interpreted as :

$$N'(t) - \lambda N(t) + \lambda \int_0^\infty N(t-s) ds = \lambda F(u) \quad \text{in } H \quad (8)$$

$$1-\frac{N(t)}{c}=Ih(t)\ \text{in}\ V,\quad I\in L(H,V),\quad h\neq 0 \tag{9}$$

From (6) we immediately get the asymptotic behaviour:

$$\lim_{t\to\infty}\left(1-\frac{N(t)}{c}\right)D^{k}e^{-P(t)}=0 \tag{10}$$

When the perturbation F(u) = $u^2(t)+u(t)$, in (3) we [3] obtained the following form for the solution:

$$N(t)=4tY(t)e^{2t}*D((\varphi^{-1}(t)(-\delta+H_A)*\Phi(t))^2+(\varphi^{-1}(t)(-\delta+H_A)*\Phi(t))) \tag{10}$$

for $\lambda = 4$ and

$$N(t)=\lambda Y(t)e^{r_1 t}*Y(t)e^{r_2 t}*D((\varphi^{-1}(t)(-\delta+H_B)*\Psi(t))^2+(\varphi^{-1}(t)(-\delta+H_B)*\Psi(t))) \tag{11}$$

for $\lambda \neq 4$ and $\lambda > 0$, where we denoted:

$$\Phi(t)=2H(t)\int_0^\infty \tau e^{2\tau}\varphi(t-\tau)d\tau+\varphi(t)\qquad,\ \varphi\in U=D^k(p;H^1(R))$$

$$A(t)=4H(t)(Y(t)te^{2t})$$

$$\Psi(t)=\lambda H(t)\int_0^\infty\int_0^\infty e^{r_1\sigma+r_2\tau}\varphi(t-\tau-\sigma)d\tau d\sigma+\varphi(t)$$

$$B(t)=2\lambda H(t)(Y(t)e^{r_1 t}*Y(t)e^{r_1 t})$$

$$H_A=-A(t)-A(t)^{*2}-\dots-A(t)^{*n}-\dots$$

$$H_B=-B(t)-B(t)^{*2}-\dots-B(t)^{*n}-\dots$$

For another particular case, when F(u) satsfy the functional equation $F(u+\theta\varphi)=F(u)+\theta H(\varphi)$, $u,\varphi\in U$ and $\theta\in(0,1)$ and $d\alpha(s)=G(s)ds$, by the same method, the solution has the form:

$$N(t)=\frac{c}{\lambda}\delta * Vp\left(\frac{1}{tY(t)G(t)}\right)$$

3. NUMERICAL APPROXIMATIONS I

One of the most promising method of approximating the solution for the control problem (3) - (4) seems to be the Galerkin method.

Assume V is a separable space and let an internal Hilbert approximation for V be $\{V_h, P_h, R_h\}$, $h \in N$ which is stable and convergent [4], [5] , and such that $V = \overline{\bigcup_h V_h}$

This can be done if we take , for instance $w_1, w_2, \ldots, w_h, \ldots$ a free and total sequence in V and we denote by V_h the subspace of V generated by $w_1, w_2, \ldots, w_h$. P_h is the canonical injection $V_h \to V$ and R_h is the ortogonal projection of V on V_h .
This way , to each V_h , $\forall k \in \mathbb{Z}$, there corresponds the space $\mathcal{D}^k(p; V_h)$ where the weight p satisfies the required property. It is obvious that:

$$\mathcal{D}^k(p; V_h) \subseteq \mathcal{D}^k(p; V)$$

$$\overline{\bigcup_{h=1}^{\infty} \mathcal{D}^k(p; V_h)} = \mathcal{D}^k(p; V)$$

Let us consider the state equation

$$\left(N'(t) - \lambda N(t) + \lambda \int_0^\infty N(t-s)\,ds, v\right)_H + \left(1 - \frac{N(t)}{c}, v\right)_V = \left(h(t) + \lambda F(u), v\right)_H \qquad (12)$$

As we have mentioned , the optimal control $u^* \in V_{ad} \subset V$ -V_{ad} a closed convex subset of $V = U$ - exists and is unique , U = space of controls . We set $V_{ad}^h = V_{ad} \cap V_h$ and define $u_*^{\xi_h} = R_h u_* \in V_{ad}^h$. To the projection $u_*^{\xi_h}$ and for $v_h \in V_h$, there corresponds the approximate state equation

$$\left(\frac{d}{dt} N_h^{\xi_h}(t) - \lambda N_h^{\xi_h}(t) + \lambda \int_0^\infty N_h^{\xi_h}(t-s)\,ds, v_h\right)_H + \left(1 - \frac{N_h^{\xi_h}(t)}{c}, v_h\right)_{V_h} = \left(h(t) + \lambda F(u_*^{\xi_h}), v_h\right)_H \qquad (13)$$

with $v_h \in V_h$ in the sense of scalar distributions. By [4], [5] , this equation has a unique solution $N_h^{\xi_h}(t)$ with an appropriate G ,

$$N_h^{\xi_h} \in \mathcal{D}^k(p+G; V_h) \cap \mathcal{D}^k(p; H) \qquad (14)$$

and $N_h^{\xi_h} \to N$ for $h \to \infty$ in the norm of the state space.
The approximate perturbation functional is defined by the expression:

$$J_h^{\xi_h} = J(u_*^{\xi_h}, N_h^{\xi_h}) = \left(1 - \frac{N_h^{\xi_h}}{c}\right)\int_0^\infty N_h^{\xi_h}(t-s)\,ds - F(u_*^{\xi_h}) \tag{15}$$

We have:

Theorem 3.1. If $\{V_h, P_h, R_h\}_{h\in\mathcal{P}}$ is an internal Hilbert approximation for V and $(u_*^{\xi_h})$ a sequence of approximations for the optimal control u_*, then $(N_h^{\xi_h})_h$, $(J_h^{\xi_h})_h$, and finally $(u_*^{\xi_h})$ are strongly convergent to $N(t)$, $J(u)$, respectively u_*.
The proof is a simple adaptation of the proof given in [4] and because of the limited space here we wont reproduce it.
If we take V=H(R) and H=L^2(R), then one can use the discretisations already set up for these spaces.

4. NUMERICAL APPROXIMATIONS II

As far as discretisation for the operator concerned, we shall make use of the finite difference algorithm for general integro-differential equations in [5]
Let $\{N_L\}_{L\in\mathcal{P}}$ be a sequence of positive real numbers converging to infinity and for every n let us take a step $h = N_L/n$. For $\ell = 0, 1, \ldots, n$ and $\xi_h^\ell \in [h(\ell-1), h\ell]$ we built the difference equation

$$\left(\frac{N_h^\ell - N_h^{\ell-1}}{h} - \lambda N_h^\ell + \lambda\sum_{\ell=0}^{n-1} N(\ell h - \xi_h^\ell)h, v\right)_H + \left(1 - \frac{N_h^\ell}{c}, v\right)_V = \left(h(\ell h) + \lambda F(u_{\ell h}^*), v\right)_H \tag{15}$$

The term under the summation sign can be taken as:

$$N(\ell h - \xi_h^\ell) = N_h^\ell - \xi_h^\ell\,\frac{N_h^\ell - N_h^{\ell*}}{h} \tag{16}$$

where for $N_h^{\ell*}$ we make an appropriate choice such that it corresponds to a value ξ,

$$\ell\,\frac{N_L}{n} - \xi_h^\ell < \xi < \ell\,\frac{N_L}{n}$$

This way we obtaine a system of $n+1$ algebraic equations:

$$\left(\frac{N_h^\ell - N_h^{\ell-1}}{h} - \lambda N_h^\ell + \lambda\sum_{\ell=0}^{n-1}\left[N_h^\ell - \xi_h^\ell\,\frac{N_h^\ell - N_h^{\ell*}}{h}\right]h, v\right)_H + \left(1 - \frac{N_h^\ell}{c}, v\right)_V = \left(h(\ell h) + \lambda F(u_{\ell h}^*), v\right)_H$$

$$\ell = 0, 1, \ldots, n \tag{17}$$

Ordering the system by N_h^ℓ, there is easy to be seen that because $h(\ell h)+\lambda F(u_{\ell h}^*)\in D^k(p;H)$, the solution for (17) exists and is unique and belongs to the Sobolev space $D^k(p+G;V)\cap D^k(p;H)$. Correponding to N_h^ℓ, we have an approximated perturbation functional

$$J_{N_h^\ell}=\left(1-\frac{N_h^\ell}{c}\right)\int_0^\infty N_h^\ell(\ell h-s)\,ds-F(u_{\ell h}^*) \tag{18}$$

and the approximated optimal control. This way, for every N_L, we have

$$N_{hN_L}^{-1},\ N_{hN_L}^{0},\ \dots,\ N_{hN_L}^{n} \tag{19}$$

$$u_{hN_L}^{-1*},\ u_{hN_L}^{0*},\ \dots,\ u_{hN_L}^{n*}$$

For $t\in(-N_L,0)$, we define $\bar N_{hN_L}^\ell$ by reflection and then define

$$N_{N_L}^n(t)=\begin{cases}0 & t\notin[-N_L,N_L]\\ \bar N_{hN_L}^\ell & t\in[-h(\ell-1),-h\ell]\\ N_{hN_L}^\ell & t\in[h(\ell-1),h\ell]\qquad \ell=0,1,\dots n\end{cases} \tag{20}$$

and the same for $u_{N_L}^{n*}$. The perturbation functional becomes:

$$J_{N_L}^n=J(u_{N_L}^{n*})=\left(1-\frac{N_{N_L}^n(t)}{c}\right)\int_0^\infty N_{N_L}^n(t-s)\,ds-F(u_{N_L}^{n*}) \tag{21}$$

From Theorem 2.1. in [5] follows:

Theorem 4.1. The approximate solutions for the stated Optimal Control Problem are strongly convergent to the optimal solution in the corresponding spaces:

$$N_{N_L}^n(t)\xrightarrow[n\to\infty]{}N(t)\quad\text{in } D^k(p+G;V)\cap D^k(p;H)$$

$$J_{N_L}^n\longrightarrow J\qquad\text{uniformly},\quad U_{N_L}^n\xrightarrow[n\to\infty]{}U\quad\text{in } U$$

5. PENALTY METHOD

Let us returne to the system (3) - (4). Instead of minimizing the perturbation functional under the restraine (4), one might consider the problem of minimizing a "so cold" penalty functional:

$$J_\varepsilon(N,v)=\left\|\left(1-\frac{N(t)}{c}\right)\int_0^\infty N(t-s)\,s^n ds-F(u)\right\|^2_{V;p+\sigma,k}+\gamma\|v\|^2_U$$

$$+\frac{1}{\varepsilon}\left\|N'(t)-\lambda N(t)+\lambda\int_0^\infty N(t-s)s^n ds-\lambda F(u)\right\|^2_{H;p,k}$$

defined over $Y\times U$, where

$$Y=\left\{N \mid N'(t)-\lambda N(t)+\lambda\int_0^\infty N(t-s)\,s^n ds-\lambda F(u)\in D^k(p;H)\right\}$$

with the norm $|||N|||^2_Y=\|\cdot\|^2_{V;p+\sigma,k}+\|\cdot\|^2_{H,p,k}+\|P(t,D)N-\lambda F\|^2_{H;p,k}$
The penalty problem consists in : find(N_ε, u_ε) such that

$$J_\varepsilon(N_\varepsilon,u_\varepsilon)=\inf_{N\in Y,\,v\in U_{ad}\subset U} J_\varepsilon(N,v)$$

The infimum exists and is unique, and $(N_\varepsilon,u_\varepsilon)\underset{\varepsilon\to 0}{\longrightarrow}(N,u)$

REFERENCES

1. Oster,G: 1977,Moder Modeling of Continuum Phenomena,Lect.in Applied Math.vol.16,Am.Math.Soc.Provid.
2. Simionescu,Cl.L.:1978,Simulation of Control Systems,IMACS,North Holland Publ.Comp.
3. Simionescu,Cl.L.:1979,Mathematics and Computers in Simulation,XXI,pp.178-181,North.Holland.Publ.
4. Simionescu,Cl.L.:Information Processing 77,pp 925 -928,IFIP,North-Holland Publ.
5. Simionescu,Cl.L.:1977,Optimization Techniques,PartI I,pp.370-379,Springer-Verlag
6. Volterra,V: 1931,Lecons sur la Théorie Mathematique de la Lutte pour la Vie,Gauthier Villars,Paris

SYSTEMS ANALYSIS IN THE MARITIME INDUSTRY

Ernest Koenigsberg

School of Business Administration
University of California, Berkeley
Berkeley, CA 94720

INTRODUCTION

Few people today, or even yesterday, would think of the maritime industry as being a leader in technological innovation. Leadership, to laymen, implies new and bold machines or devices which show promise for the future. In the maritime industry innovations of that type--hovercraft, hydroplanes, ground effect machines, and super submarines--have proved to be uneconomic for the movement of cargo, and nothing has followed the early fanfare.

In its slow and plodding way the industry has actually been a leader in technological and operational innovation, at least in the period 1955-1975. The trade literature often refers to the "container revolution" and "tanker revolution" in discussing the changes in general cargo and liquid bulk transportation. Dramatic changes have occurred in other areas of maritime transportation. The evidence would indicate that the radical changes in transportation technology were a result of the growth of trade, but to some extent the technology has driven the growth rate.

Starting about 1950-1955 we can list the major technological changes.

1) The containerization of the "liner trades" and the demise of the "flexible" tramp ship.

2) The development of the "super-tanker" and very large bulk vessels.

D. G. Lainiotis and N. S. Tzannes (eds.), Applications of Information and Control Systems, 505-511.

3) The development and use of a wide range of specialized ships:

a) parcel tankers;
b) LNG, LPG, and Methane ships;
c) large ocean going barges;
d) LASH and SEABEE barge ships;
e) automobile transportation ships.

4) The development of port facilities to accommodate the special vessels and enhance their economic effectiveness.

5) The use of timely weather data to decrease travel time and improve the safety of vessels.

Systems analysis has played a role in most of these technological and operational developments. We emphasize the difference between technological and operational developments. Super tankers are a result of a technological change in the design of non-rigid structures. Containerization, on the other hand, is a result of a change in the operation of ships as well as changes in design. (The early experiments were carried out by carrying containers as deck cargo on conventional liners and then by modifying conventional ships to carry containers.) We can find examples of the use of systems analysis in the following areas:

a) container ship design, selection and operation;
b) design and operation of container terminals;
c) scheduling of dedicated or partially dedicated fleets (LNG, Bauxite, etc.);
d) scheduling of tanker and bulker fleets;
e) design and operation of ocean-going tugs and barges;
f) design of ships for minimum lifetime costs.

In this presentation we can only present a few examples and only in outline form. A larger set of examples is included in the bibliography.

SELECTION OF CONTAINER SIZE(S)

One of the earliest applications of systems analysis to the maritime industry was the work done for Matson Navigation Company by Weldon (1958). One of the important problems in the design of a maritime container system is the selection of the container dimension. Weldon carried out his analysis using the data for the Hawaii-West Coast U.S. trade.

The study recognized that there were a number of factors which influence the selection of container dimensions. First, there are the limits set by state and local authorities on road operations:

1) vehicle width and height;
2) vehicle load limits (or maximum axle loads);
3) operation of articulated vehicles.

Then there are constraints set by design limits on hatch sizes in ships and by design limits on cranes. Lastly, the container size will depend on cargo characteristics:

1) shipment size;
2) cargo density.

The study showed that the width was restricted to 8 feet to meet all highway requirements, but that since the container trailer would have to be redesigned the height could be as much as 9 feet and the vehicle would still meet the highway height limits. The major analysis dealt with container length. The results are shown in Figure 1, which shows the effects of shipment size and cargo density on container length.

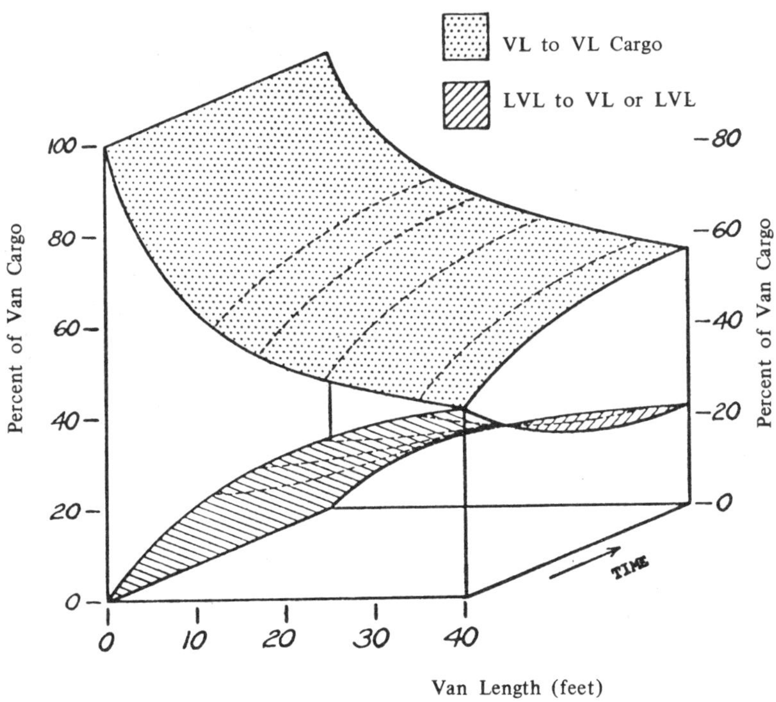

Figure 1. Van Shipments vs. Van Size

The work was then extended to select the container size which results in lowest cost considering all factors--utilization of ship space, container cost, container handling cost, container stuffing cost, and ground transportation cost. The results, as shown in Figure 2, clearly demonstrate the advantages of the 24 foot container in the U.S.-Hawaii trade.

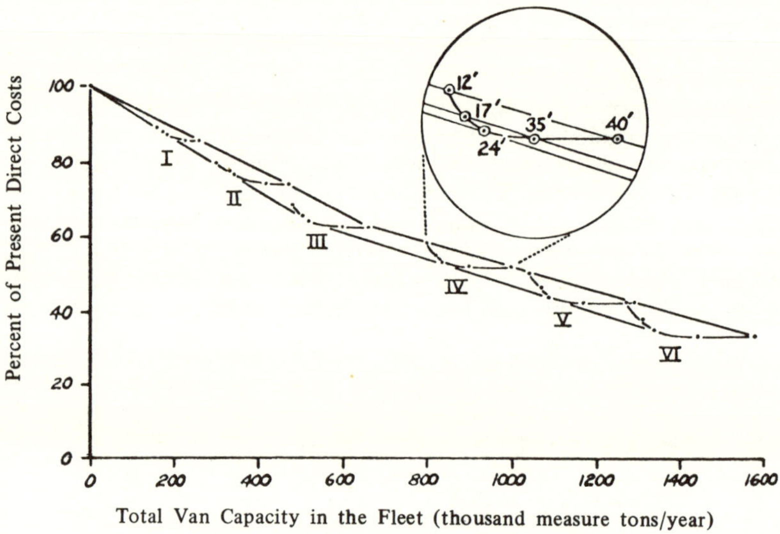

Figure 2. Annual Direct Operating Costs

As a follow-up note: the International Standard Organization (ISO) selected a set of standard international container sizes without any analytic study. In their infinite wisdom they selected a set of containers 8 feet wide, 8 feet high, and lengths of 10, 20, 30 and 40 feet. There are almost no 30 foot containers (a good size in world trade) because there is no use at all for 10 foot containers. At this time very few new containers meet the ISO height standards.

I should point out that the best sizes for today's distribution of cargo density and shipment sizes and the revised highway regulations in the U.S. (and EEC) would yield different results. In trades where articulated "double" over-the-road vehicles can be used, the container size would tend to be 8 feet wide, 9 to 9.5 feet high, and 27½ or even 30 feet long. In trades restricted to a single trailer, the optimum size would probably be 40 to 48 feet long.

The alloted space does not allow me to go into other maritime applications in any detail. The bibliography includes several cases which cannot be included here. We can review, however, the applications of systems analysis to ship allocation, scheduling, and planning.

ALLOCATION, SCHEDULING, AND PLANNING

i) Simple scheduling--e.g., selecting and allocating ships in dedicated service in simple port-to-port (or multiport cycles) operations--has been treated using queue-like models. Kaplan, et al., have used a simulation model to plan the operations of fleets of LNG ships. The problem is to select the number of ships and how many berths and other port facilities are required to move a specific tonnage between a set of ports using vessels of a given size. The model was also used by making ship size a parameter for analysis.

Koenigsberg and Lam (see also Koenigsberg and Meyer) used cyclic queue models to examine the same problems. These analytic models are much simpler and produce results which are almost identical to those obtained by the more complex and more expensive simulation models.

ii) Scheduling with some complexity--e.g., scheduling a mixed liner fleet in a service with a small number of port alternatives or scheduling a small bulker fleet with a small number of alternate ports. This type of scheduling and allocation problem has been treated using linear programming (mixed integer or all integer methods) and by heuristic models. Sorensen, et al., use a simulation model which applies good heuristics and some "look-ahead" features to allocate ships to ports and cargo to ships which meets some set of good operating standards. Sullivan and Koenigsberg treat the same problem using a linear programming model. The results in both cases were quite similar and lead to significant changes in fleet schedules.

We note here that the linear programming model is a mid-range planning model for allocating ships to routes given some general characteristics of cargo flow demands. The simulation model is a short range planning and operations model to be used for scheduling fleets over the next 30 to 60 days. Kaiser Aluminum Corp. uses a heuristic simulation system for scheduling a small bauxite fleet (which is supplemented by short term or voyage charters).

iii) Scheduling with a high degree of complexity--e.g., a large fleet of tankers (both owned and chartered) engaged in international liquid bulk and/or bulk and ore trades--has been treated by Appelgren, among others. There are several companies which "operate" fleets of 50 or 60 tankers, bulkers, and/or

O/B/O vessels. These ships have no fixed service, but carry cargoes as they develop from the business of the company, long-term agreements to carry cargo from port-to-port for other companies and cargoes of opportunity. The terms of carriage are somewhat loose and the payoff from diverting ships (and from chartering in and chartering out) are quite large. The problem of scheduling a fleet of as few as 20 mixed vessels for only a 60-day period is extremely complicated. Major companies are using programming and network methods to attack these problems.

These areas are only representative of the uses of systems analysis in the maritime industry. The number of applications reported in the open literature is not too large, say, as compared with the air transport industry, but they have contributed to the design, development and operation of improved maritime systems.

BIBLIOGRAPHY

Containerization

Weldon, F.: 1958, *Cargo Containerization in the West Coast - Hawaiian Trade*, Operations Research, 6, pp. 649-670.

Edmond, E. D. and Maggs, R. D.: 1976, *Container Ship Turnaround Times at U.K. Ports*, Maritime Policy Management, 4, pp. 3-19.

Allocation Scheduling and Planning

Koenigsberg, E. and Lam, R.: 1976, *Cyclic Queue Models of Fleet Operations*, Operations Research, 24, pp. 516-529.

Kaplan, M., Wentworth, R. C., and Hischier, R. J.: 1972, *Simulation and Optimization of LNG Shipping Systems*, ASME Transaction Petroleum Mechanical Engineering and Pressure Vessels and Piping Conference, New Orleans, September 17-21.

Koenigsberg E., and Meyer D.: 1980, *An Interacting Cyclic Queue Model of Fleet Operations*, Logistics and Transportation Review, 16.

Williams, N.: 1970, *Linear Scheduling*, in Proceedings of NATO Conference on Applications of Mathematical Programming, Cambridge, 1968, pp. 70-75.

Sullivan, W. J. and Koenigsberg, E.: 1970, *Mixed Integer Programming Applied to Ship Allocation*, in Proceedings of NATO Conference on Applications of Mathematical Programming, pp. 273-280.

Olson, C. A., Sorensen, E. E., and Sullivan W. J.: 1969, *Medium Range Scheduling for a Freighter Fleet*, Operations Research, 17, pp. 565-582.

Gilman, S. and Williams, G. F.: 1976, *The Economics of Multi-Port Itineraries for Large Container Ships*, Journal of Transportation Economics and Policy, May.

Conley, J. H., Farnsworth, R. S., Koenigsberg, E., and Wiersema, V.: 1968, *A Linear Programming Approach to the Total Movement of a Homogeneous Product*, Transportation Science, pp. 289-302.

Appelgren, L. H.: 1968, *A Column Generation Algorithm for a Ship Scheduling Problem*, Transportation Science, 3, pp. 53-68.

Appelgren, L. H.: 1971, *Integer Programming Methods for a Vessel Scheduling Problem*, Transportation Science, 5, pp. 64-78.

McKay, M. D. and Hartley, H. O.: 1972, *Computerized Scheduling of Sea Going Tankers*, Naval Research Quarterly, pp. 255-264.

AUTHOR INDEX

SUBJECT INDEX